New Food Product Development

Global Strategies and Practices for Successful Innovation

Fourth Edition

Holger York Toschka

CRC Press
Taylor & Francis Group
Boca Raton London New York

CRC Press is an imprint of the
Taylor & Francis Group, an **informa** business

Designed cover image: CravenA Shutterstock contributor

Fourth edition published 2025
by CRC Press
2385 NW Executive Center Drive, Suite 320, Boca Raton FL 33431

and by CRC Press
4 Park Square, Milton Park, Abingdon, Oxon, OX14 4RN

CRC Press is an imprint of Taylor & Francis Group, LLC

First edition published by CRC Press 1994
Second edition published by CRC Press 2004
Third edition published by CRC Press 2011

Library of Congress Cataloging-in-Publication Data
Names: Toschka, Holger Y., author.
Title: New food product development : global strategies and practices for successful innovation / Holger York Toschka.
Description: Fourth edition. | Boca Raton, FL : CRC Press, 2025. | Revised edition of: New food product development : from concept to marketplace / Gordon W. Fuller. 3rd edition. Boca Raton, FL : CRC Press, c2011. | Includes bibliographical references and index.
Identifiers: LCCN 2024038276 (print) | LCCN 2024038277 (ebook) | ISBN 9781032698489 (hbk) | ISBN 9781032724584 (pbk) | ISBN 9781032724591 (ebk)
Subjects: LCSH: Food--Marketing. | Food--Research. | New products--Marketing. | Product management. | Food industry and trade--Technological innovations.
Classification: LCC HD9000.5 .F86 2025 (print) | LCC HD9000.5 (ebook) | DDC 338.1/90688--dc23/eng/20250116
LC record available at https://lccn.loc.gov/2024038276
LC ebook record available at https://lccn.loc.gov/2024038277

ISBN: 9781032698489 (hbk)
ISBN: 9781032724584 (pbk)
ISBN: 9781032724591 (ebk)

DOI: 10.1201/9781032724591

Typeset in Times
by KnowledgeWorks Global Ltd.

Contents

About the Author

Holger York Toschka, Ph.D., was until recently Senior Director of Research and Development Foods at Unilever PLC, a British multinational fast-moving consumer goods company. Holger's responsibilities in the past years included driving general food science and technology, innovation and product maintenance, in local, regional and global food product portfolios. He has 30-plus years of food industry management and leadership experience in the areas of packaged, branded food development, food science and technology, biosciences, human nutrition, project and program management, including manufacturing and sourcing, margin improvement, simplification, regulation and sustainability. During his time with the company, Holger was heading product development in more than 20 countries on the local, regional and global scale, leading to taking responsibility for hundreds of new product innovations, renovations and refreshments. He can look back to an end-to-end experience in the packaged food sector, ranging from science detection, product and concept development, shaping of product manufacturing and safety up to activation and market maintenance. Besides many years in leading roles in the global R&D centres, Holger spent more than seven years in Southeast Asia, in functions where he was leading foods R&D for all markets in Southeast Asia, Australia and New Zealand, sitting on various local and regional innovation boards and managing multinational teams of 100+ managers, experts and technologists. Alongside his work in overall food research and development, Holger spent a couple of years in ice cream and refreshment R&D and food manufacturing and sourcing. In his time with Unilever, he had the opportunity to provide many presentations and lead talks on international conferences in Europe and Asia, as well as provide lectures at local universities about food science and technology, nutrition, health food and sustainability. Holger earned his Ph.D. in biochemistry and molecular sciences with an emphasis on gene expression and regulation at the Free University of Berlin and the Max Planck Institute for Molecular Genetics. He has published several scientific articles in the field of food science and molecular biology, holds several patents in the biosciences and recently helped to edit a book about the "science of the Thai cuisine". Holger has been married for many years to Ute, has four adult children and right now lives again in Southern Germany. Today he works as a global food and innovation consultant.

Preface
Why this book?

Writing such a book today requires a good portion of optimism and confidence that people will read it. In my case, the target audience I have in mind are higher-degree students or those taking their first job in enterprise making, commercial food product and process development, academia, institutions, specialist labs and research facilities, and from enthusiastic startups to big multinational fast-moving consumer goods (FMCG) food companies.

One key ability in product innovation – especially in food innovation – is to recognize how and why food and drink trends are moving and then linking this to the right time frame! A promising trend is ideally detected in its early stages, but with a granular understanding of how it will grow or fade. Responding too late limits the market opportunity; others will have taken the lion's share of the new growth. The "salted egg" aroma gained explosive popularity in some Asian restaurant dishes in the middle of the last decade. Many food makers interpreted this aroma direction as rather a "niche savoury dish", and short term, as more of a hype! But five years later, salted egg was widely used in condiments, dressings, sauces, salads, soups, meals, buns, cakes, chips, ice cream and beverages.

Being too early isn't good either. Quiche outside France, fondue outside Switzerland, matcha tea or the seasonal superfoods either needed many years to establish themselves or had a short lifespan.

For a food maker with an existing business, "doing nothing" or "being wrong" are both not good options. For a bigger company, jumping on a short-term hype rather than a trend means it is too late and will not earn its money back. That's due to the time needed to bring an idea to commercial launch.

In a sense, this book continues the editions of Gordon W. Fuller's book *New Food Product Development*. His third – and last – edition was published in 2011 and provided at the time a solid perspective about food product development principles and the factors shaping them. It guided the reader well through the process from initial idea search, phase prototyping and testing up to market launch. The mostly Western North American view gave valuable orientation for students in food science and technology focussing on these markets.

Since then, the world – and especially the world of foods, the most vital part of FMCG – has changed dramatically. Vegan or vegetarian was a lifestyle choice, functional food a niche serving the rationalist, and sustainability more an exciting topic in academic discourses. Markets have grown and segmented, some parts are much more consolidated, distribution chains are global and products are international. The life of most consumers has changed. Some changes were gradual, like global urbanization bringing hundreds of millions of people from the countryside into the cities. Most jobs there were more mental than physical. The lifestyle changed, work intensity increased and expectations towards good food were reshaped.

And while this was happening for many years at a rather constant pace, something nobody had expected shaped it radically. The global pandemic and lockdowns, followed by civil unrest, sourcing and distribution challenges and price inflation, redefined new food product making.

New solutions must fit lifestyles and individual circumstances. And they must be better, for the consumer and the planet, but also real "foody", genuine, natural, simple, authentic, cleaner and reliable. Sometimes better means more local and reliable, but in other situations, better means more exotic, international, inspiring and surprising.

So, nearly 15 years after the last edition, this new version could not just be a simple update. The scenery has changed dramatically, and consumers have accepted that uncertainty is unavoidable. The future is less predictable. Living in the "here and now" with a good portion of day-to-day focus is a balance between enjoying life now and the physical and emotional resilience for themselves and the ones they care for. The "food makers" had to respond to this, within their business models, which are often built on economies of scale and efficiency. This is not so easy, as new food items have to hit the market much more quickly now. More products are coming to market much more quickly, taking share and market growth from those already there. This shortens the product life cycle, and new ideas have to hit the markets more quickly and more often, while fully complying with many more sophisticated standards, ranging from safety, traceability, environmental impact, data privacy, convenience and availability. But not only has speed to market accelerated. The market itself, trends, barriers, new technologies, regulations and taxation move forward more quickly. A popular bonmot, often quoted in the last few years, says: "Change has never been this fast. And it will never be this slow again"!

This is good news for people in the business of making new food products. More needs, more opportunities, but probably also more stress. Hence, it is truly healthy to understand the mechanisms behind the innovation scenes and have good confidence about the tools that will help to manage these situations successfully.

Last, but surely not least, science and technologies have made major steps forward. Breakthroughs in such diverse areas as bioscience and data technology, genome editing (Crispr Cas) and neural computer science (AI) have opened up the technical possibility to make radically better, more superior food products. Health food was one trend in some major markets in 2010; today it dominates all major markets. It is no longer an option to choose, but a "*conditio since qua non*" – a non-negotiable! And while in the past it may have been sufficient to "add" a single "good ingredient" to an existing food format and make a claim, it nowadays requires a more holistic approach to create competitive solutions. Just adding a functional "goody", like a vitamin or a mineral, brings hardly any advantages and is often not even possible. First, the food product needs at least to be cleared properly from any perceived "baddie", like sugar, salt, fat or artificial ingredients. Preferably, the food product matrix boosts the "effect" of the "active": Increased bioavailability, low glycaemic load, better satiety or satiation, or a portion of the daily fruit and vegetable recommendation.

With all of that happening, it was difficult for me to just update a next fourth edition of Fuller's book. Bringing just some novel data, fresher references and maybe a few new anecdotes and stories would likely miss the key points.

But his work truly helped me to organize the content, and I took learnings from the structure of the book. The old chapters "What Is New Food Product Development?", "Where to Find Successful Ideas" and "How to Go to Market" remain crucial questions in any product development process, but the tools have been developed further. Doing fair justice of the diversity of the food sector is nearly impossible, but his idea to explain separately the characteristics of food service and the business-to-business (B2B) ingredient sector makes even more sense today as they differ so much from the retail consumer market.

Other building blocks like the options for the "manufacturing strategy", the "legal and regulatory environment" and the "idea testing" are still essential elements. Here the change is more in the "how to do it" rather than in the "what to do".

Not only does R&D have a different role, but the activity of business partners in an organisation has changed, so it's essential to recalibrate roles and responsibilities in new food product development.

The interaction between R&D and the food safety and quality department, protecting consumers, brands and companies is today also more proactive. New hazards, challenges and wider and more open markets shifted the thinking to "safe by design" and, hence, into product developers' responsibility. At the same time, quality assurance (QA) has more influence on the next new food product through their observational skills.

I worked for many years in a food multinational in different roles spanning from basic science and technology, partnerships building, new product development to manufacturing and innovation steering in various local, regional and global roles. That influenced my thinking and convinced me to give the managerial elements of foods R&D work a broader space.

New chapters will address the tools available for the food industry and manufacturers to select, sharpen, fine-tune and support new food product launches and build robust idea portfolios. Most companies are not working at any given time only on the next launch. The innovation pipeline needs constant feeding to ensure a continuous flow of new product ideas, while the short- and medium-term risks and opportunities require reflection. Hence, its necessary in the conversation about the end-to-end process on new product development to shift the perspective from a single product to the overall portfolio management. Therefore, after discussing contemporary food project management tools, more insights on portfolio management are included.

The dramatically changing predictions of the changes in the environment, especially in the food sector, require much more attention than in the edition 15 years ago. The whole sector is not the defendant; it is the enabler of better solutions. Sustainable sourcing of all ingredients, environmentally friendly manufacturing, reduced carbon footprint, better waste management and an overall circular economy are critical contributions also from those designing and making such new products.

Finally, no book about new food products can end without an educated guess about the trends in the next years. On one side, I said extraordinary changes in consumer and customer trends happen more quickly and more often and the world of foods post-Covid is not the same as it was before! This phenomenon is sometimes called the "VUCA", an acronym standing for a world full of volatility, uncertainty, complexity and ambiguity. So, the extra challenge on top for food developers is to help predict success in unpredictable, uncertain and complex market situations.

Luckily, not everything will change: The importance of taste, total sensorial impression, convenience and the price-value equation remains critical for success. Just the functional "health and wellness" aspect has gained a lot of momentum lately.

I believe that we are in a special moment of time, in which science and technology can bring powerful solutions for key consumer needs. Therefore, I built an extra chapter, where I reflect on key aspects of enablers for areas called "personalized nutrition", "performance food", "mental health", "plant based", "mimicking meat, fish and dairy", "gut health", "biotics" and more.

My own professional career and the experiences I was allowed to make forced me to take a global perspective on food product development. This required me to think about tensions and synergies between local cuisines, culture and culinary preferences, and the motivation to develop products broadly to apply efficiencies and scale in such a way that consumer prices remain attractive. Local market dynamics still rule, but globalization is accelerating, and most relevant players aim to be successful in much more than just one market. So, a view on globalization in product design, its scope, but also boundaries, is included. This includes concepts like "core product and formulations design" and where it can and where it can't be used. "Glocal" in food is probably much more than a buzzword. For the international food business, it's the mandate to find the balance between the success of global formats like burgers, trusted food aspects like "locality", and facts about how and where the food industry can cut unnecessary "food mileages". To state it already here: "Design once, deploy everywhere" may remain an important goal but will always find its limitation.

In terms of style, I do not want to write a "lecturing" book. For me, "storytelling" was the better way to "onboard" new team members, helping them to land smoothly in a new and pretty different environment, where they wanted to leave their "footprint".

But to offer a little help to those educating and training students in food science and technology, I included case studies in each chapter. Some are short and crisp; others are a bit more complex and longer. Some of them show that what is better and what is worse only becomes obvious once concrete details are considered.

Acknowledgements

Towards the end of a more than one year journey, it is time to say, "Thank You"! What started as the containable idea to help make a next edition of Gordon W. Fuller's book *New Food Product Development: From Concept to Marketplace* became a complex undertaking for which I needed enormous support. His book gave me the guidance and structure to stay on track throughout that long period. The tonality inspired me to shift from fact-based lecturing more towards storytelling. While the perspective here could only be mine, many old and new friends and colleagues helped me with additions, corrections, advice and insight. At least a few I want to mention here.

Starting with my former colleague at Unilever, Bert Debets. Thanks for taking the time to discuss the quality and safety dimension of food development and sharing your view on some of the latest tools. Marnix van Amerongen, you sharpened my view about the differences between branded food maker and the food ingredient industry. Wouter Musters did that for the Food Service sector. Garmt Dijksterhuis, for providing input into the psychological side of food and liking. Aprajita Mahendiratta, Apra, you helped me to understand the full potential of consumer technical insight work as the missing link between product development and consumer. Boontarika Chanvorachote, Nui, it was so helpful we could walk together again through the mechanics of project and portfolio management in FMCG. Thanika Daechanantasilp, kop khun krap for the discussion about food packaging and the nitty gritty of plastic recycling.

During this excursion, I also had the opportunity to make some new contacts. Professor Stephan Martin, thanks a lot for sharing all your superior knowledge about diabetes, diet and lifestyle. It sharpened my understanding of future food product principles. Professor Daniel Wefers, for the balanced view on industrial food products and some of the myths on spelt. I am grateful to H.J. Yoon, for the possibility to calibrate my view on consumer trends and products with telling insights from Korea. The reconnection after many years with my old colleague Professor Reinhard Kohlus turned out to be a fruitful excursion about food process sciences.

A huge thank you to the publishing team, namely, Laura, Steve, Meeta and especially Lucy. At the time of writing this, we have not finished it all. But now, already, your dedication and hard work helped me to manoeuvre through the exercise. Steve, without your persistence I wouldn't have started the book.

The next bundle of thanks goes to my family, Hannah, Max, Jan, Sandrine, Fabian, Jasmin and Joel, for the patience, the curiosity and all the discussions we had about food and eating. Max, you get an extra one for your effort to capture the essence of what and how AI will change product development.

Finally, my beloved wife Ute! Not only for all your support – but someone must also be responsible for the courage to start this adventure and to fuel the endurance to finish it.

What Is Food and What Is Packaged Food?

1

INTRODUCTION

Depending on which book or website you open, the huge importance of food and eating is very differently explained. Nutrition Scientists may rationally emphasize the physiological dimension and condense the essence of food into a "substance consisting essentially of protein, carbohydrate, fat, and other nutrients used in the body of an organism to sustain growth and vital processes and to furnish energy. . . " (The Editors of Encyclopaedia Britannica, 2024). And nutritionists would wish to put more emphasis on the relation between food and health, may be dramatically, like Anne Wigmore, an early 20th-century raw food advocate, "The food you eat can be either the safest and most powerful form of medicine or the slowest form of poison" (White, 2022). Moreover, while there were times when nutrition science appeared to be more of an ideology than an academic discipline, today's science is progressively more able to provide personal, situation-specific guidance. "A calorie is no longer a calorie". Today we know that calories from distinct food items have different impact on a persons body.Various physiological factors, ranging from gender,age, genetic set up over stress level to sleep patterns shape how we use and metabolize food energy.

A chef, proud of his or her culinary skills, could add the advice of Jacques Torres, a famous French pastry chef and chocolatier, that "life is too short, so start with dessert" (Roses, 2023). They all would highlight different aspects about food, like "the love of cooking", "the foundation of genuine happiness" (*Our History | Auguste Escoffier School of Culinary Arts*), an art, a passion, a song or the "ultimate giving" (*Top 25 Quotes by Jamie Oliver*) and "Gastronomy is the greatest form of therapy that anyone can be exposed to" (*Top 25 Quotes by Marco Pierre White*). For them, feelings, emotions, pleasure, art, culture and heritage are the most important. Practical skills, technology and science are the enabler of those feelings.

The discipline of "food history" aims to understand how people have grown, cooked and eaten food and what that meant for culture and tradition, as food and eating also are ways of expressing oneself, connecting with others and shaping one's own identity. And others would add that such an academic approach would need to be complemented by a deeper understanding of "culinary history", searching for the origins and evolution of recipes and cooking (The Hungry Historian, 2023).

Even sociologists study food in a search for the true core and look at food in relation to history, progression and the future development of society, with all the "medical, ritual, spiritual, ethical and cultural, environmental and labour applications" (Wikipedia contributors, 2024a). Food is local, incorporates culture and cultural history and maintains people's perceived identity. Within families and communities, traditional meals and cuisine are passed down from one generation to the next, and immigrants take their eating tradition to their new homes and pass them to their offspring. They may have been the first to see how the role of food in life is changing over time, between cultures and people and even with a single day of an individual.

DOI: 10.1201/9781032724591-1

I could list more professions and academic disciplines keen to discover the essence of food, but I guess the point is clear. As food is so ubiquitous on our planet, fulfils so many diverse needs, has so many roots and histories and can mean numerous things to different people, knowingly or unconsciously, it is no real surprise that new food products are always emerging. And in the most developed food markets nowadays, already more than 50% of all food eaten is industrially processed and packed. The different facets of food and rapidly changing living conditions, urbanization, cultural conflicts, generation shifts, employment, personalization, labour intensity, family life, pandemics, economic turbulence and climate emergencies constantly offer new opportunities to fulfil the needs of the consumer better. What was leading yesterday is mainstream today and lagging tomorrow. Any enterprise with offerings on the market today must be constantly aware of a competitor launching a better solution and taking a share. Not trying to develop new food products is simply not an option.

It is therefore no surprise that per year about 30,000 new food and beverage products are entering the markets (George, 2019). Most of them are not successful; failure rates of 50%–90% are reported (Rutkowski, 2022).

All these new products come to an existing $10 billion (2023) total food market that is expected to grow to $12 trillion by 2027 (Statista, 2024a). The packaged food part of it is about 3–3.5 trillion U.S. dollars, or around one-third (Maximize Market Research Pvt Ltd., 2023).

All these new products need to be imagined, prototyped, designed, tested, scaled, manufactured, registered, marketed and sold, while at the same time the company portfolio on the market needs constant refreshment, redesign or value improvement. Altogether, this forms the lifeline of the food consumer packed goods (CPG) industry, and compared to the total market 30,000 new food products do not sound too much (NielsenIQ, 2021).

And of course, not all these ideas have made it to the market (yet). Even so, some analysts state that given the up to 90% failure rate, too many have made it to the market; some of those not launched are "recycled" afterwards.

More and more new food products need more food professionals developing them and ideally the technology, skill and knowledge to increase the success rate. And in many countries food and agronomy are the leading industrial sectors, so no wonder, more than 1500 universities globally offer nutrition and food studies (EduRank, 2024) and compete for the most successful professional training. Traditionally, universities like Davis, Cornell and Purdue in the United States and Wageningen, Reading, Hohenheim and Milan in Europe are leading the list. But more and more universities in China, Malaysia, Singapore, Japan, Korea and South Africa are climbing up the ranking. Each of these has thousands of nutrition and food science publications. And remarkably, all of Asia has more universities with nutrition science departments than the rest of the world.

All of them are providing high-level contemporary knowledge, ensuring full understanding of biological, chemical, physical and behavioural sciences and leading knowledge in processing, preservation, quality evaluation, public health aspects and utilization of foods.

This book does not want to compete here! It is instead an attempt to explain the principles of profitable new food product development, how to find opportunities, convert them into successful launches and manage all the hurdles and difficulties on the way to the market. The focus will be less on all the important technical details needed to make a commercial food product. There are numerous books and courses out there to learn how to best make fortified staples or condiments, nutritionally more valuable snacks, wholesome ready meals, great-tasting ice cream, low calorie, clean label, natural ready to eat, vegan or dairy-free meat analogues or dairy alternatives, socially responsible and ecofriendly replacements and super functional food. But particular aspects of the "making of" these products will be covered here, when recent and relevant changes in consumer perception or enabling technology can make a true difference. Throughout the course of the book, many product design and compositional details will be presented and discussed, not to teach the basics, but to show how new trends and opportunities in the consumer space and science and technology enable the translation into a successful new food product.

After the end of the academic path, many students move into their first positions in safety, product or process design, nutrition, manufacturing, regulatory or general R&D and become involved in commercial food product development. In the new context, some things are different. Science and technology are no longer everything. Professionals with very different academic backgrounds share their view of the world and become colleagues, subordinates or bosses. In the "fast-moving consumer goods" (FMCG) sector, the marketing function often is in the lead. Marketers see a world full of opportunities and focussed on those facts that support the idea for the future. A full and deep 360-degree evaluation of a situation, reviewing all strengths, weaknesses, opportunities and threats that technicians and scientists are used to for them is too time consuming and missing the opportunity. They are driven by the wish to make the next launch, not to thoroughly analyse the case. R&D people should do the full analysis, but not forget to mention that this is to maximize the chances of the next launch and not an academic attitude.

Fruitful cooperation with business partners is the key to successful new product development. And teamwork with R&D plays a critical role. Hence, a bigger part of the book will deal with the innovations such as how entities and companies are looking for them and, once they find promising candidates, how they evaluate them to separate the "good from the bad and the ugly".

Some of the methods used here are not dedicated food tools; they often come from other industry and innovation management, often from IT or the pharmaceutical industry. For the food business, they often have customized elements.

A beginner in the food industry, starting as an assistant or junior manager, will often realize how different the new job is. All they learned is important, but not all is needed. Firstly, they are no longer primarily a knowledge receiver, but sender, explainer and decision supporter. The potential of a new technology or science, the design of an experiment, the capture of all data often require decisions based on food science knowledge. But a scientific judgement is only part of the business process; it can create or support an idea or show hurdles and threats. So, it's important for people in food R&D to understand the thinking and actions of colleagues from other functions, have a good understanding where their views and opinions are essential and have a feeling how they land their arguments. Formulas and calculations are often not a good way to tell an exciting story in R&D, while in finance they may be the final judgement. So parts of this book explain the work of other functions in new food product development. The focus is then on those elements that are important for R&D, because what is learned from them can be used in the further development process.

In a wider sense, this could help build an efficient and effective career for those wanting to climb up the ladder and have the opportunity to leave better and bigger footprints.

The development from cradle to grave for packed food is generally not a quick shot. It normally takes months, sometimes years, before a product or product range becomes launched and even longer before it becomes a commercial success or even a sustained performer in the market. So, new food development is more about the answer to trends and less about reaction to hypes. And in that development process there are several distinct phases and various steps in between, where specific signals need to be understood and addressed accordingly.

New food is never an undertaking for the single function in a company. Neither R&D, marketing, buying, sourcing, manufacturing, quality, finance or legal and regulatory can do such a job. In successful innovation and product launch, the different functions have their specific roles and responsibilities. And while this then is a true team approach, the decision taken at the necessary stages needs to build on these special roles and responsibilities. We will have a look at those. These functions and their contributions can come from inside the company or outside throughout the total process or a particular element or limited time. All these approaches have their own strengths and weaknesses and will be reviewed. Knowing them helps to make the right decisions!

Companies, teams and individuals normally do not shape a single new product or product range. One project may have one team but several formats or products or markets. Individuals in the innovation chain often work on several teams and contribute to very different offerings. A product launch after long and hard work is not the end! More often, it is just the beginning! After launch, the product performance needs to be monitored, evaluated from various sides, optimized and maybe repaired, or just adapted to emerging market realities. And if it's a stable success, however that is defined, new variants, range extension, new market segments or completely new markets are waiting for action.

Finally, and to make the point, bigger players in the food sector cannot focus on one or a few new products at a given time. They need to manage plans for a longer time span, normally a couple of years, for different markets, with different risks and chances in a pretty unpredictable environment. Hence, numerous ideas and projects of varying maturity need to be managed through time. This type of portfolio management will be introduced as well.

Changing lifestyles, increasing consumer awareness and sometimes more affluent consumers keep creating demand for new innovative food products. And while there are good reasons to treat the food sector as a conservative one compared to others, the past years have shown the importance of new developments for the health and growth perspective of the food industry. A recent review by Guiné et al. looked at 146 different articles about the role of innovation in food and concluded that "... in the food industry, like any other industry, product and process development is considered a vital part of a smart business strategy. Failure to develop new and improved products relegates companies to competing solely on price which favours the players with access to the lowest cost inputs" (Guiné et al., 2020).

In this review, product – and in particular packaging and the processing – have been highlighted as the key areas and leading keys for commercial success.

As mentioned earlier, trends in foods are changing – sometimes quickly! So, it makes sense to dedicate a special chapter to the big trends in food product development and the new insights and knowledge in science and technology enabling those. This is especially true, as some big trends seem to grow bigger roots and grow robustly.

At the time of writing, in summer 2024, there is no reason to assume that the popularity of vegan and vegetarian diets will decrease. So, the consumer demand for sensorially satisfying plant-based meat alternatives may get more traction. The contribution of science and technology here is not limited to the realization of such products through specific grains and lentils, specifically designed plant proteins, texturizers and stabilizers, etc. Science finds more and more arguments as to why eating less animal protein is good for the healthy life of an individual, a better future for the planet and a more regenerative agriculture.

The rise of non-dairy milk alternatives and the creation of new functional beverages with health benefits beyond hydration remain popular. The technical developments in cold pressing or fermentation bring new opportunities.

Recent technical pushes in cultured meat, the farming and application of edible insects or the application of cannabidiol (CBD) promise either a regret-free meat taste and pleasure or new mental benefits in a stressful world, with calming and stress-reducing effects.

The great number of technical breakthroughs in relevant science areas are building the promise of a much more personalized nutrition, where no longer nutritional requirements on a population level are answered, but gender, life circumstances, age, acute physiological conditions or even the genetic makeup is served.

But we will look at a series of real case studies, hoping to bring alive the main insights, relevant assumptions, uncertainties and mitigation options in the decision making for new food products.

THE FOOD SECTOR

The Food and Agriculture Organization (FAO) describes the global food sector as the complete range of activities from agricultural production and processing, food distribution and retail, to food service and catering. With these components, the sector accounts for approximately between \$9.3 and \$10.0 trillion, which represents nearly 10% of global gross domestic product (GDP) at current prices (Statista, 2024b). It employs over 1.5 billion people, or 27% of the global workforce, and provides livelihoods for millions of smallholder farmers, traders, processors, retailers and consumers (FAO, 2023).

Such a size has consequences: The activities in the sector are a major driver of environmental change, it uses about 40% of the world's land, 70% of its freshwater and 30% of its energy and causes 60% of terrestrial biodiversity loss while contributing to more than 25% of its greenhouse gas emissions (IPCC, 2019).

Since 1961 the end to end (E2E) food system has increased the food supply per capita by more than 30%, doubling the crop yield per hectare through seed development, fertilizer increase (about 800%) and more intense use of water irrigation systems (more than 100%). No wonder that the nutritional situation has changed in the past 60 years: Today more than 2 billion adults are overweight or obese, while another 821 million people are undernourished and 151 million children under 5 are stunted, and 613 million women and girls aged 15–49 still suffer from iron deficiency (IPPC, 2017). And recent studies still find evidence that food demand will increase by between 35% and 56% over the period 2010–2050, driven by still ongoing population growth, economic development, urbanization and a few more reasons (Van Dijk et al., 2021). The critical challenge today is that this demand must be satisfied with minimal environmental impact and fairer access for all people.

The "old" tools, which made the success available in the past, cannot be further used for this. The focus of the new tools must be:

- Reducing food loss and waste, down from today's approximately 25%–30% of global food production
- Improving governance, such as enhancing cooperation, coordination and transparency among stakeholders and implementation of suitable policies and regulations
- Implementing more sustainable and healthy diets, less meat and more plant-based foods
- Strengthening local and regional food systems, such as urban agriculture, community-supported agriculture and short food supply chains
- Promoting sustainable agricultural practices, such as agroecology, organic farming, conservation agriculture, integrated pest management and precision farming
- More investment in agricultural research and development, especially for crops and regions that are underrepresented and vulnerable to climate change

Particularly the last aspect is often forgotten in the wider discussion. Without investment into R&D, better food systems will not emerge. Better solutions are not the result of sudden "eureka" moments, but the result of substantial investments and continuous concerted work. The average annual yield increase for key crops like wheat, maze and rice has fallen from the average 3%–6% in the 1960s to only 1%–2% recently. The investment into agronomical R&D before the "green revolution" was 6%–8% annually and dropped to about 1% in 2000. Since then, the R&D investment recovered partly, especially in India and China, and was accompanied by a doubling of the private-sector investments.

Commercially, the total food sector is forecast to still grow at a compound annual growth rate (CAGR) between 6% to 7% in this decade. This growth is driven by a still growing population on this planet and more demand for convenience, health and environmental care (Grand View Research, 2021; Statista, 2024c).

A BRIEF HISTORY OF THE FOOD INDUSTRY

Food industrialization generally refers to the process of transforming food from raw materials to standardized, mass-produced, often packaged and branded products. This process began in the late 19th and early 20th centuries as part of the Industrial Revolution. The rise of industry brought changes in the way people obtained, prepared and ate food, as well as their relationship with the land, nature and food culture.

A driver in the industrialization of food was the enclosure of the commons in Britain, which dispossessed many farmers of their land and their livelihoods. Rural people were displaced and started to migrate to urban areas in search of work. The loss of land also meant the loss of food knowledge and skills, especially for women, who traditionally had a central role in growing, gathering, preserving and cooking food. Meanwhile on the countryside, the enclosure of those commons became the starting point of large-scale agriculture and industrial farming. They quickly adopted useful new inventions in mechanization, chemical fertilizers and pesticides, and it didn't take long before they recognized the value of monocultures. This was the beginning of booming food production, but also the starting point of constant degradation of soil quality, biodiversity and animal welfare. The Industrial Revolution was based on a science burst in the West that had started in the 16th century. Around that time empirical observation, experimentation and rationality replaced the guidance given by traditional authority like the church. The new thinking quickly influenced agriculture and food production, leading to better crop yields, animal breeding, soil management, pest control and food preservation.

Those coming from the countryside to the industrial centres saw the opportunity for paid work and a better life. But in their confined urban environment, the old food supply ways through their own cultivation were no longer possible. That created the demand for affordable and convenient food to feed the increasing urban population. Food became more commercialized and commodified as new technologies enabled mass production, processing, preservation, packaging and distribution of food. New industrial food formats entered the market like canned goods, frozen foods, ice cream, breads, cakes, pies, cereals, instant mashed potatoes, pastry mixes, jams, jellies and so on.

The work in the factories was hard, with long hours, and physically demanding at low salaries, so food couldn't be too expensive, and it had to be safe, since no worker could afford to become ill. The unpacked food on urban markets was often not fresh, it was difficult to know for the buyer when and how it was made, how long it took to transport it or how it was stored. The need for food was to be affordable, safe and of constant quality.

And often the whole family, even the children, had to work to make a living. There was less time for food preparation and hours of cooking in the kitchen. The moment for a few food entrepreneurs to apply the latest scientific knowledge, allowing the sterilization of perishable food items and now offering them pre-cooked, clean and safe had arrived. That brought another dimension to the new food: Not only safe and affordable, but also more convenient after a long working day.

In England, bouillon turned into an industrial format (see the case study later), and in the early 19th century in the United States, a small company selling cheese, a local New York–based company called Kraft Bros., found a way to pasteurize processed cheese so they could avoid refrigeration, which opened the possibility to sell their product all over the country. And because at that time the town of Philadelphia had the image of a clean, premium dairy town, they chose this name for their product (Wikipedia contributors, 2024b). And nearby a company called Henry J. Heinz, founded in Sharpsburg, Pennsylvania, was successful in the business of selling ambient stable products like horseradish, pickles, sauerkraut and vinegar. In 1876 they launched a new product: Tomato ketchup! This was mass-produced, preservative-free and had an image of purity, quality and healthfulness (and maybe more importantly, tomatoes were at that time known as "love apples"). A recipe had already been invented in 1812 by the scientist James Mease, but with no idea how to make lots of vinegar to stabilize the mixture of tomatoes, sugar and spice at low cost, so it was not ready for upscale production. But that's what Heinz had.

As of today, bottled ketchup is in millions of households and is an easy way to add flavour to meals without laborious preparation. Today, Heinz is a global brand with over 150 factories in more than 50 countries. The company merged with Kraft a few years back to create an even bigger food company with more than $26 billion in turnover.

Kraft, Cadbury, Maggie, Knorr, Campbell, Quaker, Keebler, Kellogg, Quaker, Coca-Cola, Pepsi – they all started around the same time in the industrialized Western markets. What they have in common: Entrepreneurial drive, the foresight to apply new technology, a clear focus on costs, upscaling of (safe) manufacturing and distribution, economies of scale, a brand name, a quality promise and great marketing skills.

All of that became the DNA of the food industry at a time when populations grew, life expectancy increased and masses of people moved from a self-sustained countryside to the promises of the hectic urban life.

The "superpower" of these companies was a creative application of upcoming technical inventions that either were worked out over many hours in the laboratory or an inventor brought to maturity through limitlessly trialling and testing. All had the courage to apply new science and technology from other areas to make processed food.

Also needed was a way to protect the food item from all the dirt and microbial dangers: The packaging. Paper, carton, glass and cans could do the job, but they also made the food item invisible. Now the established sensorial quality inspection with finger, eyes and nose was no longer possible for the shopper. Trusted manufacturers had to put their name on the package to reassure the food quality and safety. These names became the big, trusted brands of today, with the core promise to bring clean and affordable food with a reliable, constant quality.

Today, this is no longer good enough to bind consumers to their brands. They must be supercharged with brand identity, a mission statement, unique selling points and a clear "frame of reference". And beside great taste, convenience and safety, these companies need to show social responsibility and a set of emotional benefits plus ideally new functionalities. Since then, food production has been shaped by the idea that economies of scale allow a profitable business at an attractive price for consumers.

Somewhere in the 1980s, this strategy needed rejuvenation. Something called "mass customization" was necessary to maintain the growth in the food sector. The manufacturing process had to be able to produce more personalized food products without a new price premium. A solution was the creation of "core designs" or a "chassis", applying efficient processes at scale and then making variants with minimal complexity. The "art" was to find the sweet spot between the cost-effective value and efficiency and value-adding differentiation for the consumer.

The food ingredient industry understood the question and began to offer ingredient solutions that did not disturb the high-speed, high-volume processes. Simple ways to add a variety of different tastes, aromas, colours, nutrition, pieces, shapes, forms, etc., without losing the efficiencies of scale built a successful path for the next two decades.

CASE STUDY: BOUILLON

The evolution of bouillon in the Western world in the past 250 years is a good path to follow from its origin in the countryside to basically every supermarket shelf in the world.

Originally a broth, the French "bouillon" was characterized by a translucent, flavourful liquid made from simmering meat broth and bones. This served as a base for soups and sauces and as a cooking liquid for meat, poultry, fish and vegetables.

In the beginning, bouillon was made in the kitchen in the largest quantities from using the less valuable scraps of meat and bones to prepare a tasty, nutritious and inexpensive bowl for people in the countryside. Around 1800, a few chefs in noble houses in France discovered the simple liquid for their purposes. Reduced and clarified with egg white, it became a "consommé". The chefs serving Napoleon, Tsar Alexander I and James Mayer Rothschild used it for their famous kitchen creations. Basic nutrition went premium!

But Napoleon, like other rulers and military leaders at that time, also knew that a good food supply was the key to victory for their marching troops. Sir Benjamin Thompson, later Count Rumford, a rare combination of British officer, politician, experimental physicist, weapons technologist and inventor (Wikipedia contributors, 2004c made the first version of a concentrated meat and bone extract to supply the army soup kitchens with a concentrated, dehydrated base. And during the "hungry 1840s" in Europe, German chemist Justus von Liebig was the first to concentrate meat into extracts, driven by the idea to bring the "goodness" of beef to everyone. Mass application! The result was truly delicious and convenient but too expensive. But George Giebert, a Belgian engineer who had worked with Liebig on this extract, realized an opportunity when he saw the rotting beef carcasses left in the sun after taking the skins for shoe making in Paraguay. Side stream usage made it affordable. Celebrities at that time like Florence Nightingale and the second Duke of Wellington hailed the new nutrition, and the extract even joined Stanley on his historic journey to Africa in search for Livingston. This could be seen as an early form of key opinion former endorsement and event marketing. (Great British Life Staff, 2022).

Another collaborator in Liebig's lab, Charles Gunther, brought the extract to market in England and later branded it under the name "Oxo" to underline the premium beef character. Bouillon had entered the packaged food mass market (Uyehara, 2023).

Meanwhile in Paris, the ordinary people also wanted to eat what the nobles were dining on in their salons. So, shops and restaurants offered their kitchen-made meat-based bouillons in what today would be called "economy range" (Becchi, 2023).

A bit further east in a small town called Heilbronn in Germany, Carl Heinrich Theodor Knorr was also inspired by the idea of "affordable, concentrated nutrition" for the masses. Since 1838 he had been roasting chicory to make a coffee substitute. That required a deep understanding of the drying, baking and roasting of plants to create the right aroma. It then was not a big step to drying other vegetables and spices to preserve the aroma and nutritional content. That kept him and his company busy for several years launching dry soups and later bouillon cubes, rivalled only by nearby Julius Maggi and his new ready-to-cook soups, liquid seasonings and stock cubes, which he finally launched a couple of years before Knorr. Both are today billion-dollar brands, and the bouillon – dry, granulated, powdered, pasty or wet – is still a core staple (Polaris Market Research, 2023).

From the aroma point of view (taste and smell), the stock is characterized by the savoury meaty note, signalling meat quality. But meat is expensive and not always available from the waste stream. So, German chemist Karl Heinrich Ritthausen succeeded in 1866 with glutamate production through the acid extraction from gluten. And in 1908, his Japanese colleague, Kikunae Ikeda, extracted "the savoury substance" from kombu seaweed (see Chapter 2). This was the start of another famous company called Ajinomoto and a taste enhancer called MSG.

The brands are still there, and now the global bouillon cube business is worth about $8.5 billion and stiil growing with 4%–5%. and is still growing (MarketResearchFuture, n.d.; Grandview Research, www.grandviewresearch.com/industry-analysis/bouillon-market?form=MG0AV3). The market is dominated by food multinationals like Nestle, the Swiss company that bought Liebig's Maggie company after the Second World War; Unilever, an Anglo Dutch company that merged with Knorr's brand owner Bestfood in 1990; and Ajinomoto in Asia.

THE PACKED FOOD MARKET

The historic success of packaged food is reflected in its overall size and the diversity it offers. There are many possibilities to segment it. One of the regularly used ones is a mixture of different criteria using categories, formats, food sources and consumption situations: This one uses snack foods and ready-to-eat meals but also cereals and all baked goods, dairy products, frozen foods, convenience foods, sauces and condiments, all types of beverages and or bio, vegan and organic products.

The total size in 2023 was about $3 trillion and is expected to grow annually between 6.0% and 8.55% until the end of the decade (Research and Markets Ltd, 2024). In other words, that's a 30% share of the total food market, including about $400 million for packaging material.

Immense size and still high growth rates. What is driving this? The still growing population, usually rising disposable income and the continued urbanization in emerging markets are building the foundation for this. The biggest consumer drivers are the increasing demand for variety and convenience, plus health for people and the planet.

Right now, most analysts expect the fastest growth in the snack food segment, which includes the convenient "ready meals", plus a massive shift to plant-based foods. Some scientists interpret snacking as a part of human evolution and believe they have identified a couple of genes responsible for this snacking behaviour (Nityananad, 2023).

It is not completely surprising that hunters and gatherers needed to eat something they found in nature that required little or no preparation. Our ancestors probably did not have today's idea of snacking when they were gathering and hunting and eating when food was available, but the human metabolism further evolved to maximize nutrient and energy detection and supply under the evolutionarily prevailing condition of nutrient scarcity.

The concept of "snack food" as we use it today, first came up in the 1920s in the United States, as eating food outside of the main meal eating – often sweet and energy dense – gained popularity. Then it was more culturally and not evolutionary driven. In the 1930s, the new snack foods were mainly marketed to children (Cableway Blog, 2023).

In the years to follow many countries became snacking nations, and the health consequences came into focus. Snacking as such, understood as lower amounts of food consumed more often, is not bad and can fit a contemporary urban lifestyle, but these offerings have to be more than basic cravings for sugar, salt and fat. They must be healthy snacking options. The importance of regular main meals diminishes, at least for those without kids around the table and no time to learn how to cook.

Working pressure is one reason why preparing fresh meals and cooking have lost popularity. More people are working more intensely and also wish to spend their leisure time doing more exciting things. In the 1950s in America and later elsewhere, everyone was watching television. So, a company called C.A. Swanson & Sons offered a solution called the "TV dinner" in 1953. These came straight out of the oven thanks to heat-stable aluminium with compartments for the main dish and sides, ready in just 25 minutes and nicely fitting the TV table (Wikipedia contributors, 2024b). When in the 1980s, all these popular dishes were converted into microwavable preparations, time was cut down to 5 minutes or less and accelerated the global growth of the industry even further.

Like snacks, ready-to-eat meals show a high level of convenience, and they bring extra time for other activities with the reputation of a full meal.

And the third and possibly biggest trend in the industrial food market is also not brand new. Plant-based food has been widespread in the Eastern world for centuries, encouraging vegetarianism mostly due to religious reasons. And with the beginning of the First World War, the Food and Drug Administration (FDA) tried to encourage people in the United States to make Mondays meatless, more to manage the meat supply for the troops than for health, ethical or environmental reasons. The "hippie" movement at the end of the 1960s first brought this holistic perspective on food, plant and animal welfare (Blumenfeld, 2018).

Today it's no longer only a philosophical or lifestyle view, but a sheer necessity supported by a lot of scientific evidence. Today it is a meat and dairy replacement strategy, where plant-based protein alternatives match the animal base and start to drive the meat out of the "centre of the plate".

A shared feature of the CPG market is that the food is protected by packaging. And much of the materials used for that are wasted and even pollute the environment. And plastic, a major human invention, is often functionally superior as a packaging material of food. It doesn't cost much, protects well, is easily available and is light enough to support global spreading. But today's multilayered chemical solutions are so new that nature has had no time to invent ways to integrate these materials into circular systems. So, it seems that now we must find ways to manage "the spirits we once called". The hundreds of million of metric tons produced endanger life on this planet.

REFERENCES

Becchi, M. (2023, 27 February). All About Broth: History, Origins and Anecdotes. Gambero Rosso International. https://www.gamberorossointernational.com/news/history-of-broth-the-peasant-tradition-the-french-school-and-the-massive-bouillon-cube-industry/

Blumenfeld, J. (2018, 27 August). The Plant-Based Movement: A Brief History. Delicious Living. https://www.deliciousliving.com/special-diets/plant-based-movement-brief-history/

Cablevey Blog. (2023, 18 December). The Evolution of the Snack Foods Industry. *Cablevey Conveyors*. https://cablevey.com/the-evolution-of-the-snack-foods-industry

EduRank. (2024, 29 February). *Best Nutrition and Food Science Universities in Thailand [Rankings]*. EduRank.org - Discover University Rankings By Location. https://edurank.org/chemistry/nutrition/th

FAO. (2023). *Statistical Yearbook World Food and Agriculture 2023*. Chapter 1: Economic Dimensions of Agriculture. https://openknowledge.fao.org/server/api/core/bitstreams/28cfd24e-81a9-4ebc-b2b5-4095fe5b1dab/content/cc8166en.html#chapter-1

George, M. (2019, 26 September). Did You Know? The Birth of a New Food or Beverage Product. Food Industry Executive. https://foodindustryexecutive.com/2019/09/did-you-know-the-birth-of-a-new-food-or-beverage-product

Grand View Research. (2023, 31 Juli). *Vegan Food Market Size, Share & Trends Analysis Report by Product (Meat & Seafood, Creamer, Ice Cream & Frozen Novelties, Yogurt, Cheese, Butter, Meals, Protein Bars, Others), by Distribution Channel, by Region, and Segment Forecasts, 2023–2030*. https://www.grandviewresearch.com/industry-analysis/vegan-food-market

Great British Life Staff. (2022, 5 October). Taking stock - The OXO story. *Great British Life*. https://www.greatbritishlife.co.uk/people/22606270.taking-stock—oxo-story

Guiné, R. P. F., Florença, S. G., Barroca, M. J., & Anjos, O. (2020). The Link between the Consumer and the Innovations in Food Product Development. *Foods*, *9*(9), 1317. https://doi.org/10.3390/foods9091317

IPCC. (2019). *Special Report on Climate Change and Land*. Chapter 5: Food Security. https://www.ipcc.ch/srccl/chapter/chapter-5/.

Maximize Market Research Pvt Ltd. (2023, 4 July). The global business consultancy firm. Maximize Market Research. https://www.maximizemarketresearch.com/market-report/packaged-food%20market/122151/

NielsenIQ. (2021, 19 May). Bursting with new products, there's never been a better time for breakthrough innovation. NIQ. https://nielseniq.com/global/en/insights/analysis/2019/bursting-with-new-products-theres-never-been-a-better-time-for-breakthrough-innovation/#:~:text=The%20consumer%20packaged%20goods%20%28CPG%29%20market%20is%20bursting,That%E2%80%99s%20enough%20to%20fill%20the%20average%20grocery%20store

Nityananad, A. (2023). What is Snacking Behavior? LifeDNA. https://lifedna.com/dna-blog/how-genes-influence-snacking-behaviour/

Our History | Auguste Escoffier School of Culinary Arts. (2024, 19 March). Escoffier. https://www.escoffier.edu/about/history-and-timeline/

Polaris Market Research. (2023, 3 August). Bouillon Cubes Market Size & Share, Industry Analysis 2032. Polaris. https://www.polarismarketresearch.com/industry-analysis/bouillon-cubes-market

Research and Markets Ltd. (2024). *Packaged Food Global Market Report 2024*. Research and Markets. https://www.researchandmarkets.com/report/pre-packed-food

Roses, A. K. (2023, 25 April). 23+ Dessert Quotes and Sayings for Sweets Lover [New]. Kites and Roses. https://kitesandroses.com/dessert-quotes/?utm_content=cmp-true

Rutkowski, I. P. (2022). Success and Failure Rates of New Food and Non-Food Products Introduced on the Market. *Journal of Marketing And Consumer Behaviour in Emerging Markets*, *1*(14), 52–61. https://doi.org/10.7172/2449-6634.jmcbem.2022.1.4

Statista. (2024a, 6 May). *Revenue of the food industry worldwide 2019–2029*. https://www.statista.com/forecasts/1243605/revenue-food-market-worldwide

Statista. (2024b). Food – Worldwide | Statista Market Forecast. https://www.statista.com/outlook/cmo/food/worldwide

Statista. (2024c, 10 May). Global GDP 1985–2028. https://www.statista.com/statistics/268750/global-gross-domestic-product-gdp/

The Editors of Encyclopaedia Britannica. (2024, 17 May). Food | Definition & Nutrition. Encyclopaedia Britannica. https://www.britannica.com/topic/food

The Hungry Historian. (2023, 19 October). *Food History – The Hungry Historian*. https://thehungryhistorian.com/food-history

Top 25 Quotes by Jamie Oliver (of 73) | A-Z Quotes. (o. D.). A-Z Quotes. https://www.azquotes.com/author/11079-Jamie_Oliver

Top 25 Quotes by Marco Pierre White | A-Z Quotes. (o. D.). A-Z Quotes. https://www.azquotes.com/author/47068-Marco_Pierre_White

Uyehara, M. (2023, 17 October). *The History of Bouillon Cubes*. Serious Eats. https://www.seriouseats.com/bouillon-cube-history#A%20Brief%20History%20of%20Concentrating%20Soups

Van Dijk, M., Morley, T., Rau, M. L., & Saghai, Y. (2021). A Meta-Analysis of Projected Global Food Demand and Population at Risk of Hunger for the Period 2010–2050. *Nature Food*, *2*(7), 494–501. https://doi.org/10.1038/s43016-021-00322-9

White, C. (2022, 11 October). Famous Quotes on Nutrition and Well-Being. Center for the Advancement of Well-Being. https://wellbeing.gmu.edu/famous-quotes-on-nutrition-and-well-being.

Wikipedia contributors. (2024a, 31 March). Frozen Meal. Wikipedia. https://en.wikipedia.org/wiki/Frozen_meal

Wikipedia contributors. (2024b, 1 May). Philadelphia Cream Cheese. Wikipedia. https://en.wikipedia.org/wiki/Philadelphia_Cream_Cheese

Wikipedia contributors. (2024c, 22 May). Sociology of Food. Wikipedia. https://en.wikipedia.org/wiki/Sociology_of_food

Food Myths, Misconceptions and "Tricks" 2

Today it is popular to whisper about all the "tricks" the food industry is using to maximize their profit and "fool" the consumer. In all media channels, formats are offered where "a view behind the curtain of these mighty manipulators" is provided. These narratives either aim to "demask" existing practices or prolong and rejuvenate old assumptions. The roles are generally pretty clear: Natural, unmodified food (whatever this exactly is) is good and healthy, while industrial formats are unnatural and not good and not healthy! I will try here to bring a different perspective for a few of them, examine the general assumptions and trace the roots. For this, I want to split the stories in this context into "myths and misconceptions" and "tricks".

MYTHS AND MISCONCEPTIONS

A myth captures values, beliefs and the view of the group or culture that has created them. If it is good, it entertains, offers guidance to people or even inspires them. Foods offer many such examples around particular items, ingredients or nutrition claims that vividly circulate in the public domain. Their origin here is sometimes outdated research; incomplete evidence; or just powerful personal anecdotes, generalized single situations or findings. What keeps them alive is that they have a true – or at least – feasible core.

On the contrary, a misconception is more an erroneous assumption that was never true, or at least not in the light of the latest knowledge. It may have been useful advice in the past, but now it may become misleading.

Before I look at a few ones in more depth, a few examples of misleading shortening or false generalization on food are as follows:

1. Eating eggs raises cholesterol levels! It is true is that eggs, especially the yolk, are high in cholesterol. And some studies show a statistical correlation between egg consumption and coronary heart diseases (CHDs). But newer research suggests that the cholesterol in eggs has no major impact on blood cholesterol levels (Lopez-Jimenez, 2024). The correlation between CHDs is now more placed with the way the eggs are prepared (i.e. fried) and the food items eaten with the eggs (ham, sausage, bacon, hashbrowns, etc). So, most people can eat one or two eggs per day! They are a good, affordable source of protein, healthy fats and various nutrients.
2. "Calories in = calories out". This is a typical nutritional mantra, but in its simplicity is not true, as it neglects a few factors. Physically, a calorie is well defined, but biological systems have developed flexibility in their energy supply. Our physiology has a "ancient survival reaction" to counteract energy expenditure when intake is restrained. And the underlying "metabolic rate" is also not "carved in stone", but influenced by many factors such as age, body composition, sex mass hormonal factors, catecholamines, thyroid hormones, body temperature, stress and genetics (Mihaylova et al., 2023).

 DOI: 10.1201/9781032724591-2

3. Frozen food is less healthy than fresh food! This generalization is wrong! Most frozen food items are "individually quick frozen" (IQF) shortly after harvest. That prevents freshness deterioration and nutrient loss in the time between harvest, purchase and consumption and retains a quality that market-fresh food doesn't always have. How fresh the food is depends on the crop, storage, freezing conditions and any process done with the food between the field and fork (Favell, 1998).
4. Carbohydrates cause overweight and obesity! Yes, carbohydrates can be a main source of quick energy for the body, but they can also be an indigestible bulking agent. A better guidance is a look at the quality and quantity of carbohydrates. Whole grains, fruits, vegetables and legumes are carbohydrates, providing a mix of fibres, vitamins, minerals and antioxidants. Refined carbohydrates, often in white bread, white pasta or rice, can trigger a spike in blood sugar levels that effectively blocks the use of body fat stores (Röhling et al., 2020).
5. Brown sugar is healthier than white sugar! Brown sugar is typically made by mixing granulated white sugar with molasses. That makes it browner and adds some flavour and a bit more humidity. The nutritional difference between both is neglectable. Both are a simple carbohydrate – see earlier!
6. Gluten is bad! People with celiac disease – an autoimmune disorder – should avoid gluten, as it can cause damage to the small intestine. But for the majority of the population, gluten is a good protein source coming from wheat, barley, rye and some other grains. For all those people, a gluten-free diet does not provide health benefits or prevent diseases. Actually, today's gluten-free alternatives are sometimes lower in fibre and higher in sugar and fat (Taetzsch et al., 2018).
7. Organic food is healthier than conventional food! Indeed, under basically all regulations, organic food is grown without synthetic pesticides, fertilizers, hormones, antibiotics or genetically modified organisms (GMOs). But that does not make it automatically healthier than conventional food. Both food classes have basically the same standards of quality and safety, which is set by the governmental authorities. But organic food has environmental benefits like reducing soil erosion and water pollution. Organic farming often has lower yield, and therefore higher costs.

Older misconceptions, like "fat is bad", achieved in the meantime a more differentiating view in terms of public opinion, including whether the fats are saturated or unsaturated, trans- or cis, ruminant or plant based, nuts or palm fat. Like in all vibrant sciences, knowledge moves on, and results must be revisited from time to time.

General food myths tend to have a longer history, and they are often anchored in the goodness of food and the contribution to health and wellbeing. Some well-known food myths talk about surprising, hidden benefits of that food. Whether its vitamin C fighting cancer or at least the common cold, spinach as a super source of iron, or the eternal strengths of chicken soup – the all often show a gap between popular public believes and the science supporting them.

Things are different with packaged food. The treatment process has taken away goodness or even added bad stuff. Fermentation, soaking, germination or enzymatic treatment – let alone genetic modification – inactive phytate and make many essential minerals bioavailable in cereals, legumes and nuts. Industrial cooking processes can improve folate bioavailability. These things are known to experts, but not the public! These myths find their find roots in a lack of understanding what the food item is and how it was made. Intuitively simple processes like milling, peeling, cooking and drying are known and create trust. Technically more sophisticated process technologies like sterilization, pasteurization, ohmic heating, high pressure, ultraviolet (UV) radiation, ultrasonic or food additives, etc., are less easy to understand and raise mistrust. And nobody explains them well.

Black and white statements provide no good guidance here. Most food processing influences food quality and safety. That's why it is done! And some processing methods improve safety, nutritional value, shelf life, convenience and diversity of food, while others reduce or degrade some of those characteristics. Making food safe can impact the nutritional quality.

But bold statements like "all industrially processed foods are unhealthy" (see NOVA System in Chapter 17) mask the truths and can damage the original well-meaning intent.

A third source of myths for processed food seems more social and psychological. This is understandable, as food is so much more than energy and building blocks for the body. It is linked to emotions and feelings, framed by preferences and providing identity. No wonder that it can trigger so many connections. Some people may associate processed foods with convenience, comfort, pleasure and nostalgia, while others associate items with guilt, distrust or disgust. Or they like it because it is familiar, affordable or accessible, or do not like it because it is unfamiliar, expensive or unavailable.

I have chosen a few well-known examples to trace the origin, the factors shaping the judgement and the latest scientific view. One is a typical food industry example; the others are fresh food items.

Artemas Ward and the Iron Irony

In the 1870s the German scientist Dr E. von Wolff established a method to determine the iron content in food through titration. At that time science had found ways to measure the compositions of natural materials. So, Dr von Wolf planned carefully and analysed numerous natural sources and was then able to report an impressive 35 mg per 100 g iron for spinach – a finding that made the dull, bitter, dark green leafy vegetable commodity a superb source of the so-important iron. His findings made it to the textbook level, and in 1923 Artemas Ward hailed spinach in his *Encyclopedia of Food* "as health-*giving as can be imagined*", or in today's wording, a "superfood" (Adams, 2020). During that period of economic depression, the U.S. government was looking for ways to stimulate consumption of affordable, nutritious, locally grown food. Spinach was a great candidate.

But convincing children to clear their plate from this bitter vegetable was not easy for parents. Luckily, in 1929 Elzie Segar created the comic sailor man character "Popeye", a hero protecting his lovely, but a bit helpless girlfriend, "Olive Oyl", against the attempts of brutal "Bluto". When Popeye slurped a can of spinach before the exchange of physical arguments with Bluto, he got superhuman strength and gave Olive effective protection.

And while the popularity of the comic grew, science moved on, and in 1937 Prof. Schupan declared the iron content of spinach at only one-tenth of what Dr von Wolf had published. A printing error, the foolish shift of the decimal point in the original publication? Possibly, but it is also possible that Dr von Wolf's measures had determined the iron content of his lab equipment, not only the one of spinach. Maybe Popeye would have been better off if he had licked the can! This piece of new science did not find its way into the wider public for a long time; but at least Popeye got his own statue in Crystal City, Texas, a famous spinach-growing region.

Today, of course, we know better! About 2.7 mg iron per 100 g spinach is not too impressive. And spinach also has oxalic acid, a potent iron complexing agent, able to block the absorption of the iron.

Kidney Stones From Life on Mars?

So, was all this spinach eating for nothing? No – there is no need to demystify this superfood! Popeye's reputation was restored when spinach became known as a good source of provitamin A, folate, vitamin C and low in calories. And vitamin C helps with iron absorption (on vitamin C, see the next story around Linus Pauling).

Spinach was an affordable and popular workday meal in many families after WW2. It is easy to prepare, and together with a few potatoes and a fried egg, an affordable wholesome meal.

In the meantime, nutrition science moved on and looked at whole meal, not only a single component.

Now, oxalic acid reappeared on stage as a bad guy. Basic chemistry reaction rules predict that oxalic acid reacts with the bone-building calcium from the egg, forming a indigestible complex. A classical chemical reaction was applied for centuries in industrial processes for the clearing and de-staining of all types of materials. And even worse, these complexes would precipitate in the kidneys, creating painful stones! So, that`s the final truth about spinach and the egg?

No, this was not the last turn! About 20 years ago, some scientists announced the discovery of very small living things within these kidney stones, which they called "nanobacteria", which are the nucleation point for the kidney stones (Theobald, 2005). Other scientists had their stomach pain with these "nanos", as they were seen as too small to be a living organism. For a while, an extraterrestrial origin was discussed.

While the evidence for an extraterrestrial origin softened, the debate about the nucleus of such stones continued, as mineral-protein complexes (Abrol et al., 2015), or indeed tiny living organisms, but not necessarily from Mars (Fayez et al., 2023). But beyond the exciting debate, oxalate-rich food was identified as one risk of kidney stone formation.

Approximately 1%–5% of the human population develop kidney stones during their life! The list of risk factors is led by dehydration; high sodium, protein and sugar intake; family history; obesity; different digestive diseases; surgery; metabolic disorders; or various types of infections (Akram et al., 2024). Spinach and its oxalate come much later on that list!

And eggs are not a good source of calcium. One large fried egg provides about 2%–3% of the daily value; hence, it is not likely that regular spinach consumption would endanger sufficient calcium supply.

Interestingly, rumours about spinach and calcium also exist in South Korea (Dr Hyo Jung Yoon, head of Specialty R&D at CJ Cheil Jedang Food, personal communication). Here it is the combination of spinach and tofu that is said to cause kidney stone formation. The same chemical logic for a calcium source and oxalate-containing food. The main difference to the Western world: Tofu is indeed an excellent source of calcium!

And one more act in this play! And the leading roles are still for spinach and egg! But now the myth shifts to the preparation method: Reheating! The science behind this concern: Nitrate is naturally present in the environment; it's essential and not harmful, but can convert into nitrite through bacterial activity in the mouth or the gut. And this nitrite can further react with amines or amides present to become nitrosamines, proven carcinogens in animal studies. And since fertilizers increase nitrate levels, today's grown spinach has a much higher potential to form nitrosamines together with the amines coming from eggs.

Now cancer is threat, the final boss in the fight for health! However, the true risk of eating spinach and eggs seems very low. First, spinach contains other substances, such as vitamin C and polyphenols, which are strong inhibitors of nitrosamine formation. Secondly, proteins from eggs directly bind to nitrite and then fully block the reaction to nitrosamine. Finally, the amount of nitrate in spinach varies and not only depends on growth conditions but is reduced by more than 50% by simply boiling – the standard preparation method for spinach.

So, despite all the concerns that have appeared in the last century, spinach is a healthy food choice with many health-promoting constituents (Roberts & Moreau, 2016).

The story is long and not straight but offers a few learnings. Good science is just the best knowledge at that time. Errors can still happen, but they get corrected and framed. First scientists learned how to detect and measure nutrients in food. Early experimental design became standardized, and the measurements became more accurate. Nutrition science broadened the understanding from single components to possible interactions between various food constituents, which raised new concerns. Biochemistry taught us to view the logic of isolated chemical reactions in biological systems differently. Somehow, more becomes possible and less can be fully excluded. It's a bit back to the ancient wisdom that "the dose (can) make the poison!" (Paracelsus).

A possible learning for food developers today: When designing a new food product with, for example, an extra health benefit, it needs more than the simple addition of a particular "goodie". A good understanding about the matrix effects, the total composition and maybe even the consumption situation makes the difference.

Simply fortification is not good enough. Adding vitamin C to a food product that will be heated is not the smartest idea. Unless the vitamin is there, it should act as an antioxidant and protect other minerals. And once this cooking is over, one needs to think about the process from mouth to absorption. But if the iron-fortified staples are enjoyed with a good cup of tea, the tea polyphenols complex the iron and make them unavailable for the body. By adding some milk or lemon juice, at least part of the iron can be "rescued" for the body.

And anybody who still wants to listen to the "spinach story": Please enjoy the recent controversy about nitrate oxide in spinach and a possibly positive contribution to cardiovascular health, cerebrovascular health and cognitive function (Rajendra et al., 2022).

Food myths are not generally warning against the consumption of particular food items; often they endorse food with a long kitchen history. Science is of course eager to prove such beliefs. The next examples are about this.

Vitamin C and the Common Cold

Vitamin C is a nutrient involved in immune function, collagen synthesis, wound healing and antioxidant defence. It is also said to prevent or treat the nasty common cold. An inexpensive cure could prevent the loss billions of work and school days. A good field for science to investigate.

In 1928 the Hungarian biochemist Albert Szent-Györgyi succeeded in isolating the mysterious "anti-scorbutic factor", and 4 years later the British chemist Walter Haworth described the molecular structure of vitamin C and suggested a synthesis way to produce ascorbic acid. In 1937 they both received Nobel Prizes for that. The chemical synthesis made the vitamin very affordable, so no wonder that after the Second World War, the prescription of vitamin megadose in the United States became vividly discussed in part of the medical society.

The leading name in this story is Linus Pauling, one of the very few double Nobel Prize winners in history! In 1954 he received one for chemistry for his groundbreaking research on the nature of chemical bonds and 10 years later for his restless efforts in nuclear disarmament. Profiles in science; https://profiles.nlm.nih.gov/spotlight/mm/feature/biographical-overview?form=MG0AV3

In the 1960s Pauling joined a scientific debate around "orthomolecular", a view that optimal health was based on the perfect balance of the right molecules and the right amounts in the body (Linus Pauling - Profiles in Science).

At that time, Irvine Stone, a biochemist from Wallerstein Company, successfully applied ascorbic acid in industrial applications to limit product oxidation (Wikipedia-Autoren, 2019). He convinced Linus Pauling that vitamin C not only prevents scurvy in tiny amounts but can cure "hypocalcaemia", a genetic defect in humans, forcing humans to take huge amounts of vitamin C to boost health and life expectancy. Prof. Pauling investigated deeper, made self-studies and recognized that he indeed felt much better, with fewer of these nasty colds disturbing his life quality. He told a few reporters this, and the opinion of the "big science voice" did not go unnoticed. Even the vitamin experts from the Food and Drug Administration (FDA) now asked Prof, Pauling to provide his evidence. He did that, reviewing the scientific literature and published it all in the book *Vitamin C and the Common Cold*. The book became a sensation, kicked off a controversy in mass media, and time convinced millions of people to boost their vitamin C intake through fortified products and supplements.

For many years Pauling advocated large doses of up to 10 g per day of vitamin C to prevent colds, but also cancer and other diseases. But other scientists started to criticize the design of his studies and couldn't repeat his results. Prof. Pauling himself continued to take large amounts of vitamin C daily until he died from prostate cancer in 1994 at the age of 93.

Linus Pauling's legacy is still discussed today, and a few researchers continue to explore the role of vitamin C in various health conditions, while many dismiss his theories as pseudoscience. In 2013 a Cocrane review collected the evidence from all intervention trials and concluded ". . . The failure of vitamin C supplementation to reduce the incidence of colds in the general population. . ." (Hemilä & Chalker, 2013).

Nowadays, 65–90 mg vitamin C per day are recommended, with an upper limit of 2000 mg. Above that, side effects occur, such as diarrhoea, nausea, vomiting, heartburn, abdominal cramps, headache, insomnia and kidney stones – we had that one before (Vitamin C, 2023).

Despite the unfavourable outcome of all those studies, practitioners do not discourage vitamin C supplementation, as it may help a bit, be it only through a placebo effect. And it is very inexpensive, thanks to professional chemical synthesis.

Chicken Soup and the Common Cold

If not vitamin C, something else should be able to fight that nasty infection. Already in early Greek writings chicken soup was mentioned as a remedy for the common cold. And while science was reluctant to prove it without a good idea about the active molecule, the mass media regularly brough this piece of ancient wisdom back on their pages when the season started!

Clinically, a common cold is often described as transient infection of the mucosa of the upper respiratory tract, causing an inflammatory response in the body. And as one of the most common illnesses in the world, while there is no real cure, the symptoms can be treated with various home remedies.

Time for science to investigate what the power of that old recommendation is and which role the different components could play. But also, a challenge: The planning of a clinical intervention with chicken soup brings some difficulties. The possible effect to expect is likely to be moderate at best; chicken soup has probably thousands of very different recipes, and the decision about the clinical marker to follow is difficult.

In 2000 a team from the Nebraska Medical Center took up the challenge and approached the idea that chicken soup may be a factor in attenuating the inflammatory response. The idea behind the test was that the infection causes a hypersensitive reaction in the body mediated by the migration of neutrophil cells. And if a soup could reduce or inhibit such migration, the nasty reactions may not occur. This is difficult to follow in the human body, but in *in vitro* assays it is possible to measure neutrophil migration on a plate in the presence and absence of soup (to be more accurate: Neutrophil migration using the standard Boyden Blindwell chemotaxis chamber assay with zymosan-activated serum and f-Met-Leu-Phe as chemoattractants) with 14 commercial soups and a self-created "grandma recipe". All 14 commercial samples showed a reduced migration, and a prototype "grandma soup" was second best. Dr Stephen Renard concluded that these studies showed that chicken soup "might have an anti-inflammatory activity, namely the inhibition of neutrophil migration" and "that the study may have clinical relevance" (Brown et al., 2020).

As such, no strong evidence for a beneficial effect against the common cold has been found, and so the study did not raise too much in terms of further publications. But in 2003 the *Nurse Practitioner* picked up the ball in a "short communication" using that data to report that "the synergistic properties of chicken soup may decrease the inflammatory response associated with viral illness" (Hopkins, 2003). The author also mentioned the nutritional and emotional aspects of giving a warm chicken soup. Since then, no more serious effort has been reported, but regularly with the beginning of the cold season, the more general press picks the idea up.

Chicken soup is good basic nutrition. How good may also depend on the recipe, but widespread types with some veggies can bring per cup up to 20% of recommended daily intake (RDI) of valuable protein plus vitamin A, vitamin B_6, vitamin B_{12}, vitamin E, vitamin K, iron, magnesium, phosphorus, zinc and selenium and a little bit of vitamin C. And because it is easily digestible and low in calories and fat, it is no burden for a weakened body. Just the often relatively high sodium content (up to 50% of recommended daily allowance [RDA]) can cause eyebrow raising from nutritionists. Or in more clinical words: "Chicken soup provides nutritional support, helps alleviate dehydration, and can be an effective tool for patient care management" (Hopkins, 2003).

A few more courageous publishers mention the monosodium glutamate (MSG) present in the soup and its role as an "taste enhancer", stimulating appetite in patients. But MSG has its own myth as a burden!

Taste and Health Enhancer

A classical case in the history of industrial food: MSG, or simply glutamate! It was discovered in the early 20th century by the Japanese biochemist Kikunae Ikeda at the Imperial University of Tokyo. He identified taste giving substance in local kombu as glutamic acid, one of the nonessential amino acids. Smart enough, he patented the combination with sodium and founded a company that later became known as "Ajinomoto", today one of the world's leading producers of peptides and amino acids, including MSG.

The commercial success for MSG came as a cooking aid or taste enhancer, as it was giving the typical meaty, savoury, umami notes at a very attractive price. Nowadays, it's made through the fermentation of starch, sugar beets, sugar cane or molasses. It has grown into a $7.3 billion market. (Research and Markets Ltd., o. D.)

The success story became clouded in 1968, when Robert Ho Man Kwok, a Chinese American doctor, wrote a letter to the *New England Journal of Medicine* (NEJM) reporting strange temporary symptoms after eating at a local Chinese restaurant. He described headache, flushing, sweating, chest pain, nausea and numbness in the mouth and throat and speculated about ingredients in the meal causing the reactions, possibly the soy sauce, the cooking wine, the sodium or the MSG. He encouraged readers to start their own research, as he himself was too busy with other work. A few readers reported similar cases with similar symptoms in various dishes in Chinese restaurants. All were somewhat different, so that the NEJM editors started to wonder about the significance of this obvious inconsistency. Meanwhile, a few popular newspapers picked up this debate in the clinical journal as a great piece of science. Now more people reported similar symptoms and the mystery got a name: "Chinese restaurant syndrome"! MSG was not yet mentioned as the cause.

That changed when neurologist Herbert Schaumburg started a series of very questionable MSG feeding trials with humans to pin MSG as the suspicious substance and toxin (Science History Institute, 2023). And he inspired Robert Olney, a psychiatrist at Washington University, to start experiments where he injected albino mice pups a solution with 4 mg of MSG per gram of body weight. When dissecting the animals he claimed to recognize some brain lesions. The studies were so bad, that immediately other researchers felt forced to remember that the used dose would equal about a pound of MSG per adult, that MSG is eaten and not injected under the skin and that any control was missing.

But these were the times when people in America became concerned about chemicals in their food. Just a few years earlier, the impactful book *The Silent Spring* was published. And here now was a substance invented by the Japanese, the old World War II enemy. No wonder the "holes in the brain" story made waves, and a few more ill-designed animal studies added speculations about the role of MSG in autism, obesity, cancer and blindness. Fortunately, a series of better-designed human intervention studies failed to trigger the syndrome or any consequences (Rosenblum et al., 1971).

And then the public concerns motivated the FDA, the United Nations Food and Agriculture Organization (FAO) and the World Health Organization (WHO) to once again thoroughly review the safety of MSG in all details. They unanimously concluded once again that MSG must be recognized as generally safe and present in many natural protein sources, which anyway lead to an average daily intake of 13 g of glutamate per day, to which the added MSG would add just another 0.55 g.

But the concern remained – just the reason to believe changed. In the meantime, glutamate became known as a neurotransmitter in the brain, so then it was that eating too much MSG could raise levels first in the blood and then in the brain, where functional disruption could happen. More research was needed in to prove that MSG intake was not leading to increased levels in the blood, unless taken in amounts "vastly in excess of normal intake levels" and that it would not pass the blood-brain barrier: "The blood-brain barrier effectively restricts the passage of glutamate from the blood into the brain" (Fernstrom, 2018).

As far as we know, MSG is safe, a powerful and very affordable taste enhancer, thoroughly safety tested and regulated as a food additive in Europe. But whenever consumers ask, even in Asia, where "vetsin" is used daily in many kitchens, cooks feel guilty using MSG as an unavailable taste giver and remember a "problem with the brain". Somewhere there is deep insight they cannot trust the food industry. Something so good and so cheap must have a dark catch.

And research has not stopped, but it changed direction! A team from UC Davis tested the idea using MSG as a salt (NaCl) replacer (Halim, 2020). They took four popular dishes and replaced the table salt with MSG in such a way that at least 25% sodium reduction was obtained (relevant for making a "reduced" claim). They gave the dishes to consumers and asked their opinion. The result showed that in two of the four dishes, the "MSG version" was judged superior and in the two others it came out par – quite a journey from the food street urchin, potentially massively health damaging, to a possibly powerful tool helping people to cut their salt intake without compromising taste. Or as the authors of previous study concluded, "that MSG can successfully be used to mitigate salt and sodium reduction without compromising consumer acceptance of better-for-you foods."

The next years will show if all this new research is able to correct public perception that is rooted in some awful science. In many lower-income kitchens of this world, MSG brings taste, aroma and mouthfeel to healthy, but often quite bland, vegetable dishes. But it leaves the person cooking with MSG with the unpleasant feeling that they are putting the long-term health of the family at risk. This is a concern that the table salt seldom raises, despite the much stronger scientific evidence for long-term health damage!

One point remaining is that MSG triggers meaty taste perceptions in dishes, where there is no or only little of that value-giving ingredient present.

The Ancient Goodness of Spelt

Spelt is an ancient cereal grain, related to wheat, barley and rye. Human consumption dates back much longer than the one of chicken soup, back into the Neolithic age. Since then, it has been widely grown in Europe, but was then forgotten for many years. It reappeared in the public at a moment when the nutritiousness of modern grain species became questioned. Since the mid-1940s these variants went through extensive modifications and crossbred, focusing on yield improvement varieties, later combined with reduced susceptibility to environmental stressors like diseases, insects and climate conditions. The result was a decrease in the genetic variability of that plant family combined with an assumption of the ancient nutritional properties. A pristine local breed, like spelt, held the promise of this ancient goodness, the questionable idea, that the diet in the beginning of human civilization was better than the one today.

Spelt itself is a type of wheat with a hulled grain and has a pleasant nutty, slightly sweet note. Like other whole grains, it is indeed more nutritious than fully refined grains, because the hull contains more protein, a good amount of fibre, plus essential nutrients like iron, magnesium and zinc. Like all commercially successful whole grain variants, spelt as part of a high-fibre diet can be linked to positive health. A recent meta-analysis confirmed the association between the consumption of whole grains and a decreased risk for cardiovascular disease and cancer (Aune et al., 2016).

Building on such observational studies, a different researcher team took the effort to review all evidence about the health impact of ancient grains and compared them to the dominating modern mass market variants. They found no major differences in the nutrient composition or other potentially beneficial secondary metabolites (Dinu et al., 2018), but some indications for a light improvement of antioxidant parameters, glycaemic and lipid status. Their conclusion therefore was that "the number of human intervention trials conducted to date are numerically insufficient, (and) it is not possible to conclude that ancient wheat varieties are superior to all commercial, modern wheat counterparts in reducing chronic disease risk."

So, for now, spelt is a tasty and flavourful nutritious grain, but not an almost eradicated magical superfood which now finally found its way back into the public spotlight! But there isn't much evidence that all of sudden, a hidden superpower will show up. Somewhat like vitamin C and Linus Pauling, health promises beyond the established nutrient profiles are not substantiated. As always, more good scientific work may bring new insights.

But there seems to be an underlying assumption, that a yield-driven breeding program would "unavoidably" lead to a deterioration of the nutrient profile. That can happen, but is not a logical consequence (Scharff et al., 2022). The antioxidants and other secondary metabolites we hail in these crops are parts of the "plant defence system", their protection shield against environmental stressors. Here yield improvement and nutrient profile are two sides of the same story and not a "give and take".

Belief in the "goodness of spelt" is fine and without consequence for many. Only when it becomes promoted as a solution for those truly gluten sensitive is caution needed. In the Western world about

15% of the population believe they are gluten intolerant, but maybe only 1%–2% really are (World population review, 2024). And for this minority spelt is a actually a dangerous alternative, since it has as much gluten as all the other grains. The misconception in the media went so far, that governmental health authorities were forced to issue a public warning for them (German Federal Institute for Risk Assessment, 2023).

There are many more myths about food that people believe and follow! Some originate from ancient cultures, religions or traditions, while others may have been influenced by modern opinion-formers in media, marketing or pseudoscience.

Some may be true for a few people; some are clearly more misleading than guiding. Some may be even dangerous; others probably not. An example of the latter is that some people still think that eating carrots can improve their eyesight.

A Mission for Carrots

This myth is one with a known birth date: Fall to winter 1941! The originator was the British government, and it was part of a propaganda campaign during World War II. The Germans were trying to invade England, but luckily the Royal Air Force had developed a superior airborne radar system, effective in night fights. And they had John Cunningham, nicknamed "Cat's Eyes", who was said to have shot down 19 German planes that were regularly attacking during the dark. To hide the existence of their new radar, the government launched the story that this famous John was eating lots of carrots and therefore was able to see better in darkness, so he could shoot down many planes.

The narrative was probably not only supposed to mislead the Germans but also to help citizens feel safer when moving around in the darkened cities without light (whole story: Smith 2013). No doubt, carrots are healthy vegetables and contain provitamin A. But that they improve eyesight in the dark remains a myth. But it helped to make the locally and abundantly available vegetable popular, while the Germans were blocking the shipping to England. So, where the Americans had Popeye and his spinach, the Brits got wing commander Cunningham and his carrots: Two cases of "superpower veggies".

A recurring scheme behind all those myths, and many others, is a faith that food coming from nature that is not or only minimal processed has unimaged powers. On the other hand, packaged processed food is made by the industry using dirty tricks. They either stripped of most – if not all – of the original goodness to ease their manufacturing or they have intentionally added something to the food that is increasing sales and boosting profit. And industrial food making is large-scale manufacturing, constantly changing the processes to maximize efficiency. The consequences for the food quality are not known and hard to predict. Also, industry constantly used substances like pesticides, hormones, antibiotics, additives or even GMOs to boost their profit.

Besides the tendency to make unfair simplifications, the general guidelines about food and nutrition for health are losing significance. A calorie is no longer a calorie for everybody: A strict vegetarian may be short on vitamin B_{12}, calcium, iron and zinc, while pregnant women should monitor their folate level, and triathletes better have 1 g of salt in their drink.

And while the simple recommendations from the past are becoming more and more specific (or personalized) and dependent on other factors, other "well-known stories" keep on waiting for their final verdict.

There is a "job to be done" for science, opinion-formers and industry to make food and its manufacturing "digestible" for consumers, transparently talking about the way food is grown and processed nowadays, not hiding weaknesses, and building more trust. And continuing to demystify a couple of famous food perceptions. Consumers need to come into a position where they can judge for themselves, without a feeling of being fooled. That's a long-term task, and until then it seems likely that the memorable, easily understandable stories and myths will keep on catching attention. And back to the beginning of the book: Food is not a functional, rational energy for a biological machine, but a deeply emotional perceived experience.

THE "TRICKS" OF THE FOOD INDUSTRY

Introductory Comments

Overall, the food sector – including the industry – has done an excellent job in the past decades. They developed innovative products, strong brands, effective supply chain management and robust business strategies. And they were confronted with many challenges and controversies about food safety, sustainability, nutrition, animal welfare and social responsibility. The public opinion about the food industry is therefore not uniform or static, but shaped by factors, such as media, education, culture, personal values and preferences. Some people may have a positive view of the food industry as a source of innovation, convenience and diversity, while others see them as leading causes of environmental degradation, health problems and ethical issues. The opinion about various food sectors differs: Many people have a favourable opinion about organic or locally grown food products but a very critical view about processed or fast-food products. Pragmatists are happy with French fries from fresh local potatoes!

The truth is that the food business, like any other industry, wants people to buy more or at higher price. Some of those approaches receive vivid criticism:

- The bigger the portion, the more we eat. Studies have indeed shown that people tend to consume more food when it comes in larger portions or containers (Hollands et al., 2015). Larger tableware is having the same effect. And food companies use oversized packaging to make you feel you are getting a good deal but encourage you to eat more than you need.
- The colour of food affects perception. Food companies use different colours and additives to make their products look more appealing and fresher – red dye to make meat look juicier, yellow dye to enhance the colour of cheese and butter. And it is not only the visual impression. Colour is the most sensory cue setting people's expectations about taste and aroma (Spence, 2015).
- The expiration dates are marketing tricks. It is true that food companies use expiration dates to indicate the freshness and quality of their products, even if not legally required. Normally they do that to flag the moment when the food quality deteriorates, but not when it becomes unsafe to consume. It is done to protect the brand quality perception, not to convey freshness or encourage people to throw food items away too early, so they need to buy them again, especially if they are exposed to heat, light or moisture.
- The serving sizes are unrealistic. In times of overweight, food companies tend to present their nutritional content in serving sizes that are smaller than what people really eat. For high-caloric items, a simple way is to "improve the calorie optic", as people have difficulties estimating portion size correctly. But companies are not free to define portion sizes.
- The ingredients list is unclear. Many regulations allow the use of alternative descriptions instead of using the scientific names, which could boost the artificial perception of the ingredient used. Something with a chemical name cannot be good food.. Vitamin C sounds better than ascorbic acid. And baking soda sounds much more "foody" than sodium bicarbonate.
- The sugar content is hidden. The ingredients list can indeed list different sugar types like fructose, glucose, sucrose, maltose, dextrose, corn syrup, honey, molasses, agave nectar or fruit juice concentrate. But the nutrient table should accurately sum them up.
- All additives are chemicals! Well, all substances are chemicals! But in food regulation, not all substances are made synthetically or through chemical modification. What makes an item additive is not the origin or way of making, but normally the fact that it is not consumed on its own (see the information on flavours).

"Marketing" claims

The technical, legal and regulatory aspects of food will be discussed more in Chapter 11, together with data showing how these claims are perceived and understood by consumers. This mandatory food information is regulated by laws and standards in different countries and regions to ensure that consumers are well informed, protected and not misled by false or misleading claims. Typical regulation systems are the EU Regulation (EU) No 1169/2011, the FDA Food Labeling Guide or the Codex Alimentarius, as a collection of internationally recognized standards, codes of practice, guidelines and other recommendations relating to foods.

Overpromise and Underachieve

But there are other statements, here called "marketing" claims, that are differently regulated by country or region. They are not subject to the same level of scrutiny or verification as nutrition and health claims, with their specific standards and requirements. Around those claims opportunities exist allowing the food industry to use tactics to let their products shine a bit brighter than they actually are. A few examples to highlight this:

- "*Natural, all-natural or 100% natural*" are claims found in packaged food categories – no big surprise, as many shoppers look for natural products when shopping, and many of those are also willing to pay a little extra for food having a strong, positive influence on consumers' product evaluations and choice (Berry et al., 2017). At the same time regulating authorities like the FDA in the United States have not sharply defined what "natural" is, so consumers to some extent have to fall back on their own assumptions. Then a popular pasta product with cheese can carry an "all-natural" claim, which some consumers even translate as "*organic*", "minimally processed" and "free from genetically modified organisms" (GMOs). Not completely true! In the United States the FDA has refused requests to define the term "natural" explaining that "*from a food science perspective, it is difficult to define a food product that is 'natural' because the food has probably been processed and is no longer the product of the earth*" (FDA Error, n.d.); however the agency has objected the use of the term if the food contains added colour, artificial flavours or synthetic substances. Authorities like the FDA often take the position that a claim should be consistent with consumer expectations. Consistent with the approach that consumer perception should shape the labelling rules, the FDA in started in 2015 a public consultation on the use of this term. Suggestions have been made to define different degrees of naturalness, ranging from "*100% natural*" over "*natural*" to something like "*made with natural ingredients*". Besides the ingredients used, the processes applied to manufacture are relevant. For bio-engineered material like seeds, it is already becoming difficult. Genetic modification has happened somewhere, but the result is identical to "*organic*". "Chemical processing" cannot qualify for "*natural*", but "physical" can! Drawing the line here is not always easy. Finally, processing aids are part of this discussion: These substances, like maltodextrin, are intentionally used in the processing of foods, but they are not consumed as a food up to the unintentional but technically unavoidable presence in the final product. Are such products natural?
- "*Clean label*" is another very attractive food claim and a similar case! The label term suggests to the normal consumer that a product is made with natural, simple and minimally processed ingredients. The regulatory consensus again excludes the usage of artificial colours, flavours, preservatives, sweeteners, GMOs and antibiotics. But ingredients like palm oil, agave nectar, carrageenan, stevia, almond milk or super premium "*Himalayan*" rock salt are often highly processed ingredients, and some of them have in addition a very detrimental environmental footprint.

- "*Low-fat or fat-free*": This one seems fortunately well regulated. The FDA, for example, allows such claim if less than 3 g/0.5 g of fat are in one serving. But two warnings exist: A serving is not a portion! The experts and the authorities may have ideas about serving sizes, but this varies depending on the type of food, the package size and the manufacturer's discretion. Hence, making a serving size a little smaller all of a sudden makes a product low fat. The second point is that the products can instead contain other rich sources of calories, often from simple sugars or carbohydrates. That can lead to about the same caloric content, possibly more additives and more drastic processing than the original one at a higher glycaemic response.
- "*Whole grains*": This is a statement closely associated with healthy foods. The problem starts with the fact that whole grains aren't always "*whole*." The grains may have been milled into very fine flour. So, yes, they still contain all the good ingredients of the whole grain, but the beneficial slow digestibility that is not raising the blood sugar levels is lost due to the destruction of the food matrix. And then often a relatively small amount of such whole grain is enough for the claim. The addition of simple sugars, carbs and or high-fructose corn syrups is not changing that.
- "*Other value-giving ingredients*": This is commonly a claim where it is possible for the producer to promise more than what is delivered. Examples are the popular "*veal liver sausage*", where more than 50% of the "liver content" needs to come from calf or young cattle. The rest can be pork liver, and the meat content in the liver sausage can consist of pork and/or beef or veal.

  If this amount of veal liver is still seen as too expansive, better call it: "Veal meat liver sausage". Now the sausage only needs to contain veal meat, but not veal liver. In other words, the liver in this sausage can come entirely from the pig. In that case, the consumer needs to be informed properly and the product needs to get the name "*veal meat liver sausage with pork*".

This is only a short selection of typical examples. The whole area of "*provenance claims*", generally conveying premium, is often regulated by different laws and standards differently in different countries and regions, depending on specificity and protection. The principle is to provide the true origin or provenance of the food. So, if a cheese is made in France with milk from Germany, the label must indicate the origin of the milk. If a cheese is labelled Parmesan, it needs to originate from the Parma region, be made from cow's milk and aged for at least 12 months. But outside the EU, the term Parmesan is often used for cheeses that are somewhat similar but not identical to the original. And they do not need to come from Parma any longer.

Beside trained lawyers and food experts, hardly anybody else can quickly judge what such products now actually really deliver and what is just nice storytelling or creative management of existent regulations.

Such claims trigger the peripheral, fast, intuitive, emotional and subconscious shopper response. It relies on heuristics, associations and past experiences to make judgments. The regular shopper normally does not have the time to familiarize herself with the latest legal verdicts and amendments. Their buying decision relies on the fast, intuitive, emotional and subconscious judgments to sort facts from fiction (see Chapter 11). But if then later, when there is time to listen more carefully, a critical spokesperson, friend, colleague or opinion-former in social media "demasks the promises" made on such a product as what they really mean, the general mistrust into any of such food information gets fed, and the overall confidence in the reliability of fact-based statement becomes eroded.

Just to make that point here again. There is nothing illegal here in such claims. And mostly it is not that the regulators have been somewhat tricked by smart industry lobbyists. They have just tried to allow what they believe the regular consumer expects when being faced with such claims. But what the average consumer believes and understands can differ of course.

A final example from international trade. In 2013 the EU and the United States started a broad Transatlantic Trade and Investment Partnership (TTIP), at that time covering nearly 50% of the world gross domestic product (GDP) (Wikipedia contributors, 2024), which was estimated to bring hundreds of billions of dollars to the partners. Parts of the intended agreement dealt with consumer protection and food safety. In this chapter, details on food safety, the so-called "*pathogen reduction treatments* (PRTs)", became the symbol for the food industry and their striving for dominance. A particular practice common in the United States as a means to kill bacteria such as *Salmonella* and *Campylobacter* was to soak chicken in cold chlorinated water to preserve the shelf life. This procedure was not allowed in the EU, where instead the slaughtered chicken was dipped in a 60°C water bath for a minute.

Activists managed to leak out this detail of the overall classified negotiations, creating a public outcry in Europe about the health dangers of consuming chlorinated chicken and its bad taste, which was the death kiss for the local poultry farmers and the overall suspicion that such chlorine cleaning is planned as a precedent for the wider use of additional chemicals in food processing.

It didn't help that both, the European Food Safety Authority (EFSA) and the German Federal Institute for Risk Assessment (BfR) stated that there are no safety concerns regarding the consumption. A big portion of the European consumers, guided by vocal non-governmental organizations (NGOs) rejected these "*chlorine chickens*" (German: *Chlorhühnchen*), and the entire negotiations slowed down until finally U.S. President Donald Trump stopped them for different reasons.

Superfood

The label "*superfood*" is not an invention of smart marketers in the last 20 years! Its proven first usage in the United States was more than 100 years ago. Only a few years after the first coining of a Central American state as a "banana republic", their export article became hailed for its nutrients, the versatility, ease of digestion and the very practical natural packaging. About 70 years later the FDA handed the title to the blueberry, being "the king of antioxidants", full of vitamins and minerals and low in calories (Limerick, 2020).

Then food marketing discovered the "*superpower*" of such a label, and the list quickly grew from dozens to hundreds and thousands of food items. The magic was the implicit consumption idea, that the consumption of such superfood somehow repairs other less healthy eating and drinking habits. These days a single food is rarely seen as a remedy to fix an unhealthy lifestyle with just a few portions.

Initially superfoods were typically nutrient rich, full of antioxidants, flavonoids, other secondary metabolites and micronutrients able to fight chronic diseases. Often fruit or vegetables, the list is today enriched with nuts, seeds, grains, dairy and fish. The "reason to believe" is limited scientific evidence about the presence of beneficial substances, laboratory or animal studies or intervention with very high doses, and results are difficult to transfer to regular diets. And it helps if the food item is exotic, so that the title allows for a price premium.

Superfood is a powerful marketing term but nothing that is scientifically or legally defined. A recent survey of superfood-describing websites in Europe revealed 136 food items in which the "Nutritional and healthy properties" were "*presented. . . in a very simplified manner. . . (but). . . generally not wrong*" (Cobos & Díaz, 2023). It helped that the EU prohibited general advertising with "superfood" and similar terms associating the consumption of products with health-promoting effects unless effectiveness has been confirmed through the strict approval process.

The overall relaxed scientific reactions to the "*superfood*" overpromise can be explained by the generally positive impression about fruit and vegetable consumption. A bit too much "noise" from marketing will not negatively impact health and wellbeing – only the personal budget.

And, not to forget, it's a fundamental principle in science that the best knowledge can change. Something not proven until today may become the new truth in the future. What has been demonized in the past can be redeemed in one of the next research waves, like the eggs and the fats in the cholesterol debate.

Best Before/Used By

The "*best before*" for foods has been explained as an industry trick. The train of thought is that by adding an unnecessarily early "best before" date on packaging, the food industry encourages people to throw away perfectly edible food too early and then buy more.

Now, the best before date is not a safety, but a quality, indicator. It should tell the consumer how long the food will have optimal taste, texture and nutritional value. It protects brand reputation, ensuring the product has the promised and expected sensorial quality. This date is defined by rigorous shelf-life testing and analytics, and it is meant to help consumers make informed choices about their food.

And normally no one in the food chain wants a product not consumed but thrown away. It doesn't make sense. "Artificially" shorting the time in which the product can be sold impacts the turnover negatively. It shortens distribution lengths, hence reducing the sales points, and the trade partners are not keen to have items on the shelf that can be sold only for a limited time. And the idea that consumers would happily buy a product again that they just have thrown away is largely an unsubstantiated assumption. Consumers better use their senses before putting an item just behind the "best before" date to waste.

It's becoming more difficult for consumers to differentiate between "best before" and the "use by" label. The latter one is now indeed focusing on food safety and is normally used for highly perishable food that is likely to spoil and pose a risk of making people sick. Typical items are fresh meat, fish, dairy products, fruit juices and other chilled ready-to-eat foods.

But consumer research indicates that consumers take the best before date as the safety limit and throw away good food products. It is estimated that up to 10% of the 88 million tonnes of food waste generated in the EU every year are somehow linked to best before date labelling (European Commission, 2018). As part of the EU green deal, the commission is currently planning to revise their guideline. Options under discussion are (1) the full removal of "best before" for non-perishable foods such as pasta, rice, coffee, tea; (2) total abolition of the "best before" date concept, only keeping the "*use by*" date; or (3) A better expression and presentation of the date marking concept (European Commission, o. D.).

But for this possible misleading statements on packaging, it does not seem correct to label it a food industry trick. The current regulatory framework here has no benefits for the food industry. Rather the opposite! With the end-to-end sustainability responsibility of the business, the high amount of unnecessary food waste created in consumer homes is part of the environmental, social and governance (ESG) action plan, and any regulatory guidance improving this without adding costs to the business is highly welcome.

Less Salt

Many elaborate observational studies link too much salt (sodium) intake to higher blood pressure and subsequently cardiovascular disease, kidney damage and perhaps stomach cancer or heart failure. Table salt (NaCl) contains 40% sodium and 60% chloride, both essential minerals, but in excess detrimental. Food makers have used salt for thousands of years to make food tasty and function as a taste-masking flavour enhancement and texture improvement through interactions with several food components and to preserve it. Processed food has been identified to disproportionally contribute to salt intake beyond the recommended 5 g/day. Thus, the food industry started to work intensely on salt/sodium reduction initiatives. The first challenge here is to maintain the same level of saltiness perception in food with less salt. Saltiness is not only one of the basic tastes that humans can detect but plays an important role in enhancing flavour and aroma and palatability of food in general.

Simply telling the consumer to eat less salt or to like low-salt food as much as the original one does not work. Repeatedly consumer research shows that less saltiness results in less liking, predicting less consumption and fewer sales. Even when people are told that such a product is good for their health, they often reject the idea and insist on their known taste. Therefore, just using less salt is not a viable option! A company or big brand just doing that would lose some of their loyal consumers, making them move to an alternative product or supplier. Or they would use their table salt to bring the products to the level of saltiness they are used to and which they like. So, that approach is an unworkable option for the food industry, and for the public health agenda an ineffective approach, not lowering the average salt intake. Thus, a "trick" is needed. Something like "same perceived saltiness, but truly less sodium!"

Today, new science and technology offer different strategies to achieve that goal by either mimicking the taste of salt or by increasing the salt intensity through stimulating other receptors in the mouth.

The experts nowadays tend to perceive the challenge of salt reduction more as "saltiness modulation". How does that make the solution more feasible? The attempts can be summarized in six different approaches:

1. Step-by-step reduction:
   That's the classical nutritional and technically least challenging approach. The idea is based on studies suggesting that saltiness perception adjusts to lower sodium levels when salt intake is reduced. Today, this "natural regulation" is discussed more controversially. However, other studies have found no significant differences in saltiness perception across age groups, or even an increase in sensitivity to salt in some. And beside those more recent doubts over the attenuation of the salt receptors, a prerequisite would be salt reduction in the total diet, not only single products.
2. Salt substitutes:
   Salt replacers are low-sodium alternatives to table salt. Saltiness is mainly transferred through a specific receptor on the tongue called, epithelial sodium channels (ENaCs), which releases a neurotransmitter triggering saltiness perception in the orbitofrontal cortex when sodium is passing through. Besides sodium, only lithium is able to trigger this. But lithium is toxic and hence not an alternative. Only the existence of a second, less elective, receptor explains, why other ions like potassium, calcium and ammonium can trigger saltiness perception too, but to a much lower extent and often with an increasing metallic side note, which sodium suppresses. As result, on average 20%–25% of the sodium present can be replaced! But salt replacers are suitable for everyone, especially those with kidney or heart problems or those taking certain medications that affect potassium levels.
3. Saltiness enhancer:
   This is a more physical approach by which hollow, more sodium crystals are used. Fewer grams of salt raised the same saltiness through better kinetics. NaCl typically crystallizes in tiny cubes, which often do not dissolve completely when eaten. So, we consume more salt than we actually taste. Physically creating more surface area per weight allows us to cut the unnecessary extra and could enable up to 50% salt reduction. But this approach only works for dry application. In the dissolved stage this approach does not, as the hollow microstructure disappears. Companies have created a patent on the ways to improve the surface-to-weight ratio (Fevzioglu et al., 2017). Similar approaches aim to lower the size of salt crystals and thereby enhance dissolution rates because of the greater surface area for rapid dissolution in the saliva, leading to increased perception of saltiness (Vinitha et al., 2021).
4. Food structuring:
   If the food consists of layered structures, the alternation of lower sodium and higher sodium layers can provide similar saltiness than the uniquely salted product. By altering the size, shape, distribution and interaction of food components, such as proteins, starches, fats and water, food structuring can influence how salt is released and sensed in the mouth. Recently gels and emulsion product structures have undergone structural food design to optimize the salt delivery to the taste buds and saltiness perception (Hebishy et al., 2023). Ideas include more porous or emulsions or gels that affect the mobility and availability of salt in the food matrix. In some bakery application or layered products like lasagne, the option exists to cut up to30% of the salt originally present. Such strategies are limited to solid food and cannot provide solutions for liquid foods like soups and sauces.
5. Cross-modal correspondence:
   This builds on the saltiness perception that other, very different sensory cues, such as colour, shape, texture or sound, can create. Studies with potato chips showed that the products were rated saltier as crispier they were (Zampini & Spence, 2004). Similar effects have been shown through the usage of Szechuan pepper and chili (capsaicin), believed to cause a chemically initiated sensation via receptors belonging to the touch system. Both spices are very prominent in Asian cuisine and have shown remarkable increase in sodium perception through their hot, prickly, pungent flavours (Wang et al., 2022).

6. Umami modulators:

   Umami is seen as the fifth distinct taste on the tongue. When the umami receptor is triggered, this raises a meaty perception in the brain. When people have consumed food having umami and salty taste, umami alone can later trigger saltiness on its own. Since more than 100 years, low-cost options exist to trigger the umami taste: Guanylate, inosinate and particularly MSG or any foodstuff containing natural MSG, like yeast extract (YEx), hydrolysed vegetable protein (HVP), fermented sauces like fish sauce and foods naturally rich in MSG, like mushrooms, tomato, Parmesan, etc. Other foods that naturally contain nucleotides (the essential building blocks of DNA and RNA that all living organisms need) can further intensify the umami perception.

   Thus, umami taste can be used as a trick to raise the saltiness perception, but unfortunately, this does not work completely. Good technical solutions nowadays allow us to reduce sodium levels by 20%–40% without changing the saltiness perception. This is an important and powerful tool to lower salt intake. This "trick" of the food industry is not applied to improve profits! That is because whatever table salt is replaced with, the alternative is more expensive, while the consumer is generally not willing to pay a small extra for such an "healthier" option. And perhaps more solutions would have been brought to "market readiness" if consumers were willing to pay a little more. This is not such an unreasonable plan, as it is possible to convince at least some consumers to pay $50, $100 or even $200 per pound for some very rare, exotic salts having the same sodium content as ordinary table salt and a few impurities (Lyons, 2022).

Weight to Volume

Did you ever wonder why a 1-L tub of ice cream weighs only 500–550 g? Or were you ever astonished how filling two scoops of premium artisanal ice cream from your favourite ice cream parlour can be?

The "trick" is called "overrun"! It describes the percentage increase in volume of ice cream compared to the amount of mix used to produce it. Or simpler: It represents how much air is whipped into the ice cream during freezing. If 1 L of ice cream mix becomes 1.5 L of ice cream after freezing, the overrun is 50%. Artisanal ice cream normally has between 20% and 50% overrun.

Overrun influences the quality of ice cream in several ways. First, it affects the texture and mouthfeel of the ice cream. The more air in the ice cream, the lighter and softer it will be, but it will also melt faster and can be perceived as having less body and smoothness. The less air in the ice cream, the denser and creamier it will be, but it will also be harder to scoop and more prone to iciness.

Second, overrun affects the flavour and nutritional value of the ice cream. The more air in the ice cream, the less flavour and aroma it will have, as the air dilutes the concentration of the ingredients. The less air in the ice cream, the more flavour and aroma it will have, as the ingredients are more concentrated. The amount of air also affects the nutritional value of the ice cream, as it changes the ratio of fat, protein, sugar and water in the product. The more air in the ice cream, the lower the fat and calorie content per serving, but also the lower the protein and calcium content. The less air in the ice cream, the higher the fat and calorie content per serving, but also the higher the protein and calcium content.

Third, and not to forget, overrun affects the cost and profitability of ice cream. The more air in the ice cream, the lower the cost of production, as fewer ingredients are needed to make a given volume of product. The less air in the ice cream, the higher the cost of production, as more ingredients are needed to make a given volume of product.

State-of-the-art industrial ice cream machines can incorporate air and form ice crystals. This process is done in a continuous freezer, where a rotating blade scrapes the frozen mixture from the walls of a cold cylinder. The freezing process determines the overrun and texture of the ice cream by controlling the amount of air and size of ice crystals in the product.

However, overrun also affects how much consumers are willing to pay for ice cream. Consumers often say that they prefer higher-quality ice cream with lower overrun and that they are willing to pay for it. Therefore, manufacturers balance the possible overrun with aroma-intensive ingredients in the solid phase to achieve an organoleptically satisfying result.

So, is industrial ice cream just a cheap trick to cut cost of the materials and increase profitability? A simple "yes" would probably not be the adequate answer. The frozen dessert is a popular treat and enjoyed by millions of people all over the world. The total sales were USD 73.61 billion in 2022, and the accumulated annual growth is estimated to be 5.20% (Ice-cream-market, 2024). Particularly in Asian countries where ice cream as a dessert is not part of the traditional cuisine, good growth is expected. Beyond these two major trends are affecting the ice cream world. One is the rising demand for premium and hence more artisanal ice creams offering high-quality ingredients, natural flavours, enriched with unique textures through creative toppings, etc. That's indulgence and well served through artisanal ice cream with low overrun. The second big trend is health and wellness among consumers, which has led to many launches of ice creams with added functional benefits, such as probiotics, prebiotics, antioxidants, or more guilt-free indulgence, through lower-calorie, low-fat and sugar-free options. The more ice cream is aerated, better the basis for "guilt free" indulgence is.

Consumer preference in the different markets will show how much of such "overrun" is accepted in the ice cream. Right now, leading manufacturers are working on new technologies to create the same indulgence with fewer solids.

Liquid Seasoning

Liquid seasonings are very practical to add flavour to meat dishes. They are basically known in all kitchens of the world and include all-time favourites like Worcester sauce; Maggi liquid seasoning; soy, oyster and fish sauce; teriyaki sauce; and many more. And liquid seasonings play an important role in industrial meat applications, as they are an excellent, cost-effective alternative to dry seasoning blends.

They can be applied in concentrated form, effectively infusing the meat's oil and water phase with flavour, aroma and colour. They can be customized to any flavour profile desired. And they can be handled easily since they have an excellent shelf life and are microbiologically very stable. And depending on the ingredients used, they allow "*clean label*" declaration, for example, as "*natural flavour*" or "*spice extract*".

The preservation, colouring and flavouring power of seasonings or marination is widely known. Less prominent is the fact that it is also an excellent method to tenderize meat through biological, chemical and physical methods. This improves the quality of the end product through pH lowering, for example, through the addition of vinegar or other acids, which accelerates the protein breakdown in the muscles. Smart use of different acids supports the breakdown of less valuable connective tissue, bringing even more tenderness. And proper application will considerably reduce the time required for meat to soften, the so-called aging (Rahman et al., 2023).

The old kitchen technique of marination has been optimized in industrial liquid seasoning. They can nowadays consist of flavouring agents, so natural or artificial flavours, extracts and spices and various solvents to dissolve the flavouring agents can be water, oil, alcohol or vinegar. For a good shelf-life, preservative like citric acid or sodium benzoate are added. Lecithin or mono- and diglycerides act as emulsifiers and ensure an even distribution in the different meat parts. Where needed natural or artificial colouring agents help to enhance the visual appeal of the meat. A few stabilizers like xanthan gum or carrageenan ensure consistence of the mix, and a few acidulants, like citric acid or vinegar, control the pH of the seasoning. And if in the end the taste and aroma need some extra balance, sweeteners like sugar, honey or artificial sweeteners can be added. For the final appetizing surface, treatment starches and gums are used, as they stay on the surface and create good visual appeal.

In other words, a full spectrum of great kitchen magic translated and adapted for industrial meat preparations creates visually appealing, juicy, tasty soft meat dishes at a low price, using a wide spectrum of meat. Somewhat like the making of beef bourguignon, a signature dish of the French cuisine. Just that the meat is here probably not soaked overnight with a spice and herbs mixture in well-aged medium dry red Bordeaux, but vinegar or citric acid instead.

Flavour

When we think about food, we think about taste and flavour. It's the most important reason to choose a food item from a menu or when preparing it yourself at home. Brilliant if it looks appetizing, but if it is then tasteless, the love stops. While taste can initially be explained by reactions on the tongue, flavour is so much more than just the chemical composition of food. It touches many senses: Smell, taste, touch of it in the mouth, even eye and ear. It doesn't only detect molecules but creates experience and triggers personal memories.

> *"Flavour can take you right back to your granny's kitchen or that time you went backpacking around Thailand."*
>
> (Allan, 2023)

At the same time many flavours are fragile, volatile, difficult to get and preserve and expensive. Each kitchen treatment, just storage and, of course, factory processing step impacts them. No preservation method does not change or damage the flavour profile. So, it's an incredible achievement of highly trained academic and industry scientists and experts to maintain or rebuild these sometimes also rare and expensive flavours, to add them back after the food is made.

The variety of flavours is virtually infinite, but one of the best known ones is vanilla! It should serve here as an example for the "trick" the industry is using.

Vanilla is a popular flavour extracted from the pods of vanilla orchids. Making natural vanilla involves a series of steps, including pollination, harvesting, curing and extracting. Since the orchid flowers last only for 1 day, hand pollination is needed to produce vanilla pods. That is a labour-intensive, very time-sensitive task that requires skilled workers. After 8–9 months the pods can be harvested. They are harvested when they are still green and unripe to capture all the flavour. The next step is called "curing" and is a sequence of blanching, sweating, drying and conditioning. This brings out the vanilla flavour, the conversion of glycovanillin into vanillin.

Different methods are known for curing of the pods, such as the bourbon method, the Mexican method and the Tahitian method, which result in different types of vanilla products. (By the way, bourbon vanilla got its name from the island of Réunion, formerly known as Bourbon Island, and not because of bourbon whiskey, as some people think).

This takes another 6 months before the extraction can start, for which the pods are crushed and soaked in alcohol and water. The alcohol dissolves the vanillin and other flavour compounds from the pods, creating a dark brown liquid with a strong vanilla scent. Finally, the concentrated dark brown liquid can be blended with other ingredients, such as sugar, corn syrup or other flavours.

That explains why natural vanilla is the second most-expensive spice in the world. Depending on the origin, quality and market conditions, the average wholesale price of natural vanilla in 2023 was $600 per kilogram (TRIDGE, n.d.). However, this price can fluctuate drastically depending on the supply and demand of vanilla beans. For example, in 2020, the price of natural vanilla dropped to $250 per kilogram due to an oversupply of low-quality beans and a reduced demand from the food industry amid the COVID-19 pandemic. The value of the market was nearly $1 billion in 2023 (Vanilla Market Insights, o. D.) and the total annual volume around 20,000 tonnes. But only approximately 1% of vanilla is coming from the leaves (Bomgardner, 2017).

The flavour industry provides consumers and customers with a much cheaper alternative. This artificial vanilla is made from synthetic vanillin, the main flavour compound in natural vanilla. The synthetic vanillin as such comes from various cheap sources, such as petrochemicals, wood pulp, rice bran or microorganisms. The petrochemical route works via guaiacol oxidation, nitrobenzene or chlorate; the wood pulp route extracts vanillin from lignin, then hydrolyses and oxidizes it. For a more contemporary method, rice bran, the waste product of rice milling, is used to convert ferulic to vanillin by chemical or enzymatic methods.

So, many different routes, cheap base material and efficient reactions done through classical chemistry or microorganisms, like bacteria, fungi or yeast, enable a price of $30/kg or less for the food-grade substance.

Strawberry

Natural strawberry flavour is one of the most complex fruit flavours and varies between the different cultivars, generally a complex combination of products in the different pathways of the plant. One of the main compounds is furanone, which has a sweet, caramel-like smell and taste. Furanone is produced by the activity of the hormone auxin, involved in the cell wall breakdown and the release of sugars and acids. Others are methyl anthranilate, with a fruity, grape-like smell and taste; ethyl butyrate, giving fruity, pineapple-like smell and taste; linalool, which has a floral, citrus-like smell and taste; geraniol, which has a floral, but also rose-like smell and taste; and hexyl acetate, which has a fruity, but now more apple-like smell and taste. These compounds are also found in other fruits, such as apples, pears, peaches and citrus fruits.

Artificial strawberry flavour is usually made by combining different synthetic chemicals that mimic the natural compounds or creating new ones. For example, one of the most common artificial strawberry flavourings is ethyl methylphenylglycidate, also known as strawberry aldehyde. This compound has a strong, sweet, candy-like smell and taste, but it does not exist in nature. It was first synthesized in 1905 by Rainer Ludwig Claisen, a German chemist, by reacting benzaldehyde with ethyl acetate.

Another artificial flavouring in strawberries is castoreum. Castoreum has a musky, leather-like smell and taste, not necessarily a typical fruit note, but it can add depth and complexity to the flavour. It is found in many flavours, including vanilla.

The old natural extraction route was from beaver glands, but luckily castoreum can be synthesized artificially by a combination of reactions between phenols, ketones, monoterpenes, benzoic acid, benzyl alcohol, borneol and salicylaldehyde. These compounds can be derived from plant sources or synthesized in laboratories. Various chemical routes are nowadays identified to produce the substance (Kunz, 2009).

And those artificial castoreum are cheaper and easier to produce than natural castoreum, and it can also avoid ethical and sanitary issues associated with animal-derived ingredients. The beaver gland extraction flavour would, by the way, be a "natural flavour".

The exact number of artificial flavours for food is not easy to determine, as different sources may have different definitions and classifications of what constitutes an artificial flavour. For the FDA, artificial flavours are those not coming from natural sources, but the FDA does not require companies to disclose the specific ingredients of their artificial flavourings, as long as they have been approved before. Estimates range from more than 700 chemically defined flavouring substances in the United States to 1300 ingredients that are approved for use by the FDA (Tsakos, 2020).

That's big business with lots of new products and processes inspired by human creativity (for now, in a couple of years perhaps artificial intelligence [AI]) and sophisticated organic chemical knowledge. What is the reason for this enormous economic success? The use of science to design chemical routes producing substances with the same sensorial power than the natural sources, but much, much cheaper and often more resourceful.

Artificial flavours are pure chemicals mimicking the taste and smell of natural substances reasonably well. They may not capture all the nuances of the original source. Natural vanilla extract contains hundreds of different compounds that contribute to the full aroma, and many people can taste and smell the difference when they truly focus on it.

No doubt, artificial flavours are widely used in processed foods and beverages, and many people consume them frequently. In the past food makers could label their flavours as "natural identical" when the synthesized substance was chemically identical to the one food in nature (Effa, o. D). That was hiding the chemistry while rising a perception of naturalness. Today that is no longer possible.

But in some cases, consumers now have developed a preference for artificial flavours, especially if they associate them with positive emotions or memories. Artificial lemon flavour is popular example, perhaps because it reminds them of lemonade or other refreshing drinks.

Many people are concerned about the health effects of artificial flavours, as they may contain allergens, additives or preservatives. Artificial flavours are chemistry, and natural flavours are biology! Here it might be good to remember that every natural substance used in a product also consists of chemicals. So, artificial and natural flavourings can be chemically identical – they differ in their source. Flavours are tested and regulated by the authorities like FDA, so they are safe to consume.

But then are natural flavours more environmentally friendly? No, not necessarily! Natural coconut flavour, for example, comes from massoia lactone in the bark of the Indonesian *Cryptocarya massoia* tree. To get it the tree must be cut and the pieces are then subjected to a chemical extraction process, and the carbon footprint adds up during the whole transport (Dempsey, 2022).

All of this does not mean that chemical flavours are always the same as natural ones. And that the flavour industry hasn't found ways to create pleasant aromas in a different way. Some natural flavours are indeed more complex in their composition than synthetically designed ones. People with sensitive senses will spot the difference in those cases.

Next, the labelling regulation is complicated, not intuitive and not harmonized. Depending on the country where the food is produced or sold, different regulations apply. In several cases natural flavours must meet certain criteria such as being obtained by physical, enzymatic or microbiological processes and not containing any artificial or synthetic additives.

In the EU, the Flavouring Regulation (EC) No. 1334/2008 defines different types of flavourings, sets general conditions for their use, establishes a Union List of approved flavourings and source materials and stipulates labelling requirements for both flavouring manufacturers and food manufacturers. The Flavouring Regulation also specifies the criteria for using the term "*natural*" in relation to flavourings, such as the origin and composition of the flavouring part and the source material.

In other countries, like China, the regulation on natural flavours is less strict. Instead, China allows the use of glycosylated steviol glycosides (GSGs), which are derived from stevia plants but modified by enzymatic treatment, as natural flavourings. And in countries like the United States, Canada and Australia, there are also different standards and definitions for natural flavours and their labelling. Calling something a natural flavour does not mean that the flavour is coming from the expected source.

The U.S. FDA considers L-cysteine from duck feathers a natural flavouring agent, as it is coming from a natural source and does not undergo any chemical changes during the extraction process. That´s not only confusing for vegans, vegetarians or some religious groups with strict dietary guidelines on animal products but is counterintuitive for most people. And, as mentioned, consumers may be relieved to hear that the complex, musky leather note in many artificial fruit aromas is not coming from beaver glands.

The somewhat inscrutable flavour regulation can create the impression that the food industry and regulating authorities are united to not burden normal consumers with difficult facts about flavour origin and generation. They agree on a vocabulary, where all the underlying chemistry and industrial processing are not obvious and therefore cause no concern.

What then remains in public is the feeling of nontransparency and mistrust. It leaves the impression that these industrial food designers are making foodstuffs behind closed doors in non-foody, industrial processes and are not willing to explain their ways of working, as they expect the facts are too complicated for the layman to understand them properly. All this science and technology and the tough explanation do not fit with food. So, yes, artificial vanillin may be much cheaper and environmentally less questionable than the natural one, but the latter is the result of a series of simple, slow, traditional, low-tech steps. Generally, we know that an industrial strawberry yoghurt is not simply milk naturally fermented by a friendly *Bacillus* strain to which freshly cut strawberry pieces have been added. That wouldn´t be safe, would have a very short shelf life and would be expensive. The intense fruit aroma, smooth texture and pleasant sweetness, stable for 3 weeks or more at less than half a euro, requires additional functional ingredients, professional skills and efficient processes beyond old farm and kitchen food making. Nothing to be ashamed of!

So why not more transparency and openness, a willingness to understandably inform about the flavour, helping consumers make informed choices about their food consumption, which could foster trust and collaboration among the different stakeholders in the food system, such as farmers, processors, retailers, regulators, and consumers, from farm to folk? Transparency and openness in industrial food production needs to consider a few barriers and challenges that up to now hinder the disclosure and dissemination of information about food production.

Global food chains are complex and diverse, which makes it hard to trace and monitor the origin, quality of all ingredients and the final product. The lack of harmonized regulations and standards for food labelling, certification and auditing across different countries and regions makes it difficult to compare. The different actors in the food system and their interests may influence the availability and accuracy of information, for which the collection, storage, analysis and sharing will add costs. On the industry side, privacy and security concerns related to the protection of sensitive data and intellectual property rights need to be managed. Finally, consumer awareness and demand for information about food production need to be met, as these can vary, depending on the level of education, income, culture and preferences.

These things are on their way! New digital technologies, such as blockchain, Internet of Things (IoT), big data analysis and the rise of AI, are already enhancing traceability, verification and communication of information about food production. Voluntary or mandatory schemes for food labelling, certification and auditing start providing reliable and credible information about the environmental, social and health aspects of food production. The different stakeholders start to use participatory and inclusive end-to-end approaches. In many places, platforms and networks are emerging that facilitate the exchange and dissemination of information about food production among the different actors in the food system. More and more opinion-formers in the food chain start to talk about the food industry in a less biased way, providing new knowledge that can empower consumers to increase their awareness and demand for information about food production.

The Biggest "Trick" of All

There are countless stories about the food industry and their tricks to sell their products to the consumer above the true product value. The truth is that the food business, like every other business, only continues to exist if it adds value: To themselves and to others!

It is also true that the cost of food relative to income has decreased in the past centuries. The improvements in mechanization, better crop varieties and superior farming practices have significantly increased yields, which helped to lower the cost of food and so in the past centuries, many people could spend less of their disposable income on food items.

Food expenditure reflects not only the cost of living, but also the availability and accessibility of food, as well as the dietary preferences and habits of consumers. In the EU, the average household spending on food in 1960 was 32% of their total consumption. In 2019 it was only 13% (retrieved from https://ec.europa.eu/eurostat/web/main/home). The global average of food expenditure as a percentage of household income was 22.3% in the same year, with a span ranging from 6.5% (US) to 59% (Nigeria). Figure 2.1 shows the difference in the food spend on this planet.

Over time, the food sector has been very successful in making food more accessible and more affordable through advancements in technology, supply chain efficiencies and mass production, something that has been in their DNA since the beginning.

Still, food prices are influenced by a variety of factors and can fluctuate massively, due to:

- Unfavourable weather
- Changes in consumer behaviour
- Supply chain disruptions
- Political instability, affecting global grain markets and edible oil markets
- Rapid economic growth in countries like China and India creating more demand
- High fuel prices increasing the production of biofuels
- Core inflation costs for energy processing, transportation and labour

The historical perspective of the last 170 years of the prices of commodities – that's the time between now and when the food industry started to offer processed food – shows that for many food items, the indexed price (versus 1900) is much lower nowadays. Figure 2.2 shows that context.

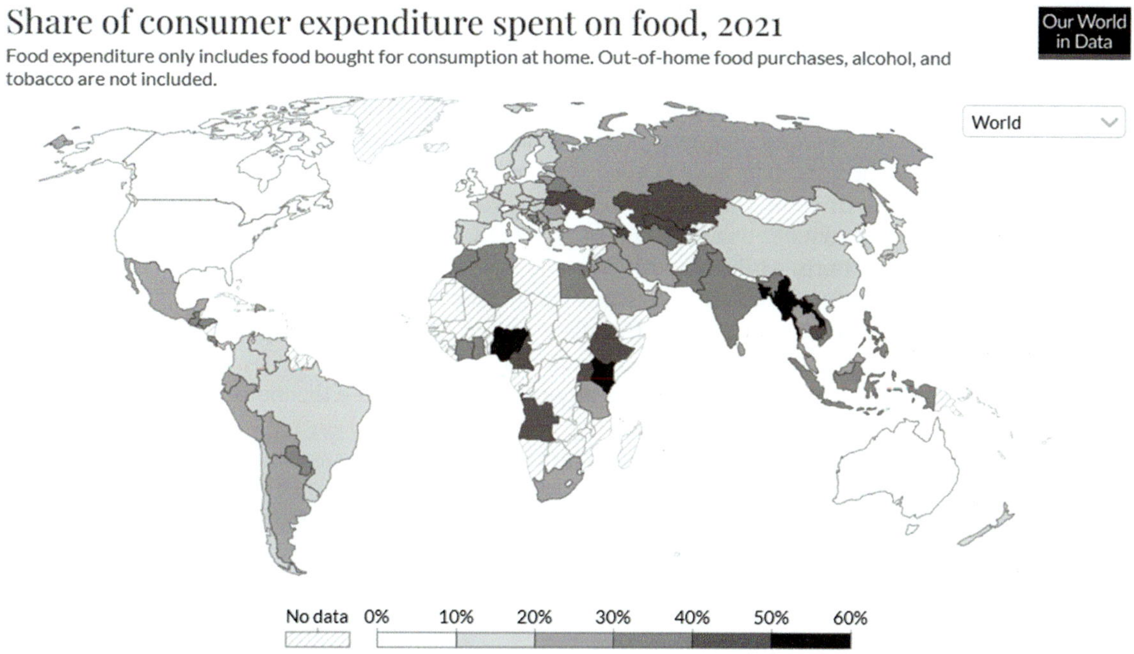

FIGURE 2.1 The share of consumer expenditure on food on this planet. (From Our World in Data, 2023.)

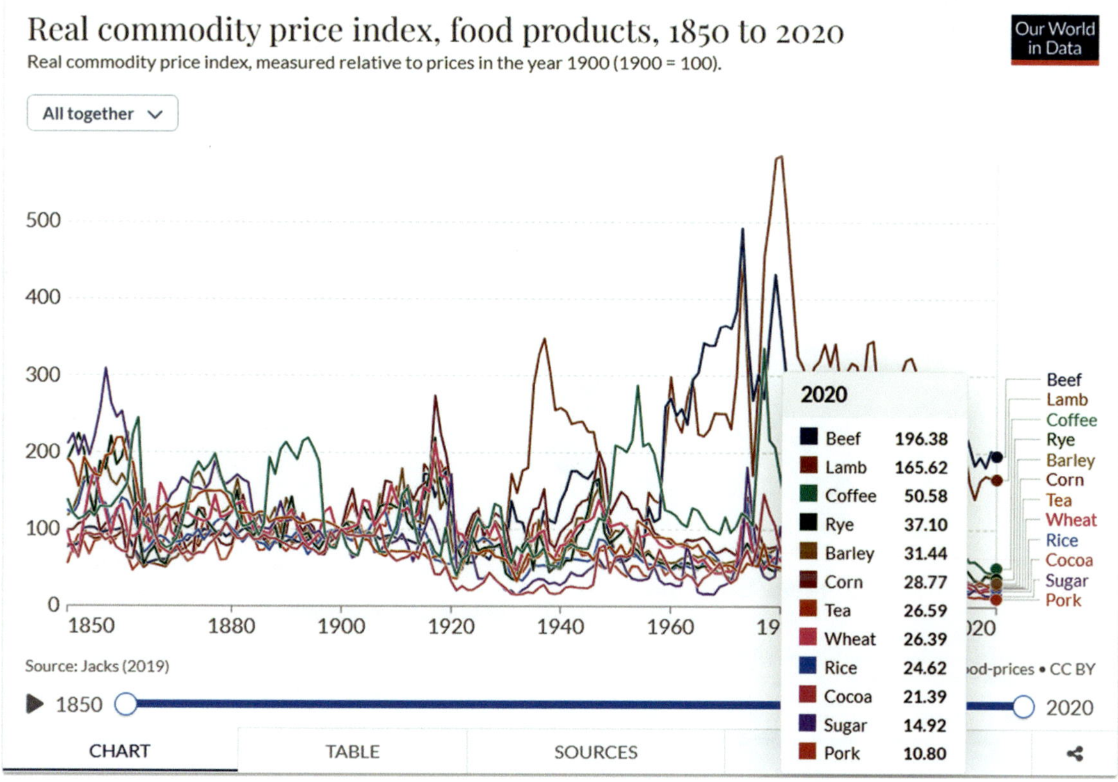

FIGURE 2.2 Commodity price index 1850–2020. (From https://ourworldindata.org/food-prices.)

The chart shows rapid and massive fluctuations over time, but it is still astonishing how inexpensive food is despite the fact that in the meantime more than 8 billion people need to be fed on this planet.

To look at this from a different perspective: In 1933, a family in the United States had to spend 25% of their household income on food; that was 21.8% for in-home consumption and 3.2% used for out of the home, like restaurants. In 2021, they only needed 9.9% of their household income and from that, already half of it was used in out-of-home consumption (Ritchie et al., 2023a). However, the United States is not the globe, and an average number can hide huge differences. So this number includes an expenditure of just 6.4% in little Luxembourg, while in South Sudan it was 59.4%.

Maybe an even better comparison in this context are the average costs of a nutrient-adequate diet throughout this world. For this, the World Bank has collected "nutrient-adequate diet data" globally. Figure 2.3 summarizes the cost difference for various countries globally.

The cost of food for the individual was generally getting lower, while the industry was getting bigger and more valuable, and the population was growing.

So, if not the availability of food that suffered in the last decades., may be a quality reduction made the food cheaper and ensured profit for the industry? The data from the FAO globally show all the same trend: Whether it is kilocalories, protein from animals or from plants, fat in most parts of the world more became available in the last 65 years, mostly achieved through yield increase.

So, the macronutrient quality did not decrease, but maybe the micronutrient profile did. Here, data here less consistent. Measurement over 50 years of the nutrient content of typical garden crops in the United States show a moderate nutrient decline for some and none for others (Davis et al., 2004). A similar

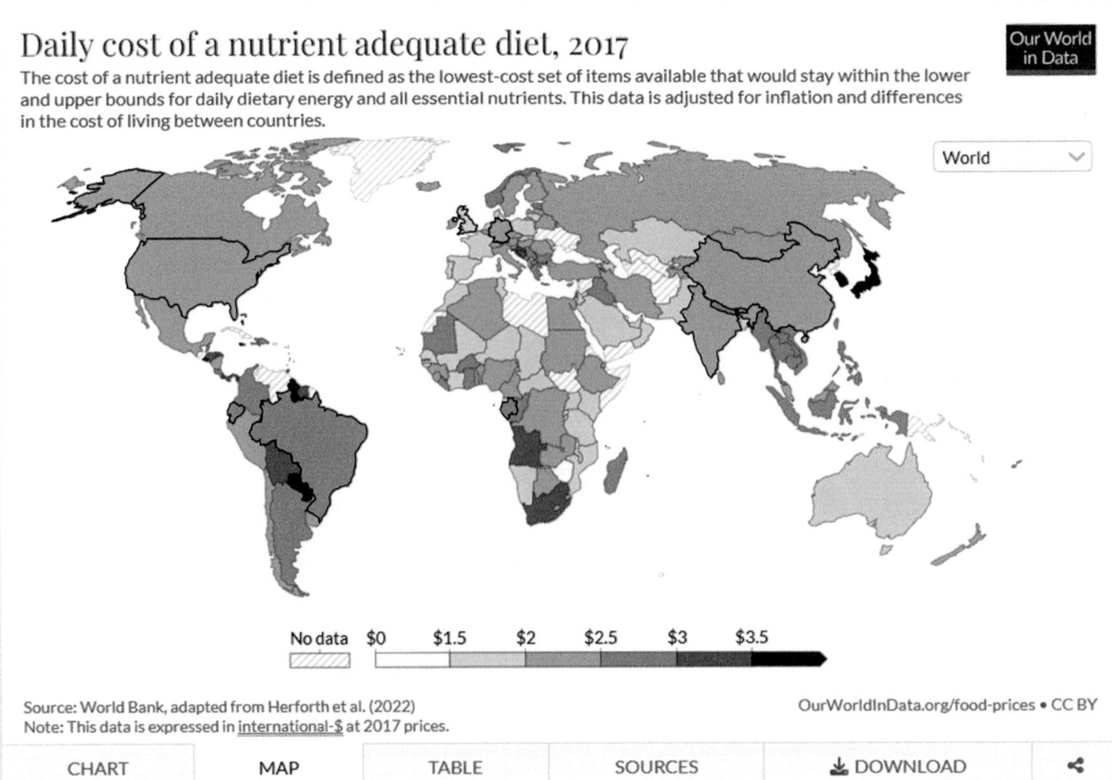

FIGURE 2.3 Cost of a nutrient-adequate diet globally. (From Ritchie et al., 2023b.)

comparison between traditional cultivars and modern high-yield crops in India indicated a higher decline in nutritional quality (Bhardwaj et al., 2024).

That indicates that there is good reason to watch the nutritional quality of crops when working on yield improvement, since the agronomical success of the past decades has depleted the soil in a non-sustainable way. This is one reason why the focus in agriculture has shifted from "sustainable" to "regenerative", where more attention is given to the conscious restoration of the soil.

But overall, the global food sector – from farm to fork – has been successful in reducing costs for the making of food products in the last centuries. Commodities are subsidized by governments, prices vary per country and within countries and average income can be very different, but the progress made in the food industry to reduce cost consistently, efficiently and effectively is impressive.

This trend is visible for most food items:

- Bread: 1 pound of white bread cost $0.09 in 1913 (inflation corrected: $1.44 in 2010); the actual cost in 2010 was $1.36.
- Milk: 1 gallon of milk cost $0.83 in 1950 (inflation corrected $9.13 in 2020); the actual cost of a gallon of milk in 2020 was $3.50.
- Eggs: A dozen eggs cost $0.52 ($5.71 in 2020 dollars); the actual cost of a dozen eggs in 2020 was $1.55.
- Meat was $0.85/lb in 1950 (inflation corrected: $9.34); the actual cost in 2020 was $5.72/lb (Jude Newsome, 2024).

Other commodities like sugar, rice, butter and burgers all show the same trend. One remarkable exception is fish. The reason is the massive decline of global wild fish stocks due to overfishing, environmental degradation and climate change (Ritchie & Roser, 2023).

No doubt, the industrialization of agriculture led to a massive conversion of small- and medium-sized farms into large industrial factories that grow, harvest, design, raise, contain and slaughter plants and animals in a predominantly efficiency-driven way. Today, globally about three-quarters of land livestock are factory farmed. And in the United States, it could even be up to 99% (Ritchie & Roser, 2024).

Most of the cost reductions in foods came from the application of new technology in agriculture, irrigation and water management, transportation and preservation, mass production or the "green revolution". And more breakthroughs can be expected in the next years from biotechnology, genetic engineering, digitalization and precision farming and AI.

The food industry made food and eating affordable and convenient, often through new ways of processing, particularly in the last 60 years. But the food quality wasn´t better at that time. Hence, simply postulating a correlation between the amount of processing and the loss of nutritional quality can be misleading (see Chapter 17). Many new processing technologies have helped to preserve food quality much better. Today food can be dried by traditional solar or shade drying, by hot air drying, freeze drying, microwave-, infrared-, drum spray combined thermal hybrid drying and many more methods. How much and which parts of the original quality are preserved depend more on the processing objectives and the right best combination of crops and technology.

The DNA of the food industry is efficiency, making food affordable and reducing costs. This is good for both the industry and consumers. But major parts of the bill have not been paid, and the environmental costs are only now becoming visible. So, a "future fit" product design must include them, reflect the full costs and, perhaps even a bit more, gives the next generations the chance of a quality of life that most of us have. And despite the technical improvements that we can hope for, it's not likely that all costs will be absorbed by efficiency increases within the industry. Some costs will be passed to the consumer – a premium most of us in the mature markets can afford.

Cutting costs is the food industries "superpower", and quantity increase is not paid with quality loss, but the full bill must include elements that have not been charged properly so far.

REFERENCES

Abrol, N., Panda, A., Kekre, N., & Devasia, A. (2015). Nanobacteria in the Pathogenesis of Urolithiasis: Myth or Reality? *Indian Journal of Urology*, *31*(1), 3. https://doi.org/10.4103/0970-1591.134235

Adams, C. (2020, 28 July). *Why was spinach once considered the ultimate vegetable?*. The Straight Dope. https://www.straightdope.com/21341870/why-was-spinach-once-considered-the-ultimate-vegetable

Akram, M., Jahrreiss, V., Skolarikos, A., Geraghty, R., Tzelves, L., Emilliani, E., Davis, N. F., & Somani, B. K. (2024). Urological Guidelines for Kidney Stones: Overview and Comprehensive Update. *Journal of Clinical Medicine*, *13*(4), 1114. https://doi.org/10.3390/jcm13041114

Allan, L. (2023, 4 May). *What is flavour?* CPD Online College. https://cpdonline.co.uk/knowledge-base/food-hygiene/what-is-flavour/

Aune, D., Keum, N., Giovannucci, E., Fadnes, L. T., Boffetta, P., Greenwood, D. C., Tonstad, S., Vatten, L. J., Riboli, E., & Norat, T. (2016). Whole Grain Consumption and Risk of Cardiovascular Disease, Cancer, and All Cause and Cause Specific Mortality: Systematic Review and Dose-Response Meta-Analysis of Prospective Studies. *BMJ*, i2716. https://doi.org/10.1136/bmj.i2716

Berry, C., Burton, S., & Howlett, E. (2017). It's Only Natural: the Mediating Impact of consumers' Attribute Inferences on the Relationships between Product Claims, Perceived Product Healthfulness, and Purchase Intentions. *Journal of the Academy Of Marketing Science*, *45*(5), 698–719. https://doi.org/10.1007/s11747-016-0511-8

Bhardwaj, R. L., Parashar, A., Parewa, H. P., & Vyas, L. (2024). An Alarming Decline in the Nutritional Quality of Foods: The Biggest Challenge for Future Generations' Health. *Foods*, *13*(6), 877. https://doi.org/10.3390/foods13060877

Bomgardner, M. M. (2017, 28 March). *The Problem with Vanilla*. Chemical & Engineering News. https://cen.acs.org/articles/94/i36/problem-vanilla.html

Brown, L. K., Miller, A., & Miller, E. (2020). Chicken Soup for the Treatment of Respiratory Infections. *Chest*, *158*(5), 2231–2232. https://doi.org/10.1016/j.chest.2020.06.062

Cobos, Á., & Díaz, O. (2023). 'Superfoods': Reliability of the Information for Consumers Available on the Web. *Foods*, *12*(3), 546. https://doi.org/10.3390/foods12030546

Davis, D. R., Epp, M. D., & Riordan, H. D. (2004). Changes in USDA Food Composition Data for 43 Garden Crops, 1950 to 1999. *Journal of the American College of Nutrition*, *23*(6), 669–682. https://doi.org/10.1080/07315724.2004.10719409

DeAlmeida, C. E., Ochoa, R., De Lima, M. C., David, M. G., Pires, E. J., Peixoto, J. G., Salata, C., & Bernal. (2014). Thermal Stress Triggers Broad *Pocillopora damicornis* Transcriptomic Remodeling, while *Vibrio coralliilyticus* Infection Induces a More Targeted Immuno-Suppression Response. *HAL (Le Centre Pour La Communication Scientifique Directe)*. https://worldpopulationreview.com/country-rankings/gluten-intolerance-by-country?form=MG0AV3

Dempsey, E. (2022, 9 November). The difference between artificial & natural flavoring might surprise you. *Utopia*. https://utopia.org/guide/do-you-know-the-difference-between-artificial-natural-flavoring/

Dinu, M., Whittaker, A., Pagliai, G., Benedettelli, S., & Sofi, F. (2018). Ancient Wheat Species and Human Health: Biochemical and Clinical Implications. *The Journal of Nutritional Biochemistry*, *52*, 1–9. https://doi.org/10.1016/j.jnutbio.2017.09.001

Effa. (o. D.). *Flavouring legislation - European Flavour Association*. https://effa.eu/eu-legislation/flavouring-legislation

European Commission (2018). *Date marking and food waste prevention*. Food Safety. https://food.ec.europa.eu/safety/food-waste/eu-actions-against-food-waste/date-marking-and-food-waste-prevention_en

European Commission. (o. D.). *European Commission - Have Your Say*. https://ec.europa.eu/info/law/better-regulation/have-your-say/initiatives/12749-Food-labelling-revision-of-rules-on-information-provided-to-consumers_en

Favell, D. (1998). A Comparison of the Vitamin C Content of Fresh and Frozen Vegetables. *Food Chemistry*, *62*(1), 59–64. https://doi.org/10.1016/s0308-8146(97)00165-9

Fayez, N., Ibrahim, M. K., Farrag, H. A., Mohamed, M. A. E. H., & Tablawy, S. Y. E. (2023). Synergistic Effect of Doxycycline and Aqueous Extract of Irradiated Khella on Structure of Nanobacteria Isolated from Kidney Stones: In Vitro and in Vivo Studies. *Cell Biochemistry and Function*, *41*(8), 1275–1294. https://doi.org/10.1002/cbf.3862

Fevzioglu, M., Kaliappan, S., Stalder, J. W., & Inc, F. L. N. A. (2017, 27 October). *US10881123B2 - Crystal morphology for sodium reduction - Google Patents*. https://patents.google.com/patent/US10881123B2/de

German Federal Institute for Risk Assessment. (2023). Spelt can also trigger allergies - low level of public knowledge about spelt being a type of wheat. In *Zenodo (CERN European Organization for Nuclear Research)*. https://doi.org/10.17590/20230202-075856

Halim, J., Bouzari, A., Felder, D., & Guinard, J. (2020). The Salt Flip: Sensory Mitigation of Salt (and Sodium) Reduction with Monosodium Glutamate (MSG) in "Better-for-You" Foods. *Journal of Food Science*, *85*(9), 2902–2914. https://doi.org/10.1111/1750-3841.15354

Hebishy, E., Du, H., Brito-Oliveira, T. C., Pinho, S. C., & Miao, S. (2023). Saltiness Perception in Gel-Based Food Systems (Gels and Emulsion-Filled Gels). *Critical Reviews in Food Science and Nutrition*, 1–18. https://doi.org/10.1080/10408398.2023.2276331

Hemilä, H., & Chalker, E. (2013). Vitamin C for Preventing and Treating the Common Cold. *Cochrane Library*, *2013*(5). https://doi.org/10.1002/14651858.cd000980.pub4

Hollands, G. J., Shemilt, I., Marteau, T. M., Jebb, S. A., Lewis, H. B., Wei, Y., Higgins, J. P. T., & Ogilvie, D. (2015). Portion, Package or Tableware Size for Changing Selection and Consumption of Food, Alcohol and Tobacco. *Cochrane Library*, *2018*(11). https://doi.org/10.1002/14651858.cd011045.pub2

Hopkins, A. B. (2003). Chicken Soup Cure May Not be a Myth. *The Nurse Practitioner*, *28*(6), 16. https://doi.org/10.1097/00006205-200306000-00005

Human Verification. (n.d.). https://profiles.nlm.nih.gov/spotlight/mm/feature/medicine

Ice-cream-market. (2024, 24 June). https://www.fortunebusinessinsights.com. https://www.fortunebusinessinsights.com/ice-cream-market-104847

Jude Newsome. (2024, 24 May). How much grocery store items cost in 1950 vs. 2020. *Ranker*. https://www.ranker.com/list/1950-food-prices-vs-today/jude-newsome

Kunz, H. (2009). The Beaver as Chemist: Total Synthesis of Enantiomerically Pure Nupharamine Alkaloids from Castoreum. https://phys.org/news/2009-02-beaver-chemist-total-synthesis-enantiomerically.html

Limerick, M. (2020, 20 October). Superfoods: The origin Story. Foods Alive Inc. https://foodsalive.com/blogs/learn/superfoods-the-origin-story

Lopez-Jimenez, F. (2024, 7 March). *Eggs: Are they good or bad for my cholesterol?* Mayo Clinic. https://www.mayoclinic.org/diseases-conditions/high-blood-cholesterol/expert-answers/cholesterol/faq-20058468

Lyons, E. (2022, 21 May). *The 10 Most Expensive Salts in the World and Why*. TheMostExpensive. https://themostexpensive.org/most-expensive-salts

Made-in-China.com. (o. D.). *Vanillin Price, 2024 Vanillin Price Manufacturers & Suppliers*. [Video]. https://www.made-in-china.com/products-search/hot-china-products/Vanillin_Price.html

Mihaylova, M. M., Chaix, A., Delibegovic, M., Ramsey, J. J., Bass, J., Melkani, G., Singh, R., Chen, Z., Ja, W. W., Shirasu-Hiza, M., Latimer, M. N., Mattison, J. A., Thalacker-Mercer, A. E., Dixit, V. D., Panda, S., & Lamming, D. W. (2023). When a Calorie Is Not Just a Calorie: Diet Quality and Timing as Mediators of Metabolism and Healthy Aging. *Cell Metabolism*, *35*(7), 1114–1131. https://doi.org/10.1016/j.cmet.2023.06.008

Our World in Data. (2023, 15 September). *Share of consumer expenditure spent on food*, 2021. https://ourworldindata.org/grapher/share-of-consumer-expenditure-spent-on-food

Promoting vitamin C. (o. D.-a). Linus Pauling - Profiles in Science. https://profiles.nlm.nih.gov/spotlight/mm/feature/medicine

Rahman, S. M. E., Islam, S., Pan, J., Kong, D., Xi, Q., Du, Q., Yang, Y., Wang, J., Oh, D., & Han, R. (2023). Marination Ingredients on Meat Quality and Safety—A Review. *Food Quality and Safety*, *7*. https://doi.org/10.1093/fqsafe/fyad027

Rajendra, A., Bondonno, N. P., Rainey-Smith, S. R., Gardener, S. L., Hodgson, J. M., & Bondonno, C. P. (2022). Potential Role of Dietary Nitrate in Relation to Cardiovascular and Cerebrovascular Health, Cognition, Cognitive Decline and Dementia: A Review. *Food & Function*, *13*(24), 12572–12589. https://doi.org/10.1039/d2fo02427f

Research and Markets Ltd. (o. D.). *Monosodium glutamate (MSG) - Global Strategic Business Report*. https://www.researchandmarkets.com/reports/4805328/monosodium-glutamate-msg-global-strategic#:~:text=The%20global%20market%20for%20Monosodium%20Glutamate%20%28MSG%29%20estimated,CAGR%20of%204.6%25%20over%20the%20analysis%20period%202023-2030.)

Ritchie, H., Rosado, P., & Roser, M. (2023a, 28 December). *Food prices*. Our World in Data. https://ourworldindata.org/food-prices

Ritchie, H., Rosado, P., & Roser, M. (2023b, 28 December). *Food prices*. Our World in Data. https://ourworldindata.org/food-prices

Ritchie, H., & Roser, M. (2023, 28 December). *Fish and overfishing*. Our World in Data. https://ourworldindata.org/fish-and-overfishing#wild-fish-catch

Ritchie, H., & Roser, M. (2024, 24 February). *How many animals are factory-farmed?* Our World in Data. https://ourworldindata.org/how-many-animals-are-factory-farmed

Roberts, J. L., & Moreau, R. (2016). Functional Properties of Spinach (*Spinacia oleracea* L.) Phytochemicals and Bioactives. *Food & Function*, *7*(8), 3337–3353. https://doi.org/10.1039/c6fo00051g

Röhling, M., Martin, K., Ellinger, S., Schreiber, M., Martin, S., & Kempf, K. (2020). Weight Reduction by the Low-Insulin-Method—A Randomized Controlled Trial. *Nutrients*, *12*(10), 3004. https://doi.org/10.3390/nu12103004

Rosenblum, I., Bradley, J. D., & Coulston, F. (1971). Single and Double Blind Studies with Oral Monosodium Glutamate in Man. *Toxicology and Applied Pharmacology*, *18*(2), 367–373. https://doi.org/10.1016/0041-008x(71)90129-3

Scharff, L. B., Saltenis, V. L., Jensen, P. E., Baekelandt, A., Burgess, A. J., Burow, M., Ceriotti, A., Cohan, J. P., Geu-Flores, F., Halkier, B. A., Haslam, R. P., Inzé, D., Lankhorst, R. K., Murchie, E. H., Napier, J. A., Nacry, P., Parry, M. A., Santino, A., Scarano, A., … Pribil, M. (2021). Prospects to Improve the Nutritional Quality of Crops. *Food and Energy Security*, *11*(1). https://doi.org/10.1002/fes3.327

Science History Institute. (2023, 18 October). *The Rotten Science Behind the MSG Scare | Science History Institute.* https://sciencehistory.org/stories/magazine/the-rotten-science-behind-the-msg-scare/

Smith, K. A. (2013, 16 November). A WWII Propaganda Campaign Popularized the Myth That Carrots Help You See in the Dark. *Smithsonian Magazine.* https://www.smithsonianmag.com/arts-culture/a-wwii-propaganda-campaign-popularized-the-myth-that-carrots-help-you-see-in-the-dark-28812484

Spence, C. (2015). On the Psychological Impact of Food Colour. *Flavour*, *4*(1). https://doi.org/10.1186/s13411-015-0031-3

Taetzsch, A., Das, S. K., Brown, C., Krauss, A., Silver, R. E., & Roberts, S. B. (2018). Are Gluten-Free Diets More Nutritious? An Evaluation of Self-Selected and Recommended Gluten-Free and Gluten-Containing Dietary Patterns. *Nutrients*, *10*(12), 1881. https://doi.org/10.3390/nu10121881

Theobald, H. E. (2005). Dietary Calcium and Health. *Nutrition Bulletin*, *30*(3), 237–277. https://doi.org/10.1111/j.1467-3010.2005.00514.x

TRIDGE. (n.d.). Empower your business with Tridge Data & Analytics. Tridge. https://www.tridge.com/data-analytics

Tsakos, L. (2020, 31 October). *What are artificial flavors?* NaturallySavvy.com. https://naturallysavvy.com/eat/what-are-artificial-flavors/

Vanilla Market Insights. (o. D.). https://www.mordorintelligence.com/industry-reports/global-vanilla-industry Vanilla Market - Size, Analysis & Trends

Vinitha, K., Leena, M. M., Moses, J., & Anandharamakrishnan, C. (2021). Size-Dependent Enhancement in Salt Perception: Spraying Approaches to Reduce Sodium Content in Foods. *Powder Technology*, *378*, 237–245. https://doi.org/10.1016/j.powtec.2020.09.079

Vitamin C. (2023, 10 August). Mayo Clinic. https://www.mayoclinic.org/drugs-supplements-vitamin-c/art-20363932

Wang, Y., Zhong, K., Shi, B., Wang, H., Liu, L., Zhang, L., Zhao, L., & Gao, H. (2022). Cross-Modal Effect of Capsaicin and Pepper Oleoresin on the Enhancement of Saltiness Perception in a NaCl Model Solution. *Food Quality and Preference*, *98*, 104542. https://doi.org/10.1016/j.foodqual.2022.104542

Wikipedia contributors. (2024, 27 April). *Transatlantic Trade and Investment Partnership.* Wikipedia. https://en.wikipedia.org/wiki/Transatlantic_Trade_and_Investment_Partnership

Wikipedia-Autoren. (2019, 14 April). *Irwin Stone.* https://de.wikipedia.org/wiki/Irwin_Stone

Zampini, M., & Spence, C. (2004). The Role of Auditory Cues in Modulating the Perceived Crispness and Staleness of Potato Chips. *Journal of Sensory Studies*, *19*(5), 347–363. https://doi.org/10.1111/j.1745-459x.2004.080403.x

Why Are New Food Products Essential?

3

Designing new food products and taking them to markets is risky, and the success of a new product is far from guaranteed. A often-quoted failure rate for new product designs (NPDs) in food and beverages of 80%–90% may be an exaggeration, but the fact that many food products or at least formats do not stay for long on the market is obvious (Rutkowski, 2022). When market agencies estimate the failure rate, they take the time span in which a product can be as a key criterion, because that is what they can measure. And why would an enterprise pull a commercial success off the market (Dijksterhuis, 2016)?

In fact, estimating the NPD success rate is more difficult. Just counting the numbers of stockkeeping unit (SKU) codes that are delisted will give a misleading picture. Companies often sell the same product (formulation) in a variety of formats, different sizes, formats, multipacks, etc. Those that are not performing well in the trade will not be ordered again and no longer produced. Operationally, managing all these different SKUs is complex and costs money. So, if some of those only have very limited incremental sales, it's better to stop producing them. The core product is then not a failure.

Other packaged food products are seasonal – they are sold for only for a limited time. Many people follow for a limited but defined time span particular dietary restrictions: Ramadan for the Muslims, Navratri for Hindus, Lent for Christians or Vassa for Buddhists. Many also follow a vegan diet for weeks or months, not necessarily due to religious motives, but to cleanse or detox the body. Trade manufacturers often produce special food products only during that time.

So, counting those delistings as well may blow up the estimated failure rate, but for many others a life span of only 12 months on market is not an immense commercial success.

So "time on market" is for the "external observer" – a valid success criterion. For the producing company, the "return on investment" – the profit after all of the costs – of a new product may be a more important one. The company will assume how long it will take to make a launch profitable. In fast-moving consumer goods (FMCG), the patience is normally rather low; even longer cases do not go beyond 3 years, and the patience already gets stressed, if the sales are not looking good 3–6 months after launch.

A new product launch often happens because the business case for this project looks better than other alternatives in the new product pipeline (see Chapter 6). Here different ideas – and different teams – compete internally. The best project is supposed to win, and the best one is often the one with the highest profit with the lowest risks. Each project team will tend to present the most promising business case, with good growth, the expected profit and the lowest investment risks.

Let's assume a new launch in the snacking area product achieves a very respectable 6% sales growth increase, 50% of it incremental to the existing range at an initial gross margin of may be 21%. But the original business case had assumed 5% incremental growth end at an initial gross margin of the calculated 22%. Compared with the business case, the launch would be considered a failure by the company, and so they delist it after 9 months. Or the launch shows the same result, and a deeper analysis of the occurred costs shows that even after distraction of costs of goods, overhead, and market expenditures, a small profit is left and the company leaves the product on the market.

A "plus" in money after the calculated time is not automatically a success. The company invested because they expected a margin better than other investment opportunities on a comparable risk level. And if the margin is too low, the project is a failure internally.

For a quick-range extension or refreshment idea, the calculation is straightforward. Buy enough ingredients and packaging material to cover the time needed to estimate the success. No need to invest

DOI: 10.1201/9781032724591-3

in new equipment that cannot be used for other products and find ways to cut the marketing costs quickly in case this is needed. Many new food launches are just "me too" products, copying something that is already there, but now under a different brand, a slightly different price or pack size or a new claim. Such small changes do not add many extra costs. But the "external observer" recognizes a new product with a "new taste" and concludes the old one was a failure. But it could just be a relaunch, making an already successful product even more successful. A relaunch – but not a failure.

The situation is more difficult for a new product, may be with new enabling technology and the potential to create a new platform of product launches for the next couple of years or markets. Success here is more difficult to estimate.

And finally, an even genuine launch failure, however measured, could create a situation that builds the foundation for the next success. A failure in NPD in foods remains a good possibility to get it right the next time. Failure is less of a problem; the associated costs are! Minimizing these costs is the better strategy. A multimillion product recall due to a weakness somewhere in the Hazard Analysis and Critical Control Points (HACCP) procedure is much more detrimental than a failed range extension.

Food companies normally do not work on one new product at a time. A good innovation pipeline needs constant refilling and the opportunity to choose. Even with a fast innovation process, say 6 months, the conditions in the market may have changed in such a way that a formerly very promising new product turns into a burden, an energy – and a money loss. So, companies need alternatives to choose from for the launch window they plan for. A couple of alternative ideas serve that purpose.

Not to be misunderstood: Whatever the real number of failures are, launching new products is risky. Why then are all companies constantly taking that risk constantly spending valuable resources, money and time?

Its simply because it's a huge market, still growing, with new opportunities, more actors and more competition looking for a profitable return. New product development and innovations are not only the lifeblood for big food companies with their mature innovation programs and structures, but also the smaller companies in the food industry are affected by the speed of change we are all facing. Recently research has started to focus on the importance of innovation for smaller companies in foods. The intensifying competition for product differentiation, the increasingly global character of food products and markets, the constantly increasing food safety requirements and all progress made in food technologies forces small food companies as well to focus on innovation. Just keeping on producing what they have made successfully in the past, maybe with a few minor, gradual improvements in the manufacturing excellence of their core products, is no longer safeguarding their business.

For private equity firms, a key reason remains the projected growth! A valuable FMCG company needs to be able to grow either the total market or their market share and expand the profit more than the rest of the market. In the business language, they are then sometimes called the "accretive growers". That separates them from the "dilutive growers", those companies, that are gaining market share but at the cost of their own margin (Dahm et al., 2023).

The vision is profitable growth! And the three routes suggested for food companies to achieve that are:

1. Strengthening the core – in most developed categories and markets
2. Expand into neighbouring sectors – either new categories or new geographies
3. Ignite breakout businesses – mobilizing resources behind bold, disruptive bets

These three principal growth areas have different characteristics and requirements. But in all areas insights are needed to create a successful new product. These are insights about the consumer needs, about the market itself, its structures and dynamics, insights about the competitors, their strengths and weaknesses and finally the company's own strengths and weaknesses. A food company may have great insights about the consumers, the market and the competitors but has only access to the necessary technology at costs that would make the new product too expensive. It lacks the essential capability and hence has no right to assume they can win the market. Figure 3.1 summarizes and shows the essential skills needed to grow successfully.

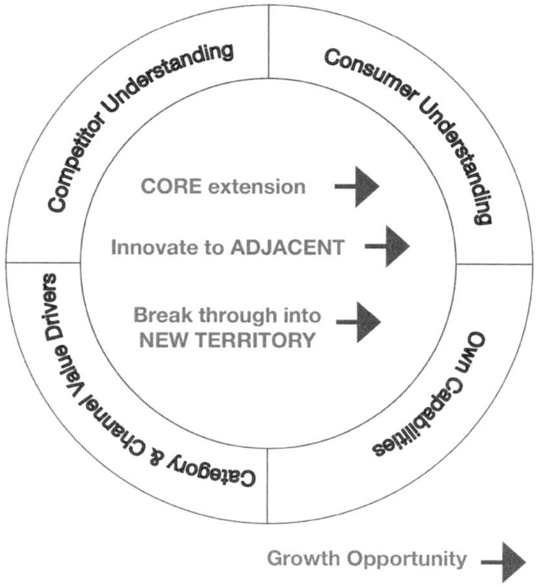

FIGURE 3.1 Key growth territories for food innovation and the four lenses for insight creation. (From https://www.mckinsey.com/industries/consumer-packaged-goods/our-insights/a-multi-lens-approach-to-finding-growth-in-consumer-goods.)

Th evaluation of each of these four lenses can be a complicated exercise. Who would disagree that knowing everything about the competitors and their abilities will give a lot of possibilities to compare one's own skills with those of others? But there is a limit to what one can know about the capabilities of the competitor. Often companies have codes of conduct going beyond what is legally possible in terms of getting insights about competitors. This allows fair competition. And the other three insight lenses are equally difficult to judge.

So, even with the best advice and deepest analysis in mind, NPD remains a sophisticated process with an uncertain outcome. Much work in the process from idea to launch aims to increase the chances of success, even if the environment is changing in between. The possibility to fail is a realistic option. But here it is more important to then fail fast and cheap. If a competitor launches a competing product faster, at a lower price or with a better story, it's important for the company to understand why that happens and what needs to change. Then taking the right learnings from failure can improve the organization's competitiveness, building the capabilities for successful future NPDs. Post-launch reviews and evaluations are structured processes to detect the weak spots in a product launch to avoid them next time (see Chapter 7). Successful new products have been recognized as critical for the business performance of manufacturing companies in today's highly competitive markets. Taking these learnings is essential, since stopping NPD in the highly innovative and accelerating food market is no option.

Hence, creating and applying suitable strategies for NPD is a top priority for business leaders. And for a couple of years researchers have been trying to uncover the drivers of successful new product performance (Vinayak & Kodali, 2014). In 1995, Cooper and Kleinschmidt conducted a broad multicompany benchmarking study. They summarized the key reasons separating solid from poor performers as ".. . high-quality new product process; a clear and well-communicated new product strategy; adequate resources for new products; senior management commitment; climate for product innovation; senior management accountability; strategic focus and synergy; high-quality development teams; and cross-functional teams" (Cooper & Kleinschmidt, 1995).

Later, a market scan in Western Europe of 12,000 new FMCG products launched between 2011 and 2013 estimated that 50%–75% of consumer-packaged goods (CPGs) failed; here understood as they did not meet their projected financial targets and were quickly taken from the market (Dijksterhuis, 2016).

I discussed the difficulty in accurately defining failure. But while different authors present different failure rates, the five key reasons for that are well captured: ". . . path-dependency; information condensations; the practices of problem definition; and window-of-opportunity effects" (Ryynänen & Hakatie, 2014).

- "Path dependency" refers to the idea that the decisions a company makes are heavily influenced by past actions and established processes.
- "Information condensation" is the process of effectively distilling and integrating knowledge from various sources.
- The "illusion of mutual knowledge and understanding in innovation" describes a cognitive bias where individuals involved in the innovation process believe they share a common understanding when, in fact, they do not.
- A good "problem definition" means a common understanding of the problem, the business context and the project objectives.
- And "window-of-opportunity effects" in innovation refers to unforeseen temporary factors allowing enterprises to accelerate. This can be technological shifts, new market demands or changes in regulations.

Fundamentally, one can distinguish two main research angles in new food product development. The first one is the technological perspective, where new technical enablers drive the idea creation process. Microwave heating, non-thermal ultra-high-pressure (UHP) treatment, plant extracts, self-heating packaging, biodegradable fibres, plant-based meat and dairy protein alternatives or CRISPR-Cas are examples. The opposite angle is the market-oriented approach, where the needs of the market drive the creative processes. Convenience, safety, naturalness, indulgence, clean labels, etc., are examples. The more recently tried consumer-led product development process seems more a variation of the latter than a radical new approach.

And while the cost for NPD and the number of new products on the market are constantly increasing, the new way of working together offers new chances to successfully compete. External collaboration cannot only fuel the product idea pipeline; it also brings new possibilities for network-building capabilities and learning (Wixe et al., 2023).

KEY DRIVERS FOR NEW FOOD PRODUCT DEVELOPMENT

About 30,000 new food products have been launched on average in the past years (MIT Professional Education, 2023). Only 2022 has been an exception, with a 12.7% decrease compared to the year before (Schwab, 2023). The pandemic, the supply, geopolitical tensions and the pressure on consumers from inflation were responsible for that exceptional trend.

But the number of food product reformulations might have been higher. To compensate for sharply increasing costs, some food manufacturers may have reduced the portion size or filling weight of their products, since that is now a possibility to absorb the oncosts or charge the inflation-tortured consumers extra; this approach is called "shrinkflation".

It is likely that this trend will change again. Maybe the 2023 and 2024 innovation rate will still be slowed down by overall instability in the world and the grey economic outlook. (Am et al., 2020). But then new unmet consumer needs for safety, convenience, health, eating pleasure and indulgence create massive new opportunities, and some of the fundamental challenges in the food system like environmental care, food waste, animal welfare and sustainability wait for better answers. At the same time food companies will intensify their efforts into producing more, better and more detailed consumer research to further reduce failures (see Chapter 8).

TABLE 3.1 General drivers for new food product development

NPD DRIVER	*FORCE BEHIND*
Product life cycle	Many products lose their profitability and attractiveness over time. They must be rejuvenated, replaced or reinvigorated in one way or another. If the product should stay on the market, it needs to be made fit for survival or replaced by a newer version. In foods the life cycles tend to become shorter, enlarging the size of the market a company is acting in or stealing the share from competitors.
New growth opportunities	Enlarging the size of the market a company is acting in or stealing share from competitors. For many financial investors, the ability of a company to grow is a key quality criterion.
New needs	Consumers, especially younger ones, demand products with less environmental impact. They request food and beverage (F&B) products that fit better their new sport and outdoor activities. Health and guilt free need to be matched by new offerings.
New behaviours	Meal moments are shifting. Some people replace traditional meals with a "good for me" snack. A warm and convenient "in between" as lunch while continuing to work. Or conscious time to prepare a gorgeous meal for friends and families on the weekend. New shopping behaviours also mean that instead half of Saturday in a supermarket, customers are doing weekly grocery on the way home in a corner shop or online. Markets are constantly changing; not reacting quickly means leaving the opportunity to the competitor.
New science and technology	Many consumer needs can now be better satisfied. Healthy soy-based beverages were for many years only growing moderately. Now, advancements in processing, flavour development and aroma masking enable products that do not have that typical beany note that consumers in some key markets do not like.
Healthier	Better food products at least "cut the baddies", like sugar, salt, calories, fat, additives, etc., to make pleasure eating non-regrettable or even "increase the goodies": Adding new, promising actives like vitamin mixes, omega fatty acids or pro-, pre-, syn- or postbiotics.
Government and regulation	Governments and influential non-governmental organizations (NGOs) push for changes food legislation to get new nutritionalor environmental guidelines. Or they just protect an important sector of local agronomy from import. Or they simply bring extra tax money.

Together with all the established "forces" driving new food product development, today's innovations all must happen faster. The world will never be so slow as it is today! Table 3.1. summarizes the drivers for new product development.

All these factors, either in isolation or collectively, are opportunities or threats for the existing business, that cannot be ignored.

The new product aims to improve the consumer's life by meeting those needs better and convincing them to try it first and later adopt it and keep purchasing it. If successful, that could either grow the market segment the product is competing in – "more people like savoury snacking" – or it takes a share of the segment from the competition. In both cases the reward for the company is growth.

New technical solutions are crucial in bringing this growth. They do not create new consumer benefits, but often fulfil existing needs better. Plant-based meat and dairy alternatives are not successful because they created a consumer wish to take better care of the planet, but because more and more people wish to contribute to a good environment and the life of farm animals. The big trend for "personalized nutrition" is not new, but now it seems possible that science and technology are able to answer the health question for the individual better. It's no longer necessary to accept the statistical risk for non-communicable disease as meal guidance. The global consequences of dietary intake remain critical to improve public health, but the individual can get personal guidance.

A different type of technical advantages is "just" improving the manufacturing efficiency. Here the product is the same, but it has been made at lower costs, and the owner can decide to improve his margin or reduce the retail price to increase market share, or do a bit of both. Or they use the money saved to compensate for oncost somewhere else in the making of the product, for example, through sugar, fat or sodium reduction.

Many of the new developments in the food ingredient sector like new flavours, aroma boosters, colourants, stabilizers and emulsifiers offer a more efficient "product making". These new solutions cut cost in the product life cycle. By finding new, efficient ways to make vanilla flavour, more consumers can enjoy these aromas in many more products. A more contemporary "win-win" example is the "double emulsion" technology. In many sauces and dressings, this cuts the costs for fat and oil by bringing in more water. But it also lowers the caloric content, allowing the pleasure of eating healthier without any compromise on the beloved sensorial properties. Radically new technology, less material costs, ultra process and healthier! Not the wrong direction for the industry to go! But here sometimes, old regulations around "standards of identity" are a hurdle. The regulations demand that something called "mayonnaise" must have at least 65% vegetable oil, while the technology nowadays allows one to create the same eating experience with less than half of it. Originally meant to avoid food adulteration, these standards nowadays make new food development more difficult. Indulgent products with less fat and fewer calories is reacting to today's nutritional challenges. The times where regulators had to ensure that mayonnaise had the full fat content in order to be real have gone by (https://www.accessdata.fda.gov/scripts/cdrh/cfdocs/cfcfr/cfrsearch.cfm?fr=169.140).

Usually in the packed food markets the products compete with each other, often in price. Just in a few key areas, like food commodities, the price and quality of food products are strictly regulated by the government. This often affects basic goods like staples and commodities. But sometimes other popular formats are included. In Thailand, the price of instant noodles was fixed at six baht ($0.23) per pack for many years. Now retailers are allowed to charge up to eight baht. And while instant noodles are a very successful food category, nutritionists would probably object to defining them as essential.

Everywhere else, the market is open for a new challenger. So having a great portfolio of commercially successful products in the market is not a moment to rest. If this market is attractive, other companies will recognize it and try to get a piece of the cake. For them every bit of it is new incremental growth and profit. So, not innovating is simply no option; even so, it can lead into a situation known as the inventor's dilemma (see Clayton Christiansen later in the chapter under "Innovation").

The future markets demand more new food product development. The helicopter view shows shelves full of brands, packaging sizes, different price points and huge array of variants. The result is tough competition among the food producers, which in turn, fuels the need to continuously develop new products. Food markets are today much more fragmented; the product differentiation of the past has created new segments and niches. But they will not fuse or amalgamate. Each product will "defend" its own territory; products from adjacent segments will look for extensions, and the owner of the product in that space will know that and react.

Companies will continue searching for unoccupied or not well-defended space in the market and a differentiating angle in the offer to consumers to land a success. The skill needed for that is a good understanding about the successful product attributes and sometimes to also find those that do not add value – something called "overdesign". Overdesign adds costs that the consumer doesn't value. A rational understanding about extrinsic and intrinsic factors influencing consumer choice is needed to address consumers efficiently.

In stores, consumers often make quick choices among multiple products with all their different extrinsic attributes, like brand, packaging, label and price, and expected intrinsic attributes, like taste, health, aroma and convenience. Experienced marketers know that when intrinsic sensory information is not available, consumers are compelled to rely on extrinsic food attributes. Recent studies in marketing and business literature show that consumers form their judgments based on heuristic conclusions about the product's extrinsic attributes in the absence of intrinsic sensory properties (Deng and Srinivasan 2013; Chai et al., 2023). While this is easy to understand, the question on "how to" make successful new products is less simply answered. Advice to "meet consumer needs", "watch the competition" or "find the sweet spot" are totally valid, but they are guideline, a frame, not the success formula.

And while the packaged food industry is constantly evolving and innovating to meet the changing needs and preferences of consumers, the space for radical, groundbreaking new innovations is getting smaller and smaller. Most successful packaged food innovations in recent years are often the shift of an existing food into a more contemporary context.

But success in innovation does not require "new food". Responding in a superior way to an existing need is sufficient. Today's busy lifestyles, less time, more convenience and personalization leave a lot of space for new products. The trend of eating smaller portions but more frequently throughout the day, rather than the traditional three-meals-a-day pattern, is often regarded as healthier. Three regular meals a day is not a physiological requirement, but a culturally adopted pattern, and it is less fitting in today's world. This is one reason why snacks are seen as one of the key growth opportunities. Formats like protein bars, nuts, dried fruits, pour-over soups and meals represent healthy snacking.

Still sometimes unexpected new formats appear. "Bubble tea" started small in the 1980s in Taiwan, where two entrepreneurs individually claimed the invention. The candidates were Lin Hsiu Hui, a product development manager at Chun Shui Tang teahouse in Taichung, and Tu Tsong-he, a teahouse owner at Hanlin Tea Room in Tainan. Both explained that they saw the beautiful traditional tea drinking culture in their country disappearing and becoming replaced by ordinary carbonated soft drinks (CSDs), especially in the younger, more agile and busy generation (Espina, 2022). So they wanted to revitalize (refresh) the highly appreciated habits of a traditional tea ceremony.

Their main inventive step was to add chewy tapioca balls to the tea, so their core formulation had four ingredients: Tea, milk or non-dairy milk, sweetener, tapioca pearls or other toppings. The tea could be anything from black, green, oolong or fruit-infused; the milk either condensed, fresh, cow- or plant-based; and the sweetener can be sugar, honey or syrup. And for the topping's tapioca, jelly, pudding or fruit could be used. Endless possibilities for product personalization, somehow brought back the private character of a "for me" moment of a tea ceremony into today's hectic urban life.

In a couple of years that built a market of about $2.3 billion in and outside East Asia, and is still growing with average annual growth of 9% in the coming years (Fortune Business Insights, n.d.). The products do not only borrow from tea culture and tradition, but they also sail on the health perception of tea, which gives the bubble drink the aura of a shield against oxidative cell damage and stress (hectic lifestyle).

For those interested in the sober health facts: A standard 500-g serving gives between 220 and 660 kcals, mainly from sugar and carbs, plus a few grams of fat, and as little as 1 g of protein. A careful nutritional advice would suggest bubble tea as a very occasional treat – surely not a regular healthy choice. But for those concerned here: It can be made with matcha tea, which has fewer calories, is lactose free and, of course, is vegan.

The third discrete force pushing new food products into the market are regulations. Markets need rules, and they are implemented to protect people's health and wellbeing. So, regulations can heavily influence the need for new products, as they may allow or forbid or newly limit usage of ingredients. Titan dioxide (E171) is a powerful colouring agent in foods (white) and was approved as an additive in Europe. But in 2021 the European Food Safety Authority (EFSA) concluded that on the base of the available data they "cannot rule out the possibility of genotoxicity", and so the European Commission banned E171 in 2022 (European Food Safety Authority, 2021). Science moves forward over time, and in food safety the precautionary principle demands action when a reasonable ground for concern of an unacceptable level of health risk exists.

More and more regulations address societal and environmental challenges (see Chapter11), trying to ensure not only a safe and good life now, but a sustainable and sufficient food supply for future generations. This is one of the reasons why increasingly more regulations focus on a responsible and resourceful sourcing. So, the end to end environmental impact of a food product becomes an increasingly more important design element in NFPD. And even if a particular tax, for example, on salt or sugar is possibly not purely inspired by the wish to improve population health, but rather a possibility to manage the government's budget, it's tremendously impacting NPD. Such regulations are a major force to reformulate existing food product ranges on a broad scale to reduce the extra tax burden of those ingredients or the extra cost coming from usage of plastic packaging.

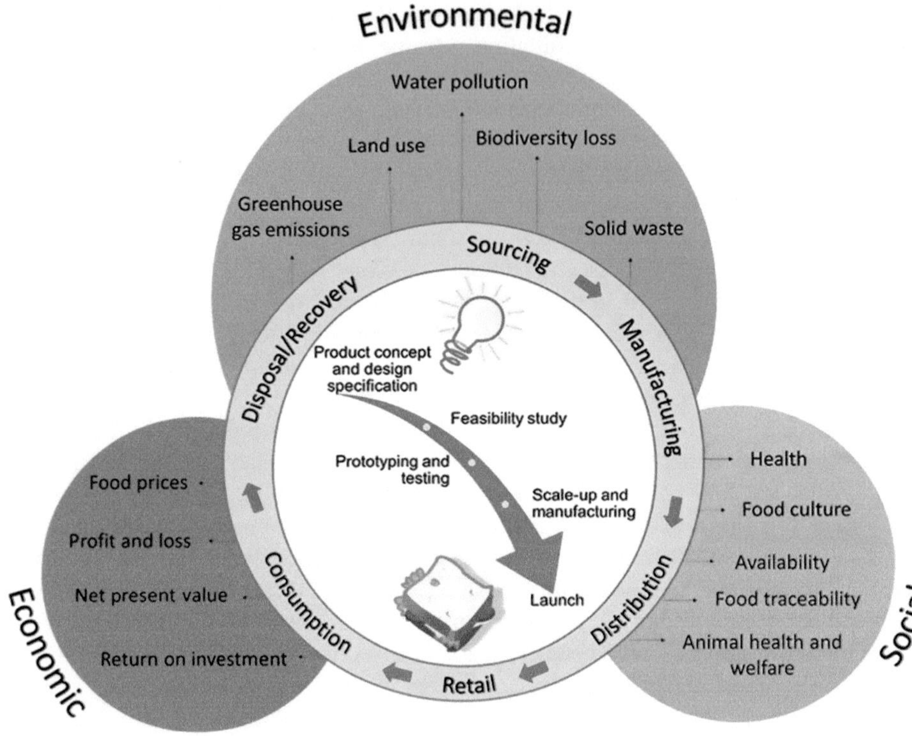

FIGURE 3.2 Incorporation of sustainability into new food product development. (From Garcia-Garcia et al., 2021.)

Most potential threats to human wellbeing or a healthy environment are caused by our behaviour, consciously or not. Long ago experts realized that many people on this planet risk their health by consuming too much salt (more accurately, sodium). Unfortunately, consumers do not read official health bulletins too often and do not want to change their taste preferences. It is then an option for the authorities to convince the food sector to adopt the problem. The companies can cut the salt usage in their products, but also they can use their core skill, marketing, to communicate to consumers much more effectively to reduce their salt intake. Following the new standards for environmental, social and governance (ESG) gives companies a "licence to operate in the future" (see Chapter 16) and the developers much more work on new food products. Now it is no longer a game of variants and range extensions, but possibly the complete overhaul of the total product range to reduce salt, sugar, calories, additives or packaging waste. The challenges in such reformulation programs can be substantial and will be discussed in the relevant chapters. Figure 3.2. captures the various key factors in contemporary NFPD.

CUSTOMERS AND CONSUMERS

Before I can talk about the different types of new food products, some minor clarifications: All the NPDs need someone to buy them. Professional analysts are usually speaking first about market size, market value, growth, etc. Only when it comes to the forces influencing the market and its dynamics, consumers and customers motivations, trends and preferences come into focus. If a product shows good growth, either more people are buying it or the current buyers buy it more often (or a combination of both). The person buying product is the customer, while the consumer is the person consuming the product or service. A customer may or may not be a consumer, and a consumer may or may not be a customer.

For example, if parents are buying a healthy kids snack for their child, the mother is the customer, and the child is the consumer. The parents have no intention to consume the snack, while the child, as consumers, would not, or cannot yet, buy it!

The motivation differs between both. The mother/father sees the next in-between hunger or appetite attack or some unhappiness in the child coming up and wants to solve the situation with something that not only tastes good but promises good nutrition and healthy growth. The child is less impressed by healthy growth perspective but focusses on direct product attributes like taste or play value. No surprising news! But for the right marketing and sales strategy, it's important to understand the difference between both and design the communication approach accordingly.

In the classical foods retail business, customers are involved differently. The typical weekend customer of a modern supermarket is coming with a shopping list and often has a clear plan of what and how much to buy. They value the convenience to find everything under one roof and the huge variety of offerings at competitive prices. They shop to save time and money and to stock up on groceries for the next week or month or to find different cuisines and ingredients that are not available at smaller stores. Special offers and value packs is a good argument for them to come. Hunger or immediate gratification is not the lead motivation, but they may also enjoy browsing the aisles, discovering new products or deals and spontaneously buying them.

In the developing and emerging markets, the "open" or "wet" markets fulfil a similar role. Customers come and buy as fresh as possible what they later in the day need for themselves and their families. The search for fresh ingredients for the daily meals without good storage options in their home and only a limited amount of cash available makes them often shop daily.

The modern trade with their supermarkets needs to keep that in mind. When the big Western chains were coming to Asia, they had to learn what "fresh" meant to the usual open market customer: Just nicely placing fish or seafood or a beautiful piece of meat on clean ice and decorating it somehow was not a good argument! Only when the stores started to sell live products did the sales take off. That's what the shoppers were used to.

The so-called "cash and carry" markets also attract people in their role as customers. Originally these were locations where the vendor could pay cash and carry the products immediately away; these are nowadays large shops where regular retail customers and food service representatives, like restaurant owners and resellers, come to stock up. They are not driven by impulse triggers, but have stock-up plans and search for economies of scale, say bigger packs at better prices.

Yet another type of retail are the cornerstone shops. Originally, these small, independent grocery stores specialized in providing small portions of daily goods including fresh, local products to their customers, placed in urban areas, where they served local customers. Cornerstone food shops cater to a diverse clientele who value quality, sustainability and community. People do not come with a long-planned shopping list; they enter on their way from to work and pick what they need right now. Bigger pack sizes and long shelf lives are not interesting, and what is bought is consumed immediately, maybe on the spot or shortly after arriving at home or work.

7-Eleven is a globally successful player in that market segment. Their customers are the people working and living around the store; food offers are often fresh, and so could be sandwiches, salads, pizzas, snacks, bakery goods, ice cream and personal care products. They adjust their offerings to the needs of the local surrounding; the company also innovates its product line by introducing new items and flavours and local preferences and trends in different markets. Products do not only differ from country to country but by community in each city.

To achieve the necessary efficiency, and hence cost structure, they follow a market dominance strategy with small shops at many corners to gain economies of scale in distribution and marketing. In their shops the customer and consumer role are closer together. People entering the store often do not have a plan on what they will buy but pick something spontaneously or come for an item they consume "on the go".

New food product development for those customers often requires the creation of new product solutions. Someone wanting a warm, healthy and filling in-between meal that can be consumed in the public transport on the way to or from work may need a spoon, fork or knife and a specially designed cup to start eating.

Impulse buying is another act of purchasing something without planning or deliberation. This one is often driven by emotions or external stimuli. For food it's a common phenomenon. While having a walk in the park, you see an ice cream parlour and immediately know that you now want at least one bowl. Ten minutes ago, that was not on your to-do list.

The colourful offerings of sweets conveniently located in appropriate pack sizes in front of the cashier are often the spots with the highest turnover in these shops. It's not unintentional that they are placed there. In a moment when parent and child are heading towards the exit to finish shopping and the child starts crying for immediate fulfilment of the suddenly awoken need, the roles of customer and consumer are clearly separated. And the strength of the pestering power of the little consumer is clearly visible. This marketing strategy is clearly not welcomed by everybody, as it may exploit people's weaknesses and encourages unhealthy eating habits.

The bigger food manufacturers know that they normally do not directly serve the consumer. They make their contracts with the retailers, often the bigger powerful retail chains. With them the manufacturers have to discuss prices and ordering volumes, category strategy and shelf space. They need to be "convinced" to place products more conveniently and visibly on the shelf, so that the products can attract the interest of the customer visiting the shop.

But for the food industry a completely different customer exists, and this is the food service sectors. Their needs, and hence the food product development, is so different from the ones in retail that they will be covered in a separate chapter. And their customer is often also the consumer. The guest visiting a restaurant is both.

And finally, for the food ingredient industry, the branded food makers or the system gastronomy are the customers. This part of the food sector knows that the packaged food industry has different needs and wants than what the final customer or consumer wants. Their offerings are defined by their corporate functionalities that the packaged food makers need to produce the food products that consumers buy. Their solutions need to fulfil the industry requirements but also meet the end consumer's needs.

The material specification for a natural colouring, like "beetroot red, betanin, E 162", would therefore first speak about aspects like "pigment purity", the composition and concentration of the non-colouring substances; the maximum values for possible contaminants, like heavy metals; and maybe define the production process before a qualification of "natural" is requested (EFSA Panel on Food Additives and Nutrient Sources added to Food, 2015). Details on this relation and the development strategies of the food ingredient industry will be discussed in Chapter 12.

PRODUCT DEVELOPMENT SYSTEMS

Product development systems in food are the processes and methods that guide the creation and improvement of food products end to end (E2E), often from the initial idea to the final launch. The impulse for new food product development (NFDP) can come in different ways, from the feedback of market research and analysis to find new consumer needs and preferences. Or by applying new technologies and techniques to enhance the quality, safety, functionality or sensory attributes of food products.

Besides the ability to collect and filter opportunities in market and science (market pull and technology push), a systematic product development approach fosters collaboration with all partners through facilitation knowledge sharing, resources and insights.

Computer-aided processes and templates can support an atmosphere of experimentation and learning and deliberate creativity, and maybe even support risk-taking and feedback. Digital laboratory journals allow researchers to record, manage and share their experimental data in an electronic format. They can easily store various types of data, such as text, images, videos, audio, graphs and tables, and enable easy search, retrieval and analysis of the data. It allows them to share data with others in real time or to export data to other platforms or applications for further processing or visualization. In food they often used to document ingredients, methods, results and experimental observations. A good template basis supports also the structured data capturing and complete documentation of experimental results.

To test and evaluate different prototypes and iterations of food products and to optimize their performance and acceptability, several commercial alternatives are available on the market like "Lab folder" (Labfolder, 2022), "Lab guru" (BioData Inc., n.d.) and many others.

New product development also requires an innovation culture and management discipline to successfully make new food products. It can be a difficult decision to step from an "ad hoc", not well-structured new product creation process to a pretty rigid experimental design and development process. The implementation will require a substantial up-front investment in time and effort before it can be used, while the "old" NPD process still must deliver results in the inbetween period. This can include behavioural change: A creative team may get its energy from trying many new things. But now, the "system" requires them to spend extra time on the detailed documentation of a failed experiment or formulations instead of quickly moving forward to the next idea. The big benefits of that experimental discipline only come later and with more captured data. The "old" mistakes will be avoided; knowledge is captured outside the "brains of the individuals"; others can build on the earlier ;results and computer applications and AI can start to model experimental results and design new solutions "*in silico*".

PRODUCT LIFE CYCLE MANAGEMENT

In a later chapter we will look more closely at modern formulation tool functionalities used today within product lifecycle management systems (PLMs). Here, the system will be reviewed for its ability to drive innovation in foods. This PLM concept began to replace older paper-based configuration management systems with computer-based software platforms that integrate computer-aided design (CAD), product data management (PDM), computer-aided manufacturing (CAM) and other tools in the early 21st century. The automotive and consumer electronic industries were early adopters understanding PLM as an innovation trigger. For foods the added value was less clear originally. In foods, safety is extremely relevant, so a huge number of real-time data about food physicochemical properties needs to be managed and combined with other data during NPD. PLM solutions must enable full traceability of exact and complete product data from raw material to finished good.

Moreover, providing a single, reliable source of unambiguous technical product description speeds NPD and helps to minimize risks due to content errors risking quality. Thus, product formulation must be certified for compliance with regulations. These differ from country to country and change constantly. A suitable PLM solution should enable companies to identify all relevant policies and regulations and comply with them in a documented form. And consumers expect new benefits and new claims, for which clear and accurate labels are required. But dedicated research and practical experience showed that even in the competitive and highly regulated food and beverage industry, an integrated PLM solution "can improve the NPD performances, affecting both firms' process management capability and coordination capability" (Pinna et al., 2018). The authors show in their paper that PLM improves the process management capability, enabling a company to have a better ability to control and improve NPD management processes. And improving the coordination capability means that the firms can coordinate NPD activities better to achieve the product development objectives.

In other words, PLM is not helping much in early innovation phases like the idea sessions or the initial feasibility check (see Chapter 7), but gets very effective during formulation management, regulatory compliance, label and artwork and overall specification management. That increases speed and minimizes error within product development. The basics of PLM are consistent and validated data of all kinds. PLM systems are normally organizing items in a unique central database and integrating templates and best practices in a single solution, something that facilitates knowledge transfer between activities and projects. Another effect of such systems is often knowledge systematization and communication effectiveness. Project team members use the form and templates to capture and share all relevant information instead of sending a huge amount of mails forward/backward with numerous attachments, which are stored elsewhere individually. Object visualization tools should lead to a more cohesive project and product representation. And higher cohesiveness can help to detect potential design problems and cut costs.

MARKET CLASSIFICATION OF NEW FOOD PRODUCTS

Naming a food product "new" can mean a lot or much less. This "new" sits in the eye of the beholder, and the use of the word depends on who speaks. And "new" on its own does not say much; it must be explained. Now "new, better formulation" is understandable. It's the well-known product, but with a change in composition, and it's easy to find out what has changed. For loyal customers it is reassurance: "Hey, I am basically the same product that you like, but now even better". And for non-users it could be a reason to try it now.

A new cleaner recipe, a new size or portion or a new convenience are other typical food "novelties". Ideas like the first microwavable crispy pizza, self-heating instant soup or the first natural blue food colouring have a different "level" of novelty. These products use new technology and often need much more development time and effort, but can build a new range or product platform. Here, the enabling technology is also presented as the "reason to believe" why this new product is better now. Kraft Heinz named their technology "360CRIP" (Kraft Heinz Company, n.d.).

But talking about the technology used to make a new product is more the exception than the rule. Probably because good food as such is traditional and perceived to be low technology. Long before the discussion about "ultra-processed food" (see Chapter 17 started, technology was mentioned in new launches, when it allowed claims like "cleaner, "more natural" "softer", etc.

New technologies in food can help to build new growth platforms, and the new technology or ingredient made by that "superior" technology gets a name (like the 1991 computer processor launch of "Intel inside").

Most new food products target the senses, making close to 50% of all innovations involving all senses and getting more sophisticated (FoodDrink Europe, 2021). Benefits for health and wellness are taking another third, when all the vegetal, natural, medical and functional features are included. And once the "single new benefits" are market standards and new solutions are not available, combinations can create a novelty: Health and pleasure, "full taste at low calories", etc. Toothpaste is a category where this approach is easily seen. What started as tooth protection, an acceptable fresh taste and a reasonable price became successively enriched by one or the other "fluoride technology" and today promises "best whitening", "gum disease protection", "care for sensitive teeth", "bad breath protection", "mouth hydration", "enamel protection" and "naturalness." It may not take long before new environmental advantages are added to the benefit list.

Back to foods: On average more beverages are newly launched than foods. It's simply often easier to bring a few new soft drink variants using the existing installed technology than to create new meals, chocolate or bakery goods.

The offerings of the markets could be viewed through the nutrition and health lens and segment along nutrient profiling based on nutrients, perceived goodies or baddies, high or low in certain items, enriched or fortified, or by the level and type of processing they received, all of which is captured in food classification systems coming from governmental organizations or NGOs, like "Nutri score" or" Nova" (see Chapter17).

Or are they viewed in the context of consumer consumption: Breakfast, lunch, dinner, in between snack, post-dinner treat, sports, outdoor, in the car, etc. The breakfast cereals once made Kellogg a very successful company. And getting Nutella out of the breakfast moment, for example, into snacking or ice cream, gives many new opportunities.

Functionality, be it health or wellness, and convenience in all its forms allow a lot of differentiation as well. Culinary dimensions are endless – once the top dishes and foods of the world have been brought to market by restaurants and gastronomy, the retail business can launch many variants for home consumption. And once a country kitchen got popular, the cuisine can be segmented into regions: Chinese becomes Szechuan, Cantonese, Mongolian, Beijing, Shandong, Hunan, Anhui, Tibetan, etc.

It requires creativity and a good story to become recognized as a new launch in these crowded markets. Breakthrough innovations are rare. "Fish fingers" were launched by Bird's-Eye in 1955 offering a better way for moms to feed their kids healthy fish, without the hassle of smelly fish or bones. And the kids got a crunchy, golden brown, fried, crispy crust that tasted like soft chicken.

A different territory of NPD comes as "new functionalities" for the branded food makers: "New resistant starch", "alpha amylases", with "PGPA", "microalgae" or the popular "alternative protein foods", with segments like plant-based foods, mycelium-based foods, precision fermentation, cell cultures, insect-based foods and components and many others.

Here the initial customer is the branded good manufacturers and the benefit must be explained differently to their customers or the consumer. Just claiming "resistant starch" is not enough: The promise to the consumer needs a tangible explanation, like bridge between the brand's messaging and the audience's trust. may promise "improved insulin sensitivity", or "lower blood sugar levels", "fewer calories", "feed for the good bacteria in your stomach".

The food market analysts do not make such sophisticated differentiations. Everything that comes with a new reason to believe is a new food product! Many of those interested take databases like "Mintel GNPD", the Global New Products Database, to look at new product launches in the more than 50 countries they are tracking. One can scan all types of subcategories, product claims, packaging attributes and ingredients. And it provides insights into product innovation and the commercial success of a product or a product category. But most excitingly, it presents all the big trends from the consumer perspective (Mintel, n.d.). Will plant-based products continue to grow? Mintel looks at their own data and reminds us that the "plant-based claim has increased by 302% between 2018 and 2022. Mintel analysts forecast that the market could grow to $160 billion by 2030". And like other data analytics and brand consulting companies, like Kantar, Nielsen, Ipsos, GfK, etc., they provide tailor-made analytical reports about markets and trends. A great source of inspiration for innovation!

New products cover a much wider range of aspects and cover much more than from the consumer perspective. These types of NPD swill be presented next.

LINE EXTENSION

A line extension is a product that brings a minor twist to the products of an established brand. It utilizes the parent brand's name and imagery and is usually in the same general product category. A line extension in food often refers to the addition of new flavours, colours, ingredients, shapes or package sizes and formats to an existing branded product or product range. Classical examples are the introduction of a new flavour of potato chips or soft drinks, juices, ice cream, or a new cereal mix, the launch of new package sizes or formats, including single-serve and family-size packages, and new formulations, such as a gluten-free or organic version versus the existing product. Technically it is often a small change in the product composition or packaging formats. Line extensions are probably the fastest possibilities for NPD. The new product still uses most, if not all, of the existing capabilities of the company; most of the ingredients have already been used before; the suppliers are known; and the supply chain knows how to make the new variant without having to deal with bigger modifications. Also, for the brand it is not a major step: It got the consumer permission to do this under the brand promise and a recognizable memory frame (logo, cloud, iconic format, etc). Some of the best-known and most valuable CPG brands started their career by calling the product they were selling: Campbell's Soup, Quaker Oats, Heinz Ketchup, Flora Maizena, etc. If a brand has "dairy" in the name, it is probably a stretch to start selling non-dairy products or products in which the consumer does not expect a dairy component. It can also work the other way around. A brand like "I can't believe it's not butter" can be many things, as long as it is not butter!

The prototyping is quickly done, especially when the "value-giving attributes" in the product brief are technically well described like "the first bite should be a burst of fresh fruity flavours, followed by a juice mouthfeel". A precise sensory description is easier than demanding something like "great taste". In the latter case, good consumer testing is needed or great trust into trend research. The further upscaling can be fast, as it normally uses existing machinery and knowledge, and the manufacturing feasibility is given. Just the new packaging and labelling need to be ordered or made and a launch slot with the trade agreed. If everything works well and the trade is willing to offer a suitable launch slot, the line extension can hit the market in 3 months. It may take longer if the extension is the introduction of a new packaging format, which differs in weight or volume. Some technical experimentation and possibly some new equipment in the manufacturing line are needed.

On the store shelf, a line extension normally appears side by side in the same category. The word "range extension" can be taken literally: The products of that brand should take more space on the shelf and get wider visibility than what the brand had before. They are additional SKUs, and they are not supposed to steal market share from the original. Instead, they should attract new consumers that have not bought the brand yet or they convince existing users to buy the brand more often. To be a success, that range extension at least must create much more value/profit than the original product may lose through cannibalization from the new range. Probably up to 50% of all new launches in food each year are basically line extensions.

BRAND EXTENSION

More complicated are the brand extensions. Often that includes the launch of new product formats, different food types or different consumption contexts. Besides the more complicated product prototyping, the conformation of the technical possibilities of such an extension requires a bigger strategic discussion as it moves the original brands with its credentials into new territory. If the brand has no "authority" in the new category, success against the established competitors will be difficult. And for the original product the risk exists that the brand image, the DNA, becomes diluted or damaged by a wrong stretch. Its therefore good to have an idea about brand "DNA" with unique characteristics that are not too closely linked to a particular product type but still does the job to be recognizable and memorable to consumers and helping to build trust and loyalty. The creation of a successful brand is a significant up-front investment, often allowing the products to carry a price premium, a value that needs to be protected.

Brand extensions are an attractive opportunity when the sales of the existing product range in a particular product category "hit the ceiling" and when future, better growth is predicted elsewhere. When dry ready meals are no longer perceived as great value for money and very convenient but more as unnatural or unhealthy, the brand owner may want to stretch into the chilled or frozen space. The brand owner has to be sure that the brand value can make that move. If the brand value is, for example, highly anchored in the "drying competence", such a move is riskier than if it lies, for example, in the unique culinary skills.

As said, in many cases the "billion-dollar food brands" started as identifiers for a particular product and an attached quality promise: "Safe and good tasting, always the same quality!" The promise then provided the consumer permission to offer new products in adjacent categories. Adjacent can have many different dimensions: A format, an eating occasion, an origin or a technology. In 1838 the chicory coffee maker Carl Heinrich Theodor Knorr started working with food drying techniques to preserve the quality of vegetables and seasonings outside the harvest season. People trusted his drying competence and so the brand relatively quickly extended into formats like dehydrated soup, meal mixes, bouillon cubes and condiments. Later the offerings in those categories migrated into different liquid technologies, as the competence of the brand had successfully conquered the culinary excellence.

Brand extension leverages the existing brand name and reputation of a company in a different product category. They expand the customer base, increase its market share and can even enhance its brand image, something sometimes needed for the "old foods brands", the reliable companion in the "good old days" but nowadays a bit boring and not up to date enough. However, brand extensions also involve risks, such as diluting the core brand identity, confusing the consumers or stimulating competition from established players in the new category. As a brand is a key asset of the company, it represents values, personality and reputation. The brand influences the way customers perceive and interact with a product and creates loyalty, trust and recognition. Basically, a brand extension must add value to the brand. The fit between the parent brand and the extension product needs to be checked carefully. The fit can be based on the similarity of the product attributes, benefits, usage situations or target markets. Successful food brand extensions are Reese's Puffs Cereal, which leveraged the popularity of the peanut butter cups to create a breakfast product. Or Sunkist's Vitamin C Tablets, which extended the orange juice brand into the health supplement category. Less successful examples are "Dr. Pepper Marinade, Zippo Women's Perfume, Frito Lay Lemonade, and Cheetos Lip Balm" (Duckler, 2024). The most prominent example of a failed brand extension in foods is probably Colgate's "beef lasagne" TV dinner launch (Harris, n.d.). A remake of the packaging is displayed in the "museum of failure" in Sweden. In hindsight the brand extension may have been so embarrassing for the company, that they said they had no memory about such an attempt to move their famous toothpaste brand into frozen ready meals. It is nowadays difficult to understand what may have convinced experienced marketers to do such a move. Building that on the taste credentials of the toothpaste seems farfetched.

The creation of a sub-brand is sometimes a possible construction; here the mother brand serves as an endorser for sub-brand, which has its own independent credential. A few examples:

- Starbucks Teavana: Famous coffee maker launching tea
- Coca-Cola Zero Sugar: A sub-brand of Coca-Cola, same great taste but no sugar
- Nestlé Nespresso: A premium machine served coffee from Nestlé
- PepsiCo Tropicana: PepsiCo's sub-brand for fresh fruit juices
- Unilever Ben & Jerry's: A eclectic sub-brand from the world's largest ice cream maker
- Mars Wrigley: The sub-brand of Mars, specializing in chewing gum
- Danon Activia: Danon's sub-brand for probiotic yogurt
- Ferrero Nutella: The famous hazelnut spread from the Ferrero Group

In contrast to typical line extensions, the brand extension can for R&D and other technical disciplines be a major undertaking. They may have to design and formulate new products and formats for which they do not have a product design and the right ingredient portfolio and need new packaging, a new manufacturing process and perhaps even the knowledge. Here, third-party product development and toll manufacturing are suitable alternatives (see Chapter 8). The brand extension can be tried without massive up-front investment into skills and machinery but will even then still need time. At least a year is normally needed between idea and launch.

REJUVENATION (REFRESHMENT)

Food product rejuvenation describes a process of updating or refreshing an existing food product to make it more appealing to consumers and keep it relevant in the market. This can involve changing the product's packaging, formulation or marketing strategy to better align with current consumer preferences and trends. Some examples of food product rejuvenation include:

- Updating the packaging design to make it more modern and eye-catching
- Reformulating the product to improve its taste, texture, environmental or nutritional profile
- Introducing new flavours or varieties to appeal to a wider range of consumers
- Changing the marketing strategy to target a new demographic or highlight new product features

The intended effect of product rejuvenation is to bring a new boost to sales, thereby extending the life cycle of an existing product by keeping it relevant in the minds of consumers. The removal of artificial preservatives, flavours and dyes from iconic formulations is a typical example and normally helps to rejuvenate the brand. Sometimes it doesn't work: In 1985, Coca-Cola introduced "New Coke," the taste reformulation of its classic soda. But the new product was not well received by their loyal users, and the company quickly reintroduced the original formula as "Coca-Cola Classic." Only this corrective action helped to rejuvenate the brand and boost sales.

Marketing theory in the 1980s sometimes differentiated the rejuvenation into "Recapture, Redesign, Refocus, or Recast" (Lazer et al., 1984):

- **Recapture** means here "product no-touch strategy". It is used when the commercial failure is not caused by product deficits. A launch may have had not sufficient marketing support, a truly bad timing or an insufficient distribution buildup. For R&D and other technical functions, this means no or low involvement.
- **Redesign** is the overhaul of an existing product. Currently this is often the improvement of the nutritional profile by cutting ingredients like salt, sugar or fat or replacing additives or other elements with a critical reputation. Sometimes the relaunch of an old product version is called redesign.
- **Refocus** in its narrow sense is targeting of different consumer groups with the same product. Again, not much technical work here if the refocus is limited to minor packaging changes like artwork. Tic-Tac is a possible example. When they lost share, that made the low-caloric content of their products the differentiator, so new consumers or those now interested in calorie intake took notice.
- **Recast** an existing product becomes modified so that it appeals to new consumer groups. Tic-Tac could also be an example, if they had modified their formulation or packaging to attract the new consumer group.

These classifications are not mutually exclusive, and they are also not used consistently. But in general, a rejuvenation, will not increase the number of SKUs on shelf, because it replaces an old one with a new one.

Generally, such launch preparation as such is not too complex. The rejuvenation/refreshment planned often does not impact the whole product design and manufacturing process, and the way to market already exists. The trade may ask some extra financial contributions to cover the limited extra work they may have with the registration and handling of the new product, but neither the trade nor the manufacturer needs to handle obsolete stocks. The refreshed variants can fill the shelf space as the old stocks are creeping out.

REFORMULATION

Food product reformulation looks at a situation from the technical perspective. An existing product or range is reformulated, so the product is redesigned. From the market or consumer perspective, this can be a refreshment or a line extension or even an innovation. And indeed, if it is additionally communicating the improvement of its taste, texture, nutritional profile, etc., it would be called a refreshment. The change of ingredients, formulation or packaging is often triggered by an identified product defect/weakness or sourcing or manufacturing problem. Here the reformulations happen silently, so with no or minimal change in the marketing mix or on packaging communication.

VALUE IMPROVEMENT PROGRAM

Another big part of the reformulation work is captured under the "value improvement program" (VIP). This characterizes systematic approaches to improve the value of one's own products or services for the company. The idea here is to identify and implement ways to at least maintain the value that the customer/consumer receives from a product or service, while reducing costs and improving efficiency.

This type of reformulation work normally targets the "bottom line", so it increases the profitability by reducing costs somewhere between product assembly and shelf. Material efficiencies through smarter formulating design, material replacement, smarter material usage or waste reduction through reduction of inefficiencies in the production process are typical examples. But also, a packaging improvement that cuts effort at the warehouse or retailer shop can be VIP if it improves the price the customer is willing to pay.

While calamities are often caused from outside the company, VIP is the result of internal processes. Often finance identifies the need to save money in the product manufacture and gives the reason for a cost-saving exercise. This can be an ad hoc request or a whole, continuously running program to optimize the cost structures in this part of the business. It requires substantial R&D resources, so sometimes prioritization is needed to balance this refreshment and innovation.

VIP programs are part of the strategy. If the financial outlook of sales and profit for the coming period does not match the financial plan of the company, a VIP becomes started. The key actors in this are procurement and R&D. R&D will have to optimize the product logic to create material savings, and procurement must find "buying leverage" through alternative suppliers or different sourcing volumes.

The impulse for a VIP program can come from procurement. Their teams have clear financial targets in the annual plan and how to buy the necessary materials and services most efficiently. But they must align with R&D as the product experts to evaluate the savings options. R&D as the designer of the product and its specification, has to confirm that the reformulation does not impact product standards. So, they sometimes are less motivated to reopen the design box to replace elements in there for products they may have just created and rigorously tested under already strict cost guidance.

An exception is the launch of an already knowingly overdesigned product, for example, due to timing pressure to catch an important launch slot. Maybe an originally planned material was not available in time, an approval is pending or line modification could not be finished.

When the "savings target" is the result of analysis of the business performance, it's often a "number given" without concrete ideas on how to get it. Then a key tool of the innovation process is used: The idea creation workshop! A session with representatives of different functions is organized, a moderator nominated and some tricks applied to get the creative juices flowing. It can be broadened by inviting external partners, especially trusted and strategic suppliers. One reason is that sometimes the cost of purchased goods is influenced by requirements of the buyer, without good understanding of the oncost triggered. A factory may have preferred to receive some good in 20-kg cartons to manage them flexibly in the factory. But after while the sales volume has increased, or the factory had delisted a product using big bag or silo delivery, and now this is free to receive bigger batches from the supplier, where such delivers is also much more efficient, saving time and lots of packaging material. For R&D this is an easy saving; ideally, they just have to rewrite the specification for the material and observe one factory trial.

Like in the innovation workshops all ideas, even the wildest, will be listed and then checked for their feasibility and a program agreed to. The savings potential of all product logic and buying leverage ideas can be calculated and summed and mirrored against the savings target. In case the end sum of the list is sufficient to close the gap identified in the business plan, the work can start. If that sum is not high enough, another creative round can start or more risky, radical ideas with bigger savings can be included. These are more drastic changes in the product design or changes to top sellers, where the volume brings the savings. These "big ticket items" are often painful for R&D. The danger that the current consumer recognized the change and dislikes it is higher, and the potential loss of sales from top sellers can make the situation even worse. Therefore, the team has to agree on clear risk mitigation plans, like rigorous consumer testing of the VIP variants, before locking in the new design.

More sophisticated VIP approaches attach risk factors to the original ideas on the list, discounting the savings potential. The big pack idea with the supplier can be straightforward based on the information at hand and therefore gets a 90% chance, while the idea to cut the vegetable content in a top seller only gets 20%. If possible, a fall-back list can be created that becomes activated when an original idea fails.

The whole program can be managed like the innovation process management, a stage gated process, where the different ideas are getting more and more of their full savings value the closer they are coming to launch, which here means implementation.

It makes good sense to create a steering team with the key functions, including marketing and regulatory, to monitor the progress being made and that is able to take a decision to get to the target in case key initiatives are not successful or delayed or a sudden opportunity appears.

By the way: As in innovation project management, the pressure is often on R&D to come up with solution as soon as possible. Once a saving is realized, the savings count, so within a reporting period having 6 months of savings is much better than only 1. But it's also true the rate-limiting steps are often the material supply. Once the new formulation is locked in, the artwork needs to change – maybe only small change in the ingredient list. And that needs new packaging material, If the packaging is not made and printed in house, the delivery of the new material will need a few weeks.

A final pool of savings are the ingredients with claims. Many food formulations use source claims, like "from Tuscany", "with extra virgin Greek olive oil", etc., and products have been launched to differentiate them from the competitors. After some years this claim may become less relevant, so it is good to check with marketing whether these "claims" are still worth the extra cost of the material.

CALAMITY

A distortion somewhere in the end-to-end product manufacture, often from outside the company, is sometimes called "calamity". This required reformulation work is unplanned and unwanted. A typical situation in foods happens if an ingredient gets "out of stock", which means that it is shortly no longer available. A supplier may not be able to provide the contracted material for political, economic or environmental reasons. Bad weather and a crop failure in a region can affect many food ingredients. Or a disturbance in the sourcing chain. Like the container displacement and cargo shortage caused by the last pandemic, for example. Or just a trade embargo. For R&D that means that if no suitable 1:1 replacement for the missed material is in sight, sophisticated reformulation will have to start, aiming to create a new version with preferably "no change" for the consumer or customer. That new reformulated product should be indistinguishable from the old one. An exception is often the legal on-pack ingredient declaration, where even very small changes in the composition change the ingredient list or the order in which the ingredients are listed. Calamity work is not a strategic intent; it's a sheer necessity to avoid that product from becoming "out of stock" and cannot be sold any longer. The latter is a very unattractive option, as turnover or shelf space lost is so difficult to gain back. And when it happens to top sellers, it's simply "no go territory".

But while the specific calamity is not foreseeable and the need to repair is not negotiable, strategic planning can happen to reduce the impact. A flexible product design, where components can be easily replaced by others, reduces the risks. A second, alternative supplier, delivering against the same or a very similar material specification, is a good fall-back option in case the first supplier cannot deliver any longer. And a dedicated R&D team experienced in such work offers an effective and quick response mechanism.

In fact, calamity situations are not rare events in big food companies. Bigger multinationals have thousands, often more than 10,000, different ingredients and raw materials with different specifications. And they may have maximized their supply security by having at least two different suppliers per specification, but still, there is nearly not a single day where not at least one ingredient somewhere is suddenly missing and not available.

The reformulation work then needs to start immediately. For R&D and procurement, a hectic period starts, searching for alternative suppliers or alternative materials that are able to deliver exchange materials or at least one is as similar as possible to the missing one.

Normally, there is also no time to test the reformulated product with consumers. And in most cases, this is also not necessary. The aim of the reformulation work is to "match the current", ideally making it the same as before. Here quite simple in-house difference testing is often sufficient to decide if the new formulation is a match or not.

On the other side, a calamity often does not affect only one formulation or SKU, but often dozens or hundreds of them, as ingredients are shared. A vegetable powder, a spice mix or a particular flavour is often used in several formulations, With the limitation in time before the SKUs are "going out of stock", it not only requires a tight project planning but prioritization as well.

R&D, especially the formulation experts, play a leading role here. First, and even that can cost quite some time, all SKUs containing the missing ingredient are listed. These commercial data need to scan for SKUs using the same formulation, like different pack sizes. Then those products still in the system but supposed to become delisted can be filtered out. Here is a need to put reformulation work into them. Or there are seasonal products in stock using that material that do not need quick reformulation, as the net production is still months away.

If this list of truly affected formulations is much higher than the time and resources available to reformulate them, a priority call is needed. R&D, as product owners, take the responsibility and cluster on product along the design logic and their similarity. Imagine if a meat flavour becomes a calamity, which is a core ingredient in different bouillon cubes, dry mixes, a couple of dry soups and some ready meals on the list. R&D can suggest the lead applications, those formulations in which the flavour is most recognizable. If a formulation change works in those, others can be swapped accordingly. This gives more time to find the alternative with the best matching characteristic. The other products in the group will get the solution implemented without rigorous testing and optimization. Sometimes those lead applications are those with the highest dose in the formulation, but the manufacturing and use by the consumer/customer can also play a role.

Instead, pure product design logic, the commercial view, can be added to. A responsible business leader or marketing colleague whose products are affected may get stomach pains if the new formulation of "his" top sellers is sent to market without any consumer check. So, often it is better to spend some extra time and money to be sure that market volume, turnover, profit or brand reputation is not affected.

For a young and ambitious food developer, the work on a breakthrough innovation may be the most attractive option. It feels more exciting to realize a new idea, create a new food product, observe the latest market tends or implement a new technology than the tedious, laborious and sometimes stressful rework on the existing portfolio. Recurring value improvement programs and calamity management are therefore not the perceived career highlights.

A big part of the business revenue is coming from the maintenance and improvement of the existing portfolio. Even a successful new launch will need a long time before it will make it to the top seller list. It creates incremental growth, but not the gross of the value. The continuous care for the product on the market is so important for the company and rightly deserves a lot of attention. Some believe that up to 40% of the available R&D workforce is on average working on this type of reformulation work.

INNOVATION

An innovation is something new and different from what is existing on the market. Food innovation is the development and commoditization of new food products, processes and services. The goal nowadays is to make healthy and nutritious offerings that are not only appealing, accessible, exciting and unique but also sustainable.

Sometimes invention and innovation are used as synonyms. That can be misleading. Invention is better described as the creation of something new that did not exist before, while innovation is the improvement, new usage suggestion or modification of something already existing. Humans did not invent fire. But using it as a source of heat and light and preparing food to make it more digestible is an innovation! Cheese as a new way to preserve milk is the innovation. The subsequent creation all those different types of cheese: Cheddar, goudas, Emmenthaler mozzarella, or brie, etc., are inventions.

Food innovations satisfy new consumer needs and preferences. Plant-based meat alternatives offer consumers wanting to reduce their meat consumption for health or environmental reasons an alternative.

Technically, innovations are characterized by scientific and technological advances or a new application of a technology. Big ticket items like biotechnology, nanotechnology, digitalization or artificial intelligence come to mind. But likewise, vertical farming is a great innovation, using controlled environments and LED lights to grow crops indoors with less water and land use.

An impulse for an innovation is not only triggered by a new consumer need or new technology developed. Social and environmental factors, culture, ethics and regulations can also cause innovation. Groups like the European Environment Agency defines social food innovations as ". . . the development and implementation of new ideas (products, services and models) to meet social needs and create new social relationships or collaborations'" (European Environment Agency, n.d.). New insect protein-based food products are truly innovative, but not necessarily fulfilling a consumer need. Sometimes consumers are not really asking for it; they instead face cultural barriers in many markets. But they can make a claim to global food security and environmental sustainability.

Innovations can also emerge when multiple stakeholders from different sectors and disciplines are coming together to create something new. The ongoing urgent debate about food waste reduction is, of course, looking for new technical solutions, But at its basic, it is a collaboration between farmers, processors, retailers, consumers and policymakers.

The debate if something new in food now is a true innovation or maybe "just" a "reformulation" or "refreshment" in foods is initially less important. If an idea offers a good opportunity to create intellectual property rights and create some strong patents, it's probably a technology-pushed innovation, as it used new science. And if a brand offers rigorous "transparency" about their product, this can be done without any new technology. At the same time, a new product format for a brand that is made by a third-party manufacturer using very mature process technology can still be an innovation for the brand. Different companies may also have different understanding when something should be called an innovation and when not. The important thing is that they identify the "character" of the idea. An innovation normally requires more checks and control points. If it's a technical innovation, milestones, proof of principal and feasibility will safeguard the maturation. Market research will ensure that the consumer need was well interpreted and the solution offered fits this. As Griffin stated some years back: "The most successful product development efforts match a set of fully understood customer problems with a cost-competitive solution to those problems" (Griffin, 1996).

And for a social innovation, the company will keep in close contact with the "problem owner" to ensure that their solution is satisfying their expectations.

Bigger innovations – and what "big" is, is company view – normally needs longer development time, more resources and even requires the addition of new skills and capabilities. Implementing new AI tools in consumer research to better predict trends means for many food companies access to those experts that can train the necessary digital networks. In the market such development should create access to a new market segment, so the cannibalization rate on the existing products is low and the incremental turnover therefore higher. A technical innovation is also not always visible for the consumer. Ways to make the same product more efficiently may need years and lot of work and aim at "no chance" for consumers. And in the space of food safety and hygiene, many new, often radical, technologies are applied to achieve more safety that the consumer would probably not recognize. Radically new innovations are sometimes called breakthroughs.

BREAKTHROUGHS

Breakthroughs, or disruptive innovations, are when new markets are created successfully or existing ones are disrupted. The current rise of plant-based meat alternatives challenges the traditional meat industry. The new products offer environmental and health benefits, but also pose a threat to the existing value networks of meat producers, processors and retailers. Or the emergence of personalized nutrition, which uses data and algorithms to tailor food and supplements to individual needs and preferences, is an attractive option. Breakthroughs are often characterized by technology, like the alternative proteins, food waste reduction, genetically modified (GM) food, restaurant digitization, nutraceuticals or maybe 3D food printing. For the established mass-market food industry, having good revenues from the existing markets working on breakthroughs can form a dilemma. They must choose between sustaining their existing market presence or eventually disrupting them by executing this breakthrough innovation. Clayton Christensen named this fundamental difficulty the "inventor's dilemma" (Christensen, 2016).

One problem is the life cycle of such an idea and its value curve. Described as an S shape, the innovation first needs time to create revenue while new buyers are coming in. Once early adopters buy in, the revenue increases and adds value for the business. And while the consumer mainstream is embracing the innovation in the background, a newer, slightly better version of the idea attracts already the first consumers. Over time more and more move to the next-generation version and the revenue of the first innovation slows down and starts to decline (see Figure 3.3).

The company with a version on the market has a dilemma. Making a better version might cut their own revenue. But they must do it before a competitor comes in and steals their share. For startups, that is an attractive position. They do not have an existing market they can lose, and a technical breakthrough can cut the market entry barriers for them.

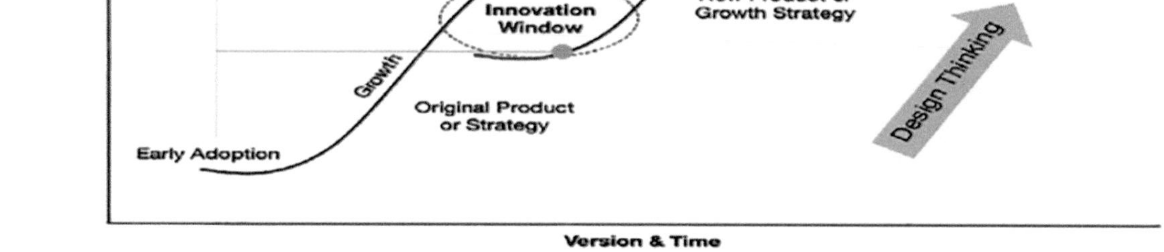

FIGURE 3.3 Product life cycle S curve and innovation windows. (From Mazouz et al., 2019.)

The problem is how to know if there will be a better version soon. If the technology has been used to the max, an invention would fail, as this has no benefit in the eyes of the consumer. So, no need to put extra effort behind such an idea. Better to stay on track after market entry and support the original invention from time to time and reminding the consumers it exists. So, the incumbent company can create a better return on investment by making low-risk variants or just campaigns from time to time. For leading FMCG brands, this is a recurring question, since other enterprises follow a value creation model in which they copy the market/innovation leader. They "match current" as much as possible, but at lower price, a benefit consumers love to take advantage of.

At this point in time, early 2024, plant-based meat and dairy alternatives seem to have reached a point where more and more consumers joint the trends, but now growth slows down and the innovators try to find the next better version that attracts consumers.

For the build of innovation pipelines and new product portfolios, this question about the future potential of such new science and technology can be very important. Whether and how a technology is capable of disrupting the current market of a company requires very good product and technology understanding – a prime knowledge domain of R&D. So, it's their responsibility to bring that into the strategy discussion.

A successful mature food business today will not prioritize effort into breakthrough innovations. Successful examples are rare, more difficult to evaluate in regular business plans and the consumer acceptance remains questionable for a long time. Food is part of the fast-moving markets, with lots of new launches all the time. So it is difficult to tell how long actually the life cycle of the product can be.

Take romanesco, a hybrid between broccoli and cauliflower, that has taken some of the share from the two original varieties but has not enlarged the overall vegetable market and will not have taken most of the share from its parents. Approximately 10 years ago cronuts, the delicious combination of croissant and donut, were hailed as one of the biggest innovations of the snacking century. Today, it is hard to find any data about their historic market share. And Ice Croissant, a croissant filled with an ice cream bowl inside for heating in the oven, was launched end of the 1990s by one of the biggest ice cream and frozen companies in Europe. Technically, exciting! A warm croissant fresh out of the oven with a still cool ice cream filling without any chemical tricks, just using physics, where the airy layered pastry structure protected the ice cream from the heat in the oven until the croissant was warm and brown. But the consumer was less excited and a bit unclear on where to put it: Not a breakfast, too sweet for a lunch and too heavy for a dessert. The breakthrough lasted two seasons.

Food is much more than nourishment: it is culture, tradition, childhood memories! It is conservative. Hence it is no surprise that real breakthrough innovations happened in the making of, the process, not the product. Nowadays, truly sodium-free saltiness, calorie-free natural sugars, no-fat or zero-calorie meals and plastic-free packaging with full plastic functionality would be true breakthroughs.

CASE STUDY: PRODUCT ONE TOUCH

Creating new products, keeping them regulatory compliant, realizing cost savings, refreshing the packaging or extending the range while bringing a new consumer-relevant claim requires a lot of time and effort. Not only the technical work on the product – it's the ordering of new ingredients, finding new suppliers, agreeing on new specifications, making new at work, testing new packaging material, trying all of it in one or more factories and testing the result with consumers/customers before presenting it to the trade to secure a launch slot.

That is not efficient. Food companies with only a few products may still work like this. They start a range extension when they see a new trend or hype, notice a new upcoming brand in the market, a competitor moves or there is a new ingredient. Bigger companies, with hundreds or even thousands of different products, all of them having a life cycle, need a different approach.

For some this much more planned approach is called "product one touch". It builds on two insights: The first one is that many of those reasons are somewhat foreseeable. Not the calamities – they remain a surprise, But the big trends in food are clearly visible: "First step is to avoid the baddies", which often means reducing salt, sugar, fat, colourants, sweeteners, flavours or other additives. This step is followed by the inclusion of more "goodies", like fibre, vitamins, minerals, unsaturated fatty acids, etc., and this concerns not only the food itself, but also the packaging. Especially the FMCG industry has to make any effort to move out of plastic packaging, or at least those materials that cannot be efficiently recycled.

And secondly the development effort and time to market for a complex change within a product are not much higher than for a simple modification, like the addition of a new flavour, a refreshed packaging or a just reduced sodium content.

The gain can be enormous, and the pain can be limited through good analysis, logical planning and focussed execution. Image a food company producing a variety of cookies, or savoury salty snacks, or dry soups and sauces that sees the trend towards more healthy, unregretted pleasure and indulgence. Just answering with a light/reduced range or range extension is not enough. Consumers turn away from the top sellers, and new regulation or taxation will impact the sales and profit of the key products. In such a situation, plans are needed to "rejuvenate" the whole core. The assortment is clustered into very similar products and formats, and rework starts for all of the at the same time. Maybe it begins with the simple butter cookies, the salty pretzels or the dry vegetable soups. Every single product gets the new basic design elements in its product brief: The maximum amount of sugar, salt, calories or a percent reduction versus current. The amount or number of additives is set, and some should get a healthy addition like vitamins, minerals and dietary fibre. All ingredients used in these product groups are listed and scanned for cost savings opportunities. With marketing is discussed which "refreshers" are needed: A few range extensions, maybe a few delisting of less well-performing products, a fresh look of the artwork, a new on-pack claim or a more sustainable and lighter packaging. All changes in product and packaging happen in the same "formulation" work, The product is only touched "once". And since the product formats are so similar, the most critical one can be identified with the manufacturing to choose them for factory trials, while the others are only simulated "*in silico*". And with market research, consumer testing slots can be booked more efficiently and with a focus on the key questions. Finally, sales receive a much broader portfolio of new products to discuss with the trade. Once the first product format wave has been modified, the next product category starts.

This structured and disciplined approach brings immense efficiency gains. The synergies in the different development steps allow work on much more products each time. The consistency of the approach – the design principles – are clearer and easier to follow, less material is wasted, ordering time is reduced and bigger volumes are created.

The time and effort saved can be used in various ways: to have more resources behind a necessary impactful innovation, or the large-scale implementation of a new technology in the factory, or for the creation of a new much broader marketing campaign to announce the new products coming. The "new story" is not about one, two or three SKUs, but for a bundle of dozens or more products.

CASE STUDY: NUTRITION ENHANCEMENT

Iin 2003 the Unilever company launched their "Nutrition Enhancement Programme" (NEP) where they created a profiling system for nutrients in product categories where too much intake causes concern, namely sugar, salt and saturated and trans fats. In the next years 10,000 products were globally reviewed by their nutrient profile and the consumer usage. And in 2010 the company decided to set out a measurable, time bond action plan, called a "roadmap", to reformulate their products where they would not meet globally recognized guidelines. Figure 3.4 shows more details.

Unilever Highest Nutritional Standards (2018)

All values presented are maximum levels unless otherwise stated

PRODUCT GROUP	ENERGY	SODIUM	SATURATED FAT	SUGARS	TRANS FAT from PHVO
Spreads and Cooking Products	NA	470 mg/100 g 600 mg/100 g salted spreads countries[1] or 1.3 mg/kcal	33 % tot fat tropical spreads 38% tot fat[2]	NA	
Emulsion based sauces	NA	750 mg/100 g mustards 2000 mg/100 g	33% tot fat or 2 g/100 g	15%en total sugars or 7% added sugar/100 g	
Water-based sauces	NA	750 mg/100 g	NA	7% added sugar/100 g	
Dairy cream alternatives	NA	1.3 mg/kcal or 100 mg/100 g	33 % tot fat or 2 g/100 g	7% added sugar/100 g	
Cream cheese	NA	675 mg/100 g	15 g/100 g	NA	
Main dishes	2 kcal/g or 700 kcal/serve	250 mg/100 g		15% total sugars	
Side dishes	2 kcal/g or 400 kcal/serve	250 mg/100 g	10%en	15%en from total sugars	
Processed meat and fish	2 kcal/g or 400 kcal/serve	800 mg/100%	5 g/100 g	NA	
Meal sauces	NA	340 mg/100 g	2 g/100 g	NA	≤1 g/100 g
bread and breakfast cereals	NA	375 mg/100 g	NA	20 g added sugars/100 g	
Small meals	400 kcal/serve	250 mg/100 g	10%en	15%en total sugars	
Seasonings	NA	265 mg/100 g	NA	NA	
Soup & Boullions	NA	265 mg/100 g	2 g/100 g	NA	
Ice cream & Water ices	110 kcal/serve	NA	3 g/serve	20 g added sugars/100g or 12 g total sugar/serve	
Savoury snacks	110 kcal/serve	300 mg/100 g	13%en	NA	
Sweet snacks	110 kcal/serve	300 mg/100 g	3 g/serve	20 g added sugars/100 g	
Beverages	NA	NA	NA	RTD tea: 5g total sugar/100 g Other: 5g added sugar/100 g	
All other products	NA	100 mg/100 g or 1.3 mg/kcal	1 g/100 g or 25% tot fat or 10%en	3 g added sugars/100 g or 15%en total sugar	

[1] Salted spread countries = UK & Ireland. Sweden, USA, Canada, Trinidad, Tobago, Grazil Chile, Colombia, Ecuador, El Salvador, Guatemala, Mexico, Panama, Paraguay, Peru, Uruguay. South and Central Africa
[2] Spreads sold in tropical areas, without chilled distribution

FIGURE 3.4 Unilever nutrition standards. Levels given are maximum levels. (From Unilever, 2018.)

With this, Unilever recognized the potential health impact of those "nutrients of concern" – sugar, saturated fat, sodium and energy – and committed to clear limits. Sugar, salt and fat have properties going far beyond their sensorial impact; they add to the appearance, texture and shelf life of foods or viscosity and make them more palatable, boosting flavour and mouthfeel. It also helps foods to last longer. But the main challenge in the reformulation is not the missing "functionality" of those ingredients, but the sensorial expectation the consumer has for that particular product or product category.

A huge effort was made to find healthy alternatives based on new compositions and alternative ingredients, like potassium as a partly sodium replacer or alternative non-artificial sweeteners, etc. Substantial long-term research programs were started with academic partners and science leaders to identify future actions that could help in the future to achieve the goals. Marketing and market research made a big effort to successfully test the new products with consumers and find exciting stories to convince them about these "better for you" products. Key opinion-formers and leading nutritionists were asked for advice and help.

Final tests in markets were sometimes frustrating! Consumers tasted the difference and/or were not interested in the "health benefit". They were often aware of the "health consequences", but their answer was "OK, but not here and not now!"; for me "taste is king".

Putting a product with somewhat compromised taste or texture on the market wouldn't make any sense. Not for the company, as they would lose sales, but also the population, as users would just shift to competitors with the old, "inadequate nutrition profile". So, progress came slowly and was laborious and expensive. But by 2021 the company moved 77% of their foods to a salt level that allows consumers to stay within the 5 g maximum daily intake. As they said ". . . this means we have removed more than 37 million tonnes of salt from our food, the equivalent of almost 15,000 swimming pools. . . " (Unilever PLC, 2024). Sugar saw a reduction " of almost 170 billion sugar cubes". And for calories and fat, time-bound targets were achieved ahead of schedule.

Once that baseline was cleared, the company set a target on "positive nutrients": For some of those nutrients published intake data were still showing a shortage. So they promised to provide 126 billion servings with at least one of the key micronutrients, like vitamin A, vitamin D, iodine, iron and zinc.

The enhancement idea was to tackle first the "nutrients of concern" and then add "positive" in the principal formulation guideline: First, "avoid the baddies, before including the goodies".

Like all of these "sustainable" commitments, you never reach the end. Doing this is a commitment to continuous improvement! So, in the meantime, the company has made new, even more stretching commitments and is working on it!

INNOVATION SUCCESS

Innovations are key to business success. Companies strive to do it; otherwise, others do it! So it is essential to be able to measure the success, which requires a tangible definition of success in innovation. Success is not well-defined phrase and can range from the creation of commercial value from novel products and services over the optimization of internal processes, the improvement of customer/consumer experiences up to the answer of how to allocate the right resources better behind the right projects in the future.

Hence, innovation goals are best linked to broader business objectives and the organization's overall mission and strategy. In the simple first step, success is the comparison of output and input. A new food product usually must create revenue, market share, customer satisfaction or maybe social impact or any combination of those. To achieve that, an investment in resources, talent, culture or processes is needed. Comparing both gives the possibility to evaluate the efficiency and effectiveness of innovation efforts.

The direct commercial success of a new product launch can be calculated. The "return on investment" (ROI) assesses the relationship between profit and the capital invested. This equation can be used prospectively and retrospectively. For the selection of the best next product launch, all ideas can be compared in their investment and expected gain. Or a product's market performance is compared to the expectation in the business case, as in a "post-launch review" to decide on urgent actions to improve the equation. A clear picture of the economic impact of NPD helps enterprises to prioritize projects that offer the highest potential for returns.

Measurable data are the basis. Finding good "key performance indicators" (KPIs) is the way for enterprises to measure success. Those do not need to be invented – many are suggested in management literature, and the "art" is just to select the few most relevant ones. A few typical indicators are listed in Table 3.2.

TABLE 3.2 Typical indicators of commercial success

COMMERCIAL KPI	*RELEVANCE*
Innovation rate	Revenue share of innovations vs. total revenue
Innovation ratio	Number of innovations vs. number of products
Market share growth	Ability to capture new customers/consumers and gain a competitive edge through novel products
Profit/revenue margin	Profit margin of new products vs. existing ones
Consumer/customer retention	Measures loyalty of consumers/customers in terms of your new products tracking how often they repurchase
Revenue from new products	Assesses the market acceptance and demand for innovations, indicating whether they meet customer needs and preferences

There are many ways to measure the direct commercial success (McKinsey & Company, 2018). In the foods business a few of them seem more important. One recurring question in the food sector is: How many innovations/NPDs do we need at all? Since food is a low-tech business and eating habits and likes are influenced by culture/tradition/safety, a metric to measure the money gained from the new launch versus existing ones on the market can be helpful. On the other side, the packaged food industry, and here the big players, are often asked to prove their longer-term "growth potential" rather than their short-term "profitability". Hence "the right number" of NPDs will depend on the food segment where innovations are planned. The soup and bouillon business may need fewer new products than the impulse sector of ice cream or soft drinks. And they may also need very different types of innovations, as they target different needs in very different situations., An elegant renovation or value improvement program will bring more cash than the investment in a new product platform that can only grow over time.

In this context of FMCG it is also necessary to understand who is buying. Are these the existing loyal consumers? And if so, what are they doing? Do they replace something else or now use it more often. And hopefully, they not replacing the company's own products (cannibalization). Or has the launch recruited new users, indicating market share gain?

More and more companies will have to make new products with a better environmental footprint. Are now more Millennials or people from Gen Z buying the innovation, or do the loyal users happily take the new offer?

A different set of KPIs looks after the innovation process and its relative strengths. The estimation of the commercial success generated can then take a shorter or longer time span. A measure like "time to market" in the fast-moving food markets is crucial as the product life cycle here is often short, and hence each week of unchallenged market presence means extra sales. In regular project planning solid assumptions are made as to when a market entry must happen. The trade often offers launch slots, which the product maker has to meet. Other food products have an inner seasonality; they are not expected throughout the year by consumers. Impulse ice cream after summer or sauce hollandaise in Western Europe after the asparagus season is not a good idea. Other launches specifically target seasonal events, like Thanksgiving, Christmas, Easter, spring break, back to school, etc.

So, yes faster is better. Many food companies therefore have innovation targets to speed up the time between initial idea and launch. That requires a deeper look into the planning/Gantt charts to identify tasks that could be shortened. That is not a target R&D can "own". But it can be "shared"! If the delivery of an ingredient takes longer than planned, supplier selection or contract details may be responsible. And if the manufacturing line is not ready or the marketing campaign not ready to go live, R&D may have had no responsibility. But if a prototype needs to be reworked and retested several times before being finalized, the R&D function has found space for improvement.

For the longer-term innovation perspective or the creation of a new product platform, the number of patents, patent applications, publications or academic collaborations are a relevant KPI. They would drive a healthier longer-term innovation perspective, and then they are typical R&D targets. For those companies with bigger research departments or strong scientific collaborating with external partners, patent targets are useful, driving long-term value. Measuring that for a group of highly productive formulators, constantly filling the product pipeline with line extensions, could distract them from their tasks.

And another set of KPIs take an even wider view. They look after the "longer-term health" of the organization, including their ability to create valuable ideas. They are often summarized under "innovation culture". KPIs here often need to "sense" the atmosphere. Just counting the number of people working in R&D, their academic grade or years of experience is not sufficient for improvement. Surveys, group discussions and questionnaires deliver more substance. That should be done over a longer period and repeatedly, and results must be discussed with all involved and targets jointly agreed. A few of those more indirect KPI and their meaning are listed in Table 3.3.

As said, there are numerous suggestion for KPIs available in management literature, websites or blogs that can be used to measure the innovation performance or potential (McKinsey & Company, 2018; https://digitalleadership.com/blog/how-to-measure-innovation/; https://www.mplinnovation.com/post/measuring-innovation-success).

TABLE 3.3 A selection of "wider" indicators of innovation success

WIDER KPIS	*RELEVANCE*
Time to market	Speed is money, especially in FMCG. More a process target than a project one.
Innovation pipeline value	Potential revenue from products in development or planned. It reflects the strength and diversity of the innovation portfolio.
Innovation culture index	Evaluates the degree of innovation culture looking after leadership support, employee engagement, collaboration, risk tolerance, learning, etc.
Iterative learning and continuous improvement	Part of the culture, but relevant to understand the organization's ability to refine/modify innovation strategies and adapt to changing market dynamics.
Number of patents/applications/publications/collaborations, etc.	Typical science or breakthrough measures. Patents and publications are output KPIs. Number of partners, etc., are input.
Innovation index	More a composite score combining different multiple indicators of innovation performance. Can be helpful to benchmark the company's abilities with those of others.

Single KPIS can be combined into "scorecards", or if they reflect the totality of business activities a "balanced scorecard". Here, it's advisable to focus on a few key ones. Too many diverging KPIs simply cost too much time to collect. That can distract and confuse the innovation machinery, and the original business intent becomes diluted. The teams often perceive this as control and performance measurement tools, which can cause rejection. But a few well-set KPIs are good guidance for teams to keep focus and embed their work properly into the business context.

Targets should fit well together, like incremental turnover and innovation value. And another important aspect here is the time over which the performance is judged. If the innovators in a company are in their totality working on ideas, spanning a typical FMCG time horizon of 3.5 years, with a couple of launches and uptake of new ideas every year, a 3-year rolling comparison is helpful to measure progress annually without losing the longer-term innovation power.

And if a new food launch uses a new differentiating technology and is not performing well after a couple of months, is that then a failure? Did the consumer miss the intended benefit, or do they just need a little more time and encouragement? A frozen food launch of a "less salt" range is not growing market share while costing a little more? Is it time to delist?

The setting of innovation KPIs is normally not done by R&D on their own. It comes as a query from the business leadership. And therefore, it's critical that R&D fully understands which KPIs are chosen and why and how they are measured and what success would look like. Essentially, they are signposts on the route to more competitiveness, not individual targets. Wrongly chosen or doubtfully implemented, they distract from value or become "boxes to tick".

REFERENCES

Am, J. B., Furstenthal, L., Jorge, F., & Roth, E. (2020, June 17). Innovation in a Crisis: Why It Is More Critical than Ever. McKinsey & Company. https://www.mckinsey.com/capabilities/strategy-and-corporate-finance/our-insights/innovation-in-a-crisis-why-it-is-more-critical-than-ever

BioData Inc. (n.d.). *Lab Management Software | Laboratory System | Labguru*. https://www.labguru.com/?msclkid=f9d436a23293101a4d1c50beeb30a1a5&utm_source=bing&utm_medium=cpc&utm_campaign=BING%7CROW-EN%7CSRC%7CBRD%7CLabguru%7CEXM&utm_term=labguru&utm_content=Bran)

Cai, W., Chen, T., Ren, T., & Qing, P. (2023). Watch Out for the Healthy Trap! How Food Flavor Influences Healthiness Evaluations and Consumption. *Food Research International*, *173*(2). https://doi.org/10.1016/j.foodres.2023.113379

Christensen, C. M. (1997). *The Innovator's Dilemma: When New Technologies Cause Great Firms to Fail*. Boston, MA: Harvard Business School Press.

Cooper, R. G., & Kleinschmidt, E. J. (1995). Benchmarking the Firm's Critical Success Factors in New Product Development. *The Journal of Product Innovation Management*, *12*(5), 374–391. https://doi.org/10.1111/1540-5885.1250374

Dahm, J., Enger, C., Grimmelt, A., Lowrie, J., Marquis, A., Murphy, R., Ometto, M., Timelin, B., & Wittmann, N. (2023, July 26). A multi-lens approach to finding growth in consumer goods. McKinsey & Company. https://www.mckinsey.com/industries/consumer-packaged-goods/our-insights/a-multi-lens-approach-to-finding-growth-in-consumer-goods

Deng, X., & Srinivasan, R. (2013). When Do Transparent Packages Increase (or Decrease) Food Consumption? *Journal of Marketing*, *77*(4), 104–117, https://doi.org/10.2307/23487406

Dijksterhuis, G. (2016). New Product Failure: Five Potential Sources Discussed. *Trends in Food Science & Technology*, *50*, 243–248. https://doi.org/10.1016/j.tifs.2016.01.016

Duckler, M. (2024, May 9). Clearing the confusion between line extension and brand extension. *FullSurge*. https://www.fullsurge.com/blog/clearing-the-confusion-between-line-extension-and-brand-extension

EFSA Panel on Food Additives and Nutrient Sources added to Food. (2015). Scientific Opinion on the Re-Evaluation of Beetroot Red (E 162) as a Food Additive. *EFSA Journal*, *13*(12), 4318. https://doi.org/10.2903/j.efsa.2015.4318

Espina, D. (2022, June 22). *History of Bubble Tea | Who invented Boba*. BubbleTeaology. https://www.bubbleteaology.com/history-bubble-tea-who-invented-boba/

European Environment Agency. (n.d.). *Reimagining the food system through social innovations*. https://www.eea.europa.eu/publications/reimagining-the-food-system-the

European Food Safety Authority. (2021, May 6). Titanium dioxide: E171 no longer considered safe when used as a food additive. *European Food Safety Authority*. https://www.efsa.europa.eu/en/news/titanium-dioxide-e171-no-longer-considered-safe-when-used-food-additive

FoodDrink Europe. (2021). *Data & Trends 2021 EU food and drink industry*. https://www.fooddrinkeurope.eu/wp-content/uploads/2021/11/FoodDrinkEurope-Data-Trends-2021-digital.pdf

Fortune Business Insights. (n.d.). *Industry-reports/bubble-tea-market*. https://www.fortunebusinessinsights.com/industry-reports/bubble-tea-market-101564.

Garcia-Garcia, G., Azanedo, L., & Rahimifard, S. (2021). Embedding Sustainability Analysis in New Food Product Development. *Trends in Food Science & Technology*, *108*, 236–244. https://doi.org/10.1016/j.tifs.2020.12.018

Griffin, A. (2012). Obtaining customer needs for product development. In M. D. Rosenau, Jr. (Ed.), *The PDMA Handbook of New Product Development*, 211–230. New York: John Wiley & Sons Inc. https://doi.org/10.1002/9781118466421.ch13

Harris, H. (n.d.). *What a viral fake lasagne taught me about failure*. https://www.prospectmagazine.co.uk/ideas/philosophy/39799/what-a-viral-fake-lasagne-taught-me-about-failure

Kraft Heinz Company. (n.d.). *Kraft Heinz Aims to Revolutionize the Microwave with 360CRISP, Its First Ever Innovation Platform that Delivers a Pan Like Crisp in Seconds*. https://news.kraftheinzcompany.com/press-releases-details/2023/Kraft-Heinz-Aims-to-Revolutionize-the-Microwave-with-360CRISP-Its-First-Ever-Innovation-Platform-that-Delivers-a-Pan-Like-Crisp-in-Seconds/default.aspx

Labfolder. (2022, August 2). *Learn more about how Labfolder ELN can benefit your lab*. https://labfolder.com/features/

Lazer, W., Luqmani, M., & Quraeshi, Z. (1984). Product Rejuvenation Strategies. *Business Horizons*, *27*(6), 21–28. https://doi.org/10.1016/0007-6813(84)90088-0

Mazouz, A., Alnaji, L., Jeljeli, R., & Al-Shdaifat, F. (2019). Innovation and Entrepreneurship Framework within the Middle East and North Africa Region. *African Journal of Science, Technology, Innovation and Development*, *11*(6), 699–710. https://doi.org/10.1080/20421338.2019.1573959

McKinsey & Company. (2018, October 8). *How to take the measure of innovation*. https://www.mckinsey.com/capabilities/strategy-and-corporate-finance/our-insights/how-to-take-the-measure-of-innovation

Mintel. (n.d.). *2024 Global Food and Drink Trends*. https://www.mintel.com/food-and-drink-market-news/global-food-and-drink-trends/

MIT Professional Education. (2023, October 2). Why 95% of new products miss the mark (and how yours can avoid the same fate). https://professionalprograms.mit.edu/blog/design/why-95-of-new-products-miss-the-mark-and-how-yours-can-avoid-the-same-fate/.

Pinna, C., Galati, F., Rossi, M., Saidy, C., Harik, R., & Terzi, S. (2018). Effect of Product Lifecycle Management on New Product Development Performances: Evidence from the Food Industry. *Computers in Industry*, *100*, 184–195. https://doi.org/10.1016/j.compind.2018.03.036

Rutkowski, I. (April 2022). Success and Failure Rates of New Food and Non-Food Products Introduced on the Market. *Journal of Marketing and Consumer Behaviour in Emerging Markets*, *1*(14), 52–61. DOI: 10.7172/2449-6634.jmcbem.2022.1.4

Ryynänen, T., & Hakatie, A. (2014). "We Must Have the Wrong Consumers" – A Case Study on New Food Product Development Failure. *British Food Journal*, *116*(4), 707–722. https://doi.org/10.1108/bfj-08-2012-0215

Schwab, P. (2023, September 11). Food Innovation on the Decline (−12.7%): Trends in 2024. Market Research Consulting. https://www.intotheminds.com/blog/en/food-innovation-trends-2024/

Unilever. (2018). *Unilever highest nutritional standards*. https://assets.unilever.com/files/92ui5egz/production/0c9437ea9360553258a6f081d4b5933f958d6e9f.pdf/hns-external-2018.pdf

Unilever PLC. (2024, May 16). Setting standards for the future of food. *Unilever*. https://www.unilever.com/news/news-search/2021/setting-standards-for-the-future-of-food/

Vinayak, K., & Kodali, R. (2014). The Relationship between NPD Innovation and NPD Performance: the Moderating Role of NPD Best Practices in Indian Manufacturing Industry. *Measuring Business Excellence*, *18*(2), 39–59. https://doi.org/10.1108/mbe-03-2013-0017

Wixe, S., Nilsson, P., Naldi, L., & Westlund, H. (2023). Correction to: The Role of Collaboration and External Knowledge for Innovation in Small Food Firms. *The Annals of Regional Science*, *70*(1), 157. https://doi.org/10.1007/s00168-021-01093-8

New Product Development Essentials

4

The second part discusses the different organizational options to set up R&D expertise and teams. Internal organizations, like functional departments and innovation centres, are compared with external expertise and various forms of open innovation.

This part of the book discusses the different prerequisites that a successful new product development enterprise needs. Nothing guarantees the commercial success of a product or service launched in the food and beverage market, but a couple of key factors can massively influence the chances. The focus here are the roles and responsibilities R&D has in that context, which can be versatile, sometimes proactive, impulse giving, sometimes more reactive and sometimes a sounding board. For members of R&D teams, it is good to be aware of that so they can reflect and act. That's a "must have" for food innovation. Technical professionals in R&D that want to leave a footprint in the new food development arena should not only understand the different strategic parts, but influence them and appreciate the roles of colleagues from other functions. The chapter starts with strategies, those for business and for R&D, and tries to outline the sequence and interdependency of both. Strategy here means those plans and actions an enterprise is doing differently than their competitors. Operational efficiency is another critical factor, a different one, as here activities are compared with those of the market rivals with the intent to do it better. This can be "benchmarked". In the past doing things better was seen as a sufficient strategy in fast-moving consumer goods (FMCG), but increasingly both a differentiating strategy and operational efficiency are necessary – one alone is not sufficient. Strategy comes first.

THE BUSINESS STRATEGY

A business strategy is a plan describing the goals and actions of an enterprise. It states what the company wants to achieve, how it will achieve it and why they think they are successful. The strategy integrates all the internal and external factors affecting the company, including the needs and expectations of the customers, competitors and stakeholders. The full strategy is often structured by a "vision", a "mission" and a "values statement", plus sometimes "objectives" and "key performance indicators" (KPIs).

It is not so easy to see differentiate those aspects, but some features can be used to make it more tangible.

The world's biggest branded food manufacturer, Nestlé, communicates the vision "To be the leading Nutrition, Health, and Wellness Company" (Mission & Vision, n.d.; Nestlé Global, 2021). This is a long-term aspiration. It's the guidance for employees and should inspire them to work toward the greater goal. Interestingly, in this case the vision does not mention explicitly food, which can be understood as the aspiration to go beyond offering nutritious products and covers initiatives and practices for more holistic nutrition, health and well-being.

DOI: 10.1201/9781032724591-4

Overall, the task of the vision is to guide strategic decisions and long-term planning, so, for example, decisions where to invest and where not. It guides the mission statement, which in Nestlé's case says, "Nestlé's purpose is to enhance the quality of life and contribute to a healthier future."

At first sight, this may sound similar! But it portrays the shorter-term doing and offers a plan of how the vision can be achieved. And while the vision is primarily for employees, the mission also wants to speak to consumers, customers, and investors. And expanding nutrition, health, and wellness into an "enhancement of quality of life" suggests the intention to address the environmental and social challenges associated with its business. This now sets a clear direction for the operations, gives an idea about a focus in product and service development and the importance of sustainability practices in sourcing and manufacturing, including efforts in research and development.

The "values" then add information about the way of working, how the enterprise will run their business. It sets the principles and beliefs that characterize the culture and identity of the business. Nestlé's values are "Our values are rooted in respect. A respect for ourselves. For others. For diversity. And for the generations who will follow in our footsteps" (Nestlé Global, 2021; Nestlé Global, n.d.b).

Sometime companies use all three elements to amalgamate the "business purpose" out of them. In Nestlé's case: "Our purpose is to unlock the power of food to enhance quality of life for everyone, today and for generations to come". That is essence and direction of the business.

As said, such statements give information about the way an enterprise plans to move from today's reality towards the desired future. That is a development activity that needs a plan. And the plan often needs research and development.

So, people in R&D must understand the business strategy. It the plan that is transmitted by the product and services of the future that R&D creates. They must bring product differentiation; understand pricing and distribution; and contribute to sourcing, innovation, manufacturing and sustainability.

And a good strategy better not be too broad and vague, trying to please everyone. It must also clarify the things the business will not focus on, to give focus. How narrow or broad such a strategy is depends also on the size of the enterprise and the business they are in. In the food and beverage (F&B) business, a high innovation rate is important, but ideas need direction. Global players have a wider strategy; others focus their approach more:

- *Niche strategy*: It targets specific segments of the market with unique needs or preferences, specifically offering products or services to cater to exactly those needs. A company can focus on outdoor sports nutrition, DIY healthy vegan snacks or ultra-local sourcing. Supporting R&D teams here often have leading specialized skills.
- *Cost leadership strategy*: Here the focus is to offer products at a lower price, while matching the perceived quality of the market leader. They may achieve this through different profitability, lower profit margins, leaner fixed costs, like overhead structures, better economies of scale or more efficient operations. R&D teams focus on technical product insights and technology and less on the identification of new consumer needs.
- *A differentiation strategy*: This involves offering products that have distinctive features or benefits that set them apart from competitors'. For example, a company may use high-quality ingredients, innovative recipes or attractive packaging to create unique value proposition for consumers/customers. Those companies normally need strong R&D development partners, often as an overhead function.
- *A diversification or expansion strategy*: Expanding into new markets or product categories that are related or unrelated to the existing ones. For example, a company may enter new geographic regions, add new product lines or acquire other businesses to increase its market share and revenue. Start-ups belong here as well.

In all those cases a viable business strategy needs input from R&D. It is important to sharpen the business strategy. For a cost leadership strategy, great market observation tools, detailed competitor understanding and sophisticated cost benchmarking tools are essential. The development team must quickly find those new product launches in the market that will be a success. They must have a firm view of how they can make a comparable product at a lower price. Product and process designers should be able to copy fast to be the first follower. But a company who is an innovation leader would not so much "scan the market": They would want the capability to find new market trends and turn their observations into insight and insight into action.

The company leadership provides the business priorities, defines the markets, maybe relevant segments and brings strategic options and tools, like a suitable brand, a particular way to market, important strategic relations and factories, and they provide the resources to work on those plans. It needs R&D input to make these products and services fitting to the business or just finding promising ways to move the business into new markets through new platforms and disruptive innovations. They clarify how much it would take to achieve a planned goal. In the past some companies have critically underestimated the effort it takes to launch a food product with new attractive health proposition under a new food regulation (see Chapter 11). Maybe they have not discussed deep enough with expert functions what is needed to finish the way. Or these expert functions were too optimistic about the path to go? Did they underestimate the capabilities of competitors? Or did something disrupt the market?

Similarly, how can the company leadership commit to sustainability, regenerative agriculture, recyclability, plastic reduction, healthier and more nutritious products, a differentiating premium taste and proven superior quality without an in-depth dialogue with those teams that will have to create the solutions promised in the strategy?

THE R&D STRATEGY

R&D's key role is to advance the corporate plan by driving innovation, so R&D strategy reacts to company strategy. The corporate strategy defines the business and the key battlegrounds. That's the question on what the company's purpose is and "what" it wants to bring to market. Next then is the question of "how" to bring it to market. Here R&D has to provide their judgement. The company wants to come up with ambient stable meals, soup concentrate, frozen vegetables or smoothies. R&D would need to design the products and answer which technologies, skills and resources are needed for that. R&D has the expertise to such questions. And if not all necessary is in place yet, they help to make a time-bound plan and when and how to close the gap. They train the existing R&R team, recruit new experts, start technology scouting, build an open innovation partnership or search for a third-party manufacturer. R&D back-translates the product planned into the enabling skills and technology.

The company wants to introduce a new range of chunky liquid foods with fresh herbs. Is then "inline" or "in pack" sterilization technology better? Does it need air drying or lyophilization? How big will the chunks be with the different technologies, how fresh and colourful the herbs, etc.? Only R&D and technology can answer those questions.

R&D works for most of their time on solutions for the future. If all technical prerequisites for the new product are already available in company, it's the lead time that sets the timeline. If major parts are missing, maybe new recyclable packaging material, new suppliers, a new technology or just legal confirmation about market conformance, that launch maybe delayed by a couple of years.

Not all food companies drive the market. The market leader is the first one to offer a new product and that has to find a consumer base. Such company better is top in finding consumer preferences. The follower can copy the "first mover product" as good as possible, but at lower costs. That can be because they have access to a more cost-effective supply structure, less overhead or lower marketing costs. Different R&D skills for different food companies in the same market segment are required.

Bigger retailers often use that approach. They create their own brand, also known as a private label, and put it aside the established market leader at lower costs. They may not have to finance R&D or a supply chain, living with less overhead structure. Ones they found a co-maker, a company with no interest in having their own brands on the market, they are in business. And if they can offer them substantial volumes, because they have a big turnover, they can expect a good price from the producer. This is a typical product development strategy of discounters.

Likewise, for R&D the tasks in the business strategy are formulated as targets, objectives and scope, basically measurable results with time-bound commitments. Ideally those targets are "SMART": Specific – Measurable – Achievable – Realistic – Timely.

That helps R&D to take good decisions. Where useful the results can be benchmarked and compared with others to measure the outcomes and impacts of R&D activities ("best in class"). This allows progress tracking and improvement or corrections. Various indicators and metrics can try to measure the value of R&D from different angles:

- The number of products launched, or the number of projects started or finalized. Depending on the business strategy patents, publications, collaborations and third-party funding prototypes, successful factory trials can be important KPIs. A cost leader does not need to spend much time on their own patents or publications, while the numbers and time of factory trials indicate how fast they can come to market with a "product copy" of the innovation leader.
- Financial aspect like return on investment, how much business success has been created with how much R&D investment, is attractive to determine the performance. Those KPIs measure current performance, but they are not very suitable to predict the abilities in the future.
- If the market is changing, new skills and capabilities are critical, so an R&D strategy can be new skill recruitment, extra training and education, new external collaborations or even acquisitions. They are also specific for R&D, but focus not directly on the current business performance, by building the future capabilities.
- Commercial numbers like top-line and bottom-line growth, market share or product profitability, material savings and consumer and customer satisfaction. Data directly connecting R&D effort with short-term business success. Often this is more a share target, since the success depends also on other functions like procurement, marketing and manufacturing. If a new recruitment wave or training cycle is not working, R&D has taken the decisions. Other functions were not much involved here (HR perhaps).
- Societal benefits resulting from the R&D work and output. Maybe a company wants to cut their carbon footprint, reduce the amount of plastic they use, make the material recyclable, cut the sodium, or sugar content of all products. Such targets may not directly improve the company's financial performance, and they may even not have built new skills, but they can be critical for the success of the longer-term plan. Most of the environmental targets a company has are in this area.

A R&D strategy that is not well connected with the business strategy bears the danger that the research and development program lacks the necessary granularity, clarity and confidence to materialize the company's ambitions. Nobody plans for a R&D ending up isolated from corporate priorities, disconnected from market needs and out of sync with the speed of business. Again, the "art" sits less in the "what" or "what not" to do, but in the "how" to do it. Management literature gives some advice on how to build a robust strategy (Brennan et al., 2020). Key elements are as follows:

- Understanding the barriers to R&D success
- Choosing the right elements for the strategy
- Pressure testing them before ratifying it

Has R&D those skills and capabilities that make the difference? Is the investment commitment sufficient to create the skills needed? Application of AI is highly demanded nowadays. But the infrastructure and staff are expensive. Being able to hire one or two machine learning engineers is not enough. These people need context, data, infrastructure – additional time and money!

Future business plans have to be bold, and the R&D strategy must match that ambition. A good R&D strategy is stretching, but still feasible. And to be realistic, the real constraints and opportunities need to be discussed and accepted.

The growing confidence in the availability of simulation and automation technologies created a steep drop of experimentation costs and accelerated R&D throughput. The pace of corporate innovation is further spurred by the increasing emergence of broadly applicable technologies, often coming from other industrial sectors such as digital and biotech.

The percentage of the turnover that becomes reinvested into R&D is called R&D intensity. The branded food industry comprises mostly low-tech sectors with a low percentage of turnover invested into R&D. In 2021, the highest share with about 20% of total turnover of global research and development spending was made within the software technology producing industry, followed by health with 17% and food, as part of FMCG, at 1–5% (Statista, 2024).

That does not mean their R&D departments are small. The top F&B companies have turnover between $80 billion and $3 billion each (Food engineering, 2020). To illustrate: In 2022, Nestlé's global turnover went towards $100 billion, and while the annual research and development expenses was below 2%, that was still a $1.777 billion investment into R&D (Nestlé Global, n.d.a). This can be seen as low compared to current investments in pharma, with spendings of up to 40% of their annual turnover, but it's well above the average of 0.8% for food and beverages (Skillicorn, 2019).

Hence even at a reinvestment level of only 1–2% of turnover into R&D that still creates big funds for R&D, and current corporations are only one part of the equation. And in the food ingredient (FI) industry the R&D levels are often significantly higher – at lower double digits. These trillion of dollars a year that companies spend on R&D is matched by the public sector. Well-funded start-ups, meanwhile, are developing and rapidly scaling innovations challenging established business models or steer industry growth into new areas. A recent analysis by the European Union indicated that these investment do pay off (Pedro et al., 2018).

What are these funds used for? Primarily to make new products and services to stay ahead of their competitors. But also to constantly improve existing processes:. Making new packaging solutions to reduce waste, finding new ingredients, making existing options healthier or developing ways to improve production efficiency. Different companies naturally differ a lot here. A new startup or mature company will put most of the available funds in new product or process management. A global player with dozens or hundreds of factories and thousands of products will have to put much more effort into the maintenance of their existing portfolio.

Beside the directly NPD linked activities like product formulations, nutritional calculation, process and packaging development, labelling designs and regulatory compliance new production processes that meet quality at lower costs are important. This includes researching new equipment that increases safety standards while minimizing labour costs and minimizing waste throughout the production line and the exploration of automation solutions reducing errors and increasing efficiency. A R&D strategy investing in the creation of a new platform or the extension of an existing range is possible. Big food companies may do all of it; smaller ones must choose their battles.

Today's foods R&D focus is innovation, efficiency and sustainability. Innovations are the fuel of the business creating growth, efficiency helps to create those funds for the growth and sustainability is the licence to be able to do that profitably in the future.

R&D departments are not the only organization focussing on new product development. Marketing, consumer science, quality assurance and others are as well.

Mature companies often prioritize incremental NPD. Here, incremental projects account for more than half of an average company's R&D investment, even though bolder bets and more radical reallocation of the innovation portfolio could potentially deliver higher rates of success. But this has higher risks. Adding a few percent of turnover to existing ranges may lead to predictable near-term return.

Focusing innovation solely around the core business may enable a company to coast for a while – until the industry suddenly passes it by. A mindset that views risk as something to be better avoided rather than faced and managed can be unwittingly reinforced by how the business case is measured. Transformational projects at one company faced a higher internal-rate-of-return hurdle than incremental R&D, even after the probability of success had been factored into their valuation, reducing their chances of securing funding and tilting the pipeline toward initiatives close to the core.

And there is surely a good relation between the amount of money spent and the commercial success. But even with outcomes clearly measured, the often-lengthy period between initial investment and finished product can obscure the R&D organization's performance. Yet, this too can be effectively managed by tracking the overall value and development progress of the pipeline so that the organization can react and, potentially, promptly reorient both the portfolio and individual projects within it.

Big food multinationals may have many brands, and huge product ranges, with many products long established in the market. Many of those do not need new R&D or much NPD effort. The products have a stable consumer base, and the product design is optimized. They make a robust commercial contribution and need just basic maintenance support, until an unexpected situation forces R&D to work on the product design. Calamities like an out-of-stock supplier or regulatory changes can cause this. This majority of the product offerings provide the resources for the company to invest into new trends and future science and technology to create innovation platforms and help to ensure future profit. Such new areas are normally more research intense and hence require a disproportional investment into research and development, particularly if this new territory should be protected by powerful entry barriers for the competition, like intellectual property rights to block competition to jump on the bandwagon.

So, the budget must be available to productively execute a good R&D strategy. But the correlation is less direct than one might expect. In 2018 PwC published an analysis in which they described the key factors differentiating top innovators from the rest beyond their actual R&D investment (Jaruzelski et al., 2018). They mentioned six key characteristics:

"1. They closely align innovation strategy with business strategy.
2. They create company-wide cultural support for innovation.
3. Their top leadership is highly involved with the innovation program.
4. They base innovation on direct insights from end-users.
5. They rigorously control project selection early in the innovation process.
6. They excel at each of these first five characteristics and have been able to integrate them to create unique customer experiences that can transform their market."

And the authors put their emphasis on the point that hitting the first five ones is by far not sufficient. The success comes with the right mixture of thorough attention to strategy, culture, top-level involvement, consumer insights and execution throughout the full innovation process. With that the solution is less "a plug and play" module companies can use to have better R&D, but an enterprise-specific built company culture.

While that is for every food company to find out for themselves, the role the cross functional team and their organization have in successful new product development is much more general.

THE IMPORTANCE OF CROSS FUNCTIONAL TEAMS

Cross-functional or multidisciplinary or interdisciplinary teams consist of members from different departments with versatility in their expertise and the ability to collaborate affectively to achieve a common goal. Today they are a core component of modern project management. The model gained wide acceptance with the rise of agile and lean methodologies, emphasizing collaboration, knowledge sharing and reduced hand-offs between workflow steps.

Generally, cross functional teams seem to improve project outcomes, with higher completion rates, better budget adherence and increased goal achievement. They are particularly effective for addressing complex challenges. Setting up a team in that way grants a much broader range of skills, knowledge and resources allows a team to tackle a wider range of challenges faster and look at opportunities from different angles. In contrast to the old-fashioned top-to-bottom hierarchy where employees are dependent on the decision-making of one person in charge, cross functional work allows for a much more self-organized environment. They are said to increase an organization's ability to respond to market changes by bringing together individuals from different departments, functions and disciplines, leveraging collective knowledge and skills. For a couple of reasons cross functional teams promise better results than more departmental set-ups in the FMCG industry:

Innovation and Adaptability: Cross functional teams recognize changes in the environment faster and respond quicker. A team with members from different departments looks at the environments from diverse perspectives. That makes decisions more solid.

Speed and Efficiency: Before decisions can be taken enabling technology and actual consumer and customer expectations need to be brought in. In FMCG these changes happen fast. Being able to judge quickly which developments are relevant and which ones less so requires functional expertise and the chance to mirror the findings quickly against the assessment of other functions. Cross functional teams are not necessarily faster in decision making.

Resource Allocation: Simply the setup of a cross-functional team gives enhanced access to the resources the project requires. Preferably the different team members have a clear commitment from their line managers of how much time and effort they can invest into the project.

Strategic Alignment: In the FMCG industry, integrated, cross functional teams allow a better realization of strategic agenda of the business. Category marketing may look differently at market opportunities than channel marketing or sales.

Challenging the Status Quo: In an organization with "classical" departmental setup (necessary for different reason than a team setup) cross functional teams seem more likely to challenge the status quo and current ways of working. They may even act as change agents.

Employee Engagement: Team membership or even team leaders empower people. That increases the engagement of the individuals, offers new develop perspectives and can help to retain talent. These teams build a sense of cohesion and collaboration, and employees feel more connected to the organization as a whole.

Therefore, members of cross functional teams tend to perform better, they "go the extra mile" and results may hit the market more efficiently, they "bring more rubber to the road".

A few basic principles set a functional team up for success:

- Every project needs to have an end-to-end accountable leader. That does not mean the person has to take all decisions, but should know which decisions are necessary when, where and how.
- That leader needs well defined measurable goals, committed resources, a clear understanding about his/her decision power and a transparent time plan.
- Team members need to be dedicated to the success of their project. Often the departmental line manager is not part of the team and may see different needs and priorities throughout the business year. Preferably results of the project and the individual contribution should be an important part in the performance evaluation of individuals.
- The project needs a governance structure where it can constantly re-evaluate to best meet the market requirements in a quickly changing market.

So, easy and logical guidelines for cross functional team are the secret recipe for success? Actually, no! Like most things in life, there is a second side, a consequence.

In a detailed study of 95 teams in 25 leading corporations, nearly 75% of cross functional teams were classified as dysfunctional (Tabrizi, 2024). Five criteria were used to measure success: (1) Staying in the planned budget, (2) remaining on schedule, (3) sticking to the specifications, (4) meeting customer expectations and/or (5) maintaining alignment with the company's corporate goals. In that research teams often failed in three out of these five.

Likewise a study by Sethi et al. analysed the innovative power of cross functional teams in 141 cross-functional product development teams and concluded "that innovativeness is positively related to the strength of superordinate identity in the team, encouragement to take risk, customers' influence, and active monitoring of the project by senior management" (Sethi et al., 2001).

Typically, when implementing a cross functional team, this is the odd one out in the typical departmental organization in a company. That can bring irritation. And therefore, beside the focus on selecting the most motivated, open-minded individual, covering all appropriate functional areas and having excellent communication skills, the question on how to be embedding them into the rest of the organization and the leadership support is critical.

While the team must have the authority and the accountability to accomplish the task they got, the surrounding management structures must provide adequate resources and support for the team in terms (money, time and human resource guidance).

Good communicate skills are important to manage innovation in a cross functional team effectively, but unclear governance and accountability weaken the team. The business leadership must provide the direction and decision-making and can even act as an example when senior leaders show their ability to make difficult and complex decisions.

Sometimes the beginning of a new cross functional NPD project is seen as a light entry. The different department heads agree to delegate limited resources to the team. And the requester, often the project leader, doesn't exaggerate the resource needed, because she/he wants to start. But the initiative may get some characteristics of street poker! The department head pays the basic fee, the initial resource request. The "game" starts, and the next cards show the opportunity and the extra resources needed. "Players" need to decide if they want to continue feed the pot or wait for the next game.

In practice, the commitment for sufficient resources, time and money from the beginning onwards is difficult. Processes like the stage gated process (see Chapter 7) help as they force the team to estimate their resource needs for the next phase, promise the "delivery" and come back to the sponsor. Then they explain their case, ask for more or other resources, get those eventually granted and move on into the next phase.

For budget-controlled process that is a feasible approach. But one needs to be aware that this can harm the team's motivation; it signals a lack of trust and importance. The team may feel like guinea pigs and less like a powerful innovation machine. Then individuals may decide to reduce their commitment, particularly in moments when the team has to take risky or uncomfortable decisions. A too stretching target on science and technology for R&D, a massive line modification with unclear operational efficiency or simply not enough money to do the necessary consumer work are classical factors leading to disengagement of team members in R&D, supply chain and marketing.

A good way to manage all the different expectations is a living connection into the organization. Here the project leader should interact with department leads about the project goals and the ongoing progress. In these stakeholder interviews the team learns to understand the thinking of functional departments, show themselves as having good visibility for the project and build a better understanding and extra support.

If the implementation of cross functional innovation teams is new for a company, a driving force from within helps, like a few senior people infusing the innovation thinking and keeping it alive. And the team needs someone explaining the new technologies and best practices regarding innovation creation. This expert or moderator can come from outside and lead the process. This person can generally not fulfil the task within the company to align the team with the strategy of the organization and ensures the harmony with the ongoing business by managing noise and disruption created by the team.

Comparison between industrial sectors shows differences in the team setup. High-technology sectors like information and communication technology (ICT), medical pharma, etc., often start a project with the mixture technical experts, transforming later parts of the team into a new one, where marketing, sales and legal regulatory are joining. Or even handing the result of the cross functional technical team to hierarchical marketing group for further execution. Apple's iPhone team was initially a pure technical team from design, engineering, software development and operations.

In typical foods launch projects this is not the case. Firstly, because most of the new product ideas are coming from an identified consumer need. It's therefore useful to have someone in the core team, ensuring that the original driver stays at the core of the solution. Secondly, the relatively short timelines. Success in FMCG is about being fast (but not wrong). A sequential process from technical to marketing to sales would take too much time. Only longer-term science innovation projects, like the development of a salt-free taste enhancer or a plastic free functional packaging, start with technical teams that regularly update "the business" about their progress and receive feedback.

Nowadays cross-functional teams have become ubiquitous in the foods business because companies need to speed innovations to market. "The teams are like arteries, connecting parts of the body, enabling the whole organism to renew itself", as Behnan Tabrizi wrote. "That's why it's so important for leaders to pay attention to the way cross-functional teams are set up and how well they work: when they don't function, the organization's arteries harden. When they do, goals are met, and the organization is ultimately more successful" (Tabrizi, 2024).

Preferably a company would look for the perfect innovation team. With the ability to fill an innovation pipeline, maybe even build an innovation strategy, entrepreneurial skills reaching from the ability to be a visionary, but playful and experimental, up to the inner drive to progress projects keen to get reach a result, while having the inner strength to take calculated risks. These people and teams normally do not just visibly appear somewhere in the organization and just need to be picked; they need to be trained and supported to grow.

THE ROLE OF THE PROJECT LEADER

Many companies in the food industry are marketing driven, and the way they manage their new product development is on a project basis. A new product launch, a new manufacturing line, a new factory becomes a new project and is end to end (E2E) managed with the help of contemporary project management tools. The first key role to fill is the one of the project leaders (PLs). They do not form a sperate department, but more often are a part of the functional department pools, like supply chain, QA, marketing, R&D, market research, regulatory or nutrition.

A typical launch in FMCG can benefit from an experienced marketing person as project leader to basically "bring all the ducks in a row" and lead the project from the market angle. Here the contributions of science and technology are mirrored against the established project objectives, which have picked up market needs.

Likewise, a project aiming to implement a new process line or a whole factory is better led by someone with sufficient technical expertise, ideally someone that has to that use the machinery later on.

And the "silent" relaunch of, let's say, new recyclable packaging benefits from a packaging R&D person. And a cost-saving initiative may profit from finance leadership. All projects are somewhat different, and often the leader comes from that function where the most novelty and uniqueness can be expected.

A few more examples are in Table 4.1.

Often projects are also flagged as a "twin team" with marketing and R&D as "owning" the project. It underlines the importance of both functions in NPD. But the decision power is better clarified before. The PL shapes the idea and executes it closely together with a "his/her" team. The PL often remains the departmental team, so the respective line manager keeps the career responsibility. But the new project is now the main task of the PL, often the only task in each time frame. The PL stays in the role for the whole duration of the project which can be 4 months for a quick easy roll-out or 4 years for a new platform.

TABLE 4.1 Project types and roles in NFPD

PROJECT TYPE	*TYPICAL PROJECT LEAD*	*RATIONALE*
Range extension/ innovation existing brand/new brand/new benefit	Marketing	These NPDs require sophisticated marketing work to design the campaign or "reason to believe".
Reformulation/packaging redesign	R&D	No to minimal marketing work needed. Marketing mix with no change. (Can change, if redesign is articulated, e.g., on pack as new claim or logo.)
Value improvement (VIP)	R&D or supply chain (SC)	Especially, if focus is on material or manufacturing cost improvement. (Can change, if visible change occurs, e.g., new ingredient, etc.).
Quality improvement and consumer/customer satisfaction	Quality assurance	Internal and external project where the quality of existing product is focal point (includes market research, see DVoC).
Nutrition enhancement project	R&D (nutrition)	More and more big parts of the existing portfolio need reformulation, to reduce/cut perceived baddies like salt sugar, fat, MSG, etc.
New claim territory	Regulatory (SC or R&D)	Safeguarding a new claim can benefit from regulatory lead to strengthen the claim.
Market scan (new benefit/trend)	Market research or consumer insight function	To sharpen the understanding of a new consumer need and how to address it with new product or service.
Roll-out (from one market to next)	Marketing or SC	Same product in new market can require minimum technical work, depending on where product is made; SC may get lead.
Science study/academic collaboration	R&D (science or technology)	Technology needs expert leadership! Often S&T projects are running a market launch.
Export	SC or marketing	Hardly any "product touch", new artwork, new distribution and new communication in second market.

In the corresponding literature a lot of research and recommendations are available about the impact of the leadership behaviour of the project manager and the outcome in the new product development process developing new products, i.e., improving innovation speed and new product performance. Particularly the path-goal theory, originally published in 1971, has proven to be good guidance for project leadership behaviour in innovation projects (House, 1971).

In its core the theory says that a leader's behaviour is related to the satisfaction, motivation and performance of the subordinates and that a good project leader behaves in a way that complements the subordinate's abilities and tries to compensate for their deficiencies.

Particularly the "achievement-oriented leadership behaviour" has been proven to be successful in innovation processes. Here, the leader sets clear and challenging goals for the team and asks them to perform at their highest level, while showing confidence in the abilities of the team to meet this expectation. It's helpful if it is mirroring successful leadership styles of the top management.

That does not mean that other leadership styles could not work. An experienced project leader might e.g., become more "directive" in situations where the role and tasks of the team are ambiguous. Or the "participative leader behaviour" means the leader is consulting with the team members and is asking opinions before deciding.

TABLE 4.2 Key elements of a business case for NFPD

ELEMENT	*KEY FEATURE*
Business need	A description of the business need and why it is important to do it now.
Suggested solutions	At least one possible solution and how the best possible solution is identified.
Project definition	Setting the business objectives to be achieved along and the general outline of the project.
Project plan	A plan with key elements to successfully achieve your project goals.
Project scope	Clearly describing what is in scope and what is out.
Project budget	Best current estimate of costs and resources needed.
Project roadmap	Start time and timeline until launch plus important project milestones, like decision gates.
Project financials	Financial metrics describing the financial parameters.
Risk assessment	Early flagging of potential risks and the actions to mitigate them.
Project stakeholders	A list of key stakeholders available to discuss the case.

And never to forget: New product development projects with their timelines and targets can easily cause stress with the individual, so it helpful if the project leader shows concern for the team members' psychological well-being.

More directly, the project leader guides the teams to deliver high-quality projects within scope, timeline, budget, planning and overseeing all aspects of project management, including risk management, quality control and communication. The PL ensures the team has an actionable project plan, timelines, and budgets in collaboration with stakeholders, tracking project progress and ensuring that the project is delivered on time and in full (OTIF) and within budget.

The key document a project leader must deliver is the "business case", usually having the following element (Table 4.2).

The initial business case document contains the information and assumptions that are available at this early point in time. Nobody expects the PL to have crystal ball with which the future can be predicted. But a decent ability to estimate the effort needed and the chances and risks creates trust with the decision makers in the company. That maybe one reason why a typical PL normally already spent some time in the company, so it's easier to estimate what is needed and how long it will take.

Leading a project requires specific skills and competencies. These are not personal talents or compulsory parts of the academic curriculum, but rather the result of specific training. A typical training as project leader aims at a set of competencies. Some of these courses split into "skills" and "competencies", where "skills are more specific in defining exact abilities, competencies have a broader focus and comprise many skills" (Indeed, n.d.). And skills may be defined as "soft" or "hard", competencies separated into "behavioural", "functional" or "professional".

But whatever the naming conventions, the training provides the tools to lead a project in the enterprise successfully. The number of named individual capabilities can differ considerably. Some trainings identify 10 different competencies, others even up 30 – that depends on the wishes of the company leadership and maybe the consultants advising them.

The future project leader must learn to run a project technically, which includes planning and project management tools, at least the fundamentals of financial and commercial management, and basic team and people management. I have modified from the UK-based association for project management, "APM", which provides these qualifications (Competence Framework, n.d.). Table 4.3 summarizes some of the often addressed target areas and their typical training content.

People nominated for a project leader of course are not "blank" on these competencies. They have already acquired a few of them. The idea is to find the individual gaps and fill those.

TABLE 4.3 Possible training module and skill requirements for a project leader

KEY SKILL AREA	*DETAILED TRAINING CONTENT AND EXPERTISE*
Ability to structure and organize a project	Understanding the decision frame structure Governance structure Financial acumen Portfolio intention Business sustainability Business case requirement
Project preparation	Ability to secure the necessary resources Ability to manage project progression Confidence to the governance board that a project is managed well Consciousness about differences between project and the wider organization Integration of project outcome into regular business processes Develop and suggest usage of benefits and rewards
People management	Stakeholder management and general project communication Ability to resolve conflicts efficiently Leadership, empower and inspire Ability to select, develop and manage a successful team and sustain it Ensuring an inclusive environment and diverse culture Professional usage of ethics, compliance with all standards required
Project execution	Definitions of project requirements Identification of suitable solution to satisfy requirements Ensure the quality standards of all outputs Integrated project management smoothly amalgamating separate workstreams Time management Resource management, acquisition and efficient deployment and sustained utilization Budget and cost control Risk and issue management

And since all these elements are important, a quantification of the individual skill levels makes the training progress measurable. Table 4.4. shows typical skill level. The result is an individual PL candidate profiles where difference before and after the training is a measure of the effectiveness of the training.

Often it is helpful to describe the competence level in each level with a typical example and each class in particular competency with a few typical examples, so that each person that is part of the "scoring" has a common set of references.

An example for "proficient stakeholder management could be: "Uses stakeholder engagement matrix to categorize stakeholders based on their level of authority and concern related to the project" or "maintains a stakeholder communication checklist, ensuring all necessary information is conveyed to stakeholders on time in full".

TABLE 4.4 General skill levels

LEVEL		*KNOWLEDGE LEVEL PER COMPETENCY*
1	Unaware	No awareness of necessary knowledge
2	Aware	Basic awareness
3	Skills	Has working knowledge and can describe requirement
4	Proficient	Detailed knowledge of competency and capable to use and adapt it in even complex situations
5	Leading experts	Leading knowledge – can develop new concepts and solutions and train others

Before the training starts the candidate PL is asked to do a self-score, followed by the scoring from a number of colleagues, knowing that person well. The "360-degree" approach is used people in the organization "around" the candidate, direct colleagues, subordinates and line managers. That provides a view from various angles, as the behaviour of the candidate towards a direct report often differs differ from the interaction behaviour with a direct colleague or a senior manager. The 360 degree approach helps to make these behaviours visible for the candidates, their coaches and eventually also the line manager. The value of the feedback depends how well the respondents knows the candidate and how honestly they provide their feedback. It requires fairness and courage from the "judge" to help find strengths and improvement opportunities to shape a personalized training plan. The different capabilities allow different levels of personalization. Decision makers in an enterprise probably wouldn't welcome too much personal creativity in the business case preparation. But for an effective team leadership style, building on the individual strengths can be helpful.

It's not the aim to "clone" project leaders, but to enable them to lead with their individual strengths and modify their weaknesses in the training as needed.

The training volume can differ significantly; Short 1–2-day courses can get PLs "fit for purpose". Such quick infusion is sufficient for small team and projects with low complexity and low risk, like a refreshment project or line extensions. For a "full-blown" project leader a several weeklong full-days course or a weekly curriculum over 3–6 months is generally seen as more appropriate. PLs in innovations, multi-country or heavy investment projects benefit from such curriculum.

Experience on the job is another ingredient in a successful PL recipe. Understanding the business context, feeling the ambitions and vibes in an enterprise creates the frame in which individual skills and competencies can be used. As said, the leadership of an enterprise is keen to have a choice of project leader types. Development tasks can be very different. By having different PL styles in "their project leader pool" they can take sensible decision depending on the specificities of a project.

The agreement of the right "KPIs" with all stakeholders are a good first step to ensure success: Agreeing to file new patents makes much more sense when the organization is willing to spend time and money in building a new strong technical platform. But not if time to market is the most critical parameter for commercial success. Similarly, incremental turnover gives good focus for range extension with minimal cannibalization of existing product range. Plans to attack the market leader can be sharpened by aiming at "superior product performance". But for a value improvement project a KPI around "same perceived quality" is ambitious enough. The test to prove this would be here "match current" – in inhouse or consumer test. Table 4.5 shows some typical KPIs for the various types of new food product projects.

TABLE 4.5 Typical project KPIs for different kinds of NFPDs

GROUP	*POSSIBLW KPI*	*TYPICAL USAGE*
Innovation	(Incremental) turnover/profit	Range extensions, me too
	Superiority vs. lead competitor	Market leadership
	Number of patents/publications	Science projects
	Number of partners/collaborators	Collaborations
Budget	Budget variance	Resource planning and expenses
	Net income/ROI	Revenue vs. expenses
	Resource utilization	Expense effectiveness
Time	Lead time	Project with fixed launch
	Planned vs. time spent	For contracted work
Quality	Number of change requests	Minimize recycle, repair
	Consumer/customer satisfaction	Minimal complaints
Effectiveness	On time in full (OTIF) delivery	Project/key milestones
	Project velocity	In Scrum projects

The advice is not to set too many KPIs – that distracts and dilutes the project achievements. For the PL it should be more indicators – those measures that give tangible guidance for the project while it is running. They are not used to judge the project success after launch ("postmortem"). "Leading indicators" inform about the progress towards a particular goal. They are the sensors predicting the outcome and help the PL to make informed decisions.

"Lagging KPIs" then flag incidents that have already happened. They tell the how successful they were and help to prevent recurring mistakes in the future. These are often more fact based, but are less useful in project steering.

Skills, competencies, knowledge level, KPI, etc., all give the impression that new product development success can be calculated through an algorithm. Honestly, they can help a lot, but with the most important ingredient, the stakeholder management, they ensure not very much. New PLs often focus their interaction with stakeholder on the early stages of the project, to be clear about the expectation, so they can shape the project objectives. And once the project has been signed off, they are busy delivering what has been promised and forget that stakeholder position can change and that sometimes quickly. They continue to execute what has been agreed at the start. It is necessary to regularly reconnect with the key stakeholders throughout the whole project. But not only the project happens in a fast-moving environment; the whole business is in there. Even the best stakeholder cannot predict all market and business moves that can happen There are so many reasons why stakeholders must change their opinion within the project life span, especially in FMCG. The funding budget may change, so the budget owner must shift support from one side to another! The lead marketing person sees the wind in in market changing, and now a launch may not be appropriate any longer. The salesperson is close to signing a bigger contract with a key trade partner and that leaves no space for range extension. Or the supply chain director sees new risks for his factory when running the production. What originally was seen as a pushy, but promising plan, now is too risky. And what was agreed as a solid range extension without spectacular incremental growth is now no longer ambitious enough to show the growth potential of the business.

Connecting with the stakeholders in between reduces the risk of unexpected turns. And the project stays on the mind of the stakeholder and stays alive in the business agenda. Problems, alternatives, compromises and new opportunities can be discussed deeper before a final decision needs to be taken. And an important stakeholder has a possibility to discuss and reflect deeper in a 1:1 meeting prior to the gatekeeping moment with others.

A good project leader is obsessed about the success of the project. That extra energy is what makes the project special. But sometimes it needs the higher helicopter perspective to judge the commercial prospective of a new idea, even if that means stopping it. Pls are too close to see quickly that a project better changes direction or perhaps even a stop. This is their idea, their brainchild!

Stakeholders must ensure that the reason for a negative decision is understandable and well communicated. Stopping a project doesn't mean automatically the PLs has not done a good job. A sudden change in the market, maybe a competitor launch, is not a mistake of the team. Their task is to execute the project under the agreed objectives the best way

THE RIGHT PROJECT TEAM

Food enterprises, as part of FMCG work marketing driven, since they place the customer at the start of their process and use market insights to shape appropriate products. Accordingly, the project leader often comes from marketing, but gets clear leadership support from a strong technical project leader (TPL). The TPL has similar competence levels for the project and people management as the leader. Skills between both do not have to match, but they better complement. Both should split the leadership responsibilities accordingly. The fundamental ability to shape a good business case is critical marketing skill, but the technical leader better at least understand the basic commercial mechanics. The TPL will lead many activities that directly or indirectly influence the attractiveness of the business case. Possibilities to reduce material or production costs without losing quality, a manufacturing process using more line equipment that is not fully occupied or the number of trials, assays, tests certificates needed.

Good project planning skills in critical path analysis (CPM), work breakdown structure (WBS), Gantt charts or the critical chain method (CCM) benefit from thorough and disciplined data management, a task technically trained people are better equipped for.

Mastering the sometimes-sophisticated tools to move a project efficiently forward requires experience, accuracy, sequential logical thinking and a good level of patience. Maybe that's the reason why the supervision of the data management often lands in the hand of the technical lead.

Other project team members come from functions necessary for project success. In a classical organization the nomination is done by the functional line manager, based on a proposal made by the PL, where the project leader specifies the skills needed and often estimated the amount of work needed per time unit. Different units are used for that. A PL may state that he/she needs "0.8FTE the next 3 months from formulation". So, the R&D line manager has to find a capable candidate that can spend 80% of full time on that project. During the project all project team members are responsible to deliver what they have been assigned and they report their work to PL or TPL. But the team leader does not have the "line responsibility" during the project for the team members. Still, the feedback the PL is providing later to the line management for his team members about the "what" and the "how" should be the essence of the annual performance assessment for the individual.

The size of a project team in the food business varies depending on the scope and complexity of the project and the resources available for project work in the enterprise. Sometimes, a team of five to seven people is suggested as the optimal size for innovation projects: The kind of balance between collaboration, diversity and efficiency. When searching for the origin of that statement, former Amazon CEO Jeff Bezos's "pizza" rule is frequently referenced: This one says that if a team can't be fed with two pizzas, it's too large! Further calculations outside United States have clarified that two typical American pizza may serve five to eight people, and that the rules also apply to meeting sizes (Hern, 2018). A bit more scientific approach a year later used the "Bell numbers" to conclude the optimum team size. Based on the number of 1:1 interaction they suggest five people as an optimum. Explaining the argument is beyond the scope of this book, but can found in (Meyer et al., 2019).

Other papers use diplomacy instead of mathematics and conclude that "there is no one size fits all". Something I can agree too: A good team size is shaped by size context, complexity and goals. If a project requires a high degree of creativity and experimentation, a smaller team may be more flexible and agile, while if a project has a large scope, massive data input and complex deliverables, it might benefit from a more diverse resources and expertise to handle it.

Practically, the preferences of the project manager may play a role as well. Once the PL is nominated and briefed about the project objectives, he/she may have their own ideas about the team composition. One project may benefit from extra QA members to keep safety and quality aspects constantly on the team radar more closely. Others need more market research to sharply nail the consumer need. If the team reaches a size where the share vision becomes blurred, the limit is surely reached.

Launching two new variants of an existing F&B product in the home market is a minor modification in the product, perhaps only two new flavours and some formulation adjustment. For this a small core team, consisting of marketing, R&D product and packaging, is sufficient. Procurement, manufacturing and quality assurance contribution are not critical in the initial phases. They can join when needed. In other situations, early input in the design phase from those functions can reduce upscaling efforts and guarantee that product safety and quality are embedded into the product design.

In the business reality a discussion about the optimal team size has an academic connotation. Many companies allocate necessary skills in a so-called "extended team". Often the regulatory function is needed after the product is "locked" to ensure the product attributes, like the label, are compliant with all regulations. This is very different for a new product that wants to make a strong health or nutrition message. Here involvement in the early product design and concept phase is critical to achieve the project objectives.

Finance is another typical extended core team member. Their task is to build a solid business case, but they do not need to bother about product design, consumer test, factory trials or market launch preparation, etc. But a good relation to the project also has advantages. Then finance can translate the unique project features into business numbers. Those numbers in a business case are an assumption – they come from experience, based on data from the past. Sometimes a project has a potential where a detailed discussion with finance helps to express that properly in the business case.

There is a similar consideration for other "non-core "functions in new product development in foods. Market research is needed early to get consumer insights and then when it's time to test prototypes with consumers. And the sales function often comes in later when the product design is locked and the manufacturing confirmed (see Chapter 7) to help pave the way to market. But sales may also provide surprising insights from their channel that can give a product design a truly differentiating touch.

More complex projects require much bigger teams. If the project is using new science and technology for the first time, it better have representatives from science and technology in the core team to enable a smooth transition into the new product, concept and communication. Science is often organized centrally to keep critical mass in location, while the launch teams normally sit in the markets and the category marketing somewhere else. And with members placed in different locations to drive different aspect of the launch, team structures and meeting schedules can become more complex.

There may be a positive correlation between project size/complexity and team strengths, but there is also the need for involvement. Just running projects and then dropping the result in front of senior management does not stimulate adoption. Leaving the factory out until the product design is locked may be faster, but bears the danger that afterwards an efficient production is not feasible. And committing a local marketing and sales organization to launch a new project centrally developed may trigger "not invented here" symptoms. Basically, all functions needed to make the product and actively work on it join the team.

The final role needed is a project sponsor or a cross functional steering group. They are the hinge with the core business and ensure the link between the project and organization.

NECESSARY CAPABILITIES AND RESOURCES

Taking a market leader or runner-up as the example here, the necessary requirements for successful new product development are very different. A good start is the close link with the business strategy that is sharp enough to define which product ideas are in scope and which ones are clearly out. And a business sponsor or innovation board that articulates the needs and regularly reviews the progress from idea to launch.

At the heart of new food product development are normally marketing and R&D. Both functions do understand the brand and the products behind it. When they start choosing out of the endless options for the new ideas, they need to know what the brand DNA or identity is, so the brand adds (numerical) value to the product. These are the core values and characteristics that define the brand's identity and personality. The DNA should as well define the credibility limits of a brand. These values are not particular product formats, concepts like freedom, simplicity, transparency, care, etc. They need to be carefully chosen, often out of the history of the brand and they need to be honest as they are the guiding principles. Charging the DNA brand with a value the consumer does not believe will have no success. If a chocolate brand is about pure indulgence, it can as well launch ice cream. If a brand is about cooking and taking care of the family, it may as well launch a beverage, if it can be understood as nourishing and healthy.

The task for both marketing and R&D is now to find those product and services that fit the brand. A range of extensions from that perspective is a safe bet. But they carry a different risk: Cannibalization! The closer the new product is to the existing one, the higher is the risk that existing buyers will switch to the new one. Still, it can make sense, as its better consumers still buy the new product from their brand than from a competitor. But the business case for the launch will have to include assumptions about the cannibalization and may talk about incremental turnover rather than only turnover.

R&D needs to have the skills to design the product. At least they need to have the ability to brief and control the "real" product makers about the key product standards and attributes. If the R&D of an indulgence brand is not capable of designing impulse ice cream and writing the processing specifications, that's no reason not to do it! But they should be confident to translate the "branded product idea" for the manufacturers and ensure the presence of all standards. In such a case it may as well be an option to invite the "ice cream makers" right away into the innovation process.

More often NPD processes look at the product portfolio and manufacturing technology. But it needs to be different, new, better. In these cases, R&D will feel comfortable with the upcoming product brief but needs experts from the technology and manufacturing side to get a view, to which extend this may require processing upgrade or alternate processing steps. Moving from a smooth, velouté type of packed wet soup to a chunky rich soup may mean a massive change in the processing lines to get a typical, but safe product. Colleagues from the supply chain may as well know better where the new product can be made better, as they have a wider view on the processing capabilities of the industry. Safety is the reason why at one stage in the process quality assurance is joining. Once the product gets its form, they are the experts on the safety rules and can even contribute to get a "safe by design" status.

Regulatory and nutrition act in early phases as creators as well. They may see new claim opportunities, better nutrient profiles, etc., and can quickly access how much in reach such a new claim territory would be.

Many new ideas are potentially great, but they do not fit the reality. The next branded launch may be needed in three-quarters of a year to rejuvenate the brand getting on the shelf under stress from competitor. So, time is the defining factor here: The technical team will know how long they need to simply range extension and how much a novelty can be brought in without the danger of missing the launch window.

Representatives from the sales team are valuable joiners. They have the day-to-day knowledge about what the customer wants and needs. And in contrast to publicly available market information, these insights can be very specific and offer possibilities for unique solutions. These insights may not only cover new missing products on shelf but could solve problems for the customer on self or back door, e.g., with the handling or distribution.

THE ROLE OF SENIOR LEADERSHIP

An excellent cross functional team, where each carefully selected team member has great skills and is led by an experienced project leader, is still no assurance for success. The influence of the company leadership is huge. Leadership is sometimes described as "a process of social influence in which one person is able to enlist the aid and support of others in the accomplishment of a common task" (Chemers, 2014).

Different companies in the food sector need different types of innovations or new product development. With that comes the interesting research question if some leadership styles are better for some types of innovations and companies and others less so. A recent review by Kesting et al., classified innovation leadership as "directive and participative leadership, interactive leadership, charismatic leadership, transformational leadership, transactional & instrumental leadership, strategic & CEO leadership, and shared & distributed leadership" (Kesting et al., 2016). They analysed the different dimensions: people, means, effect and goals. What they found was "that different leadership styles fit differently well with different innovation types and stages". This research is helping business leaders to act innovation supportive; it's not a job profile in search for the best innovation leader. By being aware of the leadership needs in different innovation stages, a senior business leader can reflect on the most effective behaviour to drive successful innovation at that particular stage. Most FMCG enterprises do not work on one or a few NPDs, but several or dozens of projects at the same time. And for a constant "pipeline fill" they are in very different project phases. A good business leader might show consciously a style that supports the project team best in a specific way. Even in moments when the team requests the leader's guidance it's a choice to do that as classical "order-giver", as "innovation enabler" or a "servant leader" fully committed to help the teams achieve their goals.

And leadership influence starts long before the new product development process begins. The leadership is not part of the innovation or NPD team, but they must create the right environment and atmosphere to help to sharpen the project goals and accompany the creative outcome.

Company executives know best what their business needs – that is why they are in that position. During the innovation process they must find the great idea between all the good and moderate ones. The chosen product idea should not only match a consumer/customer need but must also fit the company and their brands. A wrong decision is not only a waste of time and resources and a missed opportunity but can even lead to disaster. In the early 1980s Colgate, one of the world's most famous oral care brands, decided to launch beef lasagna as a frozen TV dinner. Even today people still wonder what made the leadership believe that a toothpaste brand would bring meal credentials. But also other big FMCG companies made big mistakes by replacing old formulations for their flagship carbonated soft drinks with new ones or reworking the iconic packaging (Doering, 2017). In hindsight it makes one wonder what has driven them to upset the core of their loyal users.

For the leadership the question is less, "do we want innovation", but rather "which type of innovation can we have and do we want to have?" There are situations where it is better for the business to choose a low-risk NPD profile, for example, by refreshing an existing core range with a few variants. Or they see that a nearly complete overhaul of existing product ranges is necessary, where salt, sugar and fat reduction, additive cleansing plus more environmentally friendly packaging would serve the business better! So, they may direct most of their R&D to that task, for example, to pre-empt an upcoming new regulation or new tax that would impact the sales of the existing portfolio.

Company leadership sets fitting frames for their NPD program. But a capacity for new product development and innovation is not "just there" once a cross functional team has been briefed to work on new ideas. The foundation for good innovations must be anchored in the organization, starting from the creation of an environment wherein people and their talents can grow and where new ideas can emerge, get energy, grow and become nurtured.

Successful leaders know about the critical importance to build such an environment that encourages employees to think creatively, work passionately on projects and collaborate with others.

Before a supportive leader can stimulate ideation, the innovators need to be recruited. Here unconscious biases to hire people with interests and backgrounds like those of the people hiring sometimes lead to a high degree of homogeneity, but not necessarily great creativity. To minimize the impact of bias a strict standardizing of selection criteria, blind resume evaluations and a series of interviews by a diverse panel can help.

FMCG leaders also search for "true innovators" and describe the task as the search for a "unique set of entrepreneurial, social, and enabling skills that define the ideal innovator", as Mauro Porcini, Senior Vice President/SVP and Chief Design Officer at PepsiCo, described it once (Author Talks, 2022). They search for such "natural talents".

And once suitable innovators are recruited, leaders must ensure a culture of innovation and high-performing learning to strengthen continuous improvement as a shared value throughout the organization. It's a task of forming an environment of psychological safety, endorsing experimentation and risk-taking without the fear of failure. Some describe innovation is a natural byproduct of inclusivity, which means "when people consistently feel included and experience a sense of belonging in their workplace, they're more willing to take risks and be transparent about their ideas" (World Economic Forum, 2023).

Great business leaders stimulate creativity and innovation. They inspire other members, which grows the capacity for achievements in the organization. This is partly soft skills, but first of all organizations, creative processes in their core are chaotic and disruptive. So, the task is to have an organizational setup in place which supports the creativity of the employees and stimulates the develop of innovative projects without negatively impacting all other business processes, where creativity and "out of the box thinking are useless or detrimental.

If new product development teams are permanently asking their assigned finance partners to calculate more and more variants of a business case, this can impact the other crucial tasks of the finance team. But if the innovation team has not enough access to finance, they may not be able to find the best-telling business case numbers.

Having a keen interest in the progress of the innovation projects is part of the innovation culture. But as innovation needs empowerment, leaders should resist micromanagement. Micromanagement does not signal superb interest and great project endorsement but can easily be seen as process control and a focus on employees' weaknesses and failures. Such an unintended message has negative impact on engagement, productivity and overall work of the team. So, it's better to resist any temptation to micromanage and instead give the team greater autonomy. Such leadership style unleashes creativity, cooperativity and idea sharing and provides the impulse to experiment, as the team feels the confidence that their work is respected.

Big companies do not automatically have more potential for innovation just because they have more resources and perhaps better trained ones. On the contrary, innovation seems more difficult for them. "They are better executors than innovators, and most succeed less through game-changing creativity than by optimizing the existing businesses" (De Jong et al., 2015). They are big because they have learned so successfully how to optimize their growing business. Those proofs lead to skill in excellence in execution, but not so much about innovative power.

Leadership responsibility in innovation starts with the recruitment of suitable innovators. That requires an understanding of the innovations needed. A company with good market opportunities for range extension and refreshment may not benefit from a disruptively thinking innovation team. And a scientist keen to bring breakthrough functionality with new health food may not feel satisfied with contributions to a series of incremental range extensions. Once that match is achieved the new innovators must be armed with the essential "innovation skills", such as problem-solving, design thinking, project management, data analysis and collaboration. And to give them the space to unfold their innovative energy, they must find an appropriate atmosphere in the company, the right balance between autonomy and interdependency, giving them the possibility to drive innovations forward, without interfering too much with mature day-to-day business processes.

CASE STUDY: SEXY SAVINGS

Some months after the start of last global financial crisis, the European part of a big food multinational recognized the need to systematically scan all product categories for cost savings opportunities. Savings, or value improvement programs (VIP), are a routine management tool to improve the financial performance. Every year the business plan included a cost saving plan through material and manufacturing cost reductions to lift the profitability. The VPs of the affected functions, especially procurement, marketing, R&D and supply chain, would sponsor the restart of a cost saving initiative, setting savings target and agreeing the allocation of a particular percentage of their function team to the initiative, for R&D often involving 10–25% of the workforce. Often the "old hands" on deck would be asked to participate. They brought years of product experience and knowledge, managerial discipline and trained ways of working, as well as more crystalline business intelligence than change behaviour and new ways of thinking. The progress of the savings initiative and possible hurdles would be reported back to the sponsors from time to time.

But this time the economic outlook demanded a much stronger impulse into the organization about the critical value of cost savings for the business. The market forecast predicted a massive shift of consumer spent to more affordable economy range products, decreasing brand loyalty and a wave of new budget and no name launches. The economic situation forced many consumers to control their household spendings and seeking for opportunities to save money on daily items. Replacing FMCG premium brand product with offerings from the retailer or brands in the economy range is one of the possibilities.

For R&D and marketing, the core functions in NPD and innovation, working on cost savings was normally more a kind of second-class job, more discipline than creativity, tedious number crunching and continuous trial and error on incremental steps. More blood, sweat and tears, not the unfolding of wonderful inspiration and creativity or the brilliance of the individuals.

But for the business leadership this was the moment where the image of savings work had to change. The task ahead required more energy and new creativity from the team, more support and recognition in the company and the possibility to recruit different types of innovators for the job.

The first step was that the CEO of one of the biggest markets announced they would take the lead and act as a direct sponsor and head of the steering committee. That was unusual, since normally the CEO would not give so much time to a single project. The steering committee was enriched with different VPs, including marketing, a function normally more distant to savings work. And the core team was filled with directors from all functions. Noticeably unusual, a director from the technical function became project leader with direct access to the CEO and not marketing, as usual in a marketing-driven enterprise. The uplift in the regular project management sent a clear message into the organization and made the activity attractive for ambitious individuals.

And rather than calling out "necessary savings target" derived from the projected business plan as in regular VIP project, the team received an open brief to show options how to maximize savings without risking consumer perceived quality. This new approach encouraged the team to think big and different. The technical leadership pushed members from R&D and supply chain to "think and act as business leader", not just the functional lead of their expertise. And the formal, direct and regular interaction with the CEO gave them the opportunity to understand the FMCG business dynamics much better than in case studies in training courses.

The result after the 9-month scoping phase resulted first in a savings potential far beyond what has ever been achieved and then in a couple of scenarios, with different opportunity and risks levels agreed and share between all functions. Perhaps most importantly, the underlying ways of working, the product development approach, the concept and communication refinement created completely new ways of working between the functions and through the hierarchy. Long-standing dogma like, "VIP is not creative", "don't talk about savings to the outside" or "cost savings have to be perceived as the same" were transformed. The team developed a new understanding about their responsibility. The different functions discussed proposals for the next steps, listened to the hopes and fears of other functions, initiated test and trials to test their hypothesis and only then brought their proposals with one voice to the steering committee. First, a critical factor to successfully deliver this single project, but the whole way of working in the company changed. Detailed planning cycles created a stable frame for the innovation activities in the years to come, shifted the care for existing portfolio into the centre of attention and create completely new ways of working in R&D and marketing.

R&D ORGANIZATIONAL DESIGN

And yet again, there is no right way to organize innovation and R&D. But there are surely many ways, to get it wrong. The explosion in information and computer technology (ICT) has increased product complexity. Constantly new evolving technologies become adopted in the food industry, consumers change their preferences faster and faster, the product cycles have to accelerate and the societal, political and economic realities put innovators and hence R&D teams and departments are under extraordinary pressure. The array of product variation grows, the product portfolios expand and just the necessary care for those are already a heavy load for R&D organizations.

A few key characteristics where the new food product development differs from the R&D departments in other industries can provide robust principles. Elsewhere, new products or services show more of a component-based approach. In foods research and development is more focussed on the function the customer/consumer wants, rather than the component or technology that make the desired function work. In the past it made sense to build a departmental structure focussing on the creative exploitation of a broad game changing technology, like biotechnology, alterative preservation, taste and flavour or information technology. The individual teams could attract fresh talents, coach themselves to a higher level and start to explore options how such technology could bring value to the company. A type of future scouting was very present in science team than in the development closer to the market. The result was often a disconnect between both ends of R&D. Where one team complained about the "not invented here syndrome" of local technical teams, the others constantly advised to leave the "academic ivory tower" and face reality.

Hence, one thing that changed is that organizations, where science and development have their merits, are connected much closer, so they grouped along identified needs of consumer or customer. They are called not just called functional food or health and wellbeing, but more specifically gut health, cardiovascular disease, meal solutions, weight management, regenerative agriculture or plastic-free packaging. That naming will stimulate in the technical teams a mindset to focus on the consumer "problem" while having guidance about the science area they are acting in. Working on cholesterol lowering makes good sense for cardiovascular disease (CVD), less for gut health or convenient ready meals. The R&D leader has guidance in which academic talent pools they can recruit, while the new employees get direction for their idea and inventions. They will make use of latest science and technology, that's what they learned in the academic career, but can act in a territory that has been confirmed as a future growth area for the business.

But when it comes to the need to take up new technologies, leadership has choices: They could separate the new group/skill or expertise group into a self-standing unit, or ensure a solid link of the new expertise into the existing R&D organization, while keeping them separate or decide to fully integrate them into the existing R&D structure.

Of course, there is no simple right or wrong here! More "horses for courses"! In an efficient R&D organization, responsibilities are clear to everyone, while interfaces between and among teams are seamless and transparent. These requirements have become recently even more important, as more and teams are working remotely with technologies making that possible (Hidma et al., 2020). Some enterprises have learned it the hard way when the pandemic blocked development work at the bench, kitchen or pilot plant and when factor trials were merely impossible, others saw that the next push to trust simulation in the computer.

Any decision comes with consequences. Splitting current and new technologies has its advantages. Uninfluenced from routine ways of working, the new unit can settle, start creating new ideas and is able to realize full potential better.

This approach requires critical mass; it is resource intensive, and the budget need for such a pulse may be taken from the existing R&D teams. And it might need more time to deliver something useful. But once that has been done, the next tasks come up, as now this new idea has to be integrated into existing products and processes. The initial segregation also limits, and the transfer of capabilities and knowledge between departments may be hindered.

Integrating new technologies fully should ensure fast knowledge exchange, at least if the new technology is not so cutting edge, that the old colleagues in R&D have difficulties imagining how the findings of the new colleagues can be used. The link with the old capabilities will be smoother, and profitable results may arise earlier. The risk is that the new department results are constrained by the current skills and expertise and push the new part of the organization into existing capabilities and processes, yielding results that are far below the full potential of the technology for the business. And full integration into existing bears as well the danger that only subcritical mass is implemented, not allowing the new technology to develop properly. One nutritionist or gut health experts will have difficulties shifting the innovation thinking of a mature product development team to new platform approach.

In the end a more balanced approach, assigning a new, separate team, but thinking about good structures to embed them into the existing R&D organization, is therefore more often the chosen solution. The full functional maintenance of the existing team system a space where the new skill base can flourish, but where both can inspire each other, demands a structure with shared platforms, idea and knowledge sharing opportunities. It drives adoption if the responsibility for the new team is given to someone "higher up in the organization", a senior sponsor. That person has a full overview about the existing structure and processes to build solid links, but also the authority to ensure "open arms" and welcome plans" from the existing R&D teams. This is a leadership signal of how much this new team is wanted in the business.

A good integration plan can influence the hierarchy. A new nutritionist coming in will bring efficiency in data collection and the robust design of the nutrient table back of pack. But he/she can also bring latest insights from the science area, make suggestions on how to redesign products, so they are healthier. For that they must understand the design principles of the products made, the possible interaction of all components in the food matrix, their reactivity and options the influence that. How will the new nutrition profile change the look, taste, texture or aroma? They may even need to stretch their knowledge into food processing and packaging, just to find ways to realize the new profile by adding antioxidants, fill under inert gas conditions, changes the shelf life or apply a different packaging format.

This skill base organization is of course not just a question of internal organization. Today many more ways to use R&D knowledge are widely available (see later). But they all have their strengths and a flip side. Collaborating with external teams makes many things easier; it seems cheaper, with less up-front investment and leaves flexibility to change innovation direction faster or to follow several ideas. But it basically sets an independent base, which puts a lot of responsibility and experience on the "dockers" on both sides and their ability to create something jointly. Results are not only defined by the content development but also the ability to recognize different cultures and act accordingly.

FOOD R&D DEPARTMENT

The ability to constantly develop new products, improve existing ones or find better ways to make those is the key task of R&D in the food industry. In smaller enterprises, with a narrow product portfolio and focussed business strategy, a separate R&D team is sometimes not needed. A contract manufacturer or co-packer producing basically one food product type may not have the need to build separate R&D department. Here, it may be better to keep the total technical expertise in one group, managing the ingredient sourcing, the processing and the packing. A successful copacker must make sure that they technically leading the "making" of the product they make for others. Here, process technology can create the competitive advantage. The experts are part of the manufacturing department and help to "keep the machinery running", while perhaps looking for ways to optimize aspects or single steps. Only when a customer comes with a new product request do these experts assess the options to produce the requested variant or suggest themselves a new version. In-depth understanding of the process technology, grown by daily exposure, allows fast and confident judgement. Keeping this expert in an extra department could build barriers and limit knowledge exchange. Here it is better to nominate this manufacturing expert as the R&D contact for anybody coming from outside with a product request.

Once the moment is reached that the company experts start to work predominantly on new things, not maintenance of the current ones, classically it is time for a separate department. Overall, food R&D is a broad and interdisciplinary field and covers various aspects of food production, agronomy, processing, quality, safety, nutrition, maybe even regulatory and consumer technical insights. There is no foods R&D curriculum. R&D professionals can come from very different disciplines, technical product or process, basically all biosciences, agronomical science, measuring technologies, psychology, information and computer technology up AI developers and researchers.

The business context sets the requirements. If a company with a strong engineering department believes it lacks the potential to create new product ideas, food technologists, formulators, perhaps chefs and nutritionists, are a priority. If a food producer is mainly working with copackers, there is no need to hire process engineers or packaging experts. And an enterprise that wants to strengthen exports in the next phase may benefit from legal and regulatory food expertise.

And in a moment when most of the E2E value creation should sit in the company, a wider range of expertise is necessary. F&B innovativeness basically means the ability to create winning formulations/recipes, combined with a successful industrial realization. While the formulators, chefs and product design experts are responsible for ingredient selection, formulation and prototyping of exciting new solutions. They are driven by creativity and limited by the boundaries of industrial manufacturing.

The processing team must then take those prototypes off the bench and bring them to industrial scale. Their task is to create the space for formulator creativity and food authenticity, while keeping the ability to manufacture the final product to scale. Their "line product" ideally matches the kitchen/lab prototype, while being economically feasible and of course safe.

They approach the challenge by upscaling. But often, it's not possible to take a kitchen or bench samples and run them straight in the factory. The small-scale equipment the formulator team has used to mix, blend, stir, cut, fill, form and weigh and are different from those in a suitable factory. And the factory has no space to for long and tedious experiments to produce commercial products from ideas never tested properly.

The link with between both scales can are the pilot plant and computer simulation. In the pilot plant larger batch volumes or batches are made with technical equipment that is downsized from the critical commercial manufacturing steps. Mixing, drying, filling, freezing extrusion, cooking, dosing, forming, homogenizing, etc., can be done in many ways in the food industry. Only the use of the downsized factory machines will allow the prediction of the manufacturing feasibility at an accuracy needed to match the assumptions in the business case for the project.

In case the new food products will bc made in one factory only, it is beneficial to have the pilot plant within the factory. That facilitates logistics of the materials and allows the sharing of infrastructure. Such models work well for comakers or copackers, who basically get the order to make a specific product using their factory technology.

It is often helpful to have for the pilot plants the permission to test with consumers. This allows for more flexible testing and stepwise product optimization without spending too much material, waste creation and factory production time.

Once the product is ready it needs packaging. For many food products this primary packaging is fully integrated into the manufacturing line, and experts in the factory know how to run this step efficiently. Downsizing to pilot plant scale would be too complicated and capital expensive. Standardized formats at the end of the line, computer simulations, in-line sensors, practical experience and packaging trials as needed are the elements to manage the end-of-line outcome. This allows the test of new material, new format or sizes for smaller B2B food companies that are normally less critical. They have a few customer packaging options attached to the end of the line to fill at large scale, and the packaging material has pure storage and distribution characteristics. The regular shopper or restaurant visitor will never see it.

Some enterprises have packaging expertise in R&D, which can be part of the processing team or form a separate team. The latter is advisable when the company offers several packaging formats and serves the consumers on the market or many customers. Packaging experts will help to design the packing process jointly with the technologist. And they will interact with material and machine outside. New packaging material, less or no or a different plastic, a format change, a new distribution requirement, etc., makes the investment into packaging experts quickly profitable.

A few other functions are needed on the way to market. With the increasing internationalization and legal complexity in foods, solid knowledge in food low is important. A nutritionist ensures the correct calculation of all consume relevant information in line with the expectations of regulatory authorities. This person can be part of the formulator team to have a strong link into product design and composition. Or it can be placed together with regulatory experts in a sort of R&D compliance team. Such team is not only avoiding noncompliance cases, law infringements, competitor challenges and eventually product recall. They may indicate new claim opportunities or positive stories.

A final group of R&D people are the project leaders (PLs) or technical project leader (TPLs). Having them as a separate team does not follow the essential skill logic (formulation, processing, packaging, nutrition), but may serve the business requirement. Since many companies drive their new product development through projects, a project leader or technical project leader owns the project end to end, leads a cross functional team and reports the updates to the leadership team (more details are found earlier in this chapter).

In bigger food and FMCG companies, the overall global R&D structure can follow a design with

- Centralized R&D structure, often close to the global headquarters
- Decentralized R&D structure, where markets/categories have R&D at a location
- Hybrid structure, a central R&D plus divisional R&D structure

On a more abstract level the decision about the best R&D structure can be guided by the business model of the operation, sometimes called the "innovation game" (Mahrschid, 2022).

Big food enterprises frequently use the hybrid model. Here the vicinity to market, consumer, customer and suppliers at the periphery is maintained, while a central more basic science- and technology-focussed group drives broader product innovations, technical breakthroughs and is engaged in strategic collaboration with academic and commercial science and technology partners. Today the central units are frequently based in an innovation environment, facilitating multiple partnerships (see "Food Innovation Hubs and Food Valleys").

The fundamental split between central and local is sometime described as the "4D Model": "Define, Design, Develop, and Disseminate" or "Discover, Dream, Design, Deliver or Deploy". First captured by Gary P. Pisano from Harvard Business School, the sequence describes the fundamental jobs in R&D from background basic and experimental science to excellence in execution (Pisano, 2012).

A functioning hybrid structure requires a business size. Each units need to have a critical mass to innovate in their strategy. Central research with too few R&D people will have problems creating breakthroughs and new product platforms, writing patents and publications and building effective partnerships. And having a local R&D department able to read market expectations very well and translate centrally provided opportunities may inflate the numbers of units needed, each of them able to "do the job" they are supposed to do.

Separate R&D units of whatever type have many advantages and two disadvantages. Firstly, it allows the R&D organization to distract from the markets! The connection to customers and consumer is looser and often sporadic and ad hoc project based. Rather than spending much time observing consumers and customers and trying to distil insights out of it, their own trials and experiments, idea testing, and literature study seem more attractive. And it is close to the activities learned during the academic career.

Foods is FMCG, and such behaviour bears the danger to lose the ability to adequately translate market trends and consumer needs into new products. The balance is between closer to market and closer to science and technology. Bigger enterprises can afford to have both, part of research, often centrally, close to science and technology, suppliers, startups or academia, focussing on longer-term technical breakthrough. They do not need the constant "pulse of the market" to achieve their goals. A permanent echo from the market with all the hypes and trends may hinder staying focussed. The local R&D units need to compensate for that; they should become the best science delivered by the central units while understanding how that can be poured into competitive local products.

The second disadvantage is that this "classical setup" can separate R&D from the rest of the organization. To a good extent, the typical R&D activity scheme is different from all day-to-day business of most other functions. R&D works in the future; business operates in the here and now.

R&D gets more and more used to use their learned jargon, and their specialized lexicon makes it difficult for others to understand what's going on in that group. For them, this team is more of a "black box".

And a black box can function well if the "input" and the "output" are well defined. But successful innovation in market with accelerating change will suffer from such a sequential model.

Once could choose an approach where R&D becomes connected to the "business reality" and "market trends" at moments where the next round of NPD is needed. Now they need to be brought up to speed with professional market observation. And they need to learn how other functions like marketing or research are approaching the task. So there isn't much time to strengthen the observational skills. Afte the "consumer watch" R&D is sent back to execution and the classical "what and how" split continues! Others, marketing, supply chain or finance, say "what" the business needs. And R&D can focus on "how" that is delivered.

One option to keep the advantages of a skilled team in FMCG but connecting the team more permanently with market and trends is the integration of expert focussing on the consumer and customer needs. They would focus on the technical aspects of activities in the market, analyse competitor products, scan markets and keep a permanent link to their sister team in market research and marketing (more details in Chapter 8).

CASE STUDY: NEW SKILLS IN THE KITCHEN

A big FMCG company was successfully selling various types of products for the daily routine, ranging from home and personal care products to foods. A significant part of their food portfolio contained products needed for meal preparation and cooking, in the kitchen and on the table. This part of the business was successful for many years, and the brands were market leaders or close runners-up in many leading markets. R&D had flavour experts, ingredient specialists, former chefs of leading restaurants, very experienced formulators, leading process technologists and sensory scientist in the backup. For many years they were successfully translating new culinary trends in the different market and convenience requests into affordable new solutions. The approach to NPD was dedicated meetings with their marketing colleagues, where they regularly organized innovation workshops and creative sessions in inspiring places. The resulting ideas were evaluated, scanned and then prioritized! The marketing colleagues took the winning ideas back home and converted into colourful picture and flashy product descriptions, which they presented to experienced external marketing agencies, so that they could check the market potential and discuss internally the details of the study design. R&D was not much need in this stage but asked to help with the product description to differentiate the new ideas from solutions in the market. Or just confirm the proper use of some technical jargon.

Once the consumer study with the target group was done, the agency prepared a detailed summary where they described in lengths all technical details of the research to underline their scientific competence. The summarized the results and provided advice and on-request commercial outlooks. The debrief ended with a recommendation to "go ahead", "go, with caution and improve some aspects" or better "don't use this idea further".

The "Yes, but with some modification" advice was the one most often coming back! *There are some great nuggets in there, but this and that, we are not so sure!* This was good for the agency, as this would bring new research requests and it was not offending the customer, who surely works hard on the idea and the conceptualization.

Results were first shared with marketing, followed a bit later by a broader presentation to R&D. This gave marketing the possibility to internalize the feedback and start creating ideas of how the project could now be successfully progressed. And as the project leader came from marketing, this brought the obligation to identify "ways forward", so the idea could fully turn into the commercial success wanted.

The following cross functional discussion with R&D was often characterized "by the creative translation skills" of the marketing colleagues and the agency. They took the task to interpret the verbal feedback of the consumers into a product design language that R&D could understand and start drafting the product brief.

The R&D team was not very skilled with listening to somewhat vague consumer discussions and then extracting the essentials out of it. The more experienced R&D colleagues regarded these interviews more as a waste of valuable time. So they preferred to send their younger colleagues and instead focussed their own attention on the science and technology that would help to make the industrial product being more natural or freshly prepared. The expert talk with colleagues from the factory and external experts gave them more energy than watching rounds and rounds of consumer talks.

As a result, the commercial success coming from such new launches was moderate and often only little incremental. In general, a result, the team was accustomed to, since they knew that they were innovating in a mature market, where many similar offerings were available to consumers.

The leadership team concluded that for the quality of the R&D skills and the effort made, more return on investment would be needed to fulfil the strategy and "win more in this attractive market". So, they took advice from non-foods part of the business, where they launched successfully daily routine articles like, soap, detergents, shampoo, etc., often with commercially better success.

These businesses had not only, like all others in the company, a consumer market insight team, working closely with marketing, but as well a "consumer technical insight" team embedded in their R&D functions. Their role was to understand the behaviours, interests and mindset of consumers, but more importantly the translation of consumer talk into technical language and attributes the R&D team could fully understand and put into action. So, expert and consumer panels were trained in common attribute language, and "texture" became something different than mouthfeel, "meaty" and "umami" different things and experiences like "dish tension" an intensely disputed consumer moment.

The integration of the new CTI function into the existing "chef and formulator team" took some time and effort. The culinary chefs, until then the "final authority" in taste, had to learn the language of the consumer and accept their verdict as final. The high-level sponsorship by a SVP R&D helped a lot, and in the end, both became a powerful team, headed by a senior insight manager.

INNOVATION CENTRE

In the 1990s the idea of innovation centres (ICs) became popular in FMCG. At that stage, the main purpose was a more effective link between "market pull" and "technology push" and a more dynamic interaction between "what "and "how". In contrast to the older departmental model, here the innovation core, especially marketing and R&D, were co located in one unit. The day-to-day work was enriched through common brainstorming sessions, sometimes design sprints, innovation workshops, rapid prototyping, market reports or consumer immersions. How far the integration of different function like marketing and R&D went was part of the experimentation. ICs also serve as spaces where people with different backgrounds and academic education are coming together and focus their thinking on creative design. Since the beginning the concept of an IC has evolved over time and gone in different directions.

A light touch approach was just putting both teams physically into one space to increase and deepening the interactions. Members of one function would get insights into the work and think of the other team. Both would learn to "decode" their discipline-specific vocabulary. In food companies, the lead was given to a senior marketing through discussion around "USP", consumer "proposition", "design "thinking", etc. Other setups included the making of common business plans, activity planning and even budget planning and management. The strongest integration would be the creation of a standalone centre with one unit leader. This approach is not without difficulties, as the head of the centre would be resource and career responsible for people with very different professional backgrounds.

The companies experimenting with ICs did not switch the whole R&D and marketing structure but took part of it and reorganized them accordingly. A company could arrange a segment like "ready meals", instant snacks" or "impulse ice cream" as IC would give the new team an R&D and a marketing lead as joined leadership. The functional heads could still line report within their function to a director or VP, while the rest of the respective function would remain in their departmental structure.

Sometimes such reorganizations are combined with a geographical reorganization. Instead of local teams in lead markets, country combinations or regional teams were established. The teams were mixes of people from different countries, languages and cultures.

The setup of ICs is a signal that the company gives NPD and innovation priority and therefore is willing to make a bigger investment.

When the cross functional team gets common targets like the number of innovations, refreshments or range extensions or incremental turnover rather than function specific targets new products, a common focus is better possible. Sometimes this results in a new understanding of innovation, the R&D strategy, the specific objectives, clarified sponsorship, well-defined roles and responsibilities and measurable success factors.

It's not a training campus for technical experts to get closer to the market and consumers. But when its run well, it does create a passion for consumer needs within the technical functions. And marketing will be more regularly exposed to the principles of food "product making" and the technical, commercial and legal limitations. They get a deeper understanding of all the technical aspects. In a way, it is an attempt to bring technology push and science pull into a common arrangement.

In such IC setups, a dedicated consumer science and testing can be very valuable. While many organizations claim passionate consumer/customer-centric development, their R&D groups rarely get the opportunity to test products directly with the end user. It's a very different experience to be involved in the design of the testing protocol, thinking about the target group, prepare locations, brief testers, had out the samples and watch them closely than just reading the final agency report or joining a debriefing PowerPoint presentation.

Like other types of R&D units, ICs are not a fixed concept. They are flexible structures, adjusted to the business needs. The idea is not design protected; there were no minimal requirements to call a unit an IC. One could call the back of the garage an IC. But it always sent a strong signal about the priority of new product development and innovation to the internal and external world. The name is also not protected: If one believes that innovation centre sounds more modern or vogue than R&D department, the team can just be renamed.

Today the name innovation centre is used differently. They are described as incubators, accelerators, coworking spaces, R&D centres, etc. It's more a space, a physical platform where individuals from all types of enterprises meet and where they find resources and support for their business idea. Often it's a space or hub that offers various resources and services to entrepreneurs, startups and established businesses; a collaborative environment where individuals or organizations can come together to exchange ideas, innovate and develop their projects. They are between organizations, not within them. Sometimes even co working spaces are captured under the name innovation centre (Vbouronikos, 2023).

The big multinationals in food have also started to reorganize their R&D to become open R&D platforms. Nestlé's R+D Accelerator is a global group and provides a platform for startups, students and Nestlé employees to collaborate with Nestlé's R&D experts and access their infrastructure and technology. This one consists of 12 different sites all over the globe, each with focus on a product category or region (Nestlé Global, 2021).

Unilever's global Food R&D centre in the food valley in Wageningen is called "Hive" and constantly invites people from outside to come and share their infrastructure and collaborate with company R&D. (Plc, 2023).

The vocabulary often used to describe the ambition of these places include "culture of innovation", "entrepreneurship" and "transformation".

These places are owned by a company, but intentionally with open doors to the environment. In that sense they are a link between pure internal R&D structures and fully external services.

OPEN INNOVATION

The concept of open innovation (OI) introduced by Chesbrough who defines it "as the use of purposive inflows and outflows of knowledge to accelerate internal innovation and expand the markets for external use of innovation" (Chesbrough, 2003). It promotes the idea of using internal and external ideas and sharing them to find solutions for mutual implications and even work in sync with competitors. The beauty of the concept is that collaboration does not only make us more efficient but also helps in finding new solutions. More people, more brains, more ideas!

OI can encompass different ways in which the new product or service is developed. It does not prescribe the organization set up and governance, or how many partner are collaborating, who these partners are partners or in which phase of innovation development the collaboration happens. To capture the "classical" 1:1 cooperation between the ingredient or machine supplier and F&B maker under that roof is sometimes done, but does not capture the essence of that new approach. The latter is q simple and classical practice. A very different approach is to enter a partnership with a wider set of external partners, including academia, startups and perhaps competitors, and coordinate their contributions, organizing the innovation process and agree on all aspects of ownership.

Sometimes the benefits are described more defensively: Avoiding company blindness and circumvent wrong filtering! The underlying concern is that in an established enterprise ways of working and existing culture would only allow incomplete – or even wrong – information to feed the internal innovation process and the solutions created would therefore not be good enough.

It also makes a big difference if the OI approach is only used during a specific innovation phase, like idea generation or upscale or commercialization, or if the E2E innovation process is managed jointly.

Some authors in the management literature suggest a segmentation of OI into open, special, integrated and closed innovators. These types of OI can be evaluated against the return of investment, like required organizational requirements and transaction costs, their creative potential or ability to manage risks and the danger of information leakage (Lazzarotti & Manzini, 2009).

This provides a useful frame to shape the intended innovation partnership. Hardly any company would fully give up their internal R&D. An open innovation would be complementary with internal R&D, and not a substitute. Working with external innovators requires different managerial skills and techniques for setup, maintenance and exploitation of the partnership. The external partner cannot be expected to understand the company's innovation processes and decision making. It can be useful to set up a specific interphase team within the company managing the interaction with the partner, while the rest of the innovation machinery continues to work the way they used to within the company.

At the end the companies entering an open innovation partnership will find the way of working that is best fitting all partners and their existing structures and capabilities. And it's more used for radical, breakthrough idea searches than just incremental innovations. Here OI can be a very useful addition. The "big and bold" unsolved problems for which the company has no adequate skills is opened to outside influence, possibly looking at the problem from a very different perspective and therefore able to find a solution. The environmental impact of the sector is huge, but finding superior solutions to avoid plastic, landfill, soil erosion, deforestation, carbon footprint and improve circularity, etc., are often beyond a company's own capabilities. An investment in OI offers a new and additional opportunity. Coffee made from upcycled coffee berry husks, grain products made with spent grain from beer production or using cellular agriculture to produce animal-free cheese are recent examples.

Sometimes it's said that the switch to OI would be easier in small or medium-size companies, as they are more focussed on their business, while being more flexible in their approach and open minded. And indeed, big companies often have an established company culture and may see other options to grow their business with less risk. But they also have more resources, which makes it easier for them to set up complementary structures and manage the "innovation dilemma" (see Chapter 3).

The more open an innovation process becomes, the more distant it may be from established internal processes. Finding new ideas is one thing, testing and prototyping another one and upscaling and commercialization one more. Here the external results need to be properly connected with the business strategy and ambition. P&G's Connect and Develop Strategy is a way of innovating products by collaboration with others and commercializing them through licensing deals, so that P&G can commercially use external technologies and packages. This structure also allowed them to sell their own IPR to others: IO is here "insight out" and "outside in" (Huston, 2019).

DSM in the Netherlands is another example in the food business. They test various innovation tools like academic R&D, external and internal venturing, licensing and acquisitions to build and scale innovative ventures inside and outside. A recent famous example is Hologram Sciences, Inc., a consumer-facing company that offers personalized nutrition solutions based on individual health data (DSM Integrated Annual Report 2022, n.d.).

A "gorilla in the room" is the exploitation of new findings coming out of open innovations. The challenge lies in balancing the protection of intellectual property rights (IPRs) with the benefits of sharing and collaborating with external partners in the innovation process. Open innovation lives from an unconstrained flow of knowledge, while IPR is meant to exclude others from exploiting the finding. To handle the trade-off between openness and protection, validated formal contracts to govern the relationships are needed. This includes a view about which types of IP to protect, which parts to license or maybe release to the public domain. These considerations are not only needed for newly created "intellectual property". Innovation partners will have to fill the process with science, technology and knowledge they already have. The F&B business is largely built on mature technology adopted from other sectors, which has been improved over the years of usage, bringing huge benefits to the company, but sometimes no option to legally protect this new knowledge, since the "inventive step" was too small or the legal procedure too complicated. Sharing such knowledge in open innovation bears the danger that such a competitive advantage leaks out to others.

The difference in scale between small companies, startups and academic partners and big companies can influence the atmosphere. Big companies have big legal departments, with experienced patent attorneys and legal experts. They are costly professionals, doing their job well and drafting very detailed contracts, trying to make every possible outcome legally watertight and always with an eye on the company benefits. On the other side of the table, there are entrepreneurs, science leaders and experts, not familiar with the legal procedures and the language used. They may get the feeling that they will not get their fair share from the intended collaboration or need to spend months sending around draft contracts. For a successful collaboration, one needs to ensure that all parties benefit from it. Fortunately, academia, small and medium-size companies are now often equipped with good access to that professional knowledge.

Classical cross licensing agreements on current use or future business, open-source technologies, and IO-specific IPR management are other tools to overcome the basic tension between the need to exclusively exploit opportunities, while finding them together (StartUs Insights, 2022).

Entrepreneurship is seen as the tool to bring entrepreneurial spirit and innovation culture into the framework of a bigger company. But what is this?

FOOD INNOVATION HUBS AND FOOD VALLEYS

The impulse for the setup of food innovation hubs or food valleys does not come from commercial enterprises centres but is often initiated by governments to foster the output of that industrial sector. In 2020 the Thai government recognized the importance of food innovation and established a US$283.8 million fund to support "Food Annapolis" a project to make Thailand a global food innovation hub and leading food exporter (Jones & Pimdee, 2017). Many more countries have food innovation hubs: The

Netherlands in Wageningen, Indonesia in Jakarta, Kenya in Nairobi, India in Mahya Pradesh and others in VEA, Vietnam or Columbia (Food Innovation Hubs, n.d.). Today these are multistakeholder platforms, funded by supranational organizations, government, charity foundations, companies and private investors.

They often combine several partners with their own experts and organizations that have not worked together in the past in a systematic way. It's often more than a strategic partnership between supplier and customer. Competitors may collaborate, civil social partners may join and the government may also demand a seat at the table. Companies settling there with their R&D facilities often enjoy a couple of tax benefits, and the country hosting the locations attracts international talents.

These hubs have gained a lot of popularity in the past 10–15 years. Not only because the investment into successful science and technical research are becoming bigger and bigger, but also driven by the perception that real new developments of all the technical building blocks needed are often too complex for a single company.

These attempts for a wider E2E collaborations is driven by the insights that the future of the food sector has to be sustainable. And that requires that all contributors in the sector cooperate more and deeper to find new solutions for existing problems with a deep understanding of interdependency. Sometimes these R&D facilities are also called an IC, but they differ from the units of the past.

An example is Unilever's global Foods Innovation Centre in Wageningen. The E85M facility was placed on the Wageningen Campus, known as well as the Food Valley in the Netherlands where international food companies, research institutes and Wageningen University and Research Centre are concentrated. Nowadays the Food Valley hosts about 15,000 professionals working in food-related sciences and technological development. The whole valley is pretty flat, but the association with "Silicon Valley" serves to help create the image of a dynamic knowledge centre and intended to create conditions so that food manufacturers and knowledge institutes can work closely together from farm to fork.

Open doors, scientific lectures, presentations and shared facilities are building a campus atmosphere, attracting young talent from all over the world. For the companies located there, this creates excellent access to future talents, and for the students it allows an early and realistic view into the industry. And this hub collaborates with many similar locations all over the world. The world economic forum has launched an initiative to connect all these global hubs.

In Denmark, the Food and Bio Cluster is home to a thriving food industry that is well known worldwide for its strongholds in areas like food quality, safety, sustainability, organic food, gastronomy, ingredients and innovation technology. Food & Bio Cluster Denmark is the national cluster organization, building on the Danish food industry and agronomy. There ambition is to be the unifying platform for innovation and growth for Danish, but as well, international companies and knowledge institutions (Food & Bio Cluster Denmark, n.d.).

There are numerous new food clusters all over the world. Often, they are established where farm-based products, agronomy, feedstock, fisheries and the food industry are an essential industrial sector for the country, like in the agronomical sector in the Netherlands or Denmark, or where the food chain has been chosen as focal area in the future. Examples are "The Start-up Nation" in Israel, positioning itself at the forefront of innovation, focussing FoodTec. It covers an array of areas that includes nutrition, packaging, food safety, processing systems, cultured meat, novel ingredients, retail and restaurant tech, health and wellness and alternative proteins.

The "Singapore Food Tech Hub" is traditionally not known for its agronomy sector but wants to "triple its food self-sufficiency to 30% by 2030". With the lack of arable land, it is no wonder they are focussing on food and biotech, urban farming, artificial meat and other modern enablers (Writer, 2021).

When the food valley is attached to a university with food science, the setup gives great opportunities to cooperate with all types of industrial, governmental and society partners throughout their studies. Students have the chance to understand future places where they might work better, and the enterprises get constant access to new talent.

EXTERNAL NEW PRODUCT DEVELOPMENT

Asking an external partner to develop a new food product allows companies to access new technologies, ingredients, markets or expertise that they may not have internally. This often reduces development time, cost and risk of innovation and may provide a broader spectrum of food products that can be sold. This model works well for the big retailers in the world, when they decide to set up their "retailer own brands". For others, it is less a fundamental decision about "internal versus external" than an additional opportunity to be innovative.

A company without at least some internal R&D would have difficulties to implement what has been created outside. It's possible, but normally impractical, to hire someone to create all the formulations, make the prototypes, check how the products can be made, capture all those data in specifications, collect all product information and print it on the packaging and label. Beside that fact that all food products need maintenance. They do not simply live "happily ever after" the launch. Crop failure, supplier bankruptcy, trade barriers, new regulations, machine breakdown, consumer complaints and many more things need corrective actions on the product level by experts. An external partner may be able to handle the required changes if they are very simple because the products are simple and mainstream. But for food product with some uniqueness, it's becoming much more difficult.

At the same time new product development is happening outside the company.

The most obvious partners are the various ingredient or machine suppliers! The supplier may have started their innovation program due to the occurrence of innovations emerging in technical sectors or trends long ago. They more experts on that product or process and they need to also constantly search for new ideas to stay competitive in their market. New solutions in agriculture, biotech, food processing, packaging and ingredient technologies have been evaluated on the supplier side to deliver advantages. The food company is keen to see new ideas with the potential to gain a competitive advantage by being first in the market. Nowadays they are offering proactive solutions not only for their direct customers but driven by trends and opportunities in the consumer markets. they are acting in.

A successful flavour and fragrance house will not only work on natural aromas, meaty taste or sodium reducing actives because they got this request from their customers, but because they see the trends, which they convert into tailor-made options for the food company. They will showcase their idea and offer implementation support by making the wanted prototypes for the finished goods partner or by sending R&D experts to help with formulation, prototyping, processing or market research evaluation and optimization.

External development can also mean the that the new finished product is designed externally. This can be based on an innovation brief that has been handed over by the consumer goods company or the new product is presented to the packaged food company.

The reasons for that can vary. Often, a company with a brand has identified and new product opportunity for which they do have the skills and capabilities. Maybe they do not have the in-house manufacturing and the design skills. A brand may be strong for ambient meals, but now sees opportunities for chilled or frozen meals. All very different technologies! And if the opportunity in the market has been identified, the development must happen quickly. Things are changing in the fast-moving market and others, like key competitors, have eyes and ears as well (or the same consultants). Then there is no time to recruit and train the new skills needed.

Plant-based protein alternatives have recently created a wave of external product development requests. The companies which wanted to offer these new product ranges often quickly realized that they do not have the in-depth skills to just start to design plant-based meat products from textured vegetable protein. The plant protein source, the type of extrusion, post-texturization formulation, the choice of flavouring and colouring agents, emulsifiers, gelling agents, binders, fibres, carbs and fats all need to be selected, mixed and blended carefully to deliver competitive alternatives. This is very different from the traditional skills of a R&D team used to working with meat-based products.

Basically, all segments of the food market are manufacturers with technical and manufacturing skills to produce specific product formats. But their business model is different. They may not want to invest into the building of a brand; the marketing and sales forces needed to promote may not have a brand. Their business is to manufacture for other brands the finished products the customer wants to bring to market.

Depending on the size and ambition of those companies, they also have development teams that do understand well which product formats and variant they can make for others and which ones are not yet in scope. These teams are also used to test new raw materials in their factory for their customers.

Other companies have specialized in the product development. Often, they are providing the finished package, but an intermediate or the product in bulk still needs to be packed for retail or food service.

These companies are not small and act internationally, Companies like Solina see themselves as a "provider of customized ingredient solutions for the food industry, foodservice, butchery and nutrition sectors" and have 18,000 customers in 75 countries (About-Solina, n.d.). They provide intermediates or finished products and have the ability to pack them into different formats ranging from 10-g sachets to 1-t containers. Solina's products include sauce bases, cheese preparations, marinades, seasonings, batters, coatings and spice blends that contribute to nutritious, affordable and sustainable food concepts. Solina also offers a variety of packaging options for its products, ranging from 10-g sachets and cups to 1000-kg intermediate bulk containers. Their business model is to design new products in a category they are experienced in, not the manufacturing as such.

Others focus even more on product development for other companies with a brand. Companies like CULINEX position themselves in the earlier step of new product development offering consumer idea screening, innovation workshops, concept development, prototyping and consumer validation (CuliNEX, 2023). They will not go into contract manufacturing themselves but may provide advice for the search or in-house implementation.

And sometimes it is even difficult for a company with a problem or idea to identify a suitable innovation partner. Companies like "Nine Sigma Enterprise Solutions" connect their clients – those looking for innovation – with a wide network of solution providers (NineSigma, n.d.). That can be innovative solutions, new partners, a technology gap filler, trend scouting or new business opportunities. For that they charge clients based on the scope, duration and complexity of the projects or the value of the outcomes.

There may be many reasons why a food company decides to develop or co-develop new products with a third party. They range from extension into a new product category that may need a new manufacturing process or a different standard (allergen free, kosher, halal, etc). And before doing that up-front investment with skills, time and money, the design and manufacturing by someone else is a less risky option. The company can first test the in-market performance of the new range under their brand name. And once it starts to become a success, the "in sourcing" can be considered. Once the initial investments are done, producing in house is normally more cost effective, because the third-party manufacturer needs to have their margin as well.

In such cases the new food product development will be closely followed by the internal R&D department. That will help them to build their new skills needed for this new range and create a better understanding for a smooth transition of the product into the own factory. The company surely wants to avoid having the now internally made products not identical to the one that created the initial success. Small changes or necessary modifications in the product or packaging design can have a huge impact on the commercial performance.

EXTERNAL CENTERS

The alternative to the (re)building of internal R&D teams is to reach out to external expertise. External new food product development means creating new food products or services by collaborating with external partners. The biggest advantages of such an approach are the access to new technologies and expertise, avoiding the investment into new skills, faster speed to market and better reputation and credibility in case a reputable partner is chosen.

The consequences of innovating externally can be complicated discussions about the ownership and intellectual property, the risk of knowledge leakage and strategically dependency and loss of control over the innovation process. And working so closely with external teams requires the ability to manage the differences in culture, vision or goals of the participating parties.

Independent food development organizations provide scientific, technical and regulatory support to the food and drink industries. This builds a good alternative for companies lacking expertise in a technology field or market segment. The service offered ranges from product development, prototyping, sensory evaluation, consumer research, nutrition and health, food safety, quality and integrity, up to staff training. Those organizations have dedicated teams of experts in food science, technology, nutrition, microbiology, chemistry and law and facilities for developing, prototyping, testing, analysis and research. A well-known representatives is Leatherhead Food Research (UK) (Leatherhead Food Research, 2018). Similar enterprises can be found in many places globally, particularly in those countries where the food sector is an industrial priority. NIZO in the Netherlands ("Nederlands Instituut voor Zuivelonderzoek" or "Dutch Institute for Dairy Research"), SIFBI in Singapore (Singapore Institute of Food and Biotechnology Information") or the broader CSIRO in Australia (Commonwealth Scientific and Industrial Research Organisation) are typical examples. Some of these organizations can be used for full innovation support, ranging from idea generation until full upscaling. Interested food companies can become members and get the opportunity to tap into specific resources and get regular updates, position papers or access to networking events. This option is less for enterprises having no R&D resources at all, but those that do need specific skills and expertise where they do not want to invest immediately. A company producing ambient stable ready meals may decide to develop new products for the chilled cabinets, which is technically a different portfolio. In such case it can be advantageous to hire the needed expertise externally and test them in market. If that move is successful, the company can decide to invest into the build-up of an own skill and expertise.

Or in a company entering the home cooking market some professional expertise is missing. They may reach out for abuse testing, cooking loss or cooking instruction testing. Companies like Campden BRI would offer such help (About Campden BRI, n.d.).

The idea to collaborate in highly competitive markets with external organizations, sometimes even competitors in new food product development and innovation, has found increasing acceptance in the past years. With more and more new food product launches, increasing pressure to launch new products but a high failure rate, the possibility to mix own knowledge with new technology and skills reduced the entry barriers to new market segments and increases speed to market.

In the newer variant of such cooperative approach more and more spaces where several partners from very different parts of society and different sizes are opening up.

R&D PERFORMANCE

While there are various ways to organize R&D for NFPD, commercial enterprises will regularly review the organization they have chosen. Benchmarking is the tool of measuring the output of the R&D function.

Their performance can be simply measured in relation to the performance promised in the contract or business case, including profit margin, return on assets and return on investment. A more detailed evaluation could split the performance into three main categories:

1. The first category takes time as the lead dimension: Time to market, on-time delivery and schedule adherence. This is an important aspect in FMCG.

2. The second category focuses on cost, including total project cost, adherence to planed budget, profitability analysis, product cost or product development cost as percentage of turnover and margin analysis.
3. The last category could look at the number and size change requests during the project, including the span between initial idea and outcome (Driva et al., 2001). Here the organization would have a view about what is preferred. For a high capital investment project that is strictly gate kept, many surprising changes in direction can be negative, while quick and efficient deviations in direction in a fast Scrum project can be very valuable (see Chapter 7).

Another option to look at the performance of R&D can be into two broad categories:

1. Process outcomes that look at the effectiveness of the process in terms of teamwork and productivity
2. Product outcomes that investigate how the products performed in terms of serving the firms' strategic initiative, such as value to customers and time to market

The decision for a particular R&D organization is not binary. It can be a mixed setup, where some very fast consumer-driven market segments are served by an IC, while the more established markets are supported by a "classical" R&D department and completely new territories are managed by an OI team

REFERENCES

About Campden BRI. (n.d.). *About Campden BRI - Our history and purpose.* https://www.campdenbri.co.uk/about.php

About-Solina. (n.d.). https://www.solina.com/who-we-are/about-solina/

Author Talks. (2022, November 22). *Author Talks: PepsiCo's Mauro Porcini talks meaningful design and mentorship.* McKinsey & Company. https://www.mckinsey.com/featured-insights/mckinsey-on-books/author-talks-pepsicos-mauro-porcini-talks-meaningful-design-and-mentorship

Brennan, T., Ernst, P., Katz, J., & Roth, E. (2020, November 3). *Building an R&D strategy for modern times.* McKinsey & Company. https://www.mckinsey.com/capabilities/strategy-and-corporate-finance/our-insights/building-an-r-and-d-strategy-for-modern-times

Chemers, M. (2014). *An Integrative Theory of Leadership.* Psychology Press. https://doi.org/10.4324/9781315805726

Chesbrough, H. (2003). The Logic of Open Innovation. *California Management Review, 45*(3), 33–58. https://doi.org/10.1177/000812560304500301

Competence Framework I APM. (n.d.). https://www.apm.org.uk/resources/find-a-resource/competence-framework/

CuliNEX. (2023, November 7). *Food product development services.* https://www.culinex.biz/lp/food-product-development

De Jong, M., Marston, N., & Roth, E. (2015). *The eight essentials of innovation.* McKinsey & Company. https://www.mckinsey.com/capabilities/strategy-and-corporate-finance/our-insights/the-eight-essentials-of-innovation#/

Doering, C. (2017, June 19). 6 of the worst product failures in the food and beverage industry's history. *FoodDive.* https://www.fooddive.com/news/6-of-the-worst-product-failures-in-the-food-and-beverage-industrys-history/444783/

Driva, H., Pawar, K. S., & Menon, U. (2001). Performance Evaluation of New Product Development from a Company Perspective. *Integrated Manufacturing Systems, 12*(5), 368–378. https://doi.org/10.1108/eum0000000005714

DSM Integrated Annual Report 2022. (n.d.). *Our focus in science and innovation - DSM Integrated Annual Report 2022.* https://annualreport.dsm.com/ar2022/report-by-the-managing-board/strategy/our-focus-in-science-and-innovation.html#:~:text=We%20build%20and%20scale%20innovative%20ventures%20within%20DSM,the%20creation%20and%20development%20of%20Hologram%20Sciences%2C%20Inc

Food & Bio Cluster Denmark. (n.d.). https://foodbiocluster.com/

Food Innovation Hubs - Country Hubs. (n.d.). https://www.foodinnovationhubs.org/country-hubs

Hern, A. (2018, April 26). The two-pizza rule and the secret of Amazon's success. *The Guardian*. https://www.theguardian.com/technology/2018/apr/24/the-two-pizza-rule-and-the-secret-of-amazons-success

Hidma, A., Küchler, S., & Sandström, V. (2020, July 27). *The present-focused, future-ready R&D organization*. McKinsey & Company. https://www.mckinsey.com/capabilities/operations/our-insights/the-present-focused-future-ready-rd-organization

House, R. J. (1971). A Path Goal Theory of Leader Effectiveness. *Administrative Science Quarterly, 16*(3), 321. https://doi.org/10.2307/2391905

Huston, L. (2019, February 7). *Connect and develop: Inside Procter & Gamble's new model for innovation*. Harvard BusinessReview.https://hbr.org/2006/03/connect-and-develop-inside-procter-gambles-new-model-for-innovation

Indeed. (n.d.). *Career-development/skills-vs-competencies*. https://www.indeed.com/career-advice/career-development.

Jaruzelski, B., Chwalik, R., & Goehle, B. (2018, October 30). *What the top innovators get right*. Strategy+Business. https://www.strategy-business.com/feature/What-the-Top-Innovators-Get-Right

Jones, C., & Pimdee, P. (2017). Innovative Ideas: Thailand 4.0 and the Fourth Industrial Revolution. *Asian International Journal of Social Sciences (Online), 17*(1), 4–35. https://doi.org/10.29139/aijss.20170101

Kesting, P., Ulhøi, J. P., Song, L. J., & Niu, H. (2016). The Impact of Leadership Styles on Innovation - A Review. *Journal of Innovation Management, 3*(4), 22–41. https://doi.org/10.24840/2183-0606_003.004_0004

Lazzarotti, V., & Manzini, R. (2009). Different Modes Of Open Innovation: A Theoretical Framework And An Empirical Study. *International Journal of Innovation Management, 13*(04), 615–636. https://doi.org/10.1142/s1363919609002443

Leatherhead Food Research. (2018, May 23). *New product development | Leatherhead Food Research*. https://www.leatherheadfood.com/service/new-product-development/

Mahrschid, A. (2022). Brief Review of Food Industry R&D. *Journal of Agriculture & Rural Development, 2*, 25–37.

Meyer, H. A., Koal, K., Wrba, M., & Bachmann, T. (2019). Bye Bye, Two-Pizza Rule? Die Bellschen Zahlen als Komplexitätsmaß für Teamarbeit in der digitalen Produktentwicklung. *Usability Professionals*. https://doi.org/10.18420/muc2019-up-0243

Mission & Vision. (n.d.). Nestlé. https://www.nestle-esar.com/aboutus/missionvision

Nestlé Global. (2021, August 30). *Bringing innovations to life at the Nestlé R+D Accelerator, together with start-ups, studentsandintrapreneurs*.https://www.nestle.com/media/pressreleases/allpressreleases/nestle-rd-accelerator-lausanne-inauguration

Nestlé Global. (n.d.a). *Innovation, science and technology*. https://www.nestle.com/about/research-development

Nestlé Global. (n.d.b). *The Nestlé purpose is to unlock the power of food to enhance quality of life for everyone, today and tomorrow*. https://www.nestle.com/about/how-we-do-business/purpose-values#:~:text=Our%20values%20At%20Nestl%C3%A9%2C%20respect%20has%20a%20special,the%20generations%20who%20will%20follow%20in%20our%20footsteps

NineSigma. (n.d.). *Approaches - NineSigma*. https://www.ninesigma.com/approaches/

Pedro, G. D., Heinrich, H., Peter, V., Pavel, C., & Sergio, G. Y. P. (2018). *The impact of private R&D on the performance of food-processing firms: Evidence from Europe, Japan and North America*. RePEc: Research Papers in Economics. https://doi.org/10.2788/85468

Pisano, G. P. (2012). *Creating an R&D strategy*. Harvard Business Review. https://www.hbs.edu/ris/Publication%20Files/12-095_fb1bdf97-e0ec-4a82-b7c0-42279dd4d00e.pdf

Plc, U. (2023, November 10). *Unilever opens new global foods innovation centre on Wageningen campus*. Unilever. https://hive.unilever.com/news-and-events/unilever-opens-new-global-foods-innovation-centre-on-wageningen-campus/

Sethi, R., Smith, D. C., & Park, C. W. (2001). Cross-Functional Product Development Teams, Creativity, and the Innovativeness of New Consumer Products. *Journal of Marketing Research, 38*(1), 73–85. https://doi.org/10.1509/jmkr.38.1.73.18833

Skillicorn, N. (2019, August 28). *Top 1000 companies that spend the most on Research & Development (charts and analysis)*. Idea to Value. https://www.ideatovalue.com/inno/nickskillicorn/2019/08/top-1000-companies-that-spend-the-most-on-research-development-charts-and-analysis/)

StartUs Insights. (2022, September 7). *Balancing open innovation with IP rights | StartUs insights*. https://www.startus-insights.com/innovators-guide/open-innovation-intellectual-property-rights-ipr/

Statista. (2024, May 22). *Average R&D spending by industry 2022 | Statista*. https://www.statista.com/statistics/270233/percentage-of-global-rundd-spending-by-industry

Tabrizi, B. (2024, April 12). *75% of cross-functional teams are dysfunctional*. Harvard Business Review. https://hbr.org/2015/06/75-of-cross-functional-teams-are-dysfunctional

Vbouronikos. (2023). *Inside the world of innovation centers: A comprehensive guide.* JOIST Innovation Park. https://joistpark.eu/en/what-is-innovation-center/#:~:text=An%20innovation%20center%20is%20a%20physical%20space%20or,to%20exchange%20ideas%2C%20innovate%2C%20and%20develop%20their%20projects

World Economic Forum. (2023, May 17). *7 keys to foster an innovative culture at work.* https://www.weforum.org/agenda/2023/05/ideas-impact-foster-culture-of-innovation/

Writer, S. (2021, January 6). Singapore emerges as food tech hub, thanks to state support. *Nikkei Asia.* https://asia.nikkei.com/Business/Startups/Singapore-emerges-as-food-tech-hub-thanks-to-state-support

Sources of Inspiration, Finding Ideas

5

Understanding where to search for new ideas and new product inspirations traditionally fills many pages in all types of new food innovation books, blogs, articles, videos, etc. Especially around the turn of the year, the news channels are full of top trends in each food segment. For innovators looking for a check list of potential range extension, a good pool of ideas to quickly scan with target consumers is desired. The ideas are public, based on actual market observation and potentially very short lived (hype, not trend). That makes them an option for enterprises being able to quickly launch at low up-front investment with low risk. It is easier to implement in the food service industry (restaurants, etc.) then food ingredient or branded good suppliers.

For those more strategically looking for new growth opportunities, these trends are more input, not the direct solution, a personal "Eureka" moment! They are inputs, a building block of the new idea, that still needs to be created, part of the insights. Input material for a creative process, an idea session or something similar, where normally different people come together and share all those data while looking through different lenses to find the ideas that is fitting their business, is required. Such a process basically has two very different phases. The first one is wide open, trying to find everything that could lead to the next successful innovation or new product. In this stage the creativity of all participants is stimulated, and no limits are set to ensure that no unconventional idea(s) are missed. Moderators in this moment often "caution" participants to not react critically to even the strangest thought mentioned.

The next phase then puts the filters into the process. Emotions, inspirations and dreams are now sent through a filtering system that brings back the concrete reality of the company. All this is explained in greater detail in Chapter 6.

This chapter looks at the foundations of inspiration. One option to structure all those sources is to split between "market pull" and "technology push" (Katz, 1999). Market pull covers the opportunities not only in those markets, where the company already acts in, but insights from adjacent or foreign markets. In food, this is often the dominating stream of input data for an idea generation process. Marketing and market research colleagues are experts for this type of scan and may use external market agencies to sharpen the observations.

The second angle is "technology push". Here the wide field of developments in science and technology serve as initial inputs. Since many technologies in foods are adopted from other sectors, such scans would not be limited to the FMCG or food sectors. It's a much more rational process, and the data selection requires a technical understanding. From the "countless" numbers of possibilities, those must be identified that provide a tangible nucleus for the next new product. The idea should not be too narrow ("the next non-dairy smoothie"), but also not too vague ("new insights in climate change"). And it needs to solve a problem on the consumer or industry side.

In many companies' the technical functions and especially R&D gets the responsibility to bring such ideas as "digestible" inspirations into the cross functional creative process. A science push needs more justification, an idea how this could be linked to the next commercial idea. Just dropping "synbiotics", "high pressure cooking" or "glycaemic index" will not be enough for non-technical people if they cannot see the bridge to the benefit.

To avoid misunderstanding, every new science or technology push must answer a consumer need. In foods, the consumer often cannot see "the technology", but will understand the benefit. Successfully applying new science just offers the big chance to answer a need much better, and hence gain a significant competitive advantage. The frozen pizza uses cold-resistant yeast; encapsulated salt and time/temperature controlled

DOI: 10.1201/9781032724591-5

melted cheese make a tastier product, without making consumers enthusiastic about the smart technologies used (https://www.ift.org/news-and-publications/food-technology-magazine/issues/1999/august/features/marketing-pull-vs-technology-push).

Another angle can be "pure data": Statistics, facts, numbers! They provide reliable and often quantitative data, showing an opportunity as stimulus for creative thinking. Often these data relate to reliable forecast, which makes them more attractive as a thinking direction. "There are 102.000 people from Korea living in New South Wales" is a fact coming from the latest census. "And this group is expected to continue growing fast in the next years" shows a growth opportunity. This may not only be a new market – an opportunity for non-cannibalizing growth – but a reliable, fact-based and time-bound forecast.

On the next few pages, I will keep this split among those three "idea source" segments and review them a bit deeper.

PULL FROM THE MARKET

Online Sources

One of the most popular tools for new food product launches still is Mintel`s "Global New Products Database (GNPD)". It is a service that tracks new product launches in the consumer-packaged goods markets worldwide. The data sources of GNPD are mainly the product labels and packaging of new products launched in over 86 countries globally. With the help of local shoppers, the database collects actual local information. These data are clustered into 270 subcategories, 140 product claims, 200 packaging attributes and over 46,000 ingredients and new products from 86 major markets (https://www.mintel.com/industries/food-and-drink/).

GNPD analysts collect, verify and categorize this information with a standardized methodology. GNPD also uses secondary sources such as trade publications, company websites, press releases and social media to supplement the primary data and provide additional insights, trends, innovations and opportunities for product development. The database is very instrumental to monitor competitors and search for product gaps in the market.

GNPD offers a free 14-day trial for new clients who want to test its features, where they give access to all the data and reports. The core is the online database, with search functions for product subcategories, all types of product claims, various packaging attributes and all the ingredients.

The focus is on new product launches, but like many other market research agencies they also provide trendspotting. The reports cover consumer insights, competitive analysis and suggest future opportunities for different sectors and regions. Some of those reports are available for free and often provide a solid basis for opportunity spotting.

There are many more sources of data available to check on opportunities from the "market demand side". Examples are:

- *Innova Market Insights* (https://www.innovamarketinsights.com). Another global database of new food and beverage product launches but a bit more focussed on health, nutrition and wellness claims. New users can book a demo and use the dashboard, highlighting key insights and predictive analysis.
- *Euromonitor Passport* (Passport, n.d.): This is a global market research database that covers consumer markets, industries, countries and lifestyles. You can access it through your institution's subscription or by purchasing individual reports from the website or request a demo.

These are just a few well-known examples! Most of them aim to provide strategic market research reports and business intelligence consulting.

In a typical food company, the marketing team or related function would bring those market data into the innovation process, and they would as well hold the budget to pay for such service.

Kantar, YouGov, Forrester, Nielsen and Statista in principle provide a similar service. Depending on the data they own and the specific market segment the business is interested in, one or the other may be preferred. Again, the marketing colleagues would know and make that choice.

Food blogs can be another source of new ideas. Here food enthusiasts present their ideas and opinions, they complain about things they miss, weaknesses they spot and present new solutions. Refining the search is often necessary to manage the input wave: General cooking blogs, baking blogs, desserts blogs, healthy recipes blogs, kitchen and lifestyle cooking blogs, home cooking on a budget blogs, cocktails and beverages blogs, Indian food blogs, Asian food blogs, etc.

Some food bloggers are professional chefs or have a professional background in food or nutrition, while others are home cooks who share their personal experiences and recipes. Many blogs describe fresh cooked and prepared ideas, which can be a direct source of inspiration. The important step here is the transformation from the often scratch-made or kitchen-made food into an industrial solution. Just converting an exciting new dish or food item into a safe and stable packaged food format can be a too simple approach. A commercial packed product is different and often has weaknesses versus the fresh one. Another approach would be to identify the pain points the "followers" may have when they try to copy the suggestions.

Balinese cooking and dishes as an idea? New, different and complicated and many home cooks would not have easy access to all the ingredients needed for an authentic spice mixture called, for example, "basa genep": Black peppercorns, nutmeg, nuts, sesame seeds, shallots, ginger, galangal, turmeric, lemongrass, garlic, red chillies, shrimp paste, palm sugar and lime juice. Offering this as a ready-to-use dry mix or paste or sauce would allow people to fill their kitchen with a very exotic aroma. Or offer it as a rice snack pot. Doesn't sound like junk food any longer, but more like a truly exotic extra in-between treat.

The opportunity for R&D is to find those technologies that can be used to make featured items without compromising the DNA of the idea of "one-pot wonders" or bowls as sophisticated one pot dishes are interesting formats. "Healthified classics" are traditional recipes made healthier, or any kit that allows artisanal, cooking or baking at home. Closer to packaged food there is a side discussing which of the sometimes-demonized foods can be included without guilt into a good meal.

Food bloggers also communicate openly about the weaknesses of industrial products. An interested R&D person can take those comments and correct the weaknesses. And these channels can be used to check if a trend or hype is coming to an end. Maybe a team considers launching a new food with the promise of an "extraordinary antioxidant capacity". A quick blog scan will provide insights of how "hot "the topic still is, and a few minutes on the website of critical key experts will give a good idea if the expert support still exists or if some have started to get fed up with all these super stories about the super potential.

Branded interactive websites, also with chat bots, are opportunities to directly interact with brand users.

A close look at social media was always a good idea to understand consumers. Nowadays AI can do the scan and analyse trends, collect data, apply natural language processing (NLP), image recognition, machine learning algorithms and sentiment and predictive analysis (StoryLab.Ai, 2024). A practical example is described later in this chapter under "Digital Voice of the Consumer" (Patel, 2021).

SOCIAL MEDIA, BLOGS AND PODCASTS

The fastest way to detect new trends are the social media accounts, where individuals, brands, restaurants, influencers or bloggers immediately showcase new products and their personal view.

Food pictures are a common phenomenon on the internet, especially on social media platforms such as Instagram, Facebook and Pinterest. With more than 400 million posts tagged #food and 250 million tagged #foodporn on Instagram in social media, users are flooded with visual displays of food (https://theconversation.com/foodporn-people-are-more-attracted-to-social-media-content-showcasing-fatty-foods-160221).

One way to measure the popularity of food pictures is to look at the hashtags that people use to label their posts. Some of the most popular food hashtags are #foodporn, #yummy, #delicious, #healthy and #vegan (Pancer & Philp, n.d.). These hashtags reflect the various aspects of food that people want to highlight. Food hashtags can also indicate the trends and preferences of different groups of consumers, as well as the cultural and geographical diversity of food. Networks like Buzzfeed's Tasty has more than 100 million followers on Facebook and over a billion monthly views.

A good inspiration for new product development are food blogs. These websites present any type of content related to food, cooking, recipes, nutrition or gastronomy. The information provided there is not supposed to be neutral. The content of those sides represents a personal view of a food blogger, sharing personal experiences, opinions, tips or reviews on various aspects of food and cuisine. And that is their strengths; they are subjective and opinionated. That can be very insightful when it covers more hedonic aspects about what looks great, smells gorgeous or tastes good. The sites need to be read more carefully when they start to discuss scientific findings, opinions and facts, like health- and wellness-promoting ingredients and their effects. The statements made here are often far beyond what the regulation in a country, like the health claim regulation, would deem appropriate. When searching here for new health and wellbeing opportunities, a cross check with a regulation is always advisable before presenting new ideas to the business leaders.

Food blogs are extremely popular! The whole thing may have started in 1997 with Chowhound run by Jim Leff and Bob Okumura, first simply a digital bulletin for food enthusiasts to exchange information and opinions. Chowhound was influential in shaping America's culinary culture and introducing many under-the-radar restaurants to a wider audience. Leff was a jazz trombonist and food writer, while Okumura was the tech guy who built the site. They sold Chowhound to CNET Networks in 2006. Shortly after it was bought by Red Ventures, and the site was finally shut down. (https://en.wikipedia.org/wiki/Chowhound).

A somewhat different segment of the food blogger universe discusses packaged and branded food. They may be better taken with "pinch of salt", as they present opinion, not necessarily the best facts. Some of them are sponsored and may therefore not give a true review of the products. Others may have personal biases or preferences that affect their opinions. For good information, it is important to look for food blogs that have a professional tone, a clear disclosure of any affiliations or partnerships and a balanced perspective on the pros and cons of packaged branded food. Here are some examples of food blogs that meet these criteria:

- *The Food Lab*: This blog is run by a food scientist testing and various packaged branded food items, such as instant noodles, frozen pizzas and canned soups. He explains the science behind the ingredients, the cooking methods and the taste and texture of the final products. He also compares different brands and gives recommendations based on his findings (https://www.instagram.com/kenjilopezalt/).
- *Budget Bytes*: As the name suggests this blog is aimed at people who want to save money and time on cooking, but still enjoy delicious and nutritious meals. The blogger uses mostly packaged branded food items, such as beans, rice, pasta, sauces and spices to create simple and easy recipes that are affordable and satisfying. She also provides tips on how to stretch your budget, plan your meals and shop smartly (https://www.budgetbytes.com/).
- *Serious Eats*: This blog is one of the most popular and respected food blogs on the Web. It covers a wide range of topics related to food, including recipes, reviews, guides, techniques and history. The bloggers often use packaged branded food items as ingredients or as subjects of their reviews. They provide detailed and objective information on the quality, flavour, nutrition and value of the products, as well as suggestions on how to use them creatively and effectively (https://www.seriouseats.com/).

Sometimes a difference is made between a food blogger and a food influencer. In terms of content there is no sharp differentiation. An influencer may use more of an interactive tonality, while the blogger often simply shares knowledge and experience.

When looking for a good source of inspiration for a new product development or just an exciting dish to cook, a quality check is advisable. A quick first indicator of the reliability of a food blogger site is professionality in tone and writing, the open disclosure of any affiliations or partnerships and a balanced perspective on the pros and cons of the discussed aspects of that food.

For food R&D students and professionals wanting to set up their own blog, there is in the meantime numerous advice available on the internet, starting from topic choice/segment choosing to technical tips and tricks how to set up and run the website. Most important is probably to be passionate about the topic you want to talk about. And more food experts offering blogs to the public may be a great idea and a bit of a niche not really conquered. Looking at food with a professional background and talking about strengths and weaknesses, opportunities and threats can be valuable advice for consumers trying to navigate the field.

Social media listening tools such as Hootsuite Insights, Brandwatch and Sprout Social allow you to monitor social media platforms for mentions of your brand, products or industry and track changes in consumer sentiment and behaviour.

DIGITAL VOICE OF THE CONSUMER

Consumers in many countries more and more refrain from using a company structured feedback channel, like the typical care lines. Instead, they immediately put their experience on the open channels, the different types of the social media. For the company acting successfully in FMCG business, fast feedback is critical.

Having effective options to quickly and reliably receive feedback from consumers is the life blood for the FMCG industry and a great tool to understand what the consumers are waiting for. That's not new and has been known for many years. But the classical tools to hear the voice of the consumer are ad hoc interventions like interviews, online surveys, live chats, social media, website behaviour, like the famous click rates, recorded call data, online customer reviews, in-person surveys, focus groups, emails and dedicated feedback forms.

All these channels together often provide extremely generous, rich, complex and unstructured data sets. But the collection, structuring analysis and insight extraction out of these big data are very resource and time intensive and therefore slow. People have to read, understand and summarize all those sources of information.

The recent progress in data mining and machine learning enabled a new level of insight creation and product monitoring. The idea is simply to analyse large amounts of digital consumer feedback without any human reading and interpretation. All this feedback is often unstructured textual information. So, the first tool to start the data mining is the topic modelling approach. These are basically machine learning algorithms that detect latent topics running while running through all those unstructured textual documents. The second even more critical task is the validation of the obtained results. The latter can be based on semantic coherence, like the average semantic relatedness between topic words, and exclusivity, the comparison of top word occurrence between different topics are such validation tools (Barraveccia, 2022).

Unilever, a big FMCG company, engaged in this approach a few years back. The motivation was to create a fast response tool helping to centre more business decisions around their loyal consumers. And the concrete task was the building of a scalable digital voice of the consumer capability, able to analyse large, complex and unstructured consumer datasets, as they are typical in social media, product reviews and customer engagement centre to obtain real-time actionable consumer feedback (https://community.dataiku.com/t5/Dataiku-Frontrunner-Awards/Unilever-Developing-a-Scalable-Digital-Voice-of-the-Consumer/ta-p/18686).

One core element in this was machine learning–based natural language models to allow the "unsupervised learning" to identify the key themes the consumers were talking about. Learning algorithms for sentiment classification rounded out the package. The tool was quickly rolled out in over 60 markets and more than 50 own factories across the globe and is available in more than 20 languages. That helped to embed the tool into the day-to-day operations of the quality team and its partner, including R&D. The feedback is now not only available on request, for example, end of the months, but day and out, shortly after they have been made in the media.

The insights created by this approach helped to develop and launch new products based on a better understanding of the target consumer needs. And product design weaknesses could be picked up quickly. For example, when a particular packaging in the fast e-commerce sector was underperforming, the complaints were picked up quickly and corrected. For example, in the product packaging. Lastly, counterfeits became identified much faster and could be traced more successfully. New ideas for more sustainable products were generated based on the data provided.

During the COVID-19 pandemic complaints and product issues could be tracked back much better to the factories and allowed to stay close to the consumers and their behaviour.

Overall, the benefits for the company were top and bottom line: Improved product quality enabling better consumer experiences, generating more sales, but also options for cost savings.

THE COMPETITION

Knowing what the competition is doing is critical for any successful food business. Even the dominating market leader will spend time and effort to keep abreast of movements in the market. Often, it's the task of the local R&D teams to scan the market regularly, inform the business leadership and suggest reactions. In basically all market segments companies are acting with a business model to copy the market- or innovation-leading companies and their products. They watch for new launches, trending products and try to copy product and concept as good as possible in the shortest time. Since they have not spent a lot of resources on own trend analysis, idea creation, product brief development and consumer testing, these companies have lower overhead costs and can sustain their business on a lower net margin, so they offer a product copy soon after the original launch under their own or a retailer brand.

That's called "competitor intelligence"! Regular and structured store checks in all channels where the product category is sold often builds the basis. Shelfs are inspected and the competitor products are tasted and analysed quickly. Experience in the food category and good analytical skills are needed to clarify the details of the product composition, so it can be rebuilt. Understanding the consumer needs is not the priority. That they exist is basically confirmed by the original launch, and the communication strategy towards the target consumers is openly available. Those companies just have to ensure that their "copy" lands on the shelf at a lower price. As food and beverage are low-tech segments, the necessary production technology is freely available, and often the components can be bought from suppliers on the market. Normally such business would not invest into new manufacturing equipment to create the copy; they would normally focus their scans on products for which they have the processes or would have access to.

A food company that plans to copy a product on the market will, after sensory and compositional analyses and functional testing, run a cost benchmark to understand quickly if they are able to make the alternative at an attractive consumer price, while making their own margin.

A variant of "learning from competition" is to convert the original product into a new format. A company launches an "all in one seasoning powder" for chicken into the market. "All in one", so no sophisticated juggling with various amounts of pretty exotic herbs and spices fulfils a consumer need to conventionally prepare delicious new chicken dishes. Competing company B may see an opportunity to upgrade the seasoning into a liquid concentrate, transporting naturalness, authenticity and premiumization, because they have access to the liquid sauce manufacturing. Due to the positioning of the new concept, they can surf on the trend in the market, use the popularity of brand A launch and ask a higher price. Or the other way around: A bottled premium cooking sauce opens new culinary experiences, Korean, Chinese, Mexican, etc., but many consumers cannot afford the price asked to bring the new restaurant feeling home. A competitor may find a simple way to bring the same promise as a dry powder to the market, using a much less expensive technology.

FOOD CONFERENCES, EXHIBITIONS AND TRADE FAIRS

This is possibly the best source of new food product inspiration. The product is real, can be tested and tasted and many of them are in a single location. There is no better opportunity to experience new food products with all the senses. New product and service providers showcase and promote your food products. It allows networking with other food professionals, suppliers, distributors and media representatives and learning from their experiences and insights.

The domestic and international events offer the concentrated opportunity about new introductions and upcoming new product launches. They are another essential knowledge space for people in food development. The bigger events offer endless opportunities to search for new ingredients, visit competitor, and the latest developments in processing equipment and packaging often from all over the world. While online and print sources can deliver national news and facts very fast, these events provide a very different, multisensorial experience. Products can be touched, smelled and tasted and therefore bring a much more holistic experience.

It is impossible to only list the best-known ones, but to mention a few: Gulfood in Dubai, Anuga in Cologne, Thaifex-Anuga in Bangkok, World Food in Istanbul, Alimentaria in Barcelona or Sial in Paris as big events where the global food industry from new rural cooperatives, ingredient supplier, media channels, machine and process supplier and academia go public. Some are general; others focus on aspects or specific product groups. Today there are also fairs where only third-party manufacturers present their capabilities to those companies with good market access and suitable brands. There are sites on the internet where the events are listed, described and commented on (https://10times.com/top100/food-beverage).

Often several hundred companies are presenting their products; it can be overwhelming. A decent scan can take a couple of full days, good physical and mental condition and ideally a match plan to get the most out of it. A day at a fair is an exhausting working day. The good thing is that nobody needs a dinner at the end of the day, And the samples taken or ordered may fill a couple of days of the tasting session in the company kitchen. You may not need any dinner in the night but got samples for a quick internal tasting with colleagues the day after.

Good planning and organization help to filter the nice stuff from good and great. The events normally publish information and catalogues with all exhibitors, the company profile and their new products. That gives visiting companies and individuals the possibility to make a visiting plan. The program can be examined, and a visiting plan set up. If an enterprise can send a few people, it is beneficial to have different functions in the team. The normal suspects, marketing, R&D, procurement and manufacturing, if machine technology is present. Team members can split up broaden the coverage and meet again from time to time to compare notes. Marketing may have found an interesting product and now wants R&D to investigate more technically, procure a new supplier and needs help in the qualification, or technical advice is needed to evaluate feasibility of using material in own products.

Most of these fairs have trade shows, food trends, discussion fora, presentations on special, actual topics or cooking classes, other competitions and short creative session like hackathons. Frequently, the organizers offer the possibility for B2B meetings between interesting companies, a possibility to speak in private or under confidentiality and deepen options for a collaboration. A procurement colleague may search for third-party manufacturers in a different market or with specific manufacturing to produce a new product line. It is good if the technical people from the factory and R&D are around. They can do a first feasibility check to see how well the copacker skills are matching the product design idea. Or a marketing manager found a variety of exciting new ingredients for the range extension of a successful product line. It is always helpful if R&D is around to check the less glamorous parts of the story like shelf life, humidity, microbial limits, allergens, etc.

Visiting trade shows and fairs has a few more benefits. The softer one is the personal impression about where the markets are going and how the food provider responds to it. Sometimes a trend is well known, but in the innovations team is linked to a specific market segment. The fairs offer the opportunity to see how others are picking up the trend. A few years back "salted egg" (using salted duck egg yolks) became popular in warm savoury dishes in Asian countries. During a regional fair in Bangkok some companies started to present snacks, including jellies, cookies, cakes and chocolates tasting surprisingly good. In then took only a short time before ice cream makers realized that combined with other toppings, like dulce de leche or caramel sauce, an irresistible treat could be created. Visting fairs that are so broad open the perspective and are a great stimulus for "out of the box" thinking.

Another gain versus digital experiences is the networking opportunity. Meeting others, new or existing suppliers, representatives from completely different parts of the market or the direct competition often leads to much more insightful conversations than a simple call or video conference. And for many it's one of the few opportunities to build a network outside the country or region.

As a bonus there is the unexpected. Hardly anybody joining such fairs comes back without a couple of surprises and "eye openers". And even if those "accidental" moments do not lead directly to a new product idea, they often change the perspective about what is valuable, how things can be applied, what separates innovation from invention! A good way for fast learning.

TRADE STRUCTURES

A personal visit to the surrounding trade offers an additional perspective. It is "old school" research, but still indispensable. The unique experience here is the product of interest in the competitive environment on the shelf and the situation in which shoppers must find and choose it. One sees what the shopper sees and can get additional insights into how they shop and what creates interest. Especially for packaging developers, it is a source of inspiration to design products that differentiate them on shelf.

The bonus opportunity is to observe the shopping behaviour at the point of sale, where the shopper makes the final buying decision. R&D teams can find out if and when shoppers take time to read claims and study back-of-pack information, information that has been created with often much effort and passion, Does the more glamourous glossy print catch the eye? Or is the more natural matte packaging finishing better in transporting foodiness? Promotional campaigns and the influence on shopping behaviour can be studied. A simple visit to the trade show gives the opportunity to experience that and is something very different than the smartest AI-powered search on the internet.

The trade show itself can be split into a couple of channels and structures, and the products on sale often differ between them. Some of the common trade structures are that are often nearly forgotten in the Western markets are the differences between modern and traditional trade.

Modern trade stands for larger-scale retail operations that are organized and standardized. Typical representatives are supermarkets, hypermarkets, convenience stores and e-commerce platforms. They offer a wide assortment of products in a clean and convenient environment at competitive prices. Modern trade is growing rapidly in many regions, especially in Asia and Eastern Europe.

Traditional trades are small, independent and often family-owned groceries, kiosks, street vendors and open-air markets. Traditional trade is often still the dominant form in many developing markets in suburbs and rural areas. Traditional trade offers proximity, personal service and social interaction for consumers.

Food services or on-premises can be seen as a third type. Examples include restaurants, bars, cafes, hotels and cinemas. Here products are sold for immediate consumption – impulse buying. This is particularly interesting for beverages, snacks, confectionery and ice cream.

Each channel has advantages and disadvantages for the shoppers in terms of reach, cost, efficiency and distinct properties that can serve as idea providers for new product ideas.

The competitive battle happens at the point of sale. And like in evolution, the ones best adapted to the environment will make the best performance. Therefore a few more details about these channels come next.

HYPERMARKET

The biggest ones are hypermarkets! And here the retail store is combined with a department store and a grocery supermarket. The floor space can range from 4000 to 25,000 square metres, and they are often located in suburban areas, near highways or in shopping malls. The idea is "all under one roof". Typical international representatives are Walmart, the largest retailer in the world, with more than 11,000 stores, or Carrefour, the second largest retailer in the world, with more than 12,000 stores, followed by Tesco, with more than 6000 stores and Costco with about 800 stores in 14 countries.

To attract consumers, they often apply advanced technology to enhance the shopping experience and efficiency. For product developers these are good places to get a fast overview about the food product segment, the different types of stock-keeping units (SKUs) offered by a brand, an overview about all the main competitors and their offerings and the relative importance of a product category. A market in France will offer a huge variety of different wines or cheese, but the same chain in Vietnam may use all that space to bring the whole selection of fish sauce offerings, while the cheese tableau remains tiny. The retail prices are competitive, the business model relies on sales volume and margins are lower. Shoppers often plan their visit days before; they use the visit to stock up, but take also more time to stroll through the aisles to check for new offers. Since shoppers in tendency have more time, marketing and product campaigns can take the opportunity to interact directly with sampling and information booths.

New launches from competitors are often promoted here first. Secondary product placement can provide ideas to remind people of product usage or motivate new consumption occasion. Ideas are placing a cooking sauce in the fresh meat segment, juices with the fresh fruit asparagus with hollandaise sauce or strawberries with whipped cream sprayers.

SUPERMARKETS

They are usually smaller, on average 400–2500 square metres and often located in urban areas or near residential neighbourhoods. They may not have as much variety or depth as hypermarkets. The fresh segment may be comparably bigger than in hypermarkets. Supermarkets rely on a higher turnover, so they tend to have lower inventory. Prices can be slightly higher than hypermarkets but may offer more convenience and customer service. A visit can give a first good impression on the top selling brands, products and formats.

A variant are discounters selling at lower prices than the market average. Key features are the comparably low operating costs, so advertising, inventory, staffing and store design are kept to a minimum. Sales volume needs to be higher to compensate for lower profit margins. The product range is more limited, focusing on the most popular or essential items that appeal to a wide customer base. Discounters attract customers by offering them a clear and consistent value proposition, emphasizing the low prices, convenience and quality of the products or services. Aldi and Lidl, originally from Germany are the most successful and typical representatives.

Discounters offer "my too" products, copies of A class market leaders under their own their own brand name, summarized as "retailer own brands" (ROBs). They have both, the A class brand and their ROB version, side by side to convince customer to buy through the price difference.

CONVENIENCE STORES

Here the products that are sold quickly and at moderate costs. A convenience store is a small retail outlet that sells a limited range of FMCG products, usually in convenient locations such as gas stations, airports, train stations, residential areas and office locations. They operate for long hours, often 24/7, to cater to customers' needs at any time of the day or night. The biggest and most prominent chain is 7-Eleven, with more than 70,000 locations globally (All Top Everything, 2024). The small floor space and the simple layout allow customers to find what they need quickly and easily. Customers do not spend much time there. They come on their way to work or while heading home and just want grab food to eat now or soon. A loyal consumer base uses them to shop small amounts frequently, repeatedly as part of their daily routine.

For product developers it's good place to check the performance of smaller pack sizes and frequently bough items and to observe immediate shopping behaviours. People do not come with a shopping list; they come to get their next meal or because they forgot something. In office areas it may give good ideas about the food people take during lunch or snack breaks when they do not want to spend much time. Normally they provide heating opportunities, so that ready meals or snacks can be consumed hot. Some have even started to offer tables and seating, so that shoppers can enjoy their food there instead of on the street or public transport.

OPEN MARKETS

In the highly developed market, the open fresh market allows farmers to sell their fresh products directly from the farm to the consumer. Shoppers come here to buy fresh, unpacked regional natural food, often from bio- or organic farming. Premium prices are the usual, and longer shelf life is not expected. Often middle- to higher-class urban populations take the opportunity to enjoy freshness and vicinity. For product developers in the packed food market, a visit can bring ideas about provenience and reputation of key natural ingredients.

In the developing and emerging markets these marketplaces are often locations for lower-income groups and the D2D shopping of the rural locations. Consumers come frequently, often daily, to buy their daily consumption fresh, since they do not have good storing conditions at home and often only little cash. They are common in many parts of the world: Asia, Latin America and Africa. Where in the Western world these markets are locations for often more affluent shoppers of authentic foodstuff, in D&E these markets represent the original form of buying food. The shoppers do not have the money to regularly buy in clean super- or hypermarkets. Hygiene and freshness may be more questionable, and the shopper has to trust the seller much more. Here a branded FMCG good supplier can offer their prestigious products to the lower-income classes. That often requires different packaging material to sustain the conditions in open tropical climate better. And it requires different packaging sizes, often single portion, to allow cash-constrained shoppers to buy. Product developers can learn about all this and price value equations. Getting insights into the specific environment and its challenges gives ample opportunity to improve the product safety and robustness.

Nowadays the borders are more blurred. Not only do discounters have popular products under their own brands, but they also offer fresh food for immediate consumption in chilled cabinets beside A class brands. And convenience stores open bigger outlets with bigger packs, and supermarket chains get smaller and call themselves "express" and "to go".

The uniqueness of this source is the possibility to check products and competitors at the point of sale and to observe the shoppers and their behaviour while purchasing. One can check if a new claim on a pack is recognized and read, when consumers are open for new ideas and when they follow their own plan, what attracts them more and how the product stands up in the market on the shelf versus the competition.

The company's sales force can bring very valuable insights "fresh from the market" to new product development sessions.

ONLINE FOOD DELIVERY

The global online food delivery market has recently been estimated at USD 221.65 billion in 2022 and an expected compound annual growth rate (CAGR) of 10% until the end of the century (Grand View Research, 2023). And no doubt, the pandemic and the still increasing proliferation of smartphones have driven sales through that channel. About 70% of the sales are coming from so-called platform to consumer channels, and the remaining 30% from the restaurant to consumer.

A few major players in this sector are Uber Eats, Door Dash, Deliveroo and Just Eat. As a source of inspiration for new food product development this channel is less productive. Food delivery brings the offers of the food gastronomy sector; the key challenge is to manage food quality and safety on the way to the customer to avoid damage, contamination, spoilage and allergies. Some restaurant chains have started to use corrugated cardboard box with a thermal lincr and a vented lid to keep their products, like pizza, hot and crispy. Despite the strong growth trajectory, most delivery platforms remain unprofitable. Their cost-intensive, low-margin nature of the business remains a constant challenge (Ahuja et al., 2021). The potential for innovation is more limited. One creative approach was to combine dining and television with "taste your favourite cooking shows at home".

Food e-commerce can be separated from that. While the delivery platforms are basically taking a share from out of home restaurant eating, classic e-commerce is a channel to supply "make at home-cook at home". Here the classical retail product is ordered online and delivered at home. The innovation space is more constraint. The principle is "same, well known product in new channel". But some companies have started to design specific packaging solutions for e-commerce to minimize the risk of damage during the distribution. The advantage of this channel is the possibility to buy specialties and ethnic food products that are not available in the vicinity. Specialty shops have here the ability to create a wider reach for the products. So, for niche markets a scan may create new product ideas. A producer of culinary sauces may aim at further range extension in Europe. Here the search can show things like sweet soy sauce from coconut sugar and soybeans, very popular in Indonesia, but hardly known outside the country.

DIRECT-TO-CONSUMER BRANDS

Some brands have decided to offer their products not via the usual retail system, but only online. They can be a good source of new product inspiration as they often offer unique, innovative or niche products that cater to specific needs or preferences of the customers For practical reasons their products are relatively low weight and often have a longer ambient shelf-life, so that the distribution costs stay in balance with the product value. Typical products are ethnic food, spices and special cereals. Osom, selling starter kits for Asian dishes, and Magic Spoon, selling clean label high-protein cereal, are examples.

FOOD SERVICE, RESTAURANTS

The Food Service industry is often a great source of inspiration for food innovators. The spectrum ranges from full-service gourmet restaurants, fast-food restaurants with sit-down and take-out service, hotels, ethnic restaurants from all over the world to street vendors, all types of cafeterias and mobile to institutionalized care like hospitals elderly homes, military barracks and prisons.

A single innovator or a team searching for new product ideas would probably not start to scan all these options during the hunt for the next successful launch. But once the target segment is better defined, a visit to the leading gastronomy is not only a pleasure and a good team building exercise but also an excellent chance to shape ideas. The producer of a range of ready meals (ambient or chilled or frozen) would arrange visits to interesting ethnic kitchens to try the signature dishes before starting to translate them into industrial food.

The international slow food movement started in the mid-1980s and promoted enjoyable, conscious regional food, basically as a reaction to the spread of fast food and the loss of local culinary traditions. The movement celebrates the diversity of food cultures, promotes sustainable agriculture, local and organic food and celebrates small producers and consumers. It aims to counter the fast-food culture and its negative impacts on health, environment and society. Some food brands that have embraced the slow food idea are:

- *Whole Foods Market*: This US-based supermarket chain specializes in natural and organic foods and supports local farmers and producers.
- *Eataly*: This Italian company operates a network of high-end food markets and restaurants around the world, offering quality products from small-scale artisans and farmers.
- *Patagonia*: This company created a new product branch selling organic and regenerative food products, such as grains, beans, soups, snacks and beverages.

The long-term success of this conscious way of enjoying food made it to the top of the world most famous restaurants: Noma in Copenhagen, several times awarded as the best restaurant in the world, creates innovative and seasonal cuisine, using ingredients sourced locally and seasonally.

Most major new trends in food started in gastronomy. Nouvelle Cuisine in France and Ferrán Adriá in Spain created fundamentally new trends in foods and gastronomy bringing touch and sight, into leading cuisine, and sound has become increasingly prominent in terms of gastronomic pleasure. Other trends were the global roll-out of tapas and the globalization of ethnic and fusion cuisines, like Korean tacos, sushi burritos and Indian pizza. Or the molecular gastronomy that often manipulating the physical and chemical properties of food and creating things like edible foam.

A few years back not too many people outside Korea would be able to name three typical dishes. But nowadays bibimbap, a rice dish mixed with vegetables, meat, egg and gochujang (red chili paste); samgyeopsal, thin pork belly slices grilled on a tabletop stove and eaten with lettuce, perilla leaves, garlic, soybean paste and chili paste and kimchi; or tteokbokki, cylindrical rice cakes cooked in a spicy sauce with fish cakes, eggs, scallions and other ingredients, are known in many countries of the world.

The restaurant kitchen is one of the prime spots where new food product trends can emerge. A chef decides to experiment in his kitchen with new ingredients or unusual combinations and offers the idea to the team of staff. The skilled team uses creativity and passion to create something new and exciting. And they also do the first test market by serving their creations to the restaurant guests. Ingredient suppliers, like flavour houses and the branded packaged good industry, are constantly scouting for such new inspirations which can be converted into food products.

"Hot pot" is a Chinese cooking method using a simmering pot of soup in the middle of a dining table with a variety of raw ingredients. The ingredients are cooked in the soup and then eaten together with various dipping sauces. It's an interactive way of enjoying food with friends and family in a restaurant in many Asian countries and increasingly in the Western world.

Families and friends would love to enjoy the social atmosphere hot pot more often, but that would be expensive. Hence, soup and cooking ingredient makers have created a concentrate, liquid or dry, which enable authentic hot pot meals at home.

DATA MINING IN FOODS

Data mining is described as the process of extracting useful information from large datasets, such as customer preferences, market trends or nutritional values. It is the foundation of AI, where then machine-based systems start to perform tasks that generally require human judgement, including decision making, learning or reasoning.

Data mining in food is the practice of examining large databases to generate new information or correlations that improve all type of aspects in the food industry, ranging from primary sourcing to production, quality, safety and marketing, to customer satisfaction and waste management. Data mining techniques can be applied to analyse data from different sources, such as crop, soil, climate, consumption and sales data, to identify patterns, trends and relationships that can support decision making and problem solving. Some of the data mining techniques include association rules, clustering, classification, regression and outlier detection. A few examples:

- Association rules are used to discover frequent co-occurrences of items in a transactional database. For example, a supermarket may use association rules to find out which products are often bought together by customers, such as milk and diapers or bread and eggs. This helps the supermarket to design better promotions, cross-selling strategies or shelf arrangements. A famous seasonal example from Western European markets is the co-placement of hollandaise sauce in the fresh fruit and vegetable compartment during the asparagus season.
- Clustering is the technique that groups data points based on their similarity or proximity. A fast-food chain may use clustering to categorize its menu items based on their nutritional values, such as calories, fat, protein and carbohydrates. This can help the chain to understand the preferences and needs of its customers and offer them more personalized recommendations. Or a restaurant uses data clustering to plan the menu based on the availability of ingredients, prioritizing ingredients with shorter shelf life or those with a higher profit margin, while ingredients with a longer shelf life can be used for dishes that are less popular or have lower profit margins. This is a way to reduce food waste and increase revenue.
- Classification is a technique that assigns a label to a data point based on its features. For example, a food quality control system may use classification to detect whether a food product is fresh or spoiled based on its colour, texture, smell or other attributes. This can help the system to prevent food waste and ensure food safety.
- Regression is a technique that predicts a continuous value for a data point based on its features. For example, a food delivery service may use regression to estimate the delivery time for an order based on factors such as distance, traffic, weather or order size. Linear regression is often used to estimate the shelf life of a food product based on fundamental physicochemical parameters like water activity, pH and temperature. The dependent variable is the shelf life, and the independent variables are the water activity, pH and temperature. The model can help food manufacturers optimize the storage conditions and packaging of their products to extend their shelf life and reduce food waste. A logistic regression model can be used to predict the

probability of microbial growth in a food product based on its moisture content, salt content and preservatives. The dependent variable is the presence or absence of microbial growth, and the independent variables are the moisture content, salt content and preservatives. The model can help food safety authorities assess the risk of foodborne illness and establish appropriate standards and regulations for food processing and handling.

- Outlier detection is a technique that identifies data points that deviate significantly from the rest of the data. This technique is used in food fraud to spot anomalies in food prices, quantities, origins or ingredients that may indicate fraudulent or illegal activities. If your trusted bank calls you unprompted asking you about a particular buy you've made with your credit card, it probably signals two things: (1) The bank used outlier detection and (2) your card has been cracked! So, this technique can help the system to protect consumers and ensure food integrity.

In agriculture big data analysis has been used for many years, and the impact is impressive. The transformation that happened shifted a formerly low-tech sector into a data-driven business. By using sensors in many parts of the operation, a huge amount of data could be collected and detect surprising correlations. What once was experience and gut feel of the farmer becomes data-based insights (Talend, n.d.). Successful applications are in precision agriculture, using data from sensors, but also satellites and drones to optimize crop yield and quality, for example, by using data on soil moisture and nutrient levels, the weather conditions and pest infestation to adjust agronomical interventions like irrigation, fertilization, pest control and harvesting strategies.

LOOK AT DATA

Here comes the description of the pool of openly available data and statistics forecasting economic development of markets, growth dynamics of different consumer and customer segments and general population statistics. The importance of government politics and regulations also will be placed here. In addition to the description of the sources of such information, the handling and evaluation of the data will be considered here, including different ways of data mining and use of AI.

A good start to search for trends is "Google Trends" (Google Trends, n.d.). This is a tool that allows users to explore and compare the popularity of different search terms, topics and categories over time and across regions. It works by analysing a sample of Google's web searches to determine how often a term, topic, key word or subject is searched relative to the total number of searches. Users can enter one or more terms or topics in the search box, and Google Trends will display a graph showing the changes in interest over time, plus the geographic distribution of the interest. Additional filters are time range, category and type of search (web, image, news, etc.).

The numbers are indexed from 1% to 100%. They give basically percentages, but no absolute numbers, only the percentage of searches for a topic as a proportion of all searches. This is based on a random sample of Google searches. That means as well that a line trending downward is not saying that the total number of searches for that term is decreasing, but only that its popularity compared to other searches is declining.

One big advantage is that these data are extremely fresh – the last hour, day, week, year, etc. A look into the past can provide an idea about a future trend, but not much more. These data do not answer relevant business questions like "how big is the market?" They are more suitable to "build a story" and not so much a "fact pool". A product developer is convinced that the non-dairy ice cream is a major trend and opportunity for the company. He/she can collect the data from other searches and AI tools freely available and then use Google Trends to underline that this idea has momentum.

A newer help is the so-called SEO tool: These "search engine optimizations" are programs that were designed to support website creators to improve their website's visibility and the ranking on search engines such as Google, Bing and Yahoo.

Some of the basic features of SEO tools include keyword research, backlink analysis, rank tracking and site auditing. And that helps to identify popular keywords and phrases to target, analyse their website's backlink profile and track changes in their website's search engine ranking over time.

There are different SEO tools available, ranging from free tools with basic features to paid tools with more advanced capabilities. Popular SEO tools are Ahrefs, SEMrush, Moz and Majestic. Some good initial trend research can be done with tools like Ahrefs Keywords Explorer: This tool allows you to see how often specific keywords and phrases have been searched for on various search engines, including Google, Bing, YouTube and Amazon (Ahrefs, n.d.).

As mentioned earlier in this chapter the global market research firms such as Mintel, Nielsen, Euromonitor, Ipsos GfK and others build their insights on data. They provide tailor-made research aiming to solve a client's specific needs and can perform basically all types of customized market research. Any market segment of interest for a company will be analysed in depth to find insights that provide a growth opportunity. Depending on the specific needs, their dedicated research supports strategy development, the consumer type in basically all countries of the world, Opportunities for mergers and acquisitions, market opportunities, risks and treats, ingredient trends and short- and long-term forecasts in the industry, local and regional trends, product consumption, consumer habits and preferences. One can order details about the market size and value, the expected growths, key players, etc. Instead of focussing on food product types or segments, all the information can be collected for ingredients and their markets. The cost of the specific research support by agency depends on the scope, complexity and duration of the project; the level of customization; and expertise required. Commercial offers distinguish between syndicated research, custom research and consulting. Syndicated research is often the inexpensive one, as it's based on in-house data and reports and costs between $1000 and $20,000 per report. Custom research is tailored to the specific needs and objectives of the client and can cost up to $100,000. Consulting as the most in-depth service can reach $500,000 and more.

Beside the tailor-made reports many country and product segments reports are available. These reports are actual and very informative, they are not free of charge. But students can apply for a discount.

On their websites one can often find very insightful reports and briefings for free. They are very helpful to search consumer trends and catchy trend phrases. In terms of product and trend research, the example that follows from Euromonitor can illustrate that. The freely available trend reports can provide interesting insights into new opportunities. Typical reports would provide for example information about specific product categories, say dairy drinks with a specific proposition, maybe her probiotic and show graphics with the total number oof SKUs per product category per region with that proposition. Such report not only names market sizes and values but gives an idea how much a health proposition has been used in the particular market.

An alternative is Mintel, another leading market research company that provides insights and analysis on the food and drink industry (Mintel, n.d.a). Mintel offers different types of research support for food, such as consumer reports, market intelligence, product innovation and trends (Mintel, n.d.b).

Mintel's new product database in particular is a valuable up-to-date source of information on new products launched in a variety of markets around the world. It allows registered users to access detailed product information, such as ingredients, claims, packaging, pricing and distribution. The functionality allows to search by F&B category, market, launch date, brand, company and others. A helpful function for searchers is the possibility to compare products across markets and categories. One can select up to four products and view a side-by-side comparison of their attributes, such as ingredients, claims, packaging, pricing and distribution. Product comparisons can be filtered by attributes of interest. Figure 5.1 gives an example of a filled GNPD search mask. The user can filter product category, region, market, format ingredients, nutrient content, flavour type, claims made and other features to get an overview of the products on the market.

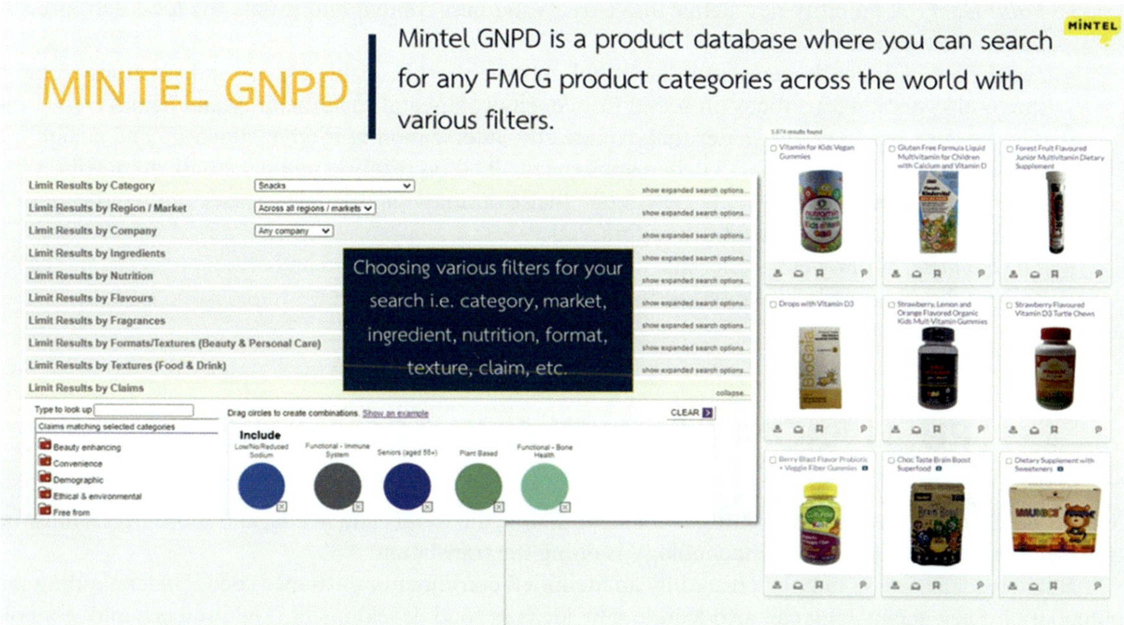

FIGURE 5.1 Mintel GNDP. (From https://www.thailibrary.in.th/wp-content/uploads/2023/07/20230717-GNPD-PR-3.jpg.)

Mintel's new product database is not free of charge, but it offers a range of subscription options to suit different needs and budgets. Users can choose from monthly, quarterly or annual plans, depending on the number of categories and markets they want to access. Users can request a free trial or a demo to explore the features and benefits of the database before purchasing.

Other great options are online platforms aggregating information on new products such as Innova Market Insights, or Food Navigator.

Beside those there are other possibilities to get updated about new product developments on the food markets. An easy option is to subscribe to trade magazines or newsletters that cover the food industry and report on the latest trends and innovations. These newsletters provide daily or weekly updates on the latest news, analysis and insights on the food sector. These platforms offer insights into consumer preferences, market opportunities and competitive analysis. By using these sources, interested people can stay updated on the new product launches on packaged food and discover new flavours, formats and benefits. Following are a few:

- *Food Dive*: A daily newsletter that covers the most important news and analysis on the food and beverage sector, from new product launches to mergers and acquisitions.
- *Food Navigator*: A weekly newsletter that delivers breaking news and insights on the global food industry, with a focus on health, nutrition, sustainability and regulation.
- *Food Business News*: A biweekly newsletter that provides in-depth coverage of the food and beverage market, including consumer trends, new product development, ingredient innovation and more.
- *Food Tech Connect*: A weekly newsletter focussing on food technology; presenting the latest developments and opportunities in food technology, ranging from e-commerce and delivery to biotechnology and alternative proteins.
- *Food Innovation Group*: A monthly newsletter with exclusive interviews and insights from leading food innovators, entrepreneurs and investors.

- *Food Hack*: A monthly newsletter that curates the best content and events for food entrepreneurs, startups and innovators.
- *Food Tank*: A monthly newsletter that highlights the most inspiring stories and solutions from the food system, with a focus on social justice, environmental sustainability and health.
- *The Spoon*: A weekly newsletter that explores the intersection of food, technology and design, covering topics such as smart kitchen appliances, 3D food printing and personalized nutrition.
- *New Food Economy*: A weekly newsletter that examines the complex issues and challenges facing the food system, from farm to fork.
- *The Counter*: A daily newsletter that offers investigative journalism and critical analysis on the politics, business and culture of food.

PUSH FROM SCIENCE AND TECHNOLOGY

Market pulls are all the insights coming from the markets, the trends and needs, and those needs must be translated into a new product. The technology is doing the translation.

But new ideas cannot only be created by an identified consumer or customer need. Understanding the potential of a new technology can also lead to new ideas in food development. Then science and technology are pushing. Testing the potential of those technical ideas with consumers is difficult. They have no experience with such a product and cannot really respond to the attractiveness or tell their buying intention. New technologies for the microencapsulation of vitamins or the creation of hollow salt crystals are not a consumer need. But using these technologies to make product which is much more stable and bioavailable vitamins or product with same perceived saltiness but much less sodium answers a big need. Then technology brings not only a new product idea, but a radial new platform, a breakthrough or even a game changer. Such innovations may need longer to be a success, as they require a behavioural change. Sugar-free drinks, alcohol-free beer and plant-based meat are a few examples where a new technology answers unmet needs.

A few more radical examples from other industries:

- *The personal computer*: The first PCs were invented by hobby engineers who wanted to explore the potential of computing, without any clear market or customer base.
- *The microwave oven*: Originally a by-product of radar research during World War II, consumers must learn the new way of heating inside-out.
- *The internet*: Originally a network of computers for military and academic purposes. Nowadays used by basically everyone.
- The smartphone, the amalgamation of various technologies, such as cellular networks, touchscreens, cameras and GPS into a single device. People first had to accept that they can make calls everywhere, every time and make it a handheld computer.

In food development, such striking examples are a bit harder to find, as the sector is generally perceived low tech. But they exist: An example is instant noodles, precooked and dried with flavouring powder and/or seasoning oil. First invented in 1958 by Momofuku Ando, a Taiwanese Japanese entrepreneur, who experimented with different methods drying and discovered that flash-frying cooked noodles in oil would create a dried noodle block that could be rehydrated quickly and flavoured. This idea did not exist in the market before the launch and led to the foundation of a famous company and brand called Nissin and nowadays a US$50 billion market, well known outside Japan and Asia.

The sous vide cooker never received the value in people's kitchens at home that it found in the food service industry globally. The cooker allows precise and consistent temperature using a constantly heated water bath, where the food is evenly cooked in a sealed plastic bag or a glass jar while submerged. This is a gently method with full retention of moisture, flavour and nutrients.

The success of technology in food is often coming when the new enabling technology is making a known product or process better, rather than a new product format.

But a new breakthrough technology is not automatically an opportunity that cannot be missed. It's the consumer who decides what will be a success. They must adopt the new, unrequested solution.

Genetically modified crops are an example coming to mind. Technically, various solutions have already been developed and tested, and some of them hold big promises about better yield, superior quality, impressive environmental stress resistance, improved nutrient profiles and more. Nonetheless, in many markets consumers robustly reject food or animal feed based on this technology due to basic concerns or fundamental mistrust.

A more recent example are edible insects: Technically a sustainable and nutritious alternative to conventional animal protein sources, but right now largely without any cultural or social acceptance or preference from consumers.

Bringing "technology push" ideas in a NFPD workshop requires an understanding of the technologies, possibilities and limitations and the translation into a product or service the consumer/customer wants. That's the job of the technical people, like R&D. If they hear about new science and technology, they can link that to existing offerings and explain how the new technology will make the product better in the eyes of the consume/customer. Many technologies successful in foods nowadays are adoptions from other sectors. An R&D person suggesting launching susceptor plates would not talk about the phenomenal heat transfer in microwave ovens and the enormous potential of the plate to absorb electromagnetic energy, but rather the option to make for the first time a crispy pizza in 2 minutes in a microwave. Or talk about a way to make restaurant-quality "ratatouille", the classical French dish of vegetables cooked in a rich tomato sauce. Prepared under vacuum ("sous vide"), the chef can keep it for long time without losing, taste, colour, nutrients, texture or aroma.

In other cases, the "technology push" remains hidden for the markets, and hence requires no behavioural change. Technology radically cuts costs in the production of a popular food product which allows the manufacturer to sell at a lower price. That doesn't require complicated explanation for the consumer.

TECHNOLOGY SCOUTING

For a long time technology scouting has become a much-debated tool to drive innovation and NPD in FMCG and foods. The big global players in the market started to use this years ago. Unilever, the big FMCG, collaborates with startups through the Unilever Foundry program to access new technologies and solutions in areas such as sustainability, health and wellness and social impact (Innovation Through Collaboration, n.d.). Nestlé partners with universities and research institutes to develop new products and processes based on scientific research and consumer insights. Procter & Gamble uses online platforms such as NineSigma and Incentive to crowdsource ideas and solutions from a global network of experts and innovators (NineSigma, n.d.).

For the sector, technology scouting is especially important as it enables them to respond to changing consumer preferences, market trends and environmental challenges. Technology scouting can involve various sources, such as startups, universities, research institutes, competitors or online platforms. Typical solutions identified so far include:

- Better and faster products and packaging through biodegradable materials, 3D printing or smart labels.
- More customized customer experience by using digital technologies, like artificial intelligence, big data and analytics.
- Expanding markets and distribution by leveraging e-commerce platforms, omnichannel sales or direct-to-consumer models.

Bagchi (2021)

It is important to note here that scouting in foods is right now not primarily used to identify new trends in the market but more to find a technical or scientific solution for an already identified need or problem. So, there is a need, but not the enabling technology. Typical searches are ongoing for sustainability, compostable packaging, better gut health actives or truly natural blue colorants. An opportunity or a threat for the company is waiting for a feasible solution that is not in sight yet. Having said this, some of these platforms and companies also promise innovation scouting (StartUs Insights, 2024).

The team connecting with such scouts are often technical functions, including R&D. The choice of solutions provided by a scouting firm need to be evaluated, checked for suitability for the company and an implementation plan needs to be made. The "docker" in the company reaches out to scouts, explaining the strategic objectives and innovation needs. Once potential solution providers are identified, they establish a relation that allows the potential partners to share all relevant information, including confidential details. Once this has ended successfully, the negotiation about the technology acquisition agreements starts and can lead to licensing, joint development or even acquisition. With the understanding of the solution, the technical function in a company starts the integration and exploitation.

The key approach in technology scouting is "TRIS", the acronym for "technology readiness information system". Here a system is segregated into small steps to identify the cause and the manifestation of a problem to find in that way the angle for the solution. It starts with the definition of the challenge for which sometimes all parts and items are listed and captured with their specific function. With the help of numerous databases, a creative person with good functional and technical skills and good customer focus would start to find new solutions for a technical problem given to him. More and more AI is taking centre stage in this process. AI can scan a large and diverse landscape of potential sources, filtering out irrelevant or outdated information and assessing the feasibility and value of the remaining technologies. The use of natural language processing (NLP) to analyse textual data from various sources, such as patents, publications, news articles, social media, etc., is often used to extract relevant keywords, concepts and trends. Machine learning (ML) afterwards can classify and rank the technologies based on their relevance, novelty, maturity, impact, etc. And at the end AI is used to visualize data and recommendation systems to present the results in a suitable and convincing form.

In a sense technology scouting could be seen as a variant of open innovation, since the search for an innovation is given to someone outside the company. But here the solution provider has normally no commercial interest in the result. They paid for the scouting they did. And the solution is a piece of science and technology, not a new solution for the market.

This scouting can therefore also be a second step in an innovation process. A new market or consumer need has suddenly been identified, but an adequate technical enabler is not available. By describing the boundaries of potential technologies, the characteristics of a solution can get sharper and thus guide the search.

PATENT DATABASES

Patent databases are valuable sources of information for new food product development. They help technically trained researchers to identify existing or emerging technologies, trends, gaps and opportunities in the food industry. Additionally, patent databases provide insights into the competitive landscape, the market potential and the legal aspects of food innovation. Often the research or development team in a company uses this source.

To get most out of a search, one needs to formulate a clear and specific research question, select the appropriate database and search strategy and analyse and evaluate the results, after which the findings can be applied for new product development. A great start is the Scopus patent database covering more than 23 million entries from the United States Patent and Trademark Office (UPSTO), European Patent Office (EPO), Japan Patent Office (JPO) and World Intellectual Property Organization (WIPO). Some examples of patent databases that can be useful for food innovation are Espacenet, Patentscope, Google Patents and Derwent Innovation (Google Patents, n.d.). The patent database is national or international, For the United States it's the USPTO (Search our Trademark Database, 2024). For international searches the IPC offers a hierarchical classification system used primarily to classify and search patent documents. The global patent platform WIPO offers more than 113 million patent documents including 4.7 million published international patent applications (PCT).

In the USPTO database an interested searcher for food patents would start in Class 426, which is specifically for food or edible material, the processes, compositions and products. This class already contains more than 50,000 patents under Class 426 in the USPTO database. On all these sites search tools are available to optimize the search. The claims are attached. Food patents of course must meet the same requirements as any other patent: Novelty, non-obviousness, utility and disclosure. A good demonstration how to do that and how to use the results are published by Otero et al. (2022).

Patent databases can help researchers to learn from the best practices, avoid duplication or benchmark their own innovations. A patent search will quickly give an idea of which solutions are available or when they found something interesting in their own experiments and see potential to protect that insight themselves.

Most patents in food are narrow in their scope, and the inventive step is not too high. Often, they claim a particular formulation with an ingredient range and an unusual result. But from time-to-time breakthrough food patents are granted:

- The microwave popcorn bag, patented by General Mills in 1981, was a breakthrough for the snack industry which created a completely new market segment.
- Aseptic packaging system, patented by Tetra Pak in 1961, enabling long-term storage and distribution of liquid foods without refrigeration or preservatives.
- Artificial sweetener aspartame, patented by G.D. Searle in 1970, which offered a low-calorie alternative to sugar and became widely used in beverages, desserts and pharmaceuticals.
- The probiotic yogurt, patented by Danon in 1987, which introduced the concept of functional foods and claimed to improve digestive health and immunity.
- The plant-based meat, patented by Impossible Foods in 2017, which mimicked the taste, texture and appearance of animal meat and appealed to the growing vegan and vegetarian market.

When looking for new innovations, the search in patent databases is a tedious undertaking. Inspiring for lateral thinkers, but often not an efficient idea search tool. Searching for "alternative protein sources" will result in hundreds of matches, often describing alternative sources, isolation and extraction and product applications – material not easily digestible in innovation workshops.

But patent searches are useful to get a good view about research intensity and which areas are "hot". And it can shape one's own research. Maybe an R&D team has done experiments and found new ways to reduce the sodium content without lowering the saltiness perception; they would scan patent databases to understand if they perhaps have a new invention at hand, or they get new inspiration to change their experiments based on what is already claimed and described. That can create new options for their own IPR or avoid infringing on others' intellectual property.

SCIENCE AND TECHNOLOG CONFERENCES

They are a bit the "high mess" for people from company R&D. Food science conferences are events where researchers, academics, industry experts and students can share their latest findings, insights, and innovations in the field of food science and technology. The number of events are enormous, ranging from very broad global topics to extremely specific areas of national interest. A good first overview can be found under the conference index (Conference Index, n.d.).

Like fairs and food exhibitions they are good opportunities for networking, collaboration and learning from each other. As a source of ideas for NPD in foods there is a bit of a mixed view: Some see them as valuable, as they can inspire new ideas, foster interdisciplinary approaches and present cutting-edge technologies, with the opportunity to directly interact with the "inventor". Others argue that food science conferences do not inspire creativity, as they often push a wider area in many talks, posters and discussions in a very limited time. Frequently people from academia, NGOs and governmental agencies present their work and scientific findings. For many of them it's a great networking event, they can meet F2F friends, collaborators and others active in their research field, creating new interactions and collaboration opportunities. For the delegates from food enterprises an option to search for new technical leads and future breakthroughs, without the need to deploy technology scouts, is available. Sometimes findings are presented that have not even been published elsewhere. And it's a good opportunity to understand research areas of interest. Representatives from food industry will as well be active and represent part of the work they are doing. These are not so much their most promising findings, but qualified contributions! They would first ensure that their new findings have been thoroughly scanned for IPR opportunities and filed before they can inform the public about it. Going public too early destroys the novelty and therefore the chance to get a legal claim exclusively for a couple of years. But their presentations as posters or talks provide a good opportunity to learn what the competition is working on for the longer run. And their presentations can contain information about an upcoming launch, since the conference is an opportunity to test the reaction of the science community.

Participation from the food industry in such conferences signals interest in the science space, the search for an open cross sector dialogue, a pool for new partners and talents and of course a sponsoring opportunity. The times where industry representatives got critical looks from academia and rather pejorative comments for the scientific contributions are largely over. Today many conferences see themselves as platforms and connectors between the sectors. Academia and independent enterprises are more open to cross sectoral cooperation and research fundings, and the industry partners must stay abreast with the latest science to shape their products and services. Many of them organize their programs in such a way that the path from scientific finding to product or service reaching the market becomes visible. Round table discussions allow speakers and audiences to discuss deeper. The increasing appearance of food startups somehow serves as well as a bridge. They often start with a scientific idea they had as the nucleus of a commercial plan, a thinking that links both sectors. And a standard element in all conference are the social events, ranging from breakfast, lunch, dinner and coffee breaks to various types of excursions offered. This offers ample opportunity to make new contacts, strengthen networks and address individuals. Needless to say, the conference locations are often top holiday destinations. A city like Bangkok is hosting more than 1000 conferences every year (Conferences in Bangkok, n.d.). This is a great opportunity to combine what´s useful with the enjoyable.

With health and wellness being the top trends in coming years, it would be negligent for branded food makers to not join and understand the current scientific debate on topics they are interested in. It's an opportunity to "test the temperature of the water", the quality of a planned claim, the robustness of the communication ahead of a new launch. Some of these scientific experts may be willing to become advisers for the company. As "opinion leaders" they influence the scientific discussion and the public. Experts can stay neutral on a new commercial launch, see it as a positive initiative or reject a product and its claim. Many of them are experts with a high public reputation, so what and how they contribute can have massive influence on the commercial success of a new product.

FOOD SCIENCE JOURNALS

There is a wide selection of print and online journals publishing the scientific and technical advantages in the academic fields of food science and technology. As an interdisciplinary field these journals cover scientific and engineering principles for production, processing, preservation, formulation, nutrition, packaging and distribution of food. Some have a very broad span; others focus on a specific topic. With the endless list of journals, it is beneficial to have some criteria at hand as to which ones are more promising to read than others, Public ranking is such a tool. Some use the "Journal Impact factor" (JIF), where the number of citations the journal received in a year is divided by the sum of published articles in the 2 years before that year (Wikipedia Contributors, 2024b). The factor provides a good orientation about the quality of the published articles, as perceived by the scientific community (that's why they cite), but not necessarily about the scientific novelty and also not about the innovativeness. It's the oldest ranking factor for journals and has received criticism in the recent past, due to limitations in research field, the 2-year window, intransparency in the way its calculated and which databases are used and an imbalance in the article types. A journal like *Nature Sustainability* may have received the highest factor because it centres around a "hot" topic, where many articles from very different research fields want to make a reference to the topic. And if the journal frequently publishes "reviews", it increases the chance to cited by others, as it itself is a summary of many research papers. Newer factors like "h-index", "SJR" and "SNIP" try to smooth out the bias (Mingers & Yang, 2017).

The Scimago score (https://www.scimagojr.com) for food science journals in 2024 was also putting *Nature Sustainability* into the pool position (Journal Rankings on Food Science, n.d.). But luckily in all those ranking the same bunch of journals appear somewhere in the top scores. The ranking as such is disputed science in itself and does not simply provide the best sources of information for an academic field chosen. A specialized journal like *Lifestyle Genomics* or a journal from the Faculty of Fisheries and Marine University in Indonesia will attract fewer readers than broad and international ones like *Annual Review of Food Science and Technology* (Annual Review of Food Science and Technology, 2017), *Trends in Food Science and Technology* (Trends in Food Science & Technology, n.d.) or the *Journal of Food Engineering* (Journal of Food, n.d.), to mention just a few. Sometimes a specialized journal can also become pushed up in the ranking, as the topic or a few leading articles are providing a popular view. Many food articles nowadays start with a reference to "sustainability"; no wonder, that good articles in that journal are used in many article introductions.

ARTIFICIAL INTELLIGENCE AND BIG DATA

I put AI and big data under the technology push stream, as the overall impact on new food products depends on the effectiveness of these technologies. AI is of course very capable of providing new insights into the market, but it is the technology that enables something new. And like in many other industrial sectors where innovations are needed, AI and big data come with the promise of being a game changer, creating faster and more successful ideas for new products.

AI and Big data are sometimes seen as two aspects of the same thing. For me, they are not! And to evaluate the potential of both, the difference is helpful.

Big data, as the name says refers to large – and maybe complex – sets of data coming from often very different sources. These data are often characterized by the three big V's: Volume, velocity and variety, characterizing the amount of data gathered, the speed in which the data are coming in and the form/formats/context in which they are accessible.

And when these data are so enormous, incompatible and differently structured (or not structured at all), machine intelligence has been used for many years to find new relations, schemes and patterns beyond the obvious. Such a search for hidden correlations is a known practice in innovation or food product development. That was seen as an "insight" created by an individual or during a discussion, even if the computer helped with the correlation of data. The reason for the insight was clear for the human creators, and they could get back and test the insight with different input data, before taking it as "valid".

Somewhere around here, digital systems got the label "intelligent". A new design concept called multilayer perceptrons (MLPs), allowed machines to "learn" from a vast amount of data, hence making them "intelligent". The MLP was introduced as a fully connected feedforward artificial neural network. Nodes became organized in a layered structure, where the first layer of nodes receives the input values of a training dataset (e.g. an image for an image classification problem), weights them and sends the result of the calculation to each node in the next layer and the next until the last layer is reached. That output layer then consists of numerical values, which represent the probability of the prediction (e.g. the probability of an object class in case of image classification). This artificial neural network requires a vast amount of data to be able to learn the underlying patterns in the data and then to solve the given problem (e.g. classifying an image that it was not trained on with the correct class). During this training process, the weights between the nodes are optimized to predict the right output.

Interestingly, the organizational similarity between how the human brain works and this latest technology breakthrough is striking (Hawkings, 2021). Similar to modern biotechnology, smart scientists invented a new construction that evolution had created millions of years back. So, with the new simplified brain-like structure and good training, digital got intelligent!

One point I want make here deals with the limitations of AI. The input-output or "black-box" nature of AI models doesn't allow humans to directly understand the inner process. So, the result may not be trusted. Even after long and exhaustive system training, the result can be so surprising that concerns appear around the limitations of the model or unrealistic underlying assumptions.

Great ideas can also emerge, by suddenly realizing something: One of the great innovations in bakery shops was to blow the air from the bakery into the sales room. The nice smell immediately caused memories of baking at home, Christmas or birthdays, because this sense triggers memories and emotions so fast. I am not sure an AI search today would come up with that solution, even when fed with all the latest details about the human senses signalling pathways. I guess, training AI on emotions and unconscious input may remain a limitation for a while.

For humans it is still difficult to fully trust the outcome that was produced in that "black box". If the AI predictions contradict a person's beliefs and experiences without providing comprehensible explanation and how the result was obtained, it is difficult to "trust the process".

In a sense it is less the objective "quality" of the AI output, but the human ability to trust the quality of the result which limits the potential of innovations suggested by AI.

Leading AI designers have recognized the problem with the black box and started to work on what they call "explainable AI". Here they use methods and techniques in such a way that the results of the AI suggested solution can be understood and conceived by human users (IBM, n.d.). But the extent to which the AI output is trusted and followed depends as well on the area where it should be applied.

In general, the adoption of a "black box" outcome in foods seems higher if the task is "technical" and does not need to predict human behaviour. One example: Once an AI system with computer vision is sufficiently trained with lots of labelled data to detect the defects in the finished food product during the production process, it makes sense for the factory to install a corresponding mechanical "pushout system" to increase line efficiency. And after a long enough successful test period, the "human supervision" of the system will be taken away, as the result can be trusted. QA will continue to take samples and check the products pushed out and may find after a while that the system identifies defective forms that were not given in the training, while there no "out of specification products" behind the pusher.

In the technical fields of food making AI has already found widespread application reducing human error, increasing safety standards, automating tasks and improving product quality. AI helps to reduce contamination in food production, leading to a safer product. Further upstream, AI can improve crop yield, accuracy in food processing, end-to-end reducing costs and minimizing waste and raising food security, while minimizing environmental impact.

A recent review on the question suggested six basic clusters where big data and AI are successfully used in the food chain: ". . . the extraction of valuable knowledge and insights, decision-making support, improvement of food chain efficiencies, reliable forecasting, waste minimization, and food safety. . . " (Rejeb et al., 2022).

More complex is the incorporation in the early phases of food innovation, like idea sessions. To predict the market success of a new food product, the "human response" is critical, as in the end enough consumers must be willing to pay the price. But human behaviour is lees uniform than a line filling process and even less predictable than the weather next week.

Solid AI applications today are less ambitious. For example, AI-based cooking tools! Once designed well and fed correctly, these applications have a few advantages on their side that may drive wider adoption. First, they answer a need! Cooking at home for your family and friends is a rising trend! That is less than old daily routine cooking, but more the "show off" when there is time and the wish to make something special or extraordinary; surprising or impressing people visiting and bringing lot of recognition. It proves much more than basic cooking skills, the ability to make something unique and with own personal touch. And that from time to time makes a good learning curve once you do it more often. The AI tools allow you to select demanding recipes, add your own ingredients and twists and even help prepare it when connected to a "smart kitchen machine", like a sous vide cooker, long-range wireless meat thermometer or smart blenders.

I will briefly explain two application areas for AI in foods, one at the very beginning of a innovation process, the cooking, and a second application covering the "downstream" supply chain.

CASE STUDY: "CHEF GPT"

"Chef GPT" is an AI-powered personal chef that can be used in very different circumstances, supporting various aspects of cooking, ranging from recipe finding, to whole plans, pairing drinks and more. It is used by professional chefs, dietitians, fitness enthusiasts or anyone wanting to improve culinary skills. The key features are:

- Pantry Chef suggests new recipes from the ingredients already in your pantry, reducing waste and saving money.
- MasterChef searches recipes and modifies them based on personal dietary requirements or personal likings to personalize cooking.
- Macros Chef suggests customized recipes, similar to what is noted earlier, with focus on individual macronutrient wants.
- Meal Plan generates personalized longer-term meals to support a balanced diet.
- Pair Perfect suggests pairing drinks for the meal.
- Mixology Maestro suggests cocktails recipes!

The basic version is free and allows the creation of a couple of recipe suggestions. After registration the user provides information about the food in storage, the kitchen equipment, their own skills and the meal context. The pro version for a monthly fee provides more functionality and unlimited use.

AI is more than an intelligent database. It contains, for example, a flavour algorithm, fed by large databases of recipes from all over the world and data about flavour compounds in a huge number of ingredients.

The forerunner in this in "Chef Watson, an AI powered recipe tool developed by IBM and first released about 10 years ago" (Bilow, 2014). The recipe selection started on the molecular level, but includes training concepts like culinary traditions.

The key element in that network was the "food pairing" concept, introduced in 2002 by the Michelin-starred chef Heston Blumenthal: It anchors in the assumption that the more aromatic compounds different foods have in common, the better they will fit together and taste. In the famous "molecular kitchen", this scientific, maybe nearly clinical, approach is used to create new surprising sensations. A little later, scientists mapped the common use of very different ingredients into a complex map, which is the base that Chef Watson was using (The Flavor Connection, 2015).

Quite often the new and surprising combinations worked, often on the flavour level. Suggestions like "peanut butter and bacon", "mango and chili" or "chocolate and chips" are not combinations the classic cookbook would come up with (https://www.scientificamerican.com/article/flavor-connection-taste-map-interactive/).

And for some it was a kind of relief that this AI chef could not precisely predict what an individual likes just by simply scanning databases and using basic algorithms and models

CASE STUDY: NEW SUPPLY CHAIN ACTS

Various countries have already installed supply chain acts aiming to protect human rights and the environment in the global food production and trade. The law regulates the responsibility of enterprises to respect human rights throughout the global supply chains. This includes the protection against child labour, the right to fair wages, the right to form unions, forced labour, discrimination, protection against land grabbing, occupational health and safety and protection against environmental violations or various standards for environmental protection.

That is already a complex task for medium-size companies and for a bigger global food company this can easily mean investigating 10,000 different suppliers, many of them having their own suppliers, all over the globe.

It is easy to imagine the "paperwork" sending all those initial complex questionnaires in all these different areas with all attachments, explanations and documentation to all of them, especially in English, French, Spanish or Mandarin or any local language. The backcombing answers then need to be translated, evaluated, compared, documented tracked and combined with improvement steps, action plans, timings, monitoring systems, etc. Later complex audits are needed or unannounced visits to confirm full compliance with the agreed standards. This is a huge and incredible complex task for procurement, that is not adding much to the business performance but costs lots of resources.

This is where new AI systems are coming. In. These are not "smart" in a human sense or surprisingly inventive, but are very capable of handling immense data sets and their relations. One origin of AI systems are the well-known language tools. With those the system can scan through all publicly available data and "listen" to all open chats in social media. Imagine workers somewhere in South Asia complaining on local social media about the blockade of a workers union or an unsafe handling with dangerous material. Someone will post a comment and add a text, picture or video. AI will recognize that and as it basically understands every language is able to interpret the situation. So, with such tools companies can much easier control that their suppliers comply with all regulations. The result are fewer audits, physical inspection and much more focussed ones – a huge increase in efficiency. And companies that go a step further in terms of transparency and corporate social responsibility (CSR) can share these date with credible NGOs to help them implement better standards or demask incompliant players.

BIG DATA IN NEW FOOD PRODUCT DEVELOPMENT

All these different sources of data are inspiration for NFPD new product ideas. But a huge amount of unstructured information as such is not creating suitable solutions. Still, most people would agree that the availability of big data has the potential to dramatically change the way new foods products are developed.

When it comes to the next layer of questions addressing the way this will happen and what is required to benefit from these potential suggestions often remain vague. The suggested advantages span from a faster innovation process over cheaper ones to a more successful one. All these elements are relevant!

As mentioned earlier, between 40% and more than 90% of all innovations in food fail, depending on how you define failure. And time is money, or shorter lead time is incremental, leading to unchallenged turnover! Being in the market 1–2 months faster than the average delivers extra sales before the market reacts. But just pushing on time without truly understanding and addressing the insights generated can be detrimental. Any NFPD is a substantial up-front investment. All opportunities to reduce market research investment. prototyping expenditures and trial costs give new degrees of freedom and are highly welcome.

Big data can help at various stages of new food product development to be more efficient and effective. Obtaining in real time consumer and market data or new trends in science and technology, while being able to flexibly match the sourcing and manufacturing options fulfils the fundamental principles of business insights, actionable insights, agility and real time (Jagtap, 2019).

Big data allows a kind of 360° approach to be used, data are internal and external and there is no particular sequence in which the material is reviewed. The ability to generate valuable business insights starts with the technology used, which allows a company to extract and analyse data. That's more than storage, computation, visualization and integration. When the underlying IT architecture can generate and analyse not only batch, but also real-time data, enterprises can continuously improve their operational performance. The "old" sequential way of trial, examine, test, improve and re-test can be skipped. Factory trials can be shorter, the learning curve is faster and less material is spent on the trial.

Less direct is the impact of the import of all potentially valuable data coming from technology (push) or market (pull). Suitable analytical tools should be able to connect developments and trends in the market with data from scientific publications and patents to identify better solutions for a known need or a new unmet need.

Figure 5.2 summarizes the holistic approach. New technical sources for new science and technology opportunities, all internal data about accessible capabilities are reviewed against all the information from the markets and the social nets.

On that level, this is still structuring the input. The expected outcome in NFPD is actionable insights or valuable outputs, combined with a prediction of the best technology, perhaps even a superior commercial strategy. And a higher level of agility is reached if all these data can be shared in time with the external environment, including suppliers, collaborators and customers.

The typical question is how a newly identified consumer or customer need can be best translated into efficient product making, including a manufacturing that maximizes the utilizing of existing machinery. These evaluations and scenarios are the moments where an idea gets a plan, where the factual requirements are getting clearer. Which technology to apply? Can existing equipment be used? Would that compromise the potential commercial success. If yes, how much? Or is the up-front investment in the new the better solution to satisfy consumer needs? The ability to quickly review in a transparent comparison different scenario can save a lot of time prior to decision making.

Then big data accelerates decision making by using inductive statistics and concepts out of large unstructured datasets, which can reveal new connections between data, hidden relationships and new dependencies and offer predictions of outcomes and behaviours. Big data cannot make a decision, but it can help decision makers to make strategic choices.

A popular quote says that "the world will never be as slow changing as it is today!" The environment is evolving faster, companies need to adapt accordingly and therefore must decide quicker. To benefit from the time and effort saved by big data usage, the decision makers must be data savvy. This means people working with big data require appropriate training so they can interpret data well. This also includes the confidence to know when sufficient data are gathered, and further data analysis will cost more time and effort than incremental information before a decision can be taken.

It's important for success to know what to decide when the data analysed stands against your "gut feeling" and all the experience made in the professional career. Has the environment changed in a way that makes it now not comparable with earlier situations? Or is there somewhere a flaw in the data, maybe

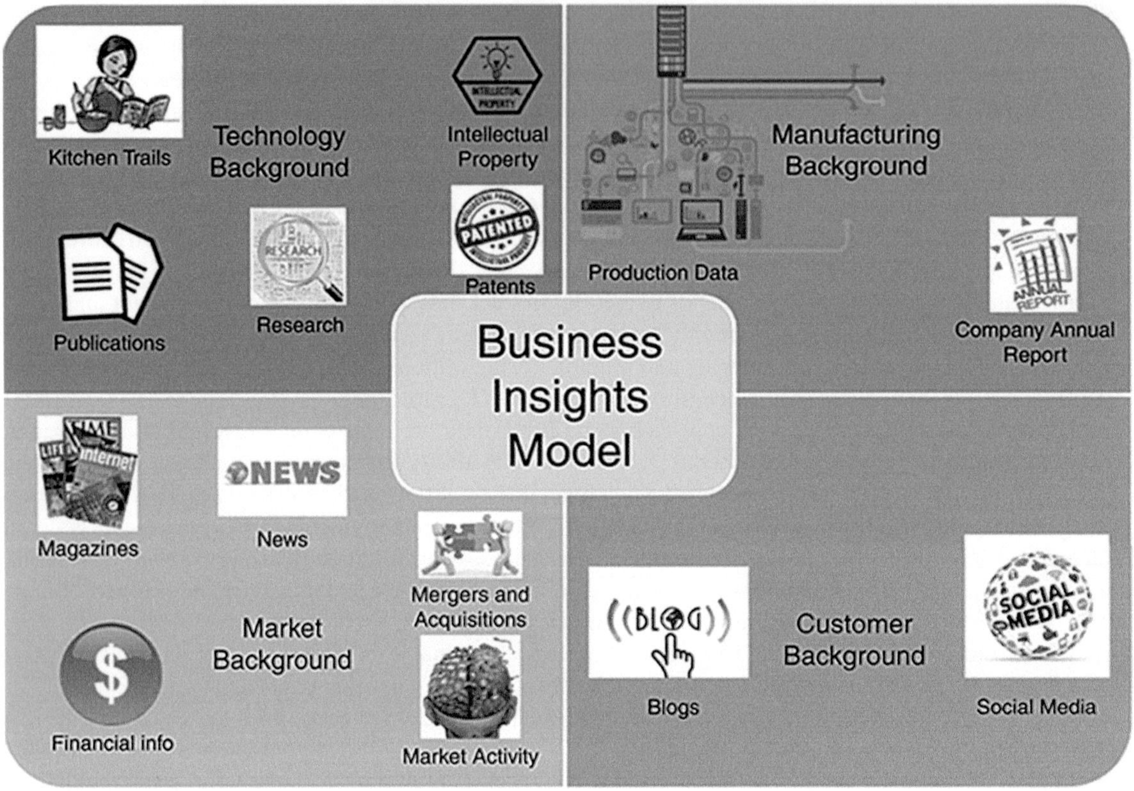

FIGURE 5.2 Big data model for food. (From S. Jagtap & I.N.K. Duong.)

wrong or missing biased input, leading to false conclusions? Crap in, crap out? Big data are not a substitute for superior business intelligence.

Efficient data processing supports are also on the way to market, as it can help to recognize potential food product quality and safety issues. So, no wonder everybody hunts for data. But there are practical limitations to the usage. The privacy and security of individuals' personal data especially related to potentially insightful information about buying habits, online shopping behaviour health, finances, or location needs to be anonymized and stored safely. Stored data are vulnerable to security breaches and cyber-attacks, and the company reputation is at stake.

Data accessibility and standardization are still limiting. Sharing potentially insightful data with other actors in the market, supplier or customer, bears the danger of giving an advantage out of hand. And of course, many datasets are stored in non-compatible formats.

Sourcing big data often results in the need to buy new tools that can process the data content. And the need to hire new experts to manage the data or train employees to use the data efficiently. Most food developers are not trained as data analysts, and data analysts normally do not know much about food product development. No wonder, some say "the cost associated with data generation is cheaper than managing it".

For the first time use of big data in the hunt for new innovations and NFPD evolutionary approaches may be more suitable. All companies have big sets of qualitative or quantitative variables that are structured or unstructured. Often, they are not machine-readable, as they are analogue or in the possession of individuals. In the past, developers often captured their formulation and bench work in lab journals, handwritten or notes in a local file. And the data R&D was using are not compatible with the manufacturing systems. Different factories had different systems – finance and procurement and marketing anyway.

In such a situation, a lot can already be gained by trying to clean the internal data and make them useable for neighbouring functions, Simplifying the necessary data interphase management stimulates harmonization and accelerates data flow.

Juts doing this can be a massive task, but the benefits can be enormous. A typical example affecting R&D is the move from older standalone systems into broad product formulation tools, like SAP-PLM. For internationally operating food companies, it is almost a "*conditio sine qua non*", a must-have.

But even that is not core big data usage! Big data are characterized by a vast amount of data generated and collected; the speed at which data are generated and must be processed; and the variety of types and formats accessible for analysis including structured, semi-structured and unstructured data. These data require special technologies, such as Hadoop, Spark or MongoDB and Cassandra as popular document-oriented NoSQL databases (Medium., n.d.).

INTERNAL SOURCES

A third angle for value input and as basis for new insights is the enterprise or company itself. In several companies, especially smaller and more specialized ones, the new food development process is an ad hoc procedure with little standardization. Reviewing how things are done internally can create a lot of value. In many food categories the sum of materials bought does not exceed about 20% of the value the consumers see on the shelf (Azanedo et al., 2020). In other words, most of the value is created during the conversion into the sellable product. No wonder that the people doing that transformation day in and day out are regularly coming up with new processes or product ideas. This goes beyond internal "constant improvement programs", basically aiming at more efficient processes, so the same products at lower costs. Company employees are consumers themselves. They naturally compare the product they are making with the other ones on the shelf while shopping. So, they can identify new market opportunities and critically review the competitiveness of their products. The ability to spot those and suggest them for new food products may have some unconscious bias Loyal employees may have a too positive view about their products and a lot of energy to suggest new ideas.

People from the sales force can help to balance such a view. Their income often depends on the success of selling a company's products. So, when talking the customers and negotiating new deals they often hear very clearly how competitive their own products are. And they are the first to pick up news from the market. New launches, changes in price and rumours about new competitors entering will bring a vivid picture about the competitiveness. Similar to QA, their responsibility ends not at the factory gate. Often, they monitor product quality and shopper reaction in the market, increasingly more by using AI, for example, in "consumer relevant quality standards; CRQS" (see Chapter 10).

REFERENCES

Ahrefs. (n.d.). *Keywords explorer by Ahrefs: Discover winning keyword ideas. at scale.* https://ahrefs.com/keywords-explorer

Ahuja, K., Chandra, V., Lord, V., & Peens, C. (2021, September 22). *Ordering in: The rapid evolution of food delivery.* McKinsey & Company. https://www.mckinsey.com/industries/technology-media-and-telecommunications/our-insights/ordering-in-the-rapid-evolution-of-food-delivery

All Top Everything. (2024, March 5). *Top 10 biggest convenience stores in the world.* https://www.alltopeverything.com/top-10-biggest-convenience-stores

Azanedo, L., Garcia-Garcia, G., Stone, J., & Rahimifard, S. (2020). An Overview of Current Challenges in New Food Product Development. *Sustainability, 12*(8), 3364. https://doi.org/10.3390/su12083364

Bagchi, S. (2021, August 17). *How-digital-Tech-is-paving-the-way-for-FMCG-innovation.* https://cxotoday.com/news-analysis/how-digital-tech-is-paving-the-way-for-fmcg-innovation/

Barraveccia, F. (2022). Digital Voice-of-Customer Processing by Topic Modelling Algorithms: Insights to Validate Empirical Re. *International Journal of Quality & Reliability Management, 39*(6). https://www.emerald.com/insight/content/doi/10.1108/IJQRM-07-2021-0217/full/html

Bilow, R. (2014, June 30). How IBM's Chef Watson actually works. *Bon Appétit*. https://www.bonappetit.com/entertaining-style/trends-news/article/how-ibm-chef-watson-works

Conference Index. (n.d.). *Conferences all over the world | Conference Index*. https://conferenceindex.org/

Conferences in Bangkok 2024/2025/2026. (n.d.). https://conferenceindex.org/conferences/bangkok-th

Fazira, E. (2023, September 6). *Functional-health-in-Asia-driving-value-in-food-innovation*. https://www.euromonitor.com/article/functional-health-in-asia-driving-value-in-food-innovation

Google Patents. (n.d.). https://patents.google.com/

Google Trends. (n.d.). *Trends*. https://trends.google.com/trends/

Grand View Research. (2023, March 31). *Online food delivery market size, share & trends analysis report by type (Platform to consumer, restaurant to consumer), by region (North America, Europe), and segment forecasts, 2023–2030*. https://www.grandviewresearch.com/industry-analysis/online-food-delivery-market-report

Hawkings, J. (2023). *A Thousand Brains: A New Theory of Intelligence*. Basic Books.

IBM. (n.d.). What is Explainable AI (XAI)? https://www.ibm.com/watson/explainable-ai

Innovation Through Collaboration. (n.d.). *Innovation Through Collaboration*. Foundry-Unilever. https://www.theunileverfoundry.com/home.html

Jagtap, S. (2019). Improving the New Product Development Using Big Data. *British Food Journal, 121*(11).

Journal of Food Engineering I ScienceDirect.com by Elsevier. (n.d.). https://journals.elsevier.com/journal-of-food-engineering

Journal Rankings on Food Science. (n.d.). https://www.scimagojr.com/journalrank.php?area=1100&category=1106

Katz, F. (1999). Marketing pull vs technology push. *IFT: Food Technology Magazine, August 1999, Volume 53, No. 8*(8). https://www.scribbr.com/citation/generator/folders/2deN3SkdQ9rd9DwOsSy8Y1/lists/3E1wNFCqePcJrVsUEsHfeY/cite/journal-article/

Medium. (n.d.). *Medium*. https://pagadala-saiteja.medium.com/big-data-analytics-a-beginners-guide-to-hadoop-spark-and-nosql-databases-8fe341816fe7.

Mingers, J., & Yang, L. (2017). Evaluating Journal Quality: A Review of Journal Citation Indicators and Ranking in Business and Management. *European Journal of Operational Research, 257*(1), 323–337. https://doi.org/10.1016/j.ejor.2016.07.058

Mintel. (n.d.a). *Food and drink industry and market insights | Mintel*. https://www.mintel.com/industries/food-and-drink/

Mintel. (n.d.b). *2024 Global Food and Drink Trends | Mintel*. https://www.mintel.com/food-and-drink-market-news/global-food-and-drink-trends/#download

The Flavor Connection. (2015). *Scientific American* (Special Edition, June, vol. 24, no. 2s), p. 8. doi:10.1038/scientificamericanfood0615-8

NineSigma. (n.d.). *Contribute to open innovation - We make innovation happen | NineSigma*. https://www.ninesigma.com/

Otero, D. M., Da Rocha Lemos Mendes, G., Da Silva Lucas, A. J., Christ-Ribeiro, A., & Ribeiro, C. D. F. (2022). Exploring Alternative Protein Sources: Evidence from Patents and Articles Focusing on Food Markets. *Food Chemistry, 394*, 133486. https://doi.org/10.1016/j.foodchem.2022.133486

Pancer, E., & Philp, M. (n.d.). *#FoodPorn: People are more attracted to social media content showcasing fatty foods*. The Conversation. https://theconversation.com/foodporn-people-are-more-attracted-to-social-media-content-showcasing-fatty-foods-160221

Passport. (n.d.). *Euromonitor*. https://www.euromonitor.com/our-expertise/passport

Rejeb, A., Keogh, J. G., & Rejeb, K. (2022). Big Data in the Food Supply Chain: A Literature Review. *Journal of Data, Information and Management/Journal of Data, Information and Management, 4*(1), 33–47. https://doi.org/10.1007/s42488-021-00064-0

Search our Trademark Database. (2024, May 15). *USPTO*. https://www.uspto.gov/trademarks/search

StartUs Insights. (2024, May 22). *Innovation scouting: Discover game-changers | StartUs Insights*. https://www.startus-insights.com/innovation-scouting/

StoryLab.Ai. (2024, April 8). *Effective Food Social Media Content Ideas [+ AI tool]*. StoryLab.ai. https://storylab.ai/examples/food-social-media-content-ideas/

Talend. (n.d.). *Big data and agriculture: A complete guide. Talend - A leader in data integration & data integrity*. https://www.talend.com/resources/big-data-agriculture

Trends in Food Science & Technology I Journal I ScienceDirect.com by Elsevier. (n.d.). https://journals.elsevier.com/trends-in-food-science-and-technology

Wikipedia Contributors. (2024a, January 19). *Chowhound*. Wikipedia. https://en.wikipedia.org/wiki/Chowhound

Wikipedia Contributors. (2024b, May 1). *Impact factor*. Wikipedia. https://en.wikipedia.org/wiki/Impact_factor

The Idea
Good, Bad or Ugly?

6

Evidently, not every idea with seemingly good market potential is always the right idea for a company. There are moments where it is better not to innovate. Longer term this is not a viable option since the food business needs growth and growth needs innovation. The task is to find the right idea. And "right" is influenced by the business context and should be identifiable from the business strategy. A new unmet need in the market can justify a substantial up-front investment into science, technology or manufacturing. Or the economic situation in the market signals that it's much better for the company to optimize their cost and profitability to stay competitive, to retain the trust of investors. Strategic analysists recognize a superior cash flow creation in the company but have doubts about the longer-term growth perspective, due to a lack of new growth platforms. Then it is time to put dedicated effort behind idea and innovation processes.

New products are coming from ideas that do not just pop up during the normal business hours. In many cases they are identified in with creativity in an ideation session.

THE IMPORTANCE OF CREATIVITY

Commercial success with new product launches in FMCG is not just a function of the amount of money and resources invested. On comparable levels some enterprises are more successful than others (Kellogg Insight, 2021). Some processes seem better to amalgamate all those "push" and "pull" and "internal" information and data into insights and are more effective in making actionable for the enterprise. When people speak about successful innovation, creativity is often intuitively regarded as a key enabler. The creativity that is a crucial part of the early innovation phase, the idea generation. Successful innovation requires additionally the tangible transformation of a creative idea into a viable solution. In a sense creativity is closer connected to imagination, while innovation links also with "implementation", the process of realizing an idea.

Creativity is the characteristic of individuals and processes to create something new and valuable. So, to be valuable for a particular business, the purpose of that business must be clear. If a company can apply a new base technology to make superior crisps, it is beneficial to first have a product-focussed creativity process, trying to find new attractive food formats in which the technology can be applied without too much complicated implementation. And if a large mining company, capable of collecting huge amounts of potassium-rich minerals as a basis for fertilizers can realize that the potassium they harvest is an attractive salt/sodium replacer in food products, this is an opportunity with more attractive margins, but also a business requiring "food-grade" operations.

Without creativity and innovativeness, there are hardly new products. But how then can stimulating input and a process to think new solutions be set up? How can a creative, essentially chaotic early phase be designed, and what are the requirements for a captured, structured and efficient forward process?

Research on useful creativity for FMCG companies highlights different aspects. There is the recognition of a sequential approach, including the role of creativity in understanding the problem and developing possible solutions and the process to choose the right one. A different perspective starts with the

DOI: 10.1201/9781032724591-6

importance of spontaneity and intuition as the source of "improvisational creativity" and how it differs from "compositional creativity", the ability to combine existing elements or ideas in novel ways.

The factual situation, like the company size, can influence the type of creativity. Some researchers see evidence that improvision is more correlated with smaller companies, while big enterprises are less inflexible and somewhat more resistant to change (Valaei et al., 2021).

And that seems logical! Food companies with a mature existing business have simply more options to grow their business compositionally, through low-risk innovations like refreshment and range extension. Adding 2%, 3% or 5% to a multimillion-euro portfolio is an attractive lower-risk opportunity. But only as long as these market opportunities exists and someone else is not changing the market, for example, through a breakthrough innovation. So, the decision of a successful business of which type of innovation to drive brings back the "innovators dilemma" (see Chapter 3 about the threat of disruptive innovations).

For new food products the human willingness to test new food items is configurated by cultural traditions, personal experiences, historical context, religious beliefs and the emotions and memories of earlier experiences. But individual curiosity and the specific situation shaped new sensorial experiences. Bringing new taste experiences into a constraint, structured and often functional moment like the everyday breakfast is probably more challenging than creating "astonishing" new experiences during a dinner in a restaurant well known for its molecular kitchen, where the guests expect to be surprised.

One can discuss which creativity is more productive for a specific business and how this than is best applied for the right innovations. But the role of individuals – the mixture of their skills and talents – and the environment and company culture remain important.

The organization of innovation centres, processes stimulating design thinking, the setup of brainstorming sessions, etc., are typical creativity-stimulating structures and processes.

Another term often used in the business context to separate success from failure is the "entrepreneurial drive". Being an entrepreneur means to search for innovations as a tool for economic success. That means ideas are sorted with firm business acumen. When starting a new business, for example, as a start-up, the business interest is often to realize a breakthrough idea to earn a part of the market. While the established food business identifies it is better for the business to just carefully start to stretch into new product groups. Some drive a speedboat; others steer a tanker. That requires different business navigation skills.

"Entrepreneurial drive" shows the strong motivation to accomplish certain things to make this business a success. And a permanent sense of curiosity helps entrepreneurs to continuously search for new opportunities. And the creativity helps them to find new ways how to do it. Passion is seen as the motor driving them.

So it is not a surprise that food companies put effort to bring more entrepreneurialism into their innovation processes, especially in those enterprises having an established business. A "purpose workshop" in that context is an attempt to connect the purpose of the company with the purpose of the individual, since this would release the entrepreneurial drive. In a sense it is the antidote to the day-to-day business routines in existing operations. Such initiatives start with reflections of the individual about their purpose, since many employees are not immediately able or willing to articulate their purpose in life. From the company honesty and authenticity are essential to be able to connect with the values of the individuals. The process must be engaging, personal and very open, not excluding potentially controversial aspects, like "care for the people and planet".

The purpose overlap between company and individual must be tangible and meaningful; they create in the individual the drive to passionately search for new growth opportunities. A good overview of how this can be achieved can be found under (McKinsey & Company, 2020).

The younger generations more and more are keen and want to understand the company values before they are willing to give their full energy. That's why many food companies have published purpose statements in the meantime:

- *Nestlé*: "Unlock the power of food to enhance quality of life for everyone, today and for generations to come" (Schneider et al., 2017).

- *Danone*: "We aim to inspire healthier and more sustainable eating and drinking practices, in line with our vision (Danone, 2024).
- *Procter & Gamble*: "We will provide branded products and services of superior quality and value that improve the lives of the world's consumers, now and for generations to come" (Gamble, 2023).
- *Unilever*: "Make sustainable living commonplace" (Unilever Plc, 2024).

And where an aligned purpose is perceived, the individual entrepreneurial increases the chance to drive superior innovation opportunities.

Earlier in this book I shared a few examples where AI can fuel innovation in food. Now we can categorize AI's potential better. For me, in situations where "compositional creativity" – the ability to recombine existing facts differently – is useful, AI will be a powerful tool. In such case, the relevant data are trained into system, where they are combined in new ways. It is easy to imagine how an AI spot can link and test all theoretical new range extensions or the next seasonal flavour. To keep it simple, the suggested top runners can be tested with consumers or compared to real-time data from social media and reflected against the latest food flavour pairing models, AI can take over the idea creation for refreshment work, at least "*in silico*".

Applying "improvisational creativity", the ability to generate novel thoughts spontaneously through unknown input might be a bigger hurdle for AI. In the human sphere these are triggered by memories, emotions and conscious and unconscious feelings. That's much more difficult to feed into the system and for AI to combine it meaningfully. The "human creator" will beat the system regularly.

IDEATION: NEW IDEA GENERATION

New product development processes in foods have often borrowed from "design thinking" (The Design Thinking Association, n.d.). This means the transformation of important consumer or customer insight into new solutions by utilizing methods originally used by designers. Such thinking is characterized by three core elements: Empathy, visualization and collaboration.

Empathy implies sensing the needs and desires of the user or consumer. Observation, interaction and immersion into customer experiences are ways to do that. There are numerous ways to awake this empathy. A direct interaction with the consumer space is much better than just the abstract discussion of an elaborate research concept. Many food companies nowadays stimulate consumer observation for their R&D people and create opportunities to directly interact with the target group and even enter their daily life, for example, through household visits or kitchen connects, where R&D people join the food preparation or cooking with their consumers.

Visualization uses the power of a visual representations of the ideas. "Seeing means believing!" Typical examples are product prototypes. It's hard to get valuable feedback from consumers about a new food product if that is only presented as written statement. A picture or, better, a multisensorial prototype immediately makes it more tangible and easier to communicate.

The third one, collaboration, implies working with a diverse group of people to generate ideas. These are the famous cross functional teams. Where possible that can be done together with consumers. Where prototypes can be improved quickly, one could improve prototypes immediately and present the improved version again. A method used in food here is "sequential recycling."

These three elements are functional elements helping to create innovative solutions and therefore they are often embedded in classical NFPD processes.

A good example of design thinking in the flavour industry is given by Michel Aubanel, from the Flavour Development of Kerry, a Irish food multinational (https://www.kerry.com/insights/kerrydigest/2020/design-thinking-and-food-innovation). He describes the finding of an industrial solution for an authentic floral cherry Sakura ice cream flavour. The traditional homemade process to get the iconic flavour into the dessert includes the making of a syrup base and an intricate cream (The Creative Adventurer, 2023).

The solution had to be a much easier process, but without losing the fragile flowery essence. The creative process focussed on the finding of the most suitable raw materials out of 10,000 possible. And it needed the combination of traditional flavour extraction expertise with modern technology.

There are many successful cases of design thinking. This one shows that a cross functional team is not always needed. A consumer/customer-centric professional with good observation skills can also lead to the next idea.

Integrating design thinking in a company's innovation machinery is a process. It may change the product only a little, while impacting the company more. In PepsiCo design thinking is no longer mainly used to design the packaging's colour, but a key element in the decision-making process, driving E2E innovation, with the core elements of empathy and genuine curiosity driving the solution of real challenges, a holistic thinking (Porcini, n.d.).

THE IDEATION SESSION

As the name implies, design thinking is just a way of shaping good ideas and not the substitute for ideation methods, the true nucleus of the idea creation. The basic ingredients for an idea creation session in food development are listed in Table 6.1.

For an fruitful ideation or brainstorming session some distance from the day-to-day (D2D) business routine is helpful. Typical functional representation for NFPD comes from R&D and marketing, supply chain, quality assurance, finance and regulatory. Human resources (HR) can join as well and take, for example, the role of a moderator. The moderator is the "master of ceremony" controlling the process, stimulating creativity and managing disturbances. A moderator does not contribute idea content.

Typically, an outside venue is booked for the session; it is great when it can happen in a stimulating environment. Some food companies run ideation sessions in a farm or a market atmosphere. That reminds participants of the origin of food, choosing, buying or eating. If you want to create food for children, why not book a location in a fun park? And in case you aim to create new kitchen or cooking products, why not book a cooking venue and invite a chef to do some creative cooking or cook yourself? An easier approach uses a room with an unusual seating order or different types of chairs. Anything that reminds the brain that this is not business as usual can stimulate creativity.

TABLE 6.1 The different elements and phases in a creative NFPD session

OBJECTIVE	*SUITABLE ACTION*
Data collection	Understanding your data, market and technology, including initial market research to find gaps, trends and consumer preferences.
Suitable Atmosphere	Nurture a creative environment: Create an environment that encourages creativity and innovation.
Functional team	Build a cross functional team: Assemble a team with diverse backgrounds, skills and perspectives, where diversity produces innovation.
Brainstorming sessions	Encourage unconstrained, free thinking, welcome all ideas and initially defer from any early judgement.
Idea generation	Use idea generation tools and techniques. Or engage a trained moderator guiding the session E2E.
Prototyping	Sometimes it is possible to create quickly the first prototypes on the spot.
Improvement	Use discussion feedback on your prototypes to iterate and improve.

Having the workshop within the company buildings is convenient and often the least costly option. The flipside is that it is more difficult to bring participants out of the D2D routines. Their engagement level can suffer. They may feel they must be approachable for all the other important business decisions of the day, so they cannot simply dedicate full time and attention to this.

The minimum requirements are that smartphones are switched off and the room can only be left for "basic needs". Even allowing people to make their calls during breaks has an influence on the engagement level.

Preferably, the idea creation session should last for more than a day. Firstly, there needs to some time for all the diverse business professionals to arrive, mentally to "be there" and "open their minds". Secondly, such sessions, often have two very different parts. There is this first phase: The ideation "pot" is filled with as many ideas as possible. Later comes the first "feasibility scan", where the wheat is separated from the chaff. Content and process in both parts differ. It is not beneficial if in the idea phase, participants begin to reject new ideas. Nor is it good to launch radically new solutions in a moment where the innovation gets clear shape.

Both parts are so different, they are better kept at a distance. Switching from unbound creativity to rational criticality does not happen with a finger flip. Announcing in the beginning of the workshop that after the coffee break the creativity should be suspended and all ideas will be critically reviewed already limits the creativity.

It's often better to announce a joint leisure/fun activity or a good dinner after the first creative working day, where all can talk and think about the ideas raised and feel rewarded for the creativity they showed. Being creative is exhaustive, as the organ that is used most consumes 20% of the total energy.

The beginning of the creative part is often an "icebreaker". And that's an important one. A quick, relaxing fun activity creates a feeling of comfort in the normally serious business context. It relaxes the atmosphere, stimulates creativity and collaboration and sets a positive tone for the session, while breaking down social barriers, creating a safe space and encouraging interaction.

Typically, not all participants know each other well, so the required introduction can be part of the icebreaker. Instead of taking time to introduce themselves with their detailed business, the moderator can ask participants as "prework" to bring one to two items they thought of, when they got the invitation to the workshop. Or they should share only fun facts about them that nobody knows. Then introduction and icebreaker are combined.

For the session, the actual location, mindset and atmosphere count! To get the creative juices flowing, more inspirations can be brought. One can put lots of food stuff from the competitors or a totally different segment on the table, showing very different environments like restaurant dining, work, mobility, etc.

A classical element is to put lots of journals on the table and ask participants to present their idea as a collage to the others: That's a visualization and one of the key steps in design thinking.

In the creative session, numerous ideation techniques are available to stimulate creativity. Examples are "mind mapping", "round robin, "storyboarding", "negative brainstorming", "prototyping", "SCAMPER" and many others for which descriptions and templates can be found on the internet (https://boardmix.com/articles/scamper-template/).

Moderators choose the tool they like, not because it's the best one (there is no best one). SCAMPER, for example, is an acronym for Substitute/Combine/Adapt/Modify/Put to different use/Eliminate and Reverse (Elmansy, 2022). It's a sequential process (see Table 6.2).

The key elements are idea stimulation, quick brain dump (non-reflected or unconstraint), sharing, explaining and building further through change of perspective. The initial focus is more on quantity than quality.

In practical terms there is also no real best practise about the number of participants. For a creative process, discussion, dialogue and interaction diversity is important. Jeff Bezos's "pizza rule" or "Bells numbers" favour smaller teams – five to seven people perhaps (see Chapter 4). But too few participants cut the potential for diverse thinking – they limit the idea potential – while too many participants can lead to undirected, repetitive and superficial discussions. Sometimes 6–10 participants are mentioned as ideal, so that everyone gets full attention and the chance to be heard. From my own experience even teams of

TABLE 6.2 The different steps in the SCAMPER ideation process

SCAMPER STEP	*OBJECTIVE*
Substitute	The exercise is to replace one element of a product (or service) with another.
Combine	Two or more elements are combined to create something new.
Adapt	Elements of the product (or service) are changed to make it more suitable for a different purpose, target group or situation.
Modify	This is the next improvement step, making the changes that improve the idea further.
Put to another use	Shifts the focus on a new use for an existing product.
Eliminate	The reduction, where the idea gets simplified and stripped to the essentials where elements from a product are removed to simplify it.
Reverse	The order of elements in a product are service are changed to see how that fits.

15 and more can work. They bring different backgrounds, skills and perspectives and that stimulates creativity. Not everybody joining from a skill base is a great lateral thinker, and a single expert, for example, from packaging, is no guarantee that the fill packaging idea potential is present. Instead, if the group is bigger, participants can be split into teams for quick ideation rounds. At the end the moderator needs to decide about the team size.

The selection of the participants can also have a hierarchical dimension. Having very senior participants, directors, VPs and SVPs joining will surely underline the importance of innovation for the company and stimulate participants to show their best engagement and "go the extra mile". But leaders may get too much attention, and others might feel less comfortable to be "naïvely creative". And senior leaders also have lots of urgent obligations during their working day, so they may only join for a while. That disturbs the atmosphere and may encourage others to adopt that behaviour and disengage in moments when they feel less involved.

A good option to combine the "best of both worlds" is to have a senior leader or sponsor joining the workshops in well-defined moments. In the beginning the person can welcome all participants and underline the business importance of new ideas in the workshop. Then the team follows the process, and the business lead comes back at end of the workshop, when ideas have been created, reviewed and evaluated. The team can present their work directly to a decision maker, who provides feedback and a perspective for the next steps.

Back to the creative session! When the first round is done, ideas are presented and discussed with the whole group. The spirit is "yes" and the intention to build them further. The famous "yes, but" phrase is forbidden. After that a clustering of all ideas can happen. Here similar and overlapping ideas are clustered together and combined under one heading. That can reduce redundancy and sharpen differentiation. A convenience idea could be linked with a packaging idea, etc., if that makes a single innovation bigger or more tangible.

For the presentation of the idea visualization, one of the core elements of design thinking can help. Seeing is believing, and it helps the participants to engage with ideas. Some professional agencies in this territory bring a draughtsman to capture ideas intuitively while team members present them.

The second part of the workshop looks at the practical likelihood of implementing the idea. It's the prioritization where most of the innovative ideas will get "killed". Members switch from creators to gate keepers. Some will now lose "their" idea. Here it is important to be transparent and clear about the selection criteria that sort ideas into "good", "bad" or "ugly". This reduces the risk that disappointed "idea owners" disengage. The selection then has an understandable purpose.

If wanted, the criteria can be weighted. Attaching a numerical value, i.e. from 1 to 5, allows prioritization. Example: A new food format requires a new manufacturing technology, which could enable a new platform with the option for a wider range of new products. This big idea may need time and money to be realized. If the business context that the senior leader has described requires ideas ready for the next launch window with the trade in a couple of months, the potentially much bigger next-generation idea may have to wait or be put aside. It is the other way around if the company needs to find new growth territories. Then, the "new technology" is the opportunity.

At this stage the "but" is not only allowed but necessary. Is the technology available or in reach? Does it need extra capital to realize it? Or do we know a copacker who makes that product for us? And will the price then still be attractive? Is there evidence that enough consumers are interested? Is it a new and just variant in the market? Is it economically attractive? How much will it cannibalize existing sales?

The scoring rationalizes the decision making and leads to a ranking of all ideas mentioned It is good, if the "innovation capacity" can be well estimated. For R&D that is the particular development work needed in different phases of the NPD. How many different prototypes or lab samples can be made? How many new raw/packaging materials/formats can be handled at that moment? How many pilot plant sessions can be made? How many of them can be upscaled? Is there enough budget to assess six different ideas in consumer testing?

At the end of the session the number of ideas that will be further worked on should be clear. The specific capacities and skills needed in the beginning of the process describe the capacity and that will define how many new ideas can be initially taken forward. During later stages of the process, ideas will be stopped and the planned capacity for those can be shifted to other ideas. But when the innovation team is already working on some older ideas on the way to market, this capacity is not available for the development of new ideas.

The actual number of ideas that can be accepted for further exploration is greatly influenced by the question of how the ideas are tested in the next stage. In foods, an early question is, of course, if the idea has sufficient market potential. This can be tested through dedicated consumer research with prototypes and storyboards – a good predictor of future market success. Or it can be quickly checked via online tests, like Toluna (Toluna, 2021). Or more "the light touch" by just analysing market research studies. Different effort – different capacity.

At the end of the workshop, it should be clear what needs to happen next for the selected ideas. A project leader and technical project leader may already be nominated, or the workshop owner summarizes the results and presents it to the decision-making body, the company or innovation board.

And the ideas that will not go forward can be put into a "parking lot". That is an idea pool that can be revisited in the next round of idea creation, which will surely come. With the selection criteria just used, the reason why ideas did not make it are evident. At this stage it takes only a little effort to capture them and add a label describing the idea and explaining, the reason why it was not chosen.

Typically, value improvement or a cost saving workshop creates more ideas to save money, than the team can work on in the next period.. Maybe the workshop has already collected enough "low-hanging fruits" to meet the expected savings target. Or some ideas seem too risky, as they may distract consumer value, and are therefore not chosen. Having those in store and recycling them when needed will enlarge the option in the next round.

The capital investment for a new product line has stopped the idea! The machinery to produce may get cheaper over time or a third-party manufacturer can be found willing to do the job. Or the necessary investment money becomes available. Or the consumer demand is regarded as too low or too high cannibalization of the current portfolio is expected. Or the creative new idea requires up-front regulatory approval, which is seen stretching out the intended launch date too far. All these good reasons can change. And if the idea is stored and the reason for stopping it is attached, the recycling in the next creative round can be efficient.

With material savings ideas, the "initial" value estimation can be robust. If an idea suggests replacing one ingredient with another, the cost difference gives the savings value. Many factors can make the calculation complicated, but the principle is simple.

For market growth ideas, the financial value is more difficult to estimate, as the size of the target group, the purchase intent, price and costs are less clear at this early stage.

HACKATHON

Hackathons as idea creation events are becoming more popular. The name comes from the software industry and is a blend of "hack" and "marathon", where "hack" means exploratory programming, not a process to get unauthorized access to digital devices and networks. It started around 1999 and is used today to describe events for rapid and collaborative development of solutions for various problems or challenges.

Hackathons usually span from a few hours to a few days. A theme needs to be pitched and ideas created and presented to a panel of judges or peers. Hackathons are hosted by various entities, such as corporations, educational institutions or communities, conferences and fairs to stimulate innovation, education and networking.

Food hackathons can be set up by a group of companies or individuals, for example, through sponsoring a side event during a big fair or conference. A single company can organize a hackathon with students from a university or a faculty or as lead sponsor of a food convention/conference.

One main difference from an "internal company ideation session" is that the question of the idea ownership can be more complicated. The terms and conditions of the events, including ownership rights, IPR, type and scope of the project and the various participants need clarification before the start. The setup may require the participants to sign a waiver or a contract that grants the organizers or sponsors certain rights or ownership over the ideas or products. This could be for various reasons: To protect their interests, to ensure quality control, to promote their brand or to commercialize the ideas.

Alternatively, participants may be allowed to retain full ownership of their ideas but ask them to grant a license or a permission to the organizers or sponsors to use them for specific purposes. More contractual attention is necessary if a hackathon is based on existing ideas or technologies that are already owned by someone else. The handling of confidential or sensitive information that is subject to non-disclosure agreements or ethical guidelines is a variant. In this case, the participants may need to respect and protect the privacy and security of such information. Last but not least, joint ownership can be agreed. Successful ideas are mostly not created by individual, but a team. Spending some thoughts on the property share based on contribution and collaboration of the participants can be important pre-work. Often that means for the participants and organizers to communicate and document their roles and contributions clearly and transparently with each other and with any other parties involved in the hackathon. This can help to avoid misunderstandings and disputes over the ownership of the ideas later. Such essential legal paperwork prior to the workshop costs extra time, which can hamper the speed of the innovation process.

The alternative is to use this tool and adopt a policy of open source or public domain for the ideas and products generated during the event. This means that anyone can use, modify or distribute them without any restrictions or obligations. With the ideas being public, perhaps rewarded with a prize as signal of the competitive setup, the uniqueness is limited.

Companies looking for commercially attractive ideas would use this approach when they see a possibility to use the idea ahead of others, when they either have a fast-track innovation paths, so they can be sure to be first in market due to their skills or maybe intellectual property rights. Or they use hackathons to create ideas or elements of ideas that will be used in combination with other unique propositions. Hackathons should foster innovation and cooperation among boundaries. Therefore, the topic chosen is often more "pre-competitive" and a wider problem angle is chosen, not directly the concrete product- or service idea. Sustainability, socio-economic issues, food security, circular economy, regional value creation, vegan food, food waste in households, restaurants, supermarkets or farms are typical topics that are wide enough to benefit from new approaches and new viewpoints. These initial ides can then be taken forward in the next, more concrete ideation sessions.

Some organizations do not like the name hackathon, or they want to pronounce an aspect of the event. They then come under names like "innovation challenge", "buildathon" "ideathon" or "creathon".

But however and wherever the ideas were created, in a next phase of development some more substantiation work is needed.

REFERENCES

Danone. (2024, January 8). *World food company - Danone*. https://www.danone.com/about-danone.html

Elmansy, R. (2022, May 23). *A guide to the SCAMPER technique for design thinking*. Designorate. https://www.designorate.com/a-guide-to-the-scamper-technique-for-creative-thinking/

Explore: Design Thinking Case Studies I The Design Thinking Association. (n.d.). https://www.design-thinking-association.org/explore-design-thinking-topics/design-thinking-case-studies

Gamble, P. A. (2023, August 16). *P&G purpose, values and principles*. Procter & Gamble. https://us.pg.com/policies-and-practices/purpose-values-and-principles/

Kellogg Insight. (2021, May 3). *When It comes to investing in product innovation, large CPG companies could learn a lot from their smaller competitors*. https://insight.kellogg.northwestern.edu/article/cpg-product-innovation

McKinsey & Company. (2020, April 22). *Purpose: Shifting from why to how*. https://www.mckinsey.com/capabilities/people-and-organizational-performance/our-insights/purpose-shifting-from-why-to-how

Unilever Plc. (2024, April 18). *Our strategy and growth action plan*. Unilever. https://www.unilever.com/our-company/strategy/

Porcini, M. (n.d.). *Creating joy through design+ Innovation*. https://design.pepsico.com/.

Schneider, M., Nestlé Executive Board, Société des Produits Nestlé S.A., & Bulcke, P. (2017). *Nestlé purpose and values*. https://www.nestle.com/sites/default/files/2021-06/nestle-purpose-values-en.pdf

The Creative Adventurer. (2023, January 27). The Best Sakura Food and Drinks to try in Japan - The Creative Adventurer. *The Creative Adventurer - A Travel Guide for Exploring the World*. https://thecreativeadventurer.com/the-best-sakura-food-and-drinks-to-try-in-japan/

Toluna. (2021, July 13). *Concept Testing: Getting started in Toluna Start | Toluna*. https://tolunacorporate.com/resource/getting-started-in-toluna-start-concept-testing/#:~:text=1%20Step%201%3A%20Review%20your%20survey%20details.%20Make,of%20the%20project%2C%20you%20are%20ready%20to%20launch%21

Valaei, N., Rezaei, S., Bressolles, G., & Dent, M. M. (2021). Indispensable Components of Creativity, Innovation, and FMCG Companies' Competitive Performance: A Resource-Based View (RBV) of the Firm. *Asia-Pacific Journal of Business Administration*, *14*(1), 1–26. https://doi.org/10.1108/apjba-11-2020-0420

Forming and Shaping the Idea

7

From a creativity session and the formulation of a good idea, it's still a long way to a successful market launch. While Chapter 6 was more about the tools to "do the right thing", this chapter focusses on "doing the things right"! The focus is now on the operational roadmap for moving a product from ideation to launch efficiently and effectively. Realizing an idea is rarely a linear, one-directional process. Throughout the maturation steps, new learnings are generated, shaping the original idea. Even the greatest inspiration will receive substantial modifications on the often-long journey to market. Foods markets are fast moving, and a market situation can change quickly. Maybe a new competitor moves in, a sudden public campaign starts, an engaging story in the press, a new health or economic crisis or just a supplier bankruptcy influences the idea. And changing an element in the mix often means that now earlier steps and decisions need to be reviewed to confirm that the change makes sense. A few typical circumstances:

- A key ingredient is no longer available, and the existing supplier needs to be contacted for a suitable alternative, which needs to be tested again.
- Or even worse, a new supplier needs to be found, who must pass the supplier qualification schemes, before it can be accredited.
- Maybe a part of the production line cannot be used any longer, even if the initial factory trial was successful.
- A competitive, differentiating "product statement/claim" turns up late in the process after regulatory review and turns out to be a potential legal risk.
- A failed shelf-life test can bear massive consequence for the product, the distribution, acceptance by the trade and the overall timing. In those moments the development needs to loop back one or more steps, find a solution and manage the consequences.

In such situations alternative plans are needed. In the latter case: Do we need to delay the launch until the "shelf-life problem" is fixed? Or will the opportunity be gone by then? Can the product be packed differently to keep the shelf life? Or is the trade and the consumer okay with a shortened shelf life? Can we perhaps even convert the problem into an opportunity by highlighting the shorter shelf life as "naturalness"? Or does this pose risk to the business case, making the original idea commercially unattractive?

And the whole development process is pressurized by the desire to innovate as fast as possible. A great market insight, a new trend, is now a great opportunity, but no guarantee it's still there in a few months.

Any useful innovation process has of course a future-oriented view and the ability and resilience to manage the unavoidable challenges. Sometimes even an adaptation of business strategy can be the right thing to do. If a company searching for the next range extensions finds a new ingredient that offers the chance for a powerful health message but the regulatory approval is missing, this might be a moment where an intended quick range extension in 6 months is converted into the plan for a new platform 1–2 years from now.

The idea is shaped by the way people think about it and the way the process is managed. People in the innovation process must be able to act in a nonlinear sequence of idea formation to shape the idea and solve problems, and the innovation management helps solve problems and channel the sharper idea into action.

In this context the concept of "design thinking" appears regularly. The essence can be captured differently. One possibility is "finding human needs and creating new solutions by using the tools and

DOI: 10.1201/9781032724591-7

mindsets of design practitioners" (Kelley & Kelley, 2013). The way it captures "design thinking" can help R&D. These product and process experts are trained to use analytic scientific thinking to sift through facts to then find patterns. A designer starts with a pattern and invents a concept to address possibilities. That's the product- and solution-oriented view; helping to avoid the risk of "falling too deep in love with the problem", rather than the solution.

A company, however successful, doesn't have the resources to put every innovative idea into practice. Deciding which idea to take and which one to leave at a given point in time is an important practice of prioritization.

A stage gated project management tool breaks up larger projects into a series of discrete stages with decision gates between them. That helps to identify knowledge gaps, capture new insights and decide what that means for the idea. It is a commonly used tool in large enterprises for more complex and risky projects.

THE STAGE GATED PROCESS

Based on research in the beginning of the 21st century on the factors separating successful innovation projects from those that fail, Robert G. Cooper, an influential innovation thought-leader, came up with a new framework (Cooper, 2008). Figure 7.1 shows the classical idea funnel, the schematic presentation of the process in which new product ideas are managed from the creation in an ideation session until launch. The funnel shape indicates that more ideas are entering the process than leaving it to market. Well defined criteria help to select the most promising idea and bring them to life.

The key elements in this approach were:

1. Cross-functional teams, composed of members from various departments and disciplines collaborating to ensure that all aspects of the project are thoroughly assessed. This team executes different tasks in each stage.

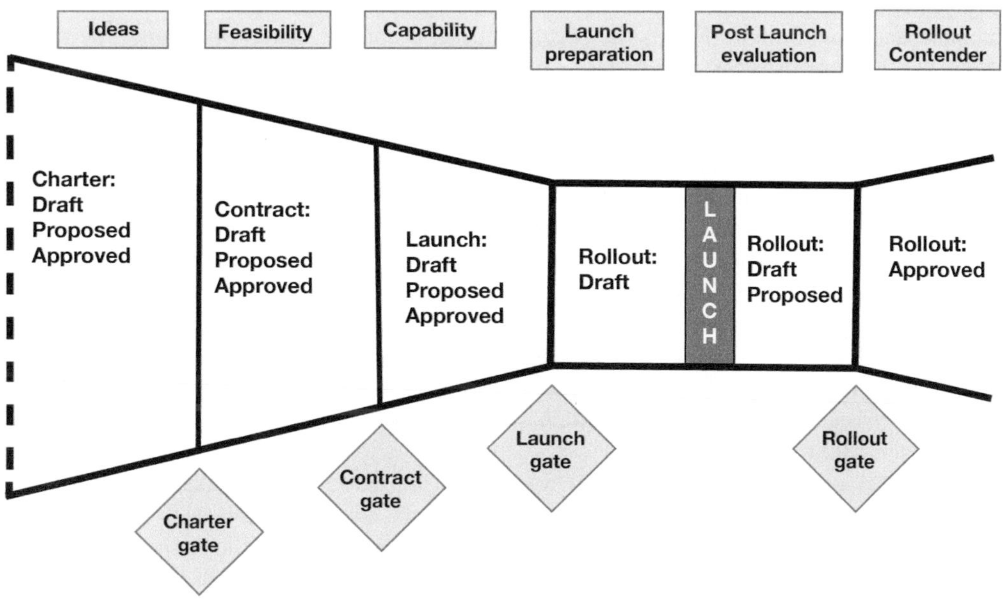

FIGURE 7.1 The stage gated process with typical phases and gates. (Modified after Sybil-Gibbs, 2014.)

2. *Stages*: The project is split into distinct phases, each representing a key step in the project's development process. Such stages could be development, feasibility analysis, design, development, testing and launch.
3. *Gates*: Formal decision points at the end of each stage. At these gates, project stakeholders evaluate the project's progress and decide whether to continue to the next stage, modify the project or terminate it.
4. *Criteria*: Transparent and specific set of criteria in each stage, outlining the specific goals and objectives that the project must meet to pass through the gate successfully.
5. *Documentation*: Comprehensive project documentation and reports at each gate to show the project's progress, challenges and any necessary adjustments

The objective is the transformation of a promising idea into an activity with extra value, normally commercial value. This value is embedded in the idea, but it needs to be measured, extracted and realized. And sometimes the extra value reaches beyond the actual idea. Even with a new product launch, a different format can strengthen the brand under which the idea goes to market and increase the intrinsic value of more products under this brand. A company selling ambient stable ready food may decide to invest into the launch of chilled meals, having fresh ingredients but a much shorter shelf life. If successful, this can raise the quality perception of the total brand, conveying more naturalness.

Or a company can decide to replace their proven and effective regular plastic packaging with a new recyclable alternative. The change requires effort and costs money and may even increase the costs of the packaging, while the consumer of the product is not (yet) willing to pay a little extra. But still, the project can be a great idea longer-term commercially; even so, the direct business case may be negative. It could pre-empt an upcoming change in regulation on plastic waste, bringing the company valuable learnings and better reputation through first mover advantage, before later the new material becomes market standard without product differentiation.

One key objective of thc stage-gated process is to identify as soon as possible those ideas with little chance of success. It shapes each idea, but it also assumes that there is more than just one or a few new ideas. Failure in NFPF is unavoidable, but early failure is much cheaper than late. And by stopping (killing) those, the assigned resources are liberated resources that can be deployed to serve better ideas.

The step-by-step process provides focus without limiting the creativity early on.

Stage-gated processes are used for various types of projects and in all industrial sectors. So, they have been adapted and streamlined towards the specific business context, while preserving the fundamental principles. The number of gates can change, and more or less effort can be requested in the different phases. Food projects as part of FMCG business tend to have a shorter funnel to allow maximum speed to market. But if the new idea requires the building of a new production line or even a factory, the gatekeepers will request very robust evidence before they release the investment.

The task remains the same: Sort out the wrong ideas as soon as possible and with the lowest effort/resources possible and maximize the success probability for those moving towards launch. The process aims to focus the available resources on those projects promising the highest payoff.

The individual enterprises decide themselves how many separate phases, what the "gate passing criteria" are and how many gates they need. In foods typically, three to four gates are installed between idea creation and launch for a step-by-step directional process. At the gates, the new evidence of the development work of the previous stage becomes reviewed against a set of predefined criteria.

The structure is a funnel, not a tunnel, which means the process is designed to take more idea initially up than it lets out later to the market.

New ideas, as output from the ideation workshop (see last chapter), customer feedback or new market research feed the funnel and become mirrored against a set of criteria such as market potential, technical feasibility and strategic fit. The ideas that are allowed to "enter the funnel" are often at least initially in the "ideation session". That means that based on the knowledge of the participants at that workshop they are regarded as valuable, but now extra and deeper analysis is needed to confirm the potential.

In this "scoping" phase, the first investigations are done to assess the technical and market feasibility. The results of that scoping are captured in a charter document where they compared with a set of agreed success criteria. With a positive outcome, the charter gate passage happens and the "feasibility phase" begins. In this phase work begins on the business case, where the market potential is confirmed and a detailed technical assessment is made, together with a financial analysis, and a detailed project plan.

At the following contract gate, the business case condenses all resource and time requirements and puts them against the business potential. With the gate passage, important contracts for machines and material are signed. At this point, failure is no longer cheap. The capability phase organizes the market entry, like the advertising, program and the negotiations with the trade – everything needed for a successful launch. Once the launch plan is signed off on, the project reaches a "point of no return". Only exceptional circumstances would now stop the idea. And after a while in the market the project performance is reviewed and subsequent actions are agreed.

Bigger companies, with more resources and ambitious growth plans, will allow more projects to enter the funnel to do dedicated project work. For a small entrepreneurial startup, the situation can be different. Here the company may simply have to quickly find the single best project, so the available resource can focus on it.

Now the different phases will be discussed in a bit more detail to bring them alive.

IDEA PHASE SCOPING

The scoping phase takes some time, ranging from a week to maybe a month. A dedicated cross functional team is nominated, in foods typically consisting of marketing, market research and R&D as a core team and supported by manufacturing, procurement and finance as extended team members. This team aims to continuously sharpen the product characteristics and to identify, manage and remove risk from the project. Typical elements are a deeper understanding of the market opportunity, including some light dedicated consumer research, and the fundamental ability to manufacturer the product, either themselves or by a co-manufacturer.

Beside a first project plan, the scoping phase should be able to show the alignment between the innovation project potential and the strategic goals of the organization, stating the expected outcomes and a plan about the necessary resources and time. It's a first "light" viability testing. On the market pull side, scoping can include "desktop work" about market size and value, consumer needs and recent competitor moves, or first product concept creation and initial light testing with the target group.

The technical aspect is also investigated. Does the idea need basic technology? Is that available and in sufficient capacity? Or do we need new technical skills and equipment, and where would that come from?

A well-proven and efficient way to conduct such investigation of all aspects is the SWOT analysis, which covers strengths, weaknesses, opportunities and threats.

SWOT ANALYSIS

This is a strategic management technique used beyond new product development grounded the situational assessment. The four words capture the thinking that for the good decision about a new idea, it's important to look into the future and critically asset the current business reality, the presence.

There a few stories about the origin of the SWOT analysis (BrightHub, 2010). Some sources trace the origin back to work of business consultant Albert Humphry in the early 1960s at the Stanford Research Institute. He formed an analytical tool to judge long-range business plans and called it "SOFT analysis", where SOFT stood for "Satisfactory" (good in the present), "Opportunity" (good in the future), "Fault" (bad in the present) and "Threat" (bad in the future). The fundamental evaluation corners were set with this: Presence (plus/minus) and Future (plus/minus).

In later discussion the idea came up to replace the "F" for faults with "W" for Weaknesses. And over the years the original presence-future axis became overlayed with an "external" and "internal" perspective (Puyt et al., 2023).

Generally, the SWOT analysis is used for strategic processes in their "own" company, like the question of which new products to develop further and which ones not. It can also be done for competitors, although the level of information available may be more limited. Beyond the pure analytical element as a tool for decision making, the analysis helps to identify weaknesses and uncover risks, against which action can be planned.

SWOT is a team working tool. Preferably, a cross functional team with representatives from all functions having a role in new food product development try to answer the questions behind the analysis. For the study the team often tries to fill a simple template, like a classical 2 × 2 matrix.

- "Strengths" are what makes the company better than the competition. Strengths are integral elements in the organization. It's fact based and present right now. It can be the quality of people working there, or particular skills, like a robust and effective innovation process, a proprietary or protected technology or special relation with main sales channel. Basically, any aspect that brings an advantage over others is a strength. Free capacity on a line with specific superior processing capabilities or access to good material stocks would be a strength as well.
- "Weaknesses" are the things that right now create disadvantages to the team or the idea. These are inherent features that can be present anywhere in the company or on the route to market. In bigger organizations specific policies or guidelines hindering an idea are probably a weakness. A company may have committed to on-pack nutrient scores, while a final formulation drives the score below the onc of a competitor product.
- Opportunities are favourable external factors that can bring extra advantage. They are available openings or chances that the team can include with additional effort or investment to make the opportunity bigger or safer. It can be a proper timing, a specific channel, special offer, an easing of restriction or a recently released application. The project realization may be in time for a big seasonal event, like Eid al-Fitr, the fast breaking at the end of Ramadan.
- Threats are often external factors negatively affecting the idea. But it can also be obstacles in the workflow, schedule, budget, or product or service. A new technology, a changing regulation, trade limitations or new competitor moves are typical threats. Rumours in the press or signals from suppliers about an upcoming product launch from a competitor would probably limit the commercial success and need to be brought into the case.

The collection of the various arguments for the alternatives can trigger a deeper discussion on specific topics, like the technical consequences attached to the creation of a new variant, the need for new technology, new suppliers, different storage and distribution pattern, new channels, a new consumer segment, leveraging a trend, regulatory uncertainty, supply chain disruption, etc. In that sense it is not a managerial decision instrument.

Some teams do not use it only to preselect a few ideas but also to discuss strategic consequences. Typical outcomes can be:

- *Strengths + Opportunities (SO strategy)*: This strategy involves using the company's own strengths to achieve more profit or a higher market share. These abilities should be expanded. It is playing on one's own strengths.
- *Strengths + Risks (ST strategy)*: With this strategy, one identifies measures that can help to use strengths to mitigate potential risks or threats. It builds a shield against risks, using strengths with conscious management of intrinsic risk and focus on minimalization of those.

- *Weaknesses + Opportunities (WO strategy)*: This strategy involves taking advantage of opportunities to compensate for weaknesses. If the market allows, a situation can be used to turn weaknesses into strengths.
- *Weaknesses + Risks (WT strategy)*: If risks are visible and the enterprise has weaknesses in this area, the strategy can avoid or counteract them with strengths or improve a weakness.

This more strategic SWOT analysis is less useful in single NPD discussions, but good for the analysis of an existing product portfolio, where it shows vulnerability of the existing stock-keeping units (SKUs) and movements in the market, resulting in an agreement of what type of NPD to initiate or speed up.

But once an idea has passed the SWOT more dedicated work for the team members starts. How strict and selective the process has to be in this stage depends on different factors. Too much selection of new variants (range extension) may not be necessary if the prototyping for consumer tests is low effort. Most ingredients may be "in stock" and the missing ones are easy to procure. But if the idea spans different technologies, like a dry ready snack, a pasty meal maker or pouch with liquid thick soup, the team can discuss if they can progress all options with the same strengths.

The strengths and weaknesses of the original idea should become clearer, and the original concept can already obtain modifications. The initial packaging idea may turn out to be difficult or unrealistic, or a similar product may have been launched by a competitor suddenly. Or a key ingredient cannot be sourced in time.

The technical members of the team will start now with some prototyping, at least working "*in silico*" on first formulations, and together with procurement start to engage with suitable suppliers for critical ingredients and drafting processing specifications. The project leader will collect the information about the basic dimensions and competencies the project entails (budget, schedule, resources needed). But it is often still a "quick scan" with "light touch". The dominant work style is desktop, with limited "out of pocket" expenditure, for example, for testing. It's not a project yet and it has no planned budget.

Typical questions for this phase are:

- What is the need that the project aims to address?
- Who are the key stakeholders and what are their expectations and requirements?
- Are there alternatives to address the problem (different solution)?
- What are the main assumptions made, and how can they be tested?
- How would success look like and how can it be measured?
- What constrains the idea and which dependencies exist?
- What are the risks and how can they be mitigated?

The amount and type of work within a scoping phase are not prescribed by the books. This reflects the requirements of the company and the project. If the innovation resources are available and the initial opportunity is a newly identified consumer need, a deeper and longer scoping of all available technologies makes good sense.

Alternatively, the company keeps it short and crisp, since the idea is close to the existing reality, without major intrinsic risks and fitting technical requirements and abilities in reach.

If the idea needs new technical solutions, technology scouting is an option. So, those technologies that help to achieve the goal need to be identified. That involves scanning of the external environment for emerging trends, opportunities and threats in an AI-supported systematic process, where different technologies are reviewed and the ability to transfer selected technologies into the organization's existing or new products or processes are assessed.

In the food industry, technology scouting can help to discover new ingredients, processes, packaging or business models that can improve the quality, safety, sustainability or profitability of food products, looking for ways to extend the shelf life of fresh produce, reduce food waste or create novel sensory experiences for consumers.

In this early phase, a first testing of the ideas selected with consumers is needed. The prioritization process in the ideation phase often is the result of an internal discussion and selection. The participants of the ideation session will have brought all their professional experience and dedication, but there is a possibility that they missed out trends and movements, so that the choices made are not fully catching the market needs or business opportunities.

Hence a quick, not too expensive check with the consumers can confirm the overall potential and provide first objective evidence. Qualitative market research will already give an idea the attractiveness in the eyes of the consumers and maybe some initial estimates about the sales potential as first building block for a business case later on. The market research tools helpful in this phase are discussed in Chapter 8.

At the end of the scouting phase all findings are captured in a document often called a "charter" or "project discovery form" that is presented to gatekeepers together with a proposal on how to move on. The document aligns the expectations and interests of the involved parties and establishes a common understanding of the project. In FMCG the document typically addresses questions like:

- Initial assessment of the business opportunity
- Business and brand fit, including fit with brand architecture
- Initial exploration of technical route to follow
- Initial concept and communication idea
- Resource plan for the next phase (feasibility)

CASE STUDY: SWOT ANALYSIS: NEW SALAD DRESSING

A strong and well-known player in the salad dressing market in Southeast Asia has identified during a creative session a few new ideas for new variants. The company has capacity for two new variants in two different packaging formats: A 20-mL pouch and a 300-mL bottle. On the market they have already Ranch Dressing, Caesar Dressing, Buttermilk Dressing, Miso, Bleu Cheese Dressing, Balsamic Vinaigrettes, Salsa, Olive Oil, Thousand Island Dressing and Italian Dressing. The new ideas were: Creamy Black Sesame, Original Wafu, Spicy Mango and Blueberry Basket!

Before entering a deeper prototyping phase and spending money on initial consumer research, the small cross functional team ran a quick SWOT analysis to get a picture based on their current knowledge. Present are R&D, Marketing, Market Research, Supply Chain, Regulatory and Quality Assurance. A set of questions was used to stimulate the discussion and start filling arguments in the matrix on the flipchart:

Strengths: What's our competitive advantage? Which of the required resources do we have already? What are our best-performing dressing product? For which dishes is it mainly used?

Weaknesses: What areas can we improve? What are our low-performing products? Where are we lacking resources? Is the size of the operation suitable for the product volumes planned?

Opportunities: What technology could we use to improve our operations? Is there scope to expand our core operations? Which new market segments can we explore?

Threats: What new regulations could threaten our operations? Press article about sesame allergenicity or competitor brand strengths? The outcome of the exercise is presented in Table 7.1).

TABLE 7.1 SWOT analysis comparison for four potential new dressing variants

STRENGHTS	*WEAKNESS*
Black Sesame (1) **Wafu (2)** **Spicy Mango (3)** **Blueberry Basket (4)**	**Black Sesame (1)** **Wafu (2)** **Spicy Mango (3)** **Blueberry basket (4)**
Unique formulation (2, 3) Fruity (4) Unique combination: Fruity, spicy (3) Good brand fit (all) Vibrant colour if transparent pack (3, 4) Authentic local cuisine (2) Incremental TO, no cannibalization (2, 3, 4) Dark brown colour signals rich aroma (2) Brand is market leader seen as innovator (all) No CAPEX need (1)	Untypical dressing flavour (2) Allergen label (1) Complex line cleaning (2) Overlap with classical savoury sauces (2) Unique ingredients: Tozazu, seaweed (2) Import material (4) W/o Stability (2, 3, 4) Shelf life (4) Niche – too spicy (3) Too sweet/sugar content (4) Processability in-house (4) Requires high CAPEX into line (3, 4) Extra costs (2)
OPPORTUNITY	*THREAT*
Black Sesame (1) **Wafu (2)** **Spicy Mango (3)** **Blueberry basket (4)**	**Black Sesame (1)** **Wafu (2)** **Spicy Mango (3)** **Blueberry basket (4)**
Leverage spicy food trend (3) Unique sourness (2) New advantageous consumer segment (2, 3, 4) Cross promotion with other spicy liquids (2, 3) Export to other markets (1, 3, 4) Health message (3, 4) Steal share from competitor (1) New HACCP may reduce cleaning effort (1) Dried natural flavours (3, 4) On trend in country (1) Fresher brand image (3, 4)	Early competitor launch (1) Not really new – me-too variant (1) Regulatory challenge health (3, 4) Perceived artificial (3, 4) Cannibalization own portfolio (1) Cost benchmark with competitor (1) Raw material cost fluctuation (4) High salt may be perceived unhealthy (2) 3PM needed (1) Supply chain disruption (2, 3, 4) Frozen storage (4) Barriers to enter new markets (1, 2)

The possible variants suggested by the innovation team were scored individually against a set of diverse criteria that the team felt confident to judge based on their diverse expertise.

By distracting the present weaknesses (–) from present strengths (+) and the future threats (–) from the future opportunities (+), the "spicy mango" variant appeared as the clear winner. "Black sesame" performed good in the present but showed clear weaknesses in the future. The team felt that "black sesame" was still attractive and it should be possible to plan actions to reduce the future threats, which have brought the variant down in the analysis. These two variants were selected for prototyping and consumer tests, while the technical team started to develop a sourcing strategy for sesame, the variant which unavoidably would contain a potential allergen, which would require long and intense line cleaning at moderate volume expectations – a situation leading to low operational efficiency and high manufacturing costs in the future.

CHARTER GATE

Gatekeepers in food innovation often first want to understand the consumer merit of the idea, the value that the target consumer gets from purchasing and using a product. Secondly, it should fit the business strategy: A company that decides to enter a premium market will understand that this needs advertising and sales money and may not have the ability to invest in new factory equipment that would bring significant material savings. And finally, at least some evidence of how the product will be made and that the resources required are available right now is necessary.

The gatekeepers will have made it clear upfront what the minimum requirements are for an idea to pass the first gate, after which normally substantially more effort and resources are needed. Typically, this can be commercial or management limits, like the minimum (incremental) turnover expected, minimum profit margin, maximum duration of the product development or maximum resource requirement. A typical list is given in Table 7.2.

And latest here, all involved project teams, project leaders and gatekeepers need to have a common understanding about the "size of the funnel", their innovation capacity. It's a waste of resource and a distraction from work if too many projects are passing onwards into phases where much more work is

TABLE 7.2 Typical questions from cross-functional gatekeepers at the charter gate

GATEKEEPING FUNCTION	*TYPICAL QUESTION FOR THE PROJECT*
Supply chain	What is the product design? Do we produce in house or third party? Do we need new processing elements/to modify processing? Do we produce the planned volumes? Do we have suppliers for all materials? Do we need extra audits (supplier, third party, new line, etc.)
R&D	Do we have all technology and skills are needed? How many (expert) resources, when and how long? What is new and different in the design? What type of experiments do we need to confirm feasibility? Have we locked the product design? What are the technical risks?
Market research	What is the market? Who are the target consumers? Which needs do we address? What type of research is planned? Is the concept tested and optimized? Which qualitative and quantitative consumer tests are needed? Budget for all tests confirmed?
Finance	What are the commercial and market risks? What are the expected sales volumes? What is the margin? How much up-front investment is needed (capital expenditure)? Do we have a draft business case with valid assumptions?
Marketing	What is the purpose and intended use? How is the strategic fit with business plans? What is the concept? Do product and concept fit? How will advertising and communication be done? How attractive are both to target consumers?

needed. On the other hand, at that first hard gate, enough projects should pass to allow further selection of the best ones. If the company has an ability to launch five new variants of a product next season, more should enter feasibility to have a choice later once more rigorous testing is done.

For the senior management doing the gatekeeping, it is important maintain the right balance between oversight of innovation projects and empowering teams to get the job done. They must set the pace and provide all support necessary for the project to be successful. And they have to take decisions. Three options they have for every project: Approve and pass gate, stop and kill project or reject and send back for additional work. While there is often a situation where the impression is that more data will make the decision clearer, gatekeepers are aware that this is a postponement of a decision, that may absorb resources used better somewhere else. Choosing the "send back" option too often can seriously impact the innovation potential of an organization. Or as Roosevelt put it in 1933: "In any moment of decision, the best thing you can do is the right thing, the next best thing is the wrong thing, and the worst thing you can do is nothing" (Brainyquote, n.d.).

While in this phase each project charter is judged on its own potential, some portfolio management may still come into action. A new product idea may have shown great consumer interest but will require lots of prototyping work or adaptation trials in the factory or analytical testing. The R&D lead may flag that there is sufficient capacity to fully test this idea but not similar ones in the coming period. Gatekeepers may then stop similar ideas requiring intense R&D work and allow instead the gate passage of another incremental project not requiring much R&D work – perhaps only a new point-of-sale activation, or a simple artwork change, something called "brand-led growth".

Some organizations have "rules of thumb" for how many original ideas should enter the next phase. If the scoping phase is not too strict, maybe because gatekeepers want to be more closely involved in idea screening (a two-edged sword, by the way), then the charter gate needs to be tougher and allow only a fraction to pass. A 20% success rate is often perceived.

FEASIBILITY GATE

In the FMCG feasibility phase, the consumer interest becomes more firmly confirmed and the technical and commercial robustness of the idea are established. Typical tasks are:

- Brand selection, if there is a choice to make
- Concept test and refinement
- Confirmation of the technical route
- Initial prototypes and refinement, aligned with the concept
- Idea about packaging design and artwork
- Rough idea about the advertising and communication plan
- Quantitative test preparation for product and concept
- Evaluation manufacturing and supply chain, first capital expenditure calculation
- Resource plan for next phase
- Initial business case

In feasibility, the evidence necessary for the next gate, the "contract document", are collected. At this stage the team should have looked at alternative project solutions compared to the idea presented at the charter gate, and the technical, operational, financial, legal and ethical requirements of the projects should have been tested and any limitations/constraints should be flagged.

In contrast to the scoping in the "idea phase", the majority of the work now is "hands-on" practical work, including product prototyping, improvement and fine-tuning, process design development, pilot plant trials and the layout of the manufacturing process and the initial test of the manufacturing. In this phase, the different workstreams progress their work. If not already done initially in feasibility the first product prototypes need to be made and tested. This can happen internally or with people outside.

The more unusual the product idea for the company is, the deeper and more thorough the testing will be. A simple line extension can be tested in an internal tasting session; an external sensory panel simply with employees in the company. Aa new core format, i.e., a plan to add a chilled meal range to the existing ambient stable one, requires more testing on the consumer and technology side. That lowers the risk of substantial mis-investment later in the process, due to a problem in the manufacturing, or contracts made to stock up materials only needed in these products or a misfit between product and brand, as Colgate experienced with their evening TV meal idea.

Towards the end of the phase R&D will have finalized the product composition and packaging formats and identified relevant suppliers of those materials needed in that launch. Both parties will have agreed on the detailed material specification, and R&D has done prototyping with that. Procurement will want an idea about the material prices based on the intended volumes needed and started price negotiations. If new suppliers are selected or the manufacturing is planned with a third-party manufacturer, an auditing process will have started involving quality assurance. The initial cost of goods can be used as input for the material costs, an important value in the business case that the idea needs.

The first upscaling of the prototypes in the pilot plant by modelling or initial factory trial must show if new manufacturing equipment is needed and the suitability of the manufacturing route. Computer-aided production trials and real production tests are used to confirm the manufacturing route. The packaging format(s) should be clear in this stage as well, and the required material should have been tested.

In collaboration with marketing and market research, the product idea is optimized. The fine-tuning after consumer feedback, the material costs and availability and the manufacturing capabilities will shape this. At the end of the phase, the product enters "design lock". In theory, new products should be designed by the consumer needs identified. But in practice, it's a balancing act. Assume a company has good "in line sterilization" capabilities, but the consumer tests for a new product indicate a slight consumer preference for "in pack sterilized" products. Is it the best to invest 1 million to build a new line or easier to tweak the communication pack a bit more to underline the benefits of the existing technology?

Beside the manufacturing skills, the product formulation needs to be locked. Based on the bill of materials, all material suppliers need to be contacted, materials sent and tested and the ingredient specifications agreed and locked. If a supplier development program is needed to achieve a particular service level, this must start in feasibility. Here the work on "speciality" ingredients normally requires more effort than the sourcing of commodities – standard materials that can be bought on the spot market.

The duration of a feasibility phase can vary significantly in foods development. The complexity and novelty for the company, availability of resources and data and the size of the testing program shape the program.

For simple line extension or rejuvenation in packaged food, feasibility can be managed in 3–8 weeks. Here, technical feasibility is mostly confirmed, since the company has designed and manufactured a similar product before. On the market side, the earlier launches have given a good understanding of the consumer wants, and the calculation of the return of investment for a successful business case can be done with good confidence.

With the technology already available, samples for consumer tests can be made without problems, either via factory trial for a larger test or with help of the pilot plant.

If the new product (range) needs technology new to the company, the feasibility phase can take 2–6 months. The testing program is generally more intense, and it remains a bit of a balancing act to decide when a product is sufficiently tested and can progress into the next phase. Foods is FMCG business – an opportunity in the market exists now, but it may be gone in 6 months. Or a competitor may come in earlier.

For the contract gate, the feasibility needs to be fully confirmed. The market potential can be compared to the investment needed.

CONTRACT GATE

The contract gate may be the most important one in the whole funnel process. Here the organization must assess its ability to execute the prototyped idea and deliver the desired outcomes. It is crucial for ensuring that the organization has the readiness and competence to innovate effectively and efficiently. Now the gatekeepers need to be assured that the proposed idea is worthwhile pursuing. It's the moment where all major investment is signed off on. So far, a good portion of resources, time and money have been invested to confirm the viability of the idea. If now the ordering of a new large-scale line for a new manufacturing equipment is necessary, the confidence normally needs to be higher than for a budget commitment for the contract of a few unique ingredients.

The document also contains a risk assessment section. Here the various risks, the probability and the impact are presented. General areas in which risks occur are:

- *Profitability*: Planned prices can be sustained in the market, the distribution target for point of sales will not be achieved, own costs are increasing, etc.
- *Technology*: The product or manufacturing design is not suitable, the "*in silico*" upscaling done so far may contain major uncertainties and manufacturing problems would block production of other products.
- *Safety*: Major upheavals such as loss of facilities, contaminations, distribution and storage shelf life.
- *Operational*: Launch used new distribution, or the outsourcing brings increasing risks associated with it.
- *Reputation*: Product liability and recalls, resulting in legal proceedings and brand damage.
- *Strategy*: Economic recession, fierce competition from retailers via private label offerings and extended promotional offers. Inability to sustain business long enough to make it a success.

The risk evaluation best includes a semiquantitative evaluation of the particular risk (low, medium, high) and the impact the materialized risk will have. The value of the risk needs to be agreed to (turnover, profit, time, people, money, reputation).

At the contract gate, the question if the product idea should be realized is answered. From now on, only minor changes in the technical mix are still possible. The total R&D effort has reached maximum and may be overtaken by supply chain and the manufacturing activities in house or out of house.

In management literature, sometimes recommendations are made that in FMCG a project should have about 40% chance to pass the contract gate. That's a good number for ideas with a short and not too heavy feasibility phase. If it's possible to make all protypes from mostly existing materials in the lab or kitchen, do the consumer research with an internal consumer panel and the manufacturing is "factory standard" or can be confirmed by "*in silico*" simulation, a company can afford to stop many projects quite late in the process.

For ideas with more substantial feasibility investment, the passage probability should be higher. Ideas that may have already struggled at the charter gate should not be "kept alive" too long. Often the relative attractiveness of a new idea in foods is too small to justify a launch. A range extension where most of the predicted sales are coming from the cannibalization of the existing range may only make sense if the underlying profitability of the new variant is much better. A typical task list is given in Table 7.3.

Here company culture is important again. A good project leader (PL) is biased; he/she will do the most to get "their" product to launch. It's this extra level of energy that drives the project forward. If now, through the great work done in feasibility the ideas show weakness, the PL should have the possibility to recommend a project "stop" to the gatekeepers. The gatekeepers now should recognize good work to confirm that this project is right now not a good one; even so a project stop is always a disappointment. But better to get off the saddle than continuing riding a dead horse.

TABLE 7.3 Typical pieces of work the various functions have to finish before the project can go to contract gate

GATEKEEPING FUNCTION	*TYPICAL TASK DONE FOR CONTRACT*
Supply chain	Confirmation of E2E manufacturing route Process design locked. Capital expenditure plan agreed, and equipment ordered When should constructions and installations start Is the contract with third party signed HACCP study finished Build demand plan based on final test results
R&D	Product and packaging design lock (if not done already) Process specification ready (technical route confirmed) All suppliers qualified and approved IPR potential assessed Claim substantiation finalized Risk assessment finalized Successful upscaling factory trial All specifications finished and artwork started
Market research	Final quantitative consumer test design
Finance	Fine-tuning business, including cash flow, NPV, IRR, ROI and payback period, to help stakeholders understand the financial value the project can bring in over a period of time
Marketing	Refinement of all key marketing mix elements Final mix test Design launch material plan Prepare local communication packages

CAPABILITY

After successful gate passage, the capability phase is entered. In the FMCG – and especially food business – the program differs from the IT and software sector, where these project management tools have been invented: "Field testing" or the launch of "beta" versions is not helpful or even dangerous. In IT and the software industry, a "god enough" prototype can be released to the market and with the help of suitable consumer feedback loops, the product will be constantly improved before the project will finally leave its nest.

The key features of food products, safety and quality, do not allow such an approach. Legal standards forbid that. To correct a "defect" in product safety means recall and will basically burn the ground for the launch of a "safer" version. And improving the quality based on early consumer feedback isn't an option in the mass market. For startups and in gastronomy the possibility exists to invite a selected group of buyer/users to tell them that they are part of the product optimization. But in most sectors of the food industry so many alternatives are already available that an initial "non-performance" will damage the overall potential.

The overall effort for R&D is coming a bit down, while marketing is busy with testing and optimizing of the marketing mix, while the manufacturing often reaches the maximum. Based on the product and process specification made by R&D, the new materials and machines are ordered, and the implementation now must wait until all material and equipment are in place. If major technical changes in the manufacturing are necessary, an experienced supply chain manger may join the team. R&D process development will work hand in hand with the factory to establish the manufacturing in the factory. The design is guided by the processing specification, a document often owned by R&D.

This document often describes all the ingredients used in the manufacturing process and the processing steps: These are the specific steps involved in the production of the food product. Details of the processing are still tested and agreed with the supply chain. Pre-mixes, batch sizes, storage and handling conditions are discussed with the factory to ensure a safe product, in line with the wanted quality standards at manufacturing conditions that reflect the "line efficiency" in the business case.

While the finished product is already locked because of the contract, the process may not be locked at this stage. The materials used for the food product production are described in the raw and pack material specification, which have been signed with the suppliers earlier. At the other end the "finished good" specifications also define the corridor of manufacturing. Typical components of the finished good specification are:

- Physical attributes, like pH, water activity, moisture, taste, flavour, colour, texture, product dimension, weight, etc.
- Microbiology, like total viable count, pathogens, yeast, moulds, spores, etc.
- Shelf life
- Packaging, including inner and outer packaging specification and sometimes including case and pellet configuration
- Shipping and handling instructions, including temperature, humidity, weight limits and pallet size

Once the product line is installed, a hazard analysis of critical control points (HACCP) is performed (see Chapter 10). The finished good specification helps to set those critical points.

At the end of this technical program the factory formally accepts the specifications and confirms thereby the full visibility.

Marketing is working on the advertising and communication program, optimizing the various elements and validating the effectiveness with the help of different tests, often involving external agencies. Finally now, the sales team becomes heavily involved. With them the channel strategy is discussed and tailored packages agreed. If the product idea is based on new consumer insights, a final quantitative test with full mix and price will happen.

All tests in this late phase are supposed to confirm earlier assumptions. The space to still make changes in the mix is limited. Changing the product too much can make the results of the consumer testing less valid. And since most large-scale lines are shaped by efficiency, major process modifications are no longer possible. A much more manual food process is less sensitive to these constrains but lacks the economies of scale.

If the idea contained some "technology push" elements, the possibility to legally protect intellectual property rights (IPRs) should be clear and patent texts should be filed.

The R&D functions for nutrition and regulatory are often active in this stage. Contact with the authorities are happening now, as product and process are locked. Where needed for market approval, the result of a shelf-life test is given to the authorities. That's obligatory in some countries for products making nutrition or health claims. In case new materials had to be ordered, those should now be in stock to support at least the volumes needed for the first production run.

Like most other phases in the stage-gated process, the length of the phase can vary, from 1 month to a year or more. Often the "technical realities" set the duration. If a new line needs to be built, commissioning and implementation need time. Or the communication package contains health and nutrition messages, which require premarket approval before they can be used. And sometimes even the supply of new ingredients, new packaging or packaging-forming equipment needs months. Sourcing a new, exciting ingredient from another country or helping an entrepreneur to scale up often costs time. But it can bring a true competitive advantage.

Once all these activities have been finalized, the project moves to the final official gate before going to market. Sometimes a company can decide to accept inefficiencies and oncosts to shorten the time to market. Perhaps some other, less efficient processes in the factory can be used, or a more expensive material, already in stock can be used instead of the originally designed one. But this approach requires good confidence that the temporary oncost can be fully captured after a given time.

LAUNCH GATE

The launch gate is the final "go" for the product or service to be ready to go to market. It is no longer the question if the idea should be launched. A launch delay or even a project stop is the exception. Only a radical change in the market conditions or the unexpected failure to finish one of the critical tasks in capability can now block the launch.

Unexpected difficulties of the supply chain to produce sufficient volumes for the necessary pipeline can cause a delay. Or the marketing and sales planes could not be finalized as expected.

The final business case should as well not show drastic changes versus earlier versions. It is usual that the latest data show some optimization options in the marketing, advertising and sales effort in one or another channel or a particular trade partner.

For R&D, this is the latest moment to confirm that all technical parameters are met. Launching a new food product successfully often requires speed. So it makes sense to not delay an earlier phase, because the final shelf-life test is not finished. Professional experience, simulation, modelling or accelerated shelf-life tests can be used to allow gate passage. But now, just ahead of the launch, the data need to be safe.

Similarly for the artwork content: Many data provided on pack are mandatory (see Chapter 11), and they must be correct. So, all the latest modifications that may have happened during the capability phase now must be "bullet proof". If the HACCP study done indicated somewhere a risk in the supply chain for a cross contamination through an allergen, this now is either excluded or the artwork needs to have the appropriate warning on the pack.

Once all boxes in the check list are ticked, the new product is ready for launch. During the production of the "launch volumes", R&D normally supports the factory, takes samples, helps with the line efficiency and defines "what is in specification" and what is not. Typical "minor defects" are described and documented, so that later on the factory has clear guidance.

POST LAUNCH GATE

The commercial launch is not the end of new product development. While the advertising in the media and the point of sale reach their activity maximum, R&D often still has to conduct important tasks.

One is linked to the activities in the market. Nowadays R&D representatives are the ambassadors of the products and the science and technology behind them. As the technical experts they are most suitable to talk about the products. Not only to an internal audience, like the marketing and sales team, so they can start wordsmithing a few catchy headlines for the public. Sales conferences and key customer meetings are an opportunity to carve out the technical "unique selling propositions" of the product: A long-lasting aroma compound, naturally sourced material from controlled farming, a smart opening/closing system, a special nutrient profile or biodegradable material. There are many stories that can be told here. At the end people are buying the product, not the advertising campaign, no matter how brilliant that may be.

Often, the technically educated product representation is also helpful in the different social channels. Or for audiences like the key opinion-formers like food vloggers or academic leads. A food launch with additional functional benefits for people or the environment requires explanation. Representation on food conferences, symposia, fairs, etc., helps to create attention and perhaps support for the launch. Conference posters, oral presentations, videos and journal articles are suitable tools that need the hand of R&D.

And in the factory the manufacturing frequently needs help with the optimization. A new product has a new process specification, where only the series of step-by-step improvements finally make the product a commercial win. If for the launch new equipment has been installed in the manufacturing line, fine-tuning and optimization can take several weeks, The R&D developers and process engineers need to work together with the manufacturing on this fine-tuning.

And long established manufacturing processes just for a variant can ask a lot of effort. Many manufacturing steps in foods use very mature, long-established technology. The competitive advantage is not coming from the ownership of the technology as such, Competitors can use it as well! It is coming from the ability to execute the manufacturing extremely efficiently, at costs better than the ones of the competitors. Being able to push up the "operational equipment efficiency" (OEE) by 1% or 2% can easily contribute to the commercial win on high-speed lines.

The post launch review is another formal milestone to evaluate the results and impacts of an NPD after it has been brought to the market. It is a way of assessing the strengths and weaknesses of the product, concept and organisation. Often, two very different purposes are included: One is the lessons learned and the best practices improvement for future endeavours. The helps the team and the company to change aspects in the innovation process that turn out to be suboptimal and keep the elements that worked well or better than expected.

The second and perhaps more urgent one is the performance analysis of the current launch. The aim here is to correct and improve those elements of the launch mix that can be improved. It's an activity within the launch. This includes the analysis of the perception and reputation of the product. The product's reputation can substantially impact its sales and overall performance, making it crucial for businesses to evaluate how consumers perceive their product.

The basis for this analysis is all data showing product market performance and indicating possibilities for further and improvement areas. The learnings are mostly for the sales force, the supply chain, marketing and R&D.

The first thing to measure the success is that the product has landed in all those places in the market where it should be according to the business. A bigger national launch with various trade partners will need a few weeks to fill all pipelines and fill the shelf.

The business case will have made assumptions about the distribution that can be achieved with the new food products. Customers want the product in the right amount at the right time. That's often measured as "on time in full" (OTIF) deliveries, an important key performance indicator (KPI) for the supply chain (sourcing units and distribution).

From there the "on-shelf availability" (OSA) can be calculated, which measures the percentage of time an item is visibly accessible for sale on shelf by consumers, where they expect it, and now, they want to buy it. The data are gathered either by a physical audit or an inventory data analysis. The flipside of this measure is the "the out-of-stock rate", which reflects the value of sales lost due to unavailability. This can be caused by supply problems between factory and shop show or the warehouse and store management. A retail outlet does not want to lose customers because an item they repeatedly ask for is not available. But they also want to optimize their shelf space, not letting to many items the consumer is not asking for occupying space.

A new product on the market should create a desire amongst consumers for the food. The product concept has to awake this, and product, packaging, point-of-sale display and the total advertising need to communicate this. Market research agencies sometimes estimate the appeal of products in development. The idea is that the broader the appeal, the greater the likelihood the consumers will try the new product.

The trial rate is something that is mainly influenced by marketing effort. This includes the attractiveness of the concept, the strengths of the brand, the choice of the communication channels and of course the amount of money spent to "make the message heard".

For the product developer, "adoption" is one of the key measures of success. Adoption is the willingness to continue buying after having tried the product once the concept has raised an expectation with consumers that the product needs to fulfil. A new launch of an established brand and product group must establish a solid consumer base, one that is willing to buy the product again, that is the repeat purchase.

A successful marketing campaign during launch will "create so much noise in the market" that enough consumers are trying the product. But this can only be for a limited time. After that shoppers should buy the product because they are convinced about the product performance.

The combination of quantitative market research data established in the "going to market" phase, average numbers of consumers, shoppers in a product group and experience in the company from earlier launches form the reference against which the new product launch has to perform. For fast-moving products, the repeat purchase is essential, where the purchase frequency reflects the "distribution of brand buyers and brand volume, across the number of purchase occasions" (MarketingMind, n.d.).

If the repeat purchase data in the post-launch evaluation, about 3–6 months after the launch, are not as expected or predicted in the business case, the product details need to be reviewed. Sometimes the marketing concept has raised expectations that the product functionality cannot fulfil. Concept and product must fit. "Overpromise and underachieve is not a solid success basis". Or it is simply the price. During product introduction, special offers have been made to encourage shoppers to try. But when the price moves to the planned price point, shoppers have got used to the promotion price and now simply wait for the "good" price to come back. Or some sensorial attributes turn out to be not fully matching expectation, as they did not show up clearly enough in the product development. Maybe the late consumer testing prior to product design lock "JAR scores" have not been interpreted well or have not been taken.

Another common reason for a product deficit, a disappointing repeat purchase, lies in the project development phase. Early in development a new food product has been tested with consumers as having a very pleasant, strong, fresh or natural taste. The concept work highlights that. But during the next weeks in the project work it turns out that the overall material costs of the product are too high to match the commercial case. The costliest ingredient is that making the unique taste. So, now they have been reduced a bit to match the business case expectation, of course, after internal tasting proving that "this little" reduction should not be perceivable by consumers.

Or the first batches of new leading ingredient(s) from a new supplier have been just a bit more intense than later deliveries, as the supplier wanted to make sure that his offer was chosen. And the signed ingredient specification that forms the physical basis of the contract has just not captured a little fluctuation in the aroma intensity of the ingredient.

A performance change during shelf life could as well be the cause. That could come from an inappropriate storage in the warehouses or during distribution. Or the new packaging is just a little bit more permeable than the old one, as it also has "lost" just a bit of protective barrier during the "cost optimization "phase of the development work.

Data directly from the point of sales (shop) or point of usage/consumption (home, on the street) have to complete the picture to identify if the underperformance is a "product design defect" or a "manufacturing design defect". Quality assurance can help with data taken under consumer relevant quality standards (CRQS) or through the digital voice of the consumer (DVoC) (Chapters 4 and 10, respectively).

There can be numerous reasons why the product is underperforming. Mostly it's not because the product developers have not done their job properly or made mistakes. So many factors or combinations of very small changes finally led to the product losing superiority. It is important in that phase that the "the defect" is analysed with an open mind. The always present "business pressure" to repair the defect should better not lead to a "finger pointing exercise" in the project team. Mistakes or suboptimal solutions happen all the time – only doing nothing, prevents that but is no option in NPF. The paradigm is to fail fast and learn quickly. It does not help the launch and undermines the ability for future team performances needed in the next projects.

These "repair options" do not sit with one function after the analysis is done. A mismatch between product and concept may be best corrected by changes on both sides. A insufficient on-shelf packaging appearance is not only the responsibility of the distribution system or the packaging design, but the result is a combination of both.

A post-launch review is not a "urgent repair mechanism" but a standard assessment. It brings learnings for future projects; of course, even new product launches in line with or above expectations of the underlying business case have the potential to become even better. In the end this is the first true reality check, nothing based on statistical models, test group reactions, estimates and simulations. And it's the moment for the team to celebrate the achievements and recognize their contributions.

FULL ROLL-OUT

Smaller local companies or startups may have a chance to test their product in a few shops or outlets or via e-commerce first before going into full scale. Bigger national enterprises can start launching in a single town or a district. And global food multinationals have the ambition to launch a newly developed food product not only in one country. That's often not enough to recover the whole costs of the development process, which may have taken months, perhaps a year, and where substantial investments have been made, particularly when the development had been flanked by intense consumer research in all phases. But they would also not launch in several countries or markets at the same time. "Such all-in," is often too risky.

In all these different upscaling situations, the enterprise would carefully review the market performance of the product in the first market. The competition is not sleeping, and they will have observed the launch. The post-launch evaluation offered the chance to improve the overall launch mix after a pressure test in the market reality.

If the manufacturing process for the next markets is not bringing new difficulties or delays, like longer-distance distribution, legal and trade barriers, regulatory approval, the need for a new manufacturing side or a suitable third party manufacturer, then a final market test of the full final mix – product and concept – is the most efficient check for the new market potential.

Often, such tests with consumers in the next market reveal further improvement options, which would make the commercial situation even more attractive. It then depends on the company's strategic intent and the key features of the food product group, in which deviations from the original mix are accepted. If new advertising material is needed, maybe because of a country with a different language, locally wanted changes in the communication package are acceptable. If the communication material has been designed centrally with a global roll-out in mind, different shots and comms packages beyond a simply voice over are time and money consuming.

Product and process developers will in such cases have tried to make their contribution for a fast global roll-out. A "chassis" approach (see Chapter 14) should have somewhat anticipated the optimization steps. Rolling out a selection of different instant noodle soups from the first country to the next ones will require some modifications. Elements like new, additional top note flavours and extra or different garnishing ingredients should not bring too much complexity and delay, if the basic chassis – maybe a 300-mL polyethylene terephthalate (PET) Pot with lid, standard instant noodles, a prepacked sauce and spice mix and a dry sachet – remains the same. A different plastic material, a different size or filling volume or a completely different formulation can have massive consequence in the manufacturing. It's always up to the roll-out team to judge which changes are unproblematic and which ones are painful. The responsible business will have to decide which compromises are acceptable and which ones are not.

POST-LAUNCH EVALUATION

Sometime after launch, the performance of the new launch in the market must be reviewed. With the introduction, the first phase of the product life cycle starts: It's the first time customers are exposed to the new product. For the launching company, this normally means a substantial investment in advertising and a marketing campaign basically ringing the bell, getting through the noise, and making consumers aware of the product and its benefits and trying it. After all the piloting and pre-launch testing, this is the phase where performance really counts.

This review should happen rather quickly, within weeks, maybe a few months. It should be some time after the launch marketing campaign has come to a stop and the product has achieved full distribution.

Normally that would be the time when the consumer had the possibility to try the new product and finds it again on the shelf to purchase it again. The ideal outcome is a continuous, repeated purchase by known users and maybe "new recruits" after the marketing campaign has come to the normal maintenance level and the market shows normal competitor activities.

The primary objective of a "post-launch evaluation (PLE)" is diagnosis and, if necessary, therapy, but not a "postmortem" analysis. At this point in time, it is not of primary interest why the product is performing as it is. The key question is "how" its performance can be improved! The PLE has the character of a gap analysis: It compares "planned versus achieved".

Both qualitative and quantitative data can be used for the evaluation. Existing commercial data build the frame to review the performance. They are the key KPI for the launch. But qualitative data like consumer or customer feedback and interviews can be used get an even better understanding of the situation. Lots of data and many aspects can be reviewed here. More data do not necessarily mean better results. The right data and good insights are often more important. A few criteria with relevance for R&D are listed in Table 7.4.

The initial trial rate after launch gives a good understanding if the mix concept and communication/product are working. If concept and communication have not been properly tested before launch, consumer and market research should be done. The trial are all attempts to encourage consumers or customers to buy the new product. This includes all types of marketing activities in all channels, like mass media, social media or campaigns at the point of sale.

If the first sales (trial) are in line with expectation, but the repeat purchase is under target, often the first action is to check product performance. One important aspect here is to evaluate if the product can stand up to the expectation raised during the launch campaign.

TABLE 7.4 Typical data collected as key performance indicators showing the performance of the new product or service in the market and their root causes

MEASURE	*POSSIBLE DISCOVERY*	*POTENTIAL INSIGHT*
Market	Initial purchase under expectation	Product not appealing. Review factors such as the product concept, packaging, display and advertising.
	Repeat purchase insufficient	Customer experience at point of sale, OK? Low adoption indicates the product performs poorly and does not meet consumer expectations. Distribution and OSA. Check product appeal on shelf, get QA and consumer/customer feedback. Target group met? Consumer need not met. New competitor launch. Shelf life or distribution stability. Control retention samples and inspect point of sale or usage.
	Market share	Poor design or functionality; difficult to open, store or use the product.
	Increasing complaints	Product falls short in comparison; negative feedback may arise.
Consumer	Inadequate packaging	Product specification too tight for manufacturing process. Quality of purchased goods changed.
	Competitor comparison	Factory staff not properly trained for product. Quality criteria not clear.
Manufacturing	Waste level	Better market scan necessary, closer market watch.

There are rarely launches in which the food product when tested is perceived as better than was expected from the campaign. Often the opposite: "Overpromise and underachieve" is the case. The food product has difficulties to meet the expectations raised, it falls short and the consumers feel disappointed. If a "low fat or fat-free" promises "same rich taste" as the full-fat version but then tastes bland and lacks flavour, the product seems to be not right. And that can be the case. The product "diagnostic" can give useful hints. Has the product been tested in market research with "just about right" (JAR) scores (see Chapter 8)? Do these show a weakness? Did the internal tasting comments show a sensorial weakness? Was the formulation or production changed during the development phase to meet the project cost targets? Is the distribution and shelf-life stability as expected? Consumer feedback from care lines or discussion in social media should help to find the root cause. This can also be done with the help of AI tools like the "Digital voice of the consumer".

Equally important is the discussion of "internal" signals. The manufacturing may complain about too high waste levels or insufficient line efficiency. R&D should now investigate jointly with the supply chain what is driving the deviation. During the first production the factory normally does not achieve the target line efficiency. In case the deviation is caused by changes in the quality of the purchased material, R&D has to review jointly with QA if the material is "in specification" and perform corrective actions as needed.

The post-launch review is not a blaming or finger pointing game; its purpose is to identify correctable weaknesses to improve the performance of the product just launched. The team should openly share "pain points" in marketing, product development and manufacturing.

Regularly, the material cost target in the business case limits the sensorial product optimization. In such a situation the R&D team knows how to further improve the product, but they cannot do it without additional costs. If the sales price and the margins are fixed, improvement is difficult. But sometimes a deep E2E cost review offers opportunities to cut some costs in one step of the process to "reinvest" then into quality somewhere else. Maybe the product packaging can cut perhaps cut one or two colours in packaging printing or the carton thickness can be reduced to generate a few cents for more or better flavour, herbs or spices. Or will longer-term investment into more accurate dosing equipment in manufacturing lead to better visible pieces or less dosing loss?

If the overall sale volumes are in line with expectations but the profit is below target, the team can suggest a value improvement program (VIP) to systematically search for savings opportunities (see Chapter 3).

The task is to find solutions to fully match the consumer/customer expectations on a lower cost level without distracting from consumer/customer perceived value. This works when the team has spent money on a product attribute that is not relevant for the consumer/customer. Sometimes this is called "product overdesign".

In very good-performing NPD teams such discussion happens before the launch. If R&D or the supply chain has serious concerns that for the money available the product may not meet the expectations raised in the marketing campaign, it's their responsibility to make that heard by the business, even if it can lead to a project stop.

In reality, things may be more complicated. A new product launch will have a quite substantial marketing investment to create the "noise" in the market about the new product, driving people in trial and repeat. If after maybe 3 months, the new food product has reached its full distribution and people had time enough to test the product and buy it again and the sales are not as expected, the business must make some important decisions. Should marketing continue to promote the new product through longer, maybe different campaigns? Or will that only be "throwing more good money" after the amount already lost? Can the product be commercially sustainably improved so that consumer expectations are better met, before the business invests more money in new/better communication?

In all these discussions, it is important that R&D fully understand the business strategy and give their educated advice on the options or limitations they see.

POST-LAUNCH REVIEW

In contrast to the post-launch evaluation, the post-launch review is a process assessing the entire process from the beginning to launch. This naturally includes the analysis of the market performance of recent product launch(es) performance, but not with the key objective to improve the business case. The objective here is to identify those elements in the whole process which led to the failure or a disappointment in the commercial success. Internal and external data are collected from all functions involved and the key facts. Important is that all stakeholders are involved in the process and that the retrospective is not becoming a shaming and blaming exercise between individuals or functions. It is a short selection of critical data to capture the learnings from the finished innovation process to improve the process in the future. A list of typical measures and insights is given in Table 7.5.

TABLE 7.5 Typical measures collected for a post-launch review to identify weaknesses made in the project

MEASURE	*POSSIBLE DISCOVERY*	*POTENTIAL INSIGHT*
EXTERNAL		
Sales volume	Consumers not buying as planned Market size wrong	Wrong price point, consumer not satisfied, benefit not clear Target group missed or misinterpreted Manufacturing, distribution problems
	Product not available as planned	Consumer feedback not positive Insufficient shelf appearance
	Competitive activity	Distribution target not achieved Product not attractive for trade Better market scan necessary, closer market watch
Profitability	Price too low Unexpected costs	Launch discount set price expectation Cost calculation incomplete Unexpected change in material or manufacturing costs. More marketing expenditure needed to create trial "Unforeseen" cost too optimistic Capital expenditure high
INTERNAL		
	Inefficient data	Wrong assumption in business case, not enough technical experimentation or consumer research
	Insufficient resources	Not enough/wrong expertise Not enough budget for tests Wrong collaboration partners Project team composition Insufficient training of people
	Insufficient time	Product concept not sufficiently tested Technical feasibility not solid Essential checks not done Launch window missed
	Insufficient project management	Team not dedicated enough Line manger interference Changing stakeholder priorities Insufficient gatekeeping Too high/low risk taking, insufficient risk mitigation

The goal of the post-launch review is to identify areas where the new food product development strategy can be improved. In case the list of improvement suggestions gets too long or complex, key principles of change management can be used:

- *People*: While the decisions are often coming from the leadership, changes affect all people in the innovation process, and they should welcome the change.
- Priority is the discussion about which improvements are the most wanted ones in this NPD process. This is often about people and their energy.
- *Brain and Heart*: Ideas for changes should be built on solid facts and data but need to be explained in an engaging, emotional way,
- *Leadership*: It's good to separate the more technical skills, like project leadership, from the soft leadership skills people would follow.

The review also sheds light on the tools used in the development process. Was the creative process open enough? Was the prioritization correct? Did we use the right data, the best tools, the suitable consumer testing, etc? Which surprises occurred, and why were they not predictable? A good analysis can show the elements in the process that can be modified to increase the chances of success next time.

All changes planned need to fit the company philosophy. If in a more hierarchical organization the leadership team decides to give their innovation teams more space in decision making, it may first be necessary to have dedicated training for the team, so they are comfortable with their new "decision power".

And finally, all these new ways of doing innovation are implemented during "normal business", since companies cannot put their ongoing product development and maintenance work to a standstill to install new ways of working.

POST-LAUNCH MONITORING

For specific food products, a long-term evaluation can provide new learning and insights. This work is sometimes called "post-launch monitoring." I will not use PLM as the abbreviation for this, as this describes for most people the wider known "project lifecycle management". This is basically the food variant of the post-marketing surveillance used in the pharma and medicinal sector to assess the safety of novel foods or functional foods or its ingredients. A functional food is food with an added health benefit, mostly conveyed by an added ingredient. Research done under this umbrella can try to measure the functionality of a substance or product or assess the safety. Functional food is normally not prescribed, and the intake is not controlled. A health promise is made, but in principle anybody can buy and consume the items without expert guidance. For functional or health food, it is important find the best dose corridor between safe and effective.

CASE STUDY: PLANT STEROL INTAKE

A good example is the addition of phytosterols (PS) to margarine in the year 2000 and a few years later in dairy products in Europe under the novel foods' regulation (Ras & Trautwein, 2017). PS inhibit the absorption of cholesterol from the blood and lead to lower plasma low-density lipoprotein (LDL) cholesterol concentrations. Unilever, at that time a global leader in yellow fats and with ample experience in dairy, decided to develop a range of products with added PS for those people with elevated LDL cholesterol levels. The idea goes back to the studies of Ancel Key in the mid-50s (full story in Chapter 17), where he reported a strong correlation between the consumption of animal (saturated) fats and coronary heart diseases. Around the same time, the pharmaceutical industry recognized the effect of PS on blood cholesterol levels.

But some cholesterol is essential for the body, and only excess leads to damage. People with moderately elevated levels would benefit from foods with PS. Others like children <5 years and pregnant or lactating women should not reduce their cholesterol intake. So, adding this ingredient to something like margarine could lead to PS intake for everybody around the breakfast table. In a long series of studies, the company confirmed it could show the efficiency and safety of a daily intake of about 2 g/day.

Next to the thorough premarket entry safety assessment. Unilever was told to run a post-launch monitoring program to confirm the product use and its known effects. A monitoring program a few years after the first launch indicated the product was indeed consumed by the target population and that average intakes were lower than expected. The findings of the PLM (post-launch monitoring) survey covering the first year of launch showed that the product was bought by the target population but that intake of spreads with added PS were lower than originally assumed. In the next years more PS-containing products received market approval, so that nearly 15 years after the first launch, the company ran another monitoring program to calculate the daily intake of PS.

In more than 80,000 households in the EU, consumer purchase behaviour was tracked to calculate the daily PS intake. The analysed data showed that PS were dominantly purchased as margarine by one- to two-person households aged 35 years plus, but not in households with children <5 years. Average intake after 15 years was still at a suboptimal level below 1.5 g/day.

Some of the key learnings out of this study were that the price premium of that product limited purchase, but also that the awareness about the cholesterol-lowering effect of PS still was low. Communication programs with health care professionals were seen as a good tool to overcome these barriers. And the data showed that the planned upper intake level was not reached, leaving a big safety space, but would also allow a better cholesterol-lowering effect with higher average consumption.

WEAKNESSES OF THE FUNNEL PROCESS

Generally, the "stage-gated process" is a good management tool for bigger activities and complex product launches. It's basically one directional – the progress is guarded by many control points, which makes it time consuming. These features are not suitable in all situations. So no wonder that scientists and business professionals have observed a couple of weaknesses in the practise of the funnel management (Hakkarainen & Talonen, 2014).

It starts with the overemphasis of the early funnel phase, the idea creation. Having many ideas is often perceived as a success. It somehow signals the "productivity" of the innovators and provides the warm feeling that the company will never be short of options to grow the business, creating the impression of rich yield. But it requires a good understanding of the business need and capabilities in this "feeding phase" to get the right ideas in. It's a push process, and a high number of initial ideas doesn't ensure quality; the new food product will not be a success. "Garbage in – garbage out"!

Tough gates and a rigid filter process help to limit the problem, so that the "bad" and "ugly" ideas can be filtered out. But terminating many projects is not without side effects. It is a waste of resources. And the dedicated team members will have put their lifeblood into the realization of "their" idea. It's "their baby", so naturally emotional and motivational factors can impact their objectivity and affect their decision making. And then there is the "sunk cost dilemma" (Chen, 2023). In its core, this is the emotional difficulty to stop a project after so much effort was already invested. In a moment when the results are not what was expected, it's hard to agree that all this time and money are lost and cannot be recovered. Maybe some more effort and different consumer research would show a different result? But when to stop "throwing more good money after the bad already spent"? Rationally, all agree that sunk costs should not affect decision making. And theoretically, everybody knows that it is the purpose of the stage-gated process to sort out most of all ideas started. It is a funnel, not a pipeline.

The business leaders therefore must ensure that a project stop is not taken as project team failure. The idea owner or project leader must be recognized for the good work they did to demask what was initially a great opportunity, but now, after a closer look and good work turns out to be not the right one any longer.

Another, different challenge for the funnel process comes from the intrinsic project individuality. Once an idea has been taken up, it will basically progress on its own. While it may have originally not been the "best" choice, another one starts to look more promising throughout the stage movement. Often the team and the gatekeepers may tend to readjust the idea and optimize it. What is not transparently happening in the process is the comparison of ideas and their needs in different development stages. One project may be in upscaling and another one as well, but the necessary process engineering can only cover one. So, should one project wait until the resources are free or better move another project further that is in product design phase? A "stage-gated" decision process manages project by project, while it is also necessary to manage them in relation to other ideas. That's the portfolio management, where all activities are compared with each other under a common set of criteria (see Chapter 9).

As said earlier, the funnel passage is a push process and ideally unidirectional. It includes the possibility to "send a project back" at a decision gate, but overall, that should not be a regular case, since otherwise all the activities in the feedback loop or recycling stream will start to clog the innovation artery. "Sending a project back" at a gate is more the recognition that some evidence is missing in the project work so far than a design principle for successful innovation.

Other innovation management schemes see continuous recycling as the core shaping project mechanism and put the continuous recycling much more into the centre of the process.

Since the COVID pandemic big FMCG companies started to transform their innovation processes. The leading idea was simply to deliver value faster through agile ways of working. And one key element was to weaken the rigid process structures and focus on delivering value. Speed becomes the lead currency.

SCRUM IN FOOD PRODUCT DEVELOPMENT

An alternative, less strict management tool for new product development is "scrum". The name appeared first in the management literature in the early 1990s and borrows its name from a successful rugby restart formation (Scrum Alliance, n.d.). For about a decade the tools have been increasingly used by big FMCG companies as a leaner process framework to manage some types of product development and other knowledge work. The methodology started like many of those "lean management" processes in software engineering but is nowadays widely used. Scrum is often described as an agile project management framework. It still has and includes meetings, roles, tools, etc., but those are more set to support the structuring and management of teamwork.

Classical project management is sometimes described as the "iron triangle" of cost, schedule and scope of a project, signalling the interdependency of these three important elements.

Teams using scrum can learn through experience, self-organize their problem solving and reflect on their successes and failures to continuously improve. The teams set up a hypothesis of how they think they can approach the task ahead, for example, a new product development idea. The team tries that out, reflects on the experience and is free to adjust as they feel appropriate.

The main difference from other project management methodologies is the iterative decision making. As a relatively new method, scrum tries to avoid the principal weaknesses of older methods, like the earlier described stage-gated process, while keeping the known strengths. Classic project management calls for project managers to look at the development as a straight process, whereas scrum divides the new product development journey into distinct segments. Agile here also means that the roles of the different team members are less rigid. Individuals that are performing the tasks are the ones taking ownership and estimating completion times.

SCRUM ROLES

The first thing a new management idea must do is to create own vocabulary. New phrases and terms signal new thoughts and breakthrough ideas. Now a "schedule" becomes a "sprint", the "project scope" is a "sprint backlog", "work structure" is a "task breakdown", "productivity" becomes "velocity" and the "launch timing schedule" is a "burndown chart". But even in scrum a few roles are defined. The classical scrum setup describes only three roles:

1. Product owner
2. Scrum master
3. Development team

The "product owner" (PO) sets the project vision/goals, defines and prioritizes the project tasks and provides expert guidance throughout the execution of a project and is finally accountable for the success or failure of a project. Originally not part of the job was the responsibility to communicate the interests or requirements to various stakeholders. But in today's reality a scrum team is only a small part of the company, but with more independence, that needs to maintain vibrant links to the rest of the business to be successful.

The core job responsibility is to develop, revise and manage the "product backlog", sometimes also called "user story". This backlog is the list of improvements must happen on the current prototype after discussion with the stakeholders in the coming "sprint" (Scrum Guides, n.d.).

The "scrum master" is the process gatekeeper, making sure that the methodology is understood and properly used. The person can also take the responsibility to link the team within the rest of the organization. The specifically trained person acts as "servant leader, facilitator, coach, manager, mentor, teacher and change agent" (Scrum.org, n.d.b) The scrum master is not directly involved in the product development but leads the innovative team through the process.

The rest is the development team. They are creating the incremental improvements on the product prototype as agreed in each segment target. They do what other team members do in a project, just that the "delivery agreement" is for the end of the current segment, the "sprint". The team also has representation from those functions that are key for the development. The primary responsibilities are performing the sprint (the work in the next segment) including self-organization of the product. The development team members present the progress in often short meetings, called "stand up". "Daily scrums" are suggested, where in very short meetings each team member reports progress. The name emphasizes the important point: As short and crisp as they can be. Not a "classical meeting", with a draft agenda, welcome, sit down, progress report discussion, conclusions, minutes and follow-up.

In contrast to older processes, all team members define the sprint goals and determine high-priority items (Ojukwu, 2021).

SCRUM PROCESS

Clear project briefs ensure alignment and speed. These are typically one-pager templates created by an initiative owner covering outcomes, milestones, resources needed and impediments. Once the brief is written, "sprint zero" starts, Here the team works to agree on the brief, challenging all assumptions and clarifying what will be delivered and by when. That starts the first sprint by the squad team.

As said, the top structure of scrum is a series of "sprints", periods with a fixed lengths of a months or less, in which the agreed work is done and reflected. As the name suggests, these periods are rather short and end with a sprint review, which leads into the next sprint.

An important milestone is the product goal, ". . . a concrete, actionable, measurable objective serving as bridge between the product vision and the sprint goals" (Scrum.org, n.d.a). A typical product goal in food product development can be a number of product prototypes, a specific consumer research, panel or market test or factory trials.

The sprint goal is the single objective for a sprint. It's a clear and actionable target but with some flexibility in terms of the exact work that is needed. The product developer in a scrum team could take the task to include two seasonal ingredients in the product prototypes. That's clear and actionable but leaves enough space to the person on how to do that.

The team agrees the sprint length, typically 1 month or less. Shorter sprints generate more learning cycles and limit risk to a smaller time frame. Speed is the lead currency. To that purpose scrum differentiates between "lead time" and "cycle time". The latter is the time between start and end of a task, while the lead time defines the moment when the task was committed. The agreement for a couple of prototypes requires to select and order new ingredients for prototypes – that's the lead time. Only part of that is the cycle time. The differentiation can help the team to identify some hidden bottlenecks; here maybe the ordering of the new ingredients takes too long!

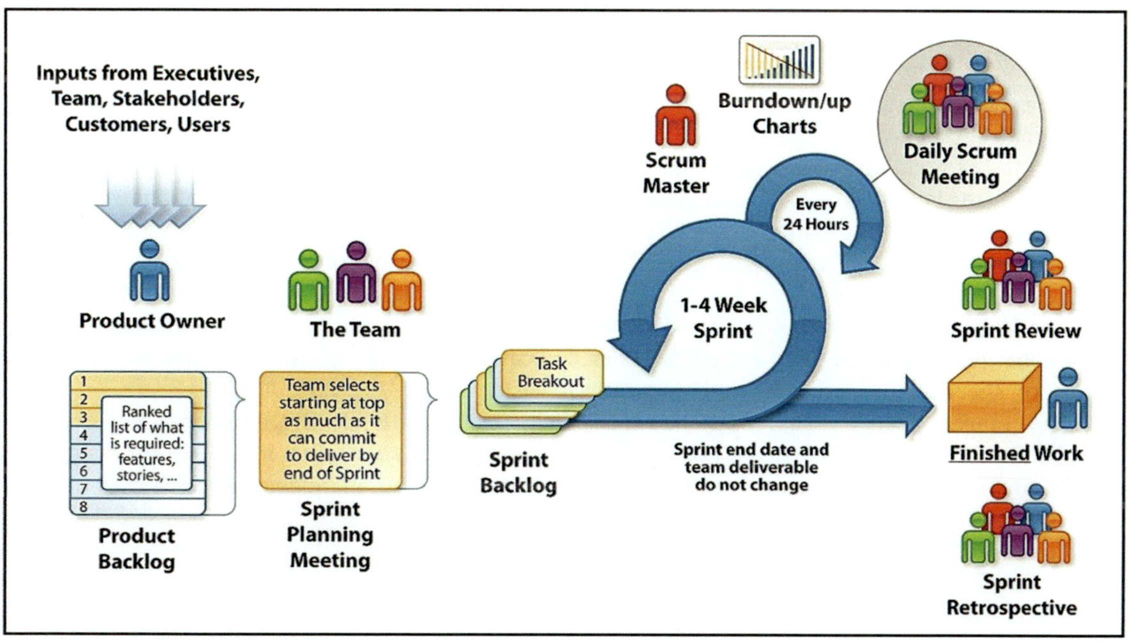

FIGURE 7.2 Schematic presentation of a typical Scrum process from the input defining the deliverables expected to the final solution.

SCRUM COMPARED TO OTHER METHODS

Beside the new vocabulary introduced, the scrum method promises to bring more focus, speed and constant improvement. That should enable a faster release of testable prototypes plus a greater ability to incorporate changes at any time. In the software industry that allowed to test useable portions of not full finished versions.

For new service ideas in the food sector like restaurant discovery or applications for food delivery systems, the scrum methodology can be used very efficiently. For the development of new food products, a few differences require attention. Overall, in food product development there is not really a "beta version" of a new food product. Every prototype that is given to a consumer or customer must comply completely with all food regulations. And new ideas that have been presented are public – that means they cannot be IP protected later on. Hence to play on the strengths of the methodology – quick testing and iterative improvement – the team and leadership should have a clear agreement about what can be tested and what can change.

For graphical and virtual testing with consumers, like "Toluna" tests or "consumer group discussion", the team can have a large degree of design freedom, based on an understanding of which types of products the company wants to produce. Testing chilled "health ready meals" with consumers is good when the company has the manufacturing and distribution system to handle that, but less so if they have only dry/ambient capabilities.

Similar in the next phase, when food prototypes are presented. The fast-learning cycles with consumers should only offer products that can be made on accessible manufacturing. Additional dosing steps, new components, exotic ingredients and beautiful garnish can easily make consumers enthusiastic. But the team often has no possibility to suggest versions that require new technology without serious discussions with the leadership team as to if the necessary investments can be made. Scrum in food works well if the team can make quick testing, followed by a fast decision based on a good understanding of which changes are possible and which ones not.

Scrum focusses more on "simultaneousness" and "adaptability" than stage-gated processes, which build more on sequencing and conceptionally a linear process, where gates are passed smoothly. Also in stage-gated processes "looping back" at a gate is possible, but more for a necessary refinement or knowledge gap closure, not to find a very different solution.

Reported disadvantages of using the scrum framework are the need for extensive training and the difficulty to apply this for large and complex projects. In scrum the solution can be very different from the original idea. Something like "scope creep" is a possible outcome, since teams are more autonomous to set new targets. A good and continuous dialogue between the team and the rest of the business can prevent a situation where the suggested solution does not fit the business requirements (Chandana, 2024). Then the intrinsic flexibility can be used to react efficiently to changes in the fast-moving markets.

Another point is the commitment of individual team members. Scrum in foods needs a truly multifunctional team. And the team members better dedicate most, if not all, of their time to the project; otherwise, speed and work density in the project will suffer.

The company reality is sometimes that a full dedication is difficult. Bigger enterprises do not run their NPD processes only on scrum. For big or more complex activities other tools, like the stage-gated process are more suitable. And many innovators have more projects or maintenance activities at the same time. Especially members from "non-core" innovation functions like supply chain, finance and legal are often loosely assigned to the team. Any urgent business priority in their functions might force them to change priorities. Now the fully dedicated project owner team must wait for the experts to find time to fulfil their assigned tasks. The team may have found in consumer research a great ready meal offering but needs the process engineer of the factory to confirm the technical feasibility, while the factory has a "priority call" like a line shut-down. If such delays happen too often, they not only slow down the process, but steal energy from the team and frustrate them.

That means the decision to introduce scrum in food product development better include ideas about the "compatibility" of scrum with the rest of the business. For a business in a first commercial idea preparation, like in a startup, scrum is a great management process for new products. The method is about prioritization and time-boxing over fixing the scope, schedule and cost of a project. Scrum in foods is more suitable for "technology push" situations, where the enabling new technology is fixed and now the most promising applications need to be found. The technological basics are non-negotiable; the solution can be anything that can be made with that technology. Such situations exist in foods as well – maybe an ingredient company has created a new protein concentrate, a new aromatic component or a new probiotic strain and is now looking for promising applications in food products. The resulting solutions are offered to customers and companies buying the ingredient. The process technology needed can be the same for all solutions. In such situations, the NPD corridor is wide, and more creative processes like scrum can bring an advantage. More difficult for "market pull" situation, where the key is often to find the technology that is best suited to fulfil the consumer need.

OTHER NPD MANAGEMENT TOOLS

IPM is not the only suitable tool at hand to manage creativity and discipline, Tools with names like "lean", "agile", "waterfall" or "kanban" are discussed in educated circles and have been used for years for several purposes. The examples listed here are not different approaches but more variants of a lighter/leaner process management in NPD. The main differences can be described as follows:

Agile is the new magic word, and more an umbrella philosophy for several methodologies. "Scrum", for example, is one execution of that philosophy.

Agile, scrum, kanban, etc., address the disadvantages of heavy planning and excessive regulation and try to nurture adaptability, knowledge accumulation and continuous improvement.

Scrum is still a process aimed to implement and build on sprints of fixed duration, while work in kanban is more supposed to happen as a continuous flow. The kanban development process is an offshoot of a Japanese production management and shares lead thought the aim to "balance demands with available capacity, and by improving the handling of system-level bottlenecks" (Wikipedia Contributors, 2024). In kanban, a project leader may start his board with the columns: Business requirements, ready to start, work, in progress, delegated to third parties and done! The business requirements are set as individual jobs to be done, which move over time from the "start" to "done" or become deleted. Such a very basic version of kanban can be made easily in Excel or Google Sheets or downloaded and can be used for different types of projects.

Scrum is batching the tasks from the product backlog to the sprint backlog. In kanban the team delivers the ask and then moves to the next one. Both share a regular structured board visualization of "the jobs to be done"!

Whatever management tool becomes chosen as the mechanism to drive efficient and effective new food product development, a critical point management with the rest of the organization. For a company growing green algae now looking for new applications of their harvest in foods, these "agile" processes are a clear option. For a grown-up food manufacturer with brands competing in the market, factories with constant supply needs and a broad product spectrum, it requires a deeper thought process to embed such an enclave of creativity and innovation in the overall business structure without disturbing the routine processes but still energizing the new team(s).

All these more "agile" operating models have two essential components in common: The dynamic front end and the stable backbone. Only both together can bring the company closer to customers, increase productivity and create the employee engagement to "go the extra mile".

The dynamic front end, the defining element of an agile organization, consists of small, cross-functional teams, often called "squads". They work to meet the specific business objectives of a project. The teams manage their own efforts; they can meet daily to quickly adjust and prioritize work, allocate new tasks and reflect on yesterday's progress. They loop back to their stakeholder, customers or consumers in regular feedback loops, as they see necessary and because they see the project changing direction. They coordinate themselves with other teams to accomplish shared goals.

The stable backbone provides the capabilities the agile teams need. The backbone includes clear rights, accountabilities and responsibilities, the necessary expertise and efficient core processes including effective links into the rest of the organization.

The agile organization moves fast, and decision and learning cycles follow each other quickly. Work proceeds in short iterations, and teams use quick testing and learning to manage risks and identifying product enhancements options. The agile organization employs next-generation technology to enable collaboration and rapid iteration while reducing cost.

As with most activities also the NPD process can be outsourced to external parties, while the respective food company focusses on their inner strengths. It is good then if the external developer knows the implementing company well; otherwise, much effort is needed to link the company reality with the new product ambition.

REFERENCES

Brainyquote. (n.d.). *Theodore Roosevelt quotes*. Retrieved June 5, 2024, from https://www.brainyquote.com/quotes/theodore_roosevelt_403358

BrightHub. (2010, December 16). *SWOT – History and evolution*. https://www.brighthubpm.com/methods-strategies/99629-history-of-the-swot-analysis/?utm_content=cmp-true

Chandana. (2024, January 29). *Scrum project management: Advantages and disadvantages*. Simplilearn.com. https://www.simplilearn.com/scrum-project-management-article

Chen, J. (2023, October 3). *Sunk cost dilemma: What it means, how it works, and example*. Investopedia. https://www.investopedia.com/terms/s/sunk-cost-dilemma.asp

Cooper, R. G. (2008). Perspective: The Stage-Gate® Idea-to-Launch Process—Update, What's New, and NexGen Systems. *The Journal of Product Innovation Management*, *25*(3), 213–232. https://doi.org/10.1111/j.1540-5885.2008.00296.x

Hakkarainen, K., & Talonen, T. (2014). The Innovation Funnel Fallacy. *International Journal of Innovation Science*, *6*(2), 63–72. https://doi.org/10.1260/1757-2223.6.2.63

MarketingMind. (n.d.). *Trial and repeat purchase - Consumer panels - Analytics | MM*. https://www.ashokcharan.com/Marketing-Analytics/~cp-consumer-panel-trial-repeat.php

Puyt, R. W., Lie, F. B., & Wilderom, C. P. (2023). The Origins of SWOT Analysis. *Long Range Planning*, *56*(3), 102304. https://doi.org/10.1016/j.lrp.2023.102304

Ras, R. T., & Trautwein, E. A. (2017). Consumer Purchase Behaviour of Foods With Added Phytosterols in Six European Countries: Data from a Post-Launch Monitoring Survey. *Food and Chemical Toxicology*, *110*, 42–48. https://doi.org/10.1016/j.fct.2017.10.003

Scrum Alliance. (n.d.). *Why is it called Scrum?* https://www.scrumalliance.org/ScrumRedesignDEVSite/media/ScrumAllianceMedia/Public%20Relations_2/What-is-Scrum-Backgrounder-2014.pdf

Sybil-Gibbs. (2014, November 1). *PPT - Innovation process management (IPM) PowerPoint presentation, free download - ID:6070228*. SlideServe. https://www.slideserve.com/sybil-gibbs/innovation-process-management-ipm

Scrum Guides. (n.d.). *Home | Scrum Guides*. https://scrumguides.org/

Scrum.org. (n.d.a). Product goal. https://www.scrum.org/resources/product-goal

Scrum.org. (n.d.b). *What is Scrum?* https://www.scrum.org/learning-series/what-is-scrum/the-scrum-team/what-is-a-scrum-master

Wikipedia Contributors. (2024, May 30). *Kanban (development)*. Wikipedia. https://en.wikipedia.org/wiki/Kanban_(development)

Continuous Improvement of Products and Concepts

8

Shaping the idea well is not enough to have a successful launch. Only the shaping and validation of the manufacturing options can ensure that the full idea comes to market. Strengths and weaknesses of in-house versus third-party manufacturing are presented. ODM and OEM are also discussed.

In the last chapter the emphasis was placed on the innovation management processes, the different phases, the work program and especially the roles and responsibilities R&D has there. This chapter is now more about the tools that are used identify improvement possibilities for a new product idea on its way to launch. Launching successfully in the highly competitive and dynamic packed food market requires a good understanding of the basic market characteristics and the ability to react accordingly:

- Generally, CPG products have short shelf lives and are subject to quickly changing consumer preferences and trends. This underlines the need of companies in this sector to constantly innovate and launch new products to meet customer demand and stay ahead of the competition. The food ingredient sector often has a longer product cycle. Their solutions are targeted to the "functional "needs of the consumer goods industry, and these needs are not that quickly changing.
- CPG products are typically sold in larger quantities but have a lower profit margins per unit than products in other sectors. That's the reason why economies of scale are so important and why the "operational efficiency" is one of the key factors to maintain sufficient profitability. The process technology applied large scale is often very mature, and this long practical expertise enables the food producers to maximize the line efficiency.
- *Brand loyalty and differentiation*: The purchase decisions for packed food products are not just based on "rational/functional reasons", but also emotional factors such as trust, familiarity, satisfaction and other emotional factors. Since food touches so many senses, it is associated with personal experience (PR Newswire, 2021). CPG companies must continuously invest in building strong brands, nurturing those brands and creating differentiated value propositions to retain and attract customers.
- *More channels, different consumers, more data*: CPG products are sold through multiple channels, wholesale for hyper- and supermarkets, corner stores, family shops, online, offline and direct-to-consumer. That's why the omnichannel strategy becomes more important. More consumer data allows better analytics and improve targeting, optimized distribution, pricing, promotion and customer engagement.

Consumer food markets are mainly pull markets. The source of innovation and competitive advantage for the businesses is therefore determined by the ability of the company to identify and satisfy unmet

DOI: 10.1201/9781032724591-8

needs faster than the competition. Here the market observation leads to insights, indicating the need and that creates a new product or service opportunity.

In FMCG needs can change quickly, so selling must stay close to the pulse of the market. Being able to do that faster than others puts a company ahead of the competition. And the quality of the research "tools" shapes the ability to get that superior understanding of what consumers really want. Superior in this context is simply "better than the competition".

The skills needed in a foods company is shaped by the business and innovations strategy. A cost leader, experienced in copying leading brands with "me too" products at a better price, will normally not invest substantially in market research skills. For them it's enough to monitor the commercial performance of new products coming to market to identify market success quickly and then start their own product development work, trying to copy these at lower costs.

A market or innovation leader, however, will put more effort into the consumer or customer research to identify first new unmet needs and then find an answer for those. The reward for that up-front effort is the "first mover advantages", often allowing a price premium.

And a classical "food ingredient" supplier will perhaps pay less attention to the consumer market movements. Their account teams will try to stay close to the consumer goods makers and listen to their changing needs, which are echoes of the market needs (see Chapter 12). If they hear that the consumer wants more plant protein–based products, clean label flavours or colourings, they work on solutions that their customers (the branded food makers) need to answer that consumer need.

Depending on type and size, the company in packaged and branded foods will have a bigger or smaller market research team. These experts select the tools to find the need and optimize the solution. Often, they are not doing that on their own, but ask specialized research agencies for support. Companies like MMR, Ipsos or GfK provide advice on suitable consumer research methodologies, make a plan and run the consumer research in the field.

Which research design is applied depends on the type of product and the phase in which the innovative idea is at that moment.

THE DIFFERENT PHASES

The consumer testing follows the same logic as the overall product and process development, starting from market segment and consumer understanding, over initial idea testing, concept and product refinement, to the final quantitative branded product and concept tests. Research in the idea phase may be just motivated by the wish to launch a new product. Later the product idea and prototype design are finished and now the fine-tuning of sensorial attributes is the focus. At the end the final version is tested, with the brand name mentioned or without, in the final pack, in home or central location and with the price or without. Finally then sales volume, frequency and penetration must be solid to plan the manufacturing and distribution process properly.

Consumer Insight Generation

Ideas are based on consumer needs. But the consumers are normally not lifting their hands and informing companies about their new needs. Research work like consumer insight aims to understand the new or sifting consumer demand. A well-known snack brand worked hard to find new snacking needs to grow their business, which was a clear expectation of their investors. They only succeeded when they realized that the "prime threat moment" of the day, the after-dinner occasion, would be saturated, since consumers have found different moments to treat themselves with a little extra. The "morning commute" was a situation that was asking for very different formats.

To understand the market quickly and efficiently two "inside" driven functional excellences have developed over time and are used in many leading consumer packaged goods (CPG) companies. These are "consumer market insight" (CMI) and the even younger "consumer technical insight" (CTI).

1. *CMI*: Consumer market insights are often data-driven observations helping the businesses to better understand their consumers/customers. The data should allow companies to make informed business decisions about how to reach and engage their target market more effectively. The main data insights pools are market research, direct consumer/customer feedback, increasingly more the social media analysis and the good old sales data analysis. By understanding the needs, preferences and behaviour of their customers, businesses can tailor their products, services and marketing strategies to better meet the needs of their target markets.

   CMI is frequently a separate functional team, and typical roles in the team are:
   - *Data Scientists*: Experts in statistical methods and machine learning to interpret complex data sets.
   - *Consumer Insights Managers*: Responsible to convert new insights into plans aligned with the business strategy.
   - *Business Intelligence Analysts*: They focus on analysing data to provide strategic recommendations for the business.
   - *Category Managers*: Experts for a particular product category with deeper insights.

   The CMI team works closely with other departments such as marketing, sales and product development to ensure that the insights lead to effective strategies and campaigns that resonate with consumers. They also stay abreast of industry advancements and leverage new technologies and methodologies to enhance their understanding of the consumer.
2. *CTI*: Consumer technical insights as a function is less known. These experts are also aiming to find new market opportunities, but they focus more on the products on the market in relation to their own ones. This encompasses the structured analysis of competitors' products and the responsibility for the design and execution of own ongoing NPD. It is the product's centric consumer research and, in a sense, the consumer voice in all R&D activities and the R&D view in all marketing plans.

Both consumer functions, CMI and CTI, start with the consumer lens. CMI normally brings the structured data analysis from the markets, trends, market sizes, the competitive landscape, shares, etc. CTI will report about functional details from their own and competitor products, like new taste and aroma variations, new fortification, new claims, shorter preparation time, less packaging or recyclable, shorter or cleaner ingredient list, resealable packaging, new ingredients, etc.

The end-to-end development of a new core food product often takes more than a year. And even a "simple" range extension requires half a year or more. That brings the necessity to regularly check back with consumers that the direction of the new food idea is still targeting an existing need in an appropriate way. And that's why it's often better to test the maturating idea on its way to market more often with consumers.

While both "insight generators" are closely cooperating, they also need an organizational understanding about the lead and end responsibility in the different testing modules. The lead suggests the plan, including activities and tests, estimates the necessary budget, controls the expenditure and reaches out to external partners, like specialized research agencies to present the idea, and discuss the research design and protocol most suitable.

Going back to the stage-gate model, the different phases from ideation to launch require different answers to successfully move the project to the next gate.

Different tests designs have evolved over time, all with their specific advantages and limitations.

After the initial ideation, in which not only the product but also to an extent the "story" (the concept (examples in Table 8.1), the reason to believe) around the product has emerged, a first reaction from the intended target group is useful; it's a first light test about the principal appeal of the idea.

TABLE 8.1 List of consumer research tools frequently used in foods

TYPE	*EXECUTION/EXPECTED OUTCOME*
Surveys	Often done via telephone or online, they allow gathering information from many people. Surveys can include questions about product preferences, purchasing habits and demographic information.
Focus groups	A smaller group of people discussing and experiencing the new product idea. A trained moderator helps to find valuable insights into consumer behaviour and preferences.
Interviews	In-depth interviews with individual consumers can provide detailed information about their opinions and experiences.
Taste tests	Consumers come to a central location and try different products and provide feedback on their preferences. It is good to develop new products or improving existing ones. An alternative is to send them the product for home tasting and usage. It is helpful when the product has to be used at home, for example, for cooking. The feedback is more difficult to capture.

PREFERENCE MAPPING

Market leaders in a food segment normally do not wait until a new trend is emerging in the market segment, where they earn their money. They know they are innovation leaders that must find the next product idea. Not the breakthrough, that is described in Clayton Christensen's classical "innovator dilemma", but more evolutionary moves, making the offer just a little bit better. For foods, sophisticated research tools have been developed to detect those opportunities. A powerful one is the "preference mapping", a technique that allows researchers to analyse data from sensory tests, such as hedonic scales, paired comparisons or ranking tests, to measure how much consumers like or dislike a product. Preference mapping uses statistical tools, such as factor and cluster analysis, or multidimensional scaling, to create a visual representation of the preferences in different consumer segments. They map also shows the relationships between consumer preferences and specific product attributes, such as flavour, texture, aroma or appearance. These are then the action points for R&D to start drafting a product brief.

In its core the method combines descriptive sensory assessments. First, a trained expert panel articulates their sensorial impression in a professional "food attribute language". Their "vocabulary" goes much beyond taste impressions like "salty" or "sour" and covers impressions of "mouthfeel" or complex aromas like "nutty", "grassy", "floral" "woody", etc.

Variations of the selected prototypes are then given to the consumers in the target segment. They represent the market and determine the sales potential; it is their preference that has to be captured accurately. Consumers are not trained in food language. So, they can be asked what they like and what not. Their vocabulary is not descriptive enough to prepare a product brief, and their ability to taste and smell differences and articulate those differs hugely. Imagine an "normal" red wine drinker aiming to capture blindly the difference between a Barolo and a Taurasi.

It's the combination of both tests, consumer preference and expert evaluation that defines the opportunity. And done with enough confidence, statistically enough consumers in the test, the analysis also provides quantitative data about preferences in the cluster, which gives information about the size and value of that targeted market segment.

Preference mapping is not new, and hundreds of papers have been published. Most of them have to do with applications for the food and beverage sector. It was invented in the late 1970s and early 1980s (Carroll & Chang, 1970). The purpose of mapping introduced then was simply the idea to reduce the complexity of sensory data by projecting products and preferences onto a two-dimensional space – that's where the "mapping" is coming from. Here, the distance between products reflects their similarity and the position of preferences indicates liking (or not). The result is a list of most preferred product attributes, which can guide product formulation work.

For a launch in a new product segment, preference mapping is less useful. For reliable feedback the consumer must be familiar with a different product to articulate preferences.

But the result offers opportunities for a line extension entering a space not yet covered in the market. Or a rejuvenation, where the existing product is modified on the basis of the preference mapping results. And indeed, when scanning the list of publications using preference mapping, often very mature market segments, like soy sauce, rice variants, fruit drinks, probiotic yogurts, peanuts or bouillon cubes are mentioned. In a sense preference mapping is a tool with higher resolution, like an UV light microscope shows more details than the visible light microscope.

Data collection and analysis depend on the specific research objectives, the products and their key attributes. Some common methods are:

- *Rating-based preference mapping*: Consumers rate each product or attribute on a scale (e.g., 1–10) based on their preference. The ratings are then analysed using statistical techniques such as factor analysis or principal component analysis to identify the main dimensions of preference and the position of each product or attribute on these dimensions in a map.
- *Choice-based preference mapping*: Consumers choose their preferred product or attribute from a set of alternatives (e.g., pairwise comparisons or best-worst scaling). The choices are then analysed using discrete choice models or conjoint analysis to estimate the utility or importance of each product or attribute for each consumer segment.
- *Sensory-based preference mapping*: Consumers evaluate each product or attribute based on sensory attributes (e.g., taste, smell, texture) using a descriptive analysis method such as QDA or CATA. The sensory data are then related to the preference data using techniques such as regression analysis or partial least squares to identify the sensory drivers of preference.

Two typical usages for preference map research are roll-outs to a new export market or an in-depth analysis of a big existing market segment. In the first case, the multinational peanut brand established in the past a great understanding about the various attributes liked by consumers in various markets. But testing some variants together with locally available samples, the company will find the best variant, understand the sensorial clusters and can so decide if they launch a product filling a "gap" in the market or if they want to go "head-to-head" with an existing product.

And doing that often enough and in sufficient markets can allow the company to predict their next innovation in an unknown market, "*in silico*", using and combining all data from older markets. AI will help here further.

The second case is more a kind of "in-depth" review of the grown market and the consumers. A local company may have been very successful for many years in making a particular product but starts to see a slow erosion in their market share, without having a visible reason for that loss, for example, a launch of a new active competitor. They decide to conduct a region/country-wide preference mapping to understand the underlying shift in the market, so they can reformulate their top seller or create a variant.

Particularly traditional formats, condiments or cooking aids like ketchup, soy sauce, mayonnaise, fish sauce, oyster sauce and mustard often have a huge market. But nobody has done in-depth research on the sensorial preferences. Local producers make their product by their processes for their customers. The next neighbour does the same. Much later and with appearance of a big national company and brand, the national market shows more spice, lighter, darker, less salty, more peppery, thicker, thinner and nutty variants. And nobody has an idea which segments are how big and if some gaps exist. The new company now can use its national distribution system to find the best-fitting variant to the right location.

The exercise starts with the collection of all samples, from leading national brands and to old, traditionally made small business samples covering the sensorial space. The next step is to characterize all the samples objectively and quantitatively. This analytical investigation covers all important physicochemical properties, like total solid content, crude ash, lipid, total nitrogen, colour value determination, pH, lipids, amino acids and nutrient profile.

The next analytical block spans the sensory space and includes the professional testing by trained individuals. More details are given later in this chapter.

For the evaluation of the samples, trained participants are asked to evaluate the acceptability of each sample by using, for example, a nine-point scale covering appearance, aroma, taste/flavour, mouthfeel, etc. One key outcome of this research arm is to determine the intensity of those sensory attributes. The "expert" description is then combined with a consumer liking and now it is possible to understand which sensorial attributes are driving the consumer liking.

After statistical analysis a type of topographical map, the "preference map", can be generated, showing sensorial attributes and the locations of samples tested in that space. This shows clusters of preferences and their relative size, which can be used to represent the market size of a preferred attribute.

Doing such a trial for a bigger market segment is a substantial undertaking. Depending on how the samples are tested (pure or in combination), the pre-work is already substantial, and trying to establish a preference map for oyster sauce would require knowing the top dishes in which the sauce is used and the most popular cooking method, Pure tasting of oyster sauce is not giving meaningful results. Then samples must be prepared, and the panellists need to be screened and trained. That may already require 4–6 months. The field work can take anything between 2 weeks and 2 months depending on size of the study and locations, and the analysis another 4 weeks before the first results are available.

An example: Years back, two leading brands in Germany both had a 45% dominating share in the market for Spaghetti Bolognese meal makers. Attempts to take shares from the other big competitor were for years not successful on both sides. Only with the preference work, it became clear that the market was consisting of two different segments of same size. One segment liked more the "sweeter, brown, fruity" sauce and the other one the tomato sourness, meaty, "red" attributes. The first one was representing the sweet spaghetti Bolognese dish for children, the other the more adult tomato meat pasta taste.

CASE STUDY: NEW INSIGHTS IN AN OLD MARKET

This case is built on a paper published by Kim, H., G., in the *Journal of Food Science.* I have asked permission to use it, as it shows so nicely where reference mapping is often used, how it is applied and how the results can be used (Kim et al., 2010).

Doenjang is a traditional soybean paste in Korea and one of the oldest and best-known condiments in the country. Like other popular condiments especially in the Asian kitchens, the traditional production involves several steps, including long-term fermentation involving different microorganisms. The modern food industry needs to simplify and streamline the fermentation to make the product commercially viable and safe. And while the industrial process differs significantly from the traditional processes, the aim is to match the sensorial properties of the traditional versions.

The research team collected five traditional and two commercial samples to map the preferences in Korea. For their descriptive panel they recruited a wider sample of students and exposed them to initial sensory screening procedure, focussing on taste and smell attributes relevant in the product category. Eight panellists were trained for 3 months on the sensorial descriptors. The doenjang samples were presented randomly in a standardized procedure.

Identically prepared samples were given to 200 consumers of different ages and judged quantitatively on a 15-point nondescriptive liking scale on odour, flavour and mouthfeel attributes.

The results of both arms of the study were statistically treated to identify significant differences and the mean values. The mean descriptive data were correlated with the consumer acceptability data to understand the relationships between the sensory characteristics and consumer acceptability. A hierarchical cluster analysis allowed to create consumer segments with a common preference pattern.

The result was 31 different attributes with different mean intensities; they appear as vectors in the two-dimensional plot and their lengths reflect the relative importance. For all attributes except "sour" the individual judgements from the panellists show consistent alignment. And the seven samples could be differentiated by their mean intensity score of the different sensory attributes. This already showed that both commercial doenjang samples were perceived as more intense in "colour"," sweetness" and "umami" than the locally produced ones, as well as other clear differences.

When the original variables were reduced to two by principal component analysis (PCA), the team saw that about 55% of all differences could be explained by a first dimension with "sweet", "MSG taste", "alcohol," and "chemical" flavour as negative attributes and "brown colour" and "bitter" taste. Another 18% in the second-dimension listed "sweet-grain" and "soy sauce" odour and "sesame leaf" flavour. In this two-dimensional plot both commercial samples were far away from the traditional ones.

In terms of consumer acceptance, both commercial samples scored the highest, which is not surprising given the assumption that both are the optimized versions of extensive consumer testing. The taste of "MSG" and "sweet" were identified as drivers of liking.

A deeper analysis of distinct areas below the main preferences indeed brought up three clusters of consumer preference, representing 38%, 36% and 26% of the consumers tested. This step showed that the overall "sweet" and "MSG" liking was actually disliked by group 2, who were looking for "bitter" taste, "soy sauce", "earthy," and "metallic" flavours with an "astringent" mouthfeel. And in the third cluster "MSG" and "sweet" again were drivers of liking for taste, and in combination with "sweet-grain,", "soy sauce" and "sesame leaf" for flavour.

This deep analysis of the doenjang market showed, first of all, that the industrial suppliers initially did a very good job in (1) finding the main drivers of linking in this very local and diverse artisanal market and (2) finding an industrially controllable process of making the paste.

What this study added was that more than 60% of the market prefers a profile not fully matched yet (if the 200 people selected are representative). That offers for the commercial supplier the opportunity to launch new variants for the preference cluster they as missing or to buy a local supplier representing that segment. The market leader would also need to watch new launches with a winning profile for the noncovered clusters coming in.

Possible limitations of the study are the geographical coverage – do the testers represent the whole national market or only a part of it? A second factor is that doenjang is a condiment, so preference testing would be better done in a dish than alone. And finally, the researchers would need to know what the share in "top dishes" in the Korean kitchen is, like jjigae, gamjatang, ssamjang, etc. The combination of these data would allow precise product development work and a robust evaluation of the commercial success.

IDEA SCREENER

For new product ideas, another test in the early phases of food product development is more suitable. As the name implies, the test is used to filter initial ideas. It includes contextualization, the intent to bring the consumption moment alive. Attention is given to the feelings, preferences and initial reactions of the consumers. This helps identify ideas that resonate positively with consumers.

Beside those, more specific criteria like market demand, profitability, feasibility or uniqueness and consumer appeal can be included. The answers can be used to generate quantitative data, allow a rating or forced rankings and give space for qualitative data, like individual comments. The ideas that resonate well can further be refined and tested again in a second round.

Through graphics and text boards one gets the consumer perception with respect to a couple of attributes like uniqueness, relevance, superiority, clarity, etc.

With open-end feedback questions, the research team can find out what respondents liked or disliked about the product ideas. The respondents are, for example, asked how unique that find the product, the quality perception they have, how much they are willing to pay and how often they would buy it.

Compared to methods like preference mapping, this method is "lighter" (less complicated, less deep and less expensive), but also several ideas can be tested for consumer appeal. It can be used to filter between similar product ideas, like variants, or only one product but with different concepts/stories. A typical example is shown in Figure 8.1, where three product ideas are tested for different angles of the idea. The results are percentage agreement numbers for all the different attributes and questions.

Several companies are offering the tools to design an idea screener online. People/companies can create their own boards and work open-ended questions, and an AI tool is converting those into quantification (like percent liking, purchase intent, etc.)

Summary of responses for each product idea </> View as JSON | Quick export | Add to export

	IS1	IS2	IS3
Name	Product idea 1	Product idea 2	Product idea 3
Product perception			
Relevance Significant differences	83% —	67% —	69% —
Superiority Significant differences	64% —	66% —	74% —
Clarity Significant differences	76% —	85% —	68% —
New and different Significant differences	82% —	77% —	67% —
Better quality Significant differences	69% —	74% —	77% —
Appealing range Significant differences	70% —	83% —	57% —
Pack appeal Significant differences	85% —	75% —	73% —
Interesting Significant differences	88% —	69% —	68% —
Uniqueness Significant differences	75% —	77% —	59% —
Likely to consider Significant differences	85% —	74% —	68% —
Comment Significant differences	87% —	83% —	75% —
Sustainable Significant differences	79% —	76% —	68% —

FIGURE 8.1 Typical result of an "idea screener test". (Taken from Conjointly, 2024.)

ROLE OF R&D

Even though the food product at this stage may only be a description on screen created by marketing, an active input from the technical side is already important. A good marketing person will try to crystallize the new, differentiation aspect out of the idea. That is necessary to articulate the reason to believe and "cut through all the advertising noise" in FMCG markets, where the consumer attention span is shorter.

The marketing colleagues need to play the drum. And they need R&D to articulate the best of what is also possible. "Underselling" the benefit of a new product is a lost opportunity or at least an inefficient use of the marketing expenditure behind the launch.

The opposite isn't good either: To catch the attention out there, the marketing colleagues will push to the maximum, but with the risk to "overpromise" the benefits of the new product. Hence, the technical team, including supply chain, nutrition and regulatory, must engage early to ensure that what will be tested with consumers can also be truly delivered in a real product.

An example: A company sees an opportunity for the launch in Italy of a probiotic, flavoured chocolate drink with low volume (shot). Marketing develops a story board and presents it to focus groups.

> "Do you love chocolate or vanilla? Do you want to boost your gut health? Then you'll love our chocolate-flavoured probiotic drinks! They are delicious, nutritious and packed with live cultures that support your digestive system. Our drinks are made with natural ingredients, no added sugar and no artificial colours or flavours. They are also gluten-free, dairy-free and vegan-friendly. Whether you need a boost for breakfast, a delicious snack or just a recharge in between, our chocolate flavoured probiotic drinks are the perfect choice. Try this convenient 50-mL PET shot bottle today and feel the difference!"

This is a good concept and probably new! The strongest argument could be the health promise. Without that the idea is flatter, a bit pale and not so new! The technical experts from R&D and regulatory could now say: *"OK, let's test it; in case it wins, we check, if it is feasible"*. No good idea! If it wins, the company wants it! And they do not look for a rock-solid argumentation as to why the product cannot carry that winning claim. The company wastes research money, and R&D "disabled" the opportunity. So, why haven't they said this earlier? And a soft early warning from the technical team that "the claim may not be possible" wouldn't have made it much better. Not achieving the claim can be taken by others as a "lack of entrepreneurial drive" or a lack of argumentation strengths versus the regulating authorities.

EARLY CONCEPT TEST

This type of test is not an alternative to the idea screener. It's more often used as an "add-on" to sharpen ideas further by exposing it to the target consumers prior to launch. This can include anything from messaging around the product to key features or ideas.

Concept testing helps you identify the following:

- Effectiveness of your concepts
- Effectiveness of your messaging and advertising
- Concept strengths and weaknesses
- Alignment of concepts with your brand
- Hidden benefits of your concept

Specialized agencies offer their service for this type of research. They can help to decide if the test is better done "monadic" or "sequential". In monadic testing the respondent is exposed to a single concept. In sequential testing multiple new products (or concepts) are surveyed by groups of participants, one after the other. Here products and concepts are typically presented in random order to prevent bias. Typical formats are:

- *Single-Concept Evaluation*: A single monadic test where the consumers or respondents are asked to complete a full evaluation of a single concept.
- *Multiple-Concept Evaluation*: This is a sequential monadic test. The respondents make a full evaluation for multiple concepts.
- *Comparative Concept Selection*: The respondents are presented a selection of concepts and askes for the one they like the most.
- *Concept Selection and Evaluation, Comparative and Monadic*: That's the deluxe version combining all elements.

The most suitable test is a balance between the wish for an in-depth, unbiased view (single monadic) and the possibility to check several concepts in less detail (multiple concepts). And, of course, a question of the budget available.

A specialized version of a concept test is "TURF", a tool frequently used to optimize product portfolios, for example, finding the right amount and type of new variants without cannibalizing the existing assortment. The test has been widely used from mid-1990s onwards The abbreviation captures the test objectives:

- **T**otal means complete or absolute.
- **U**nduplicated means it is unique.
- **R**each is the number of people reached at least once.
- **F**requency is the total number of times a person was reached.

First developed in the 1950s, as statistical tool for media planning to maximize reach and frequency of messages, it later focussed on "reach" and started added a ranking system (Thomas, n.d.). The key is "unduplicated reach", or the percentage of respondents who are activated by at least one of the suggestions offered. Through multiple choice the underlying algorithm finds the mix of new variants that adds the biggest incremental sales volume without cannibalizing existing variants.

A typical example would be ice cream: A company has three variants are on the market – let's say the top flavours of "vanilla"," chocolate" and "strawberry". Now they want to launch two more flavours out of a pool of five preselected ones, say "hazelnut", "peach", "banana", "lime" and "orange". Respondents are now asked which variant out of the offered variants they rate the highest and which one they like second best. TURF now simply finds the variant combination with the biggest total volume at minimum overlap. This avoids that the newly launched variants are getting their volume from those respondents that are switching from an already existing one on the market to a new one. The aim is of course to get new ones, not making the old ones switch (Planbeyond, 2024).

But the market reality has many more influences. A different range than the winning one could better drive purchase frequency and the opportunity for more sales than just the "incremental reach". Or a combination increases the overall market, as new consumers are coming in, that this static model would not detect. And people are buying multiple flavours of ice cream and that varies over time. Competitors, price, distribution and consumer awareness will further reduce the "isolated sales potential". This is a reason to estimate the sales potential also with the help of more expensive research.

FOCUS GROUPS

This is the most widely used qualitative technique, where small groups have a discussion guided by a moderator. Also, this is more a tool for early stages of product development: A way to get under the skin of consumers, as respondents are motivated to express their thoughts and feelings openly. The discussion can be used to select between the reactions of the group. A new product can be presented in different packaging formats and sizes, with several options of added nutrients and various taste/aroma variants. The focus group can find attractive combinations that later on can be checked quantitatively.

A typical focus group consists of 6–10 respondents and the discussion takes 1 hour or more. The moderator creates an open atmosphere, where everybody feels safe to discuss the ideas. That can be concepts boards and/or product prototypes, packaging suggestions, etc. That depends very much on the research questions. Therefore, it's important for the team to align on the key question, so moderator and stimulus material lead to the key questions.

A great location for focus groups are facilities with one-way mirrors, where the project team can observe the session. R&D should join this discussion. Imagine the focus group gets a few packaging formats presented and finally chooses one. But that one is more difficult to make and more expensive. The agency will report back which one was chosen and recommend progressing here. Only by listening to the group discussion will the team get a view if this specific format is critical for launch success and if the extra time and money are truly justified.

Such sessions can be long and tiring, but reading the agency report afterwards cannot replace listening and observing. Respondents not only articulate themselves with words. Gestures, mimics and all other types of behaviour give insights. Group dynamics and interventions of the moderator give the discussion direction. It is good practice if the cross functional team sits together immediately after the session to "compare notes", sharing observations and insights and drawing conclusions.

A variation is a "sensitized" focus group, where the respondents have received input material and questions before the session. Such sensitized groups can sometimes provide deeper insights.

The agile variant is the online focus group. It avoids travel, is less expensive and can be the only way for a focus group session when teams and consumers are very distant from each other. A global team, working from the corporate centre on a launch in several countries, will often not be able to join all local group discussions.

CONCEPT AND PRODUCT ASSESSMENT WITH CONSUMERS

A concept product test is a type of market research that involves presenting a new product idea as a testable prototype plus the concept explaining what makes this product so attractive. A good test is how good product and concept fit together. A methodology is available to highlight which elements need improvement to align both better.

If R&D joins all consumer work closely from the beginning, this often avoids the tension in this test. If earlier on, only non-technical experts and "naive" consumers have worked on the most attractive idea, the result could be a concept that the real prototypes cannot fully match. R&D involvement ensures that the technical feasibility is always present, especially as more often the key question is not if something can be made, but at which cost can it be made!

During a concept product test, participants get descriptions or prototypes of the new product, along with information about its features, benefits and potential price.

Concept product tests can be conducted using a variety of methods, including surveys, focus groups and interviews. The results of the test provide insights into consumer preferences and behaviours and guide further product development and marketing strategy.

JAR SCORES

The JAR ("Just About Right") scales are a quantitative measure to find the appropriate level for a specific sensory attribute in a food product, like saltiness or sweetness. They were invented by researchers at the Society of Sensory Professionals in the late 20th century, based on earlier work by Johnson and Vickers on optimizing sweetness in lemonade using an unstructured line scale (Johnson & Vickers, 1987). A typical JAR scales consist of 5 or 7 points, ranging from "too little" to "too much" for a given characteristic, with the middle point labelled as "just about right".

They nowadays are widely used in consumer product testing, as they can provide clear directional guidance for product reformulation or optimization. JAR scales can be used in product development in various ways, such as:

- Identifying the optimum levels of attributes in a product that maximize consumer liking or satisfaction
- Comparing different products or prototypes on their attribute levels and consumer preferences
- Evaluating the impact of changes in ingredients, processing or packaging on the attribute levels and consumer acceptance of a product
- Exploring the relationships between attribute levels and other consumer responses, such as purchase intent, loyalty or perceived quality.

Like all methods, JAR scales have some limitations. The scoring askes the consumer about the ideal levels of an attribute. Now, the consumer needs to be aware of that attribute to decide. Consumers are not trained panellists; they are often not able to articulate their sensorial experiences. Asking them if a product is "too sweet" or "not salty" enough is relatively straightforward. Most people can taste differences. "Bitter" is already more difficult. Up to 25% of the population may be "supertasters"; their "too much" may come much earlier than in the rest of the population (Rossiter, n.d.). And question about spiciness", "firmness", "crispiness", "crunchiness", "mouthfeel", "umaminess", etc., need careful wording to minimize the risk of unclear answers. For R&D the task is generally to maximize the "just about right" percentage, until a moment where a little more pushes respondents out of the middle range.

If the score for a key attribute indicates has high percentages on both ends of the scale, like "too low" and "too high", it is a polarizing score. This suggests that there is no consensus among target group consumers about the optimal level of that attribute. Product development cannot add more or take less of that attribute. Generally, 75% and more of non-polarizing score, the sum of "just about right", "slightly less than ideal" and "slightly more than ideal" are good values to progress. Less than 75% indicates that the attribute is polarizing and may need to be reformulated or can even signal two different consumer segments that cannot be served with one product (see preference mapping earlier in this chapter).

The scores are not automatically reflecting the true relevance of an attribute. But the combination with hedonic scales or preference mapping can give an idea of the relation between attribute levels and consumer choice. The "statistical penalty analysis" identifies the attributes having the most impact on consumer satisfaction.

A simple case is used as an example to see how that works. A food product is scored for sweetness, saltiness and crunchiness. The penalty analysis calculates the impact of each attribute on the overall liking of the product by comparing the JAR scores with the mean liking scores. Here the attributes that have a high percentage of consumers rating them as too high or too low become penalized, as they lower the overall liking of the product.

The JAR scale for each attribute ranges from 1 (much too low) to 5 (much too high), with 3 being just about right. The mean liking score for the product is 6.5 on a 9-point scale. The penalty for each attribute is the subtraction of the mean liking score of the consumers who rated the attribute "as just about right"

(3) from the overall mean liking score of the product (6.5). The penalty analysis would show the following results:

Attribute	JAR Score	Penalty
Sweetness	2.8	–0.3
Saltiness	3.2	–0.1
Crunchiness	4.1	–0.5

In this example the total penalty for the product is the sum of the penalties for each attribute, here –0.9. And the conclusion is that crunchiness is the most important attribute to improve, as it has the highest penalty (–0.5). Sweetness and saltiness also need some adjustment, as they still have negative penalties (–0.3 and –0.1).

Marketing and market research then have other methods to develop and test concepts and communication at a moment where the design of the product is basically locked.

THE IMPORTANCE OF CHOOSING THE RIGHT TARGET GROUP

Such thing as "the consumer" does not exist! It's not possible to aim a food product to all consumers. The company planning a new product must focus on a group of consumers with similar characteristics, preferences and behaviours regarding foods. Consumer groups in FMCG can be based on various criteria, such as demographics, psychographics, lifestyle, values, attitudes or purchase patterns. This can be health-conscious consumers, who prefer organic and natural products; impulse buyers, who make spontaneous and unplanned purchases; and bargain hunters, who look for discounts and promotions. The focus of the target groups helps to tailor the marketing strategy to meet the specific needs and wants of each segment the best way.

Probably the most important element in good consumer research is to be aware of the key target group and how this group acts and to ensure that this target group is truly present in the market research program. Marketing will know the target group very well, and the research agencies will know exactly to which group the respondents that they can recruit belong to.

And as lifestyle, values, attitude and therefore buying behaviour constantly change, target groups are changing. New ones are coming up; others get bigger or smaller. NielsenIQ, for example, published research in 2021 where they noted that "the number of consumers identified as 'newly constrained' has doubled from 23% to 46% in the four months between September and December" (NielsenIQ, 2021). Here the pandemic doubled the group of consumers in 16 different countries that felt less secure about their income prospects in 4 months from 23% to 46%.

Such data immediately shape the innovation agenda and with that the development briefs for R&D. Price leader companies in a segment may see the time to broaden their offerings to attract price-sensitive consumers with new offerings. They will look at lifestyle, age, income, etc., to target their new offerings. And market leaders may see the time to offer or strengthen their economy range, asking R&D how product costs can be reduced so that the company can compete in a lower market segment that was originally not their target group.

The FMCG foods business is mainly "market pull", so the existing needs of a group are identified and those will drive and shape the product development. The product development brief better not contain the consumer group characteristics, as they are not directly actionable in product development. The brief better contain facts and product attributes. Taking the case earlier, that could be a cost cap for the product ingredient costs or manufacturing costs allowing a lowering of the retail price, or a portion size or the number of portions increase, giving more product to consumers at the same price.

The target group will often remain the same throughout the product development, so all checks and feedback loops during the development process will be done with the same group to ensure the optimization steps were the right ones.

Let's say, a new idea is a lactose-free dairy product. Who is the target group? In a population there are those who believe they are lactose intolerant or sensitive, and this group is much bigger than the group of people that are truly lactose intolerant. But "lactose free", the core claim, is not "novel" enough, as such products are already on the market. When the research shows that 66% of that lactose-free buyers are also watching their calcium, vitamin D and protein intake, the addition of one or more of these nutrients can make the product sharper and more attractive. And when the team learns that 71% of women regularly consume lactose-free products but only 55% of men do, they can decide to focus on women as the target group (Szabó et al., 2021).

That already makes the product development brief sharper. And to address the target group even better, the analytical question could be to find out how many of those lactose-sensitive respondents regularly consume sugar-free products as well. If that target group is big enough, the development includes "low sugar", but perhaps an allowance to use sweeteners.

Companies know the target consumers of their products. When a market leader, for example, decides to rejuvenate one of their core products, they may explicitly agree with the research agency to select the test user out of a group existing "brand" or product users. This prevents the fact that during the development process the modifications made are alienating existing loyal users, those often responsible for the market share. Coca Cola's project "Kansas", the reformulation of their classical Coke into "New Coke", planned to take share from their eternal rival (Symmetry Media, 2023). And indeed, they actually attracted a few Pepsi users to switch, but unfortunately many more Coke users disliked the new taste of "their" new product.

If the rejuvenation is planned as a 1:1 replacement of the current product on the market, the final quantitative consumer test will often set "superior" as an action standard. The new packaging design, a clean(er) label approach or the sodium or sugar reduction product must "beat" the existing product performance to bring the commercial success more sales in the market. A launch could target more penetration, so more consumers buying the product. In this case the use of a "heavy user group" of the brand is not useful. If the incremental sales could come from the fact that existing consumers use the product more often, the aim is frequency. Here a split into low, medium and high users of the brand creates valuable feedback.

For important launches, like top seller relaunches, more market research money is needed. Now the research can recruit a larger consumer test group, where half of it is loyal "brand" users and the other half is "non" or "infrequent users". If the statistics are good enough, the team will get the performance with their own users and the potential to attract new users, for example, those from direct competition. In such a test the research targets could differ per consumer group: A "superior" (compared to existing) as a benchmark for existing users and a "parity" for users of competing brands. A target with "superior" for both, their own and competitors' users, is even better, but often too stretching in foods since products represent different preferences.

In all these cases it is important for R&D to stay close to the market research to ensure that particularly those product attributes that can be modified quite easily are included in the feedback. "JAR" scores are a good tool.

VALIDATING THE MANUFACTURING

Having successfully shaped a new food product in the different design phases from initial idea to final product concept and product test is critical for success, but not sufficient. The ability to commercially make that has been tested as successful and directs the late phases of the product path to the market. In theory an enterprise should decide to sell what the consumer wants. In practice, it's equally important to sell what one can make (better than others).

IN-HOUSE MANUFACTURING

The upscaling of newly created food products in manufacturing is the classical approach in the food business. A company has the possibility to make specific products or formats because they have the knowledge and the infrastructure. The new products they can develop are based on their technical capabilities and the limitations of the technology applied. A company that can make products that are sterilized in the packaging, like stand up pouches has a great technology at hand to make for example chunky soups or sauces. But they will have difficulties making top-quality classical smooth or velouté French variants, since for those type of products the aseptic in line sterilization and filling into aseptic packages is more suitable. The technological boundaries limit the possibility to fully meet a consumer demand. And hence, if a company has detected a market opportunity for such smooth soup or sauce recipes, they can decide to build up the suitable "in line" sterilization technology in-house, which requires time and capital or to find an enterprise having the needed technology. If they are lucky and identify free manufacturing capacity, they may be able to launch quickly. This outsourcing approach works well for established technologies. But for food innovations built on new, differentiating technology, this is often not possible. So, for products requiring new and innovative technologies, in-house manufacturing is the preferred option. The advantages are:

- *Quality Control*: Manufacturing in-house gives direct oversight of the entire production process. This helps to guarantee brand quality standards that are essential for consumer/customer satisfaction.
- *Cost Efficiency*: If the operational efficiency is comparable, in-house manufacturing is often cheaper, since the copacker fee can be saved. This potential to reduce logistics expenses associated with shipping rates and customs duties are another option to save money.
- *Product Availability and Flexibility*: Owning the manufacturing process offers more flexibility. In case the overall capacity is good enough, changing market demands can be answered more quickly and stock levels better managed. This is a frequent event in fast-moving goods, and is even more important if the product shelf life is relatively short. The break of the distribution chain, like in the pandemic or a unstable situation, will not immediately impact product delivery.
- *Confidentiality and Ongoing Learning*: Keeping the new product and the process in-house provides better confidentiality. Details about the new launch will not leak to the competition, and the continuous effort to optimize the processing keeps the organization creative and can even lead to more IPR.
- *Strategic Value*: Today, strategic investors see in-house manufacturing capabilities as an important asset and extra company value in synergistic mergers or acquisition opportunities. Their own manufacturing allows often a much better supply chain control, increasingly relevant in times of much more transparency and increasing food fraud.

The flipside of a decision to produce in-house are:

- *Capital Investment*: If the technology for the new product or the line capacity is not available, a significant up-front investment in infrastructure, equipment and skilled labour force is required.

- *Time*: Specifying, ordering and installing new manufacturing technologies require time and knowledge, while the market is not waiting.
- *Maintenance and Upgrades*: Just building it once is not enough. Especially food production needs constant maintenance and upgrading.
- *Limited Scalability*: Competitive food production requires high manufacturing efficiency. Once a particular scale has been installed, it is normally not possible to increase or decrease the capacity to react to fluctuating demand.
- *Labor Costs and Management*: Running your own manufacturing means searching, hiring, training and managing staff. That can be time-consuming and expensive. The local labour costs, insurances, taxes, etc., add to the financial burden.

As said, new differentiating technology may not be available from another company in the market. If a food product requires high pressure pasteurization (HPP), ultra-high temperature (UHT) processing, plant-based dairy and egg alternatives or edible packaging, the number of potential suppliers can be very limited. And waiting for others to build the manufacturing and offer it for toll production, means also delaying the launch until the new differentiating technology becomes a commodity.

If the decision for "in-house" manufacturing has been taken, R&D has to ensure that the critical features of the food product can be made with the chosen production process. Here it's less the question if the production can make the product, but how efficient it can make a product complying with all regulations, food safety standards, labelling requirements and claims made. If a new product idea is based on the promise that it contains a specific amount of value-giving ingredients, maybe a vitamin mix, premium fruits or vegetables or seafood pieces, accurate dosing might be very important. All packs of finished goods must contain the promised amount. But that should not be achieved by costly constant overdosing of those ingredients, leading to too high material losses.

Many food production lines are often high-speed lines, which ensures that food products can be made at competitive costs. Often, the operational efficiency has been optimized through many small improvements over a long time. But that often limits the flexibility in the processing steps.

In theory, innovators often say that the company needs "the technology to make the right products" and not "new food products for the technology they have". In reality, the technical experts in R&D and manufacturing will have to find ways to make the products with the existing assets without impacting the critical design elements of the product that has been successfully tested with consumers. That requires a good understanding about the attributes and creativity in the selection of ingredients and manufacturing. Maybe a premix or a different dosing sequence brings an acceptable compromise. Or ingredients with different form, size or density help to circumvent investment into new dosing equipment.

Packaging R&D will help the line operators to optimize the machine settings during production and material change. In many standard food production lines, the primary packaging is fully integrated into the line setup and often represents the "bottleneck" of the line. Here experience, good in-line sensors and great observation skills are important to bring the operational efficiency of the line to a level that was assumed in the business case.

R&D process development would work closely with the factory engineering department to optimize the E2E process. The basis is the processing specifications made by R&D and confirmed by the factory management. The lead of the product over the process helps to ensure that the originally designed product quality is maintained during the upscaling. A food company has suitable machines for their products, not suitable products for their machines.

The "definition" of the process is one of the key aspects in the handover of the product design to manufacturing. The other one is of course the product as such. R&D as the owner of the finished good specification will have to provide unambiguous reference samples for the factory. Together with the local QA, the quality of the necessary raw materials is assessed. The material intake control will assess the suitability based on agreed specifications, picture and retention samples. The assessment before the use in production reduces defects in the final product that are caused by the material used. This control is especially important in processes where direct farm-based materials are used, like fresh vegetables, cereals, legumes, etc.

Then intermediates, or semifinished goods, are identified for the same reason. Many industrial food productions have several process steps, where part of the ingredients are pre-treated together. The reasons for this can be manyfold, from simple mixing efficiencies over pretreatment to in between maturation times. The "should be" standard is also captured through samples and descriptions.

Most important are the finished good specification. These are the samples that will go to market. Beside all necessary safety checks (see Chapter 10), the food product characteristics are agreed and quality defects are defined, often split into "minor" and "major" ones. Quality assurance will regularly take a randomly defined number of samples and inspect them. Popular is a traffic light system, where particular parameters are classified as major "(red") or minor ("yellow") defects and the wanted quality aspects are "green".

Typical parameters like colour, texture, taste, aroma and shelf life are retested. The quality check is not limited to the food product as eaten or prepared but includes the consume- relevant packaging. Here the agreement on the minor defects and their tolerability must be made. Unesthetic seals, damaged outers or defects in the print quality and the presence of all mandatory information, like minimum shelf life, etc., are defined.

Once the quality of the industrially made products is defined and the factory confirmed that the food product is as designed and agreed with marketing that the product can be made with the necessary efficiency, the handover has happened.

Often it is not the question if the product can be made, but indeed at which efficiency. Line speed, process accuracy and waste levels are important criteria ensuring a sustainable routine production.

Frequently changes in the E2E process and the product can happen. The factory may not be happy with the material delivery format. Ingredients may be delivered in big pack, but now silo storage is more effective. Or part of the mix is better preblended to ensure homogeneous distribution on the final product. Or some physical parameter (size, density, form, temperature, humidity) of ingredients turn out to be difficult to manufacture. These modifications can be agreed, but always with attention to the final product. The changes should either be not visible, detachable or relevant for the consumer-perceived quality or need to be retested with the target group.

The decision to invest into one's own new manufacturing is sometimes also influenced by the financing options and tax regulations. In a country where "lease back" is an attractive option for a food enterprise, the factory can be set up and then sold to a third party but leased back. The capital invested is again available for new undertakings, while the access to competitive assets is ensured and the lease costs can be deduced from local tax.

So, R&D interaction with the manufacturing is manyfold. They bring requests for new product ideas that help the business to grow and react to market changes. This can stress the manufacturing, since the necessary manufacturing lies outside the "comfort zone" of the factory. Therefore, the "job is not done" once the new product has been successfully handed over to the supply chain. Often a longer post-launch support is needed to bring the manufacturing to an efficiency level that sustains their competitiveness.

THIRD-PARTY MANUFACTURING AND OUTSORCING

Outsourcing as such is the practice of delegating certain business functions or processes to external service providers. That can be anything from marketing, sales and R&D to financial services and manufacturing. Here only the contract manufacturing is discussed. With more and more new food products entering the market and the globalization of the food industry, the production of food by third-party manufacturers (3PMs) is increasingly attractive. This allows a strategic reallocation of resources which increases their financial performance. For the United States it is estimated that up to three-quarters of all food companies have outsourced part of their product portfolio (Aptean.com., n.d.). The main benefits of that option for food companies are:

- *Cost reduction*: A food company can save money on labour, equipment, materials and overhead expenses. By outsourcing non-core activities, food companies can cut their fixed costs and invest more in their core competencies.
- *Capacity increase*: In case the demand for a food product is bigger than the manufacturing capacity, third-party manufacturing can allow a quick expansion of the capacity.
- *New market entry*: In a situation where the product import is difficult or not profitable, companies can reach new markets by finding a local 3PM making the final product or a semifinished good from locally sourced material.
- *Capital expenditure management*: Launching a new product or an existing product in a new format is a good way to increase turnover and gain market share. Such an innovation is often associated with investment into new manufacturing processes, while the commercial return is not sure. To reduce the financial risks of such up-front investment, the initial production with a third-party manufacturer is a feasible option. As this manufacturer needs his margin, the initial sourcing costs may be higher than for an in-house operation, but in case the new product format fails, the capital expenditure is not lost. And if the format succeeds, an in-house investment and subsequent insourcing remains a good option to optimize the underlying supply chain costs.
- *Specialized expertise and advanced equipment*: External contract food manufacturers often have sophisticated equipment and industry-specific expertise that is not available within the company.
- *Internal resource and expertise focus*: By handing over basic or commodity manufacturing to external parties, the technical expertise in-house can focus on more attractive skill and capability development.

The third-party manufacturing can also bring risks and uncertainties to the contracting company and the brands:

- *Quality loss*: If the external manufacturer fails to meet quality and safety standards, low-quality or even unsafe products can come to market, in the worst case leading to a product recall.
- *Brand damage*: Inferior or unsatisfactory product quality can harm the reputation and image of the food companies and its brands, diluting the identity and differentiation of the food companies.
- *Dependency increase*: Handing the product of one's own food products to others can make a food company more dependent on the external parties for their operations and performance. Generally, the level of control and flexibility for the brand owner is reduced.
- *Order and production reliability*: 3PMs are frequently smaller manufacturers, without full back-up processes. If a risk hits their production, product supply may get interrupted affecting sales and brand reputation.
- *Knowledge transfer*: Often the outsourcing of one's own products or processes implies the transfer of specific knowledge the owner of the company has. Since food manufacturing is based on mature technology, the possibility for legal protection through patents is limited. Best practice and years of production experience can make the difference between profitable and non-profitable production. Once this knowledge has been shared, it builds the capabilities of the 3PM.

These risks can be mitigated. By establishing strict quality control processes and performing regular audits, the compliance with general food safety and brand quality standards can be safeguarded. The details in the contractual agreements, sharing of performance data and penalties for non-compliance often ensure that the 3PM will do all they can to ensure product delivery. Confidentiality agreements and non-compete clauses help to contain knowledge leakage. Such risks go beyond the product, so the mitigation plans should also span aspects of cybersecurity, preventing the leakage of data.

Outsourcing starts with the identification of a suitable partner. Some food companies have dedicated departments to scan the market for 3PMs. The expertise of the company, the reputation, location and production capabilities are key criteria before contacting the partner and asking for a proposal. If the partner is not known, due diligence – the verification of their quality and safety certifications – is the next step. Many 3PMs present their certificates on their website. Once the commercial negotiations are finished, the transition and implementation plan is needed. Beside the agreement of the production responsibilities and possible additional training for the staff, the data and system integration often requires time.

For R&D that often means the alignment on the various specifications. At a minimum all data for the final product specification need to be integrated into the data pool of the contractor, so that compliant labels, with nutrition table, ingredient content, allergen information, etc., can produced. The data management between companies often differs; here the building of an efficient, safe and reliable data transfer systems can save a lot of work. And if the customer of the contracting company, the retail organization, is also requesting detailed product datasets, an automated data flow over all interphases may be required. Some companies provide support to connect different factory systems in such configurations (Protected, 2023).

The transfer of legally compliant "finished good specifications" to the contractor is often the minimum. For the contracting company and their communication to consumers, customers and authorities, more details about the raw and packaging materials, their composition and origin and manufacturing can be useful. These data can disclose sensible and confidential information about practices and competitive knowledge of the 3PM, so that the details of the sharing need discussions. A long-term partnership helps a lot here.

Regulating authorities, like the FDA, have often established sets of robust, risk-based standards to ensure food safety in such situations where "product owner" and "maker" are not the same legal entity. Frames like the "Food Safety Modernization Act" (FSMA) contain therefore elements like reliable third-party audits by governmental agencies or private enterprises, preferably after accreditation (U.S. Food and Drug Administration, 2018).

While it can be simply cost reasons to "go external" with the food production, in innovation, a food company more often reaches out to a contract manufacturer to produce new food format for which they do not have the skills and capabilities. Reasons are the technical capabilities of the third-party company. The contracting company may have a suitable brands, channels and a sales force to bring the new product to market but not the right manufacturing. And building it up can be too resource intensive (people and money), too slow or too risky. The customer can reduce inventory and transportation costs of a finished product, get access to new markets and technologies and use competencies and marketing activities rather than production issues. This alternative is the outsourcing to a third-party company either in total or partially.

The type of agreement with a manufacturing company is sometimes split into "contract manufacturing" and "toll manufacturing" (Klinger, n.d.). This differentiation is not consistently used in the literature. The key criteria lie in the service level the third party provides. The contract manufacturer has more degrees of freedom in the production, while toll manufacturing expresses more operational involvement of the outsourcing company.

This "all inclusive" contract manufacturing includes not only the usage of factory, machines and labour but also the material stock levels, including specification and quality control of the raw materials used. The specification of the final product is agreed between both parties and sometimes even owned by the contracting company, but all specifications below the final product are managed and owned by the contract manufacturer.

Toll manufacturers also provide machinery and labour force and sometimes the raw material, but here the manufacturer normally acts under tighter guidance of the contracting company. Under this umbrella the contracting company can request to use their own (semifinished) products or ingredients in the 3PM manufacturing. This can be the case in situations where a particular item or ingredient is seen as critical for the product quality or the contracting company has better contractual conditions with the material supplier of the toll manufacturer. These types of agreements also make sense if the contracting factory has competitive knowledge in the making of a semifinished good that they do not want to share with the external manufacturer.

A food company may, for example, have great expertise in refining fats to a desired quality, which is mostly used for their own in-house production. Now a newly developed product needs exactly that fat quality, but the manufacturing design for the new product is not available within the company. In this case the toll manufacturer gets the speciality fat from the company to make the new product. A situation good for both parties. The outsourcing company can ensure that their fat is part of the formulation, without the need to disclose confidential details to others. And the 3PM does not need to invest in a perhaps sophisticated process and long learning curve to make a refined fat that the final product needs.

The advantage for the outsourcing company avoiding up-front investment and instead benefitting from the toll manufacturer's expertise and economies of scale has its price. This fee is the toll manufacturer's business model.

Toll manufactures are not smaller-size local enterprises unwilling to invest in their own brands and market access. Frequently, this can also be companies producing and selling under their own brands but also offering parts of their manufacturing skills to other companies. Some examples include the following:

- ADM, today one of the biggest food companies, provides toll processing services for oilseeds, grains, cocoa and other ingredients, sticking to their roots as a food processing and commodity trading operation.
- Cargill offers toll milling, blending, extrusion and packaging services for a variety of food products.
- Kerry provides toll fermentation, spray drying, agglomeration and encapsulation services for functional ingredients and flavours.
- Sensient provides toll extraction, distillation, concentration and crystallization services for natural colours and botanical extracts.

These food companies, often multinationals, may have started in commodity trading and processing companies. They kept their "old business", including the large-scale infrastructure and many years of experience.

Classical contract suppliers are also called "OEMs" or "original equipment manufacturers". In the past that described companies that were providing tailor-made processing solutions to a food manufacturer, like a specific transportation belt or sorting mechanism. These food "OEMs" own the manufacturing process but not the design of the product.

In this case, the R&D department of the contracting company has designed the product or at least the prototypes and then asked an outside "food maker" to produce it. The rights of the product design stay with the contracting company. How far further the rights of the contractor go depend on the level of control they take in the E2E fabrication process. Apple, for example, uses OEM, designing their own products and then reaching out to contract manufacturers, where they overlook the whole supply chain and fabrication process from components to final assembly.

In foods, traditionally more low-tech business, this split is still popular. The contractor owns the product design and the contract manufacturer the process design. The new food development and innovation process would basically run as a typical "in-house" project, up to the point when upscaling and production are starting. Here they still may advise the 3PM.

The food business now also starts with the "ODMs", the "original design manufacturer", idea. This has started in the personal care/consumer cosmetics business, where third-party manufacturers made their own new product development. Based on their in-house technical skills and the deep understanding of the technology they use, their R&D teams develop new products, which they offer companies with a strong brand as a new product. In cosmetics this collaboration has grown systematically into strategic partnerships, where the "ODM" regularly visits the branded goods company to present all their new ideas, in a sort of internal fair.

The design rights in this case are with the 3PM, but at the end all depends on the specific contract. It can be better for the branded goods company to buy the exclusivity, to own an iconic element of the product design.

Particularly in Asia this model has gained a lot of popularity. Here the local "ODM" has no suitable access to foreign markets, but they have better economies of scale, lower production costs and all the required knowledge and infrastructure. Even more sophisticated skills, like the ability to support competitive health claims through a dedicated clinical program, are provided by the ODM and help to launch new functional food products or health supplements.

In such situations the role of R&D in the contracting company is limited. They basically ensure the sourced products comply fully with the company standards and may advise marketing, procurement and sales on technical aspects of the product or "storytelling".

In summary, the role of R&D in third-party manufacturing can be very different, starting from "light touch" constellation where the supplier owns the product and process design up to fully controlled toll manufacturing, where they keep the full ownership of the full specification hierarchy, including the necessary processing. Such situations are frequent in foods, when an iconic product sells well and the market demand exceeds the capacity. To ensure that the "in-house" product cannot be distinguished from the 3PM-made one, tight control is needed. Here the risk of knowledge leakage exists. Contracting companies need a good understanding about "their" knowledge and how they can protect it in such cooperation. The toll manufacturers from yesterday can become a direct competitor tomorrow. Big food companies often have strategic documents in which they fix the level of confidentiality and ownership that they want to keep for their science and technology.

The most complicated situation is perhaps "open innovation". Here R&D teams on both sides work hand in hand to make a new product. The cooperation can start with joint ideation session or with the translation of a winning concept idea into scalable prototype. Here IP, patent and trademark rights need to be arranged in detail. Experienced legal experts in the company or from outside can help with the contractual details. This is a situation similar to "open innovation" (see Chapter 4), where both sides need trust and confidence, even if they are very different in size.

"In house" versus third party is not an "either/or" discussion. Food companies with a versatile product portfolio often have part of their offerings made by others. A relatively simple large-scale commodity process will rarely be insourced. The investment needed to install such technology at a competitive scale hardly ever pays off. And often alternative 3PMs are available, so that a "dependent situation" is not likely. On the other hand, it is very useful to use an existing 3PM to quickly bring a new product into the market without the up-front massive time and money investment to first install the manufacturing in-house. If the new food launch is a success, insourcing is a good alternative. Longer-term, the supplier margin can be saved, and the volume forecast is better predictable. Or one can even go and acquire the former 3PM, possibly a company having specific technical skills and food production certificates (Seccuro et al., 2022).

FOOD SUPPLY CHAIN THREATS

One of the most valuable goods for a food company are their brands, which carry the reputation that has been built over many years and required a lot of money. But it potentially takes only hours or days to destroy it. Company leaders are very well aware of this. Today's global E2E food supply, reaching from "from farm to fork", is often very long and complex. The ingredients for product and packaging are coming from very different sources. After production the goods must pass distribution chains of sometimes enormous lengths and complexity to finally land on the shelf in a remote place in a country with a developing and emerging market or in a supermarket in a different country. On this long and winding road many business risks are waiting. In its essence the task is to avoid ". . . all types of food safety risk or simply food fraud incidents, inappropriate sourcing of new suppliers, the usage of controversial ingredients or unmanaged regulatory compliance exposure" (FoodChain ID, 2024). The combination of a pandemic, war, economic crisis, global inflation and environmental challenges have put extra stress on the food supply chain everywhere. Nowadays it is simply not sufficient any longer to find an external third-party manufacturer able to produce the finished product as specified. The whole chain needs to be robust and resilient, able to make the product in line with all specifications, independently from agronomical harvest risks, a primary manufacturer facing technical challenges, an ingredient supplier eventually not being able to deliver the required volumes, a packaging material out of specification, a breakdown of a machine or a massive interruption in the distribution chain for the product on the way to market.

THE COSTS OF NEW FOOD PRODUCT DEVELOPMENT

There are different ways to look at the cost of new product development. One option is to sum up all costs of the steps from product idea to final launch. That clarifies the direct costs. Most easily that can be done when development contracts with external development centres are made.

These costs can differ enormously. A range extension which uses the same production process under a well-established brand with trained resources is obviously much less costly than a true breakthrough innovation within in-depth consumer work, a new processing technology, new concept and perhaps even a new brand.

A different perspective on product development costs tries to capture the costs of an "innovation capable" organization, starting from building of an idea creation and execution processes, followed by structures allowing decision making and recycling at various stages.

In many food companies the new product development costs are managed as "fixed" costs. Here, R&D is part of the "overhead costs". The money needed to hire the right people and their infrastructure, like lab, kitchen and pilot plant equipment is needed. These costs for the company are fixed and do not directly correlate with the commercial success. In FMCG foods the expenses for R&D may only reflect 1% or 2% of the full turnover, but this can still be a very sizeable budget at a scale of an annual turnover of $20, $50 or $80 billion.

Counting R&D and its infrastructure under fixed costs is often a good approach, as these costs cannot be precisely allocated to a single new product innovation. And many R&D responsibilities are routine tasks and often focus as well on the maintenance of the products already in the market.

More recent studies try to split the R&D cost and allocate them more directly to dedicated R&D effort. An approach raising interest is "activity-based costing" (ABC) allocation in open innovation models.

For those companies that do their product development with external services, most of the development work are "variable costs" – money paid for an earlier agreed to service level. Normally, these companies would brief the service company with their product request and they would revert with a fixed all-inclusive offer or a resource-based rate for labour and material per time, or a dedicated team. That can be anything between $10,000 up to $500.000 or more.

For a food company with an existing R&D department, this part of the total product development costs is often a fixed cost. And only those steps in the development that cannot be done internally require additional budget. These costs are directly related to the actual NPD and often planned as "variable costs". For this and many other reasons, it is not easy to estimate the total direct cost of a new product development in food. Some companies may limit the consumer research to internal tasting sessions and a small group discussion, while others will validate their development with a series of consumer tests, including very expensive preference mapping. Table 8.2 can therefore only give broad ranges.

So, costs can vary hugely. It depends on the quality and quantity of the research needed to bring an idea to market. An established food brand tends to test new products more intensely to minimize the risk of failure. First, the new product should not negatively affect the sales of an existing offering. Even a very

TABLE 8.2 Rough cost of various essential steps in NFPD

NPD PHASE	*COSTS $*	*COMMENTS*
Ideation session	0–20,000	Moderator, location, stimulus, knowledge capturing
Industrial design	10,000–30,000	Design workshop to bring the industrial product alive
Prototyping	5000–50,000	Iterative process possibly, with consumer customer research in between
Consumer testing	5000–500,000	Simple virtual focus group or full blown in field preference mapping and prototyping
Certification and testing	5000–50,000	Much depends on market, product class, claim and certification scheme
Packaging prototyping	0–50,000	Standard version will be made by supplier free of charge, if new moulds, etc., needed
Upscaling pilot plant	0–10,000	Incremental costs are mainly material, sometimes already available in PP or factory
Factory trial product	5000–50,000	Depends on charging model; if R&D is charged for "lost production" time, costs are much higher, otherwise material and eventually factor labour
Factory trial packaging	10.000–75.000	Packaging trials are often longer-term trials, can include short-, medium- and long-term testing

successful new product launch will deliver for many years only a fraction of the turnover an established portfolio brings.

If the launching company has the possibility to "test" their new product at a limited scale in a small market, maybe only a few shops or outlets, most of the resources for market research can be saved. The test market just needs to be a reliable predictor of the full market.

Big branded food companies regularly launch on a bigger scale! Here the pre-launch field work studies are expensive; they require substantial preparation and large numbers of participants to be representative. If the research includes a quantitative in-depth market scan, like in preference mapping, costs will raise significantly. Their own trained panel, as part of the R&D overhead costs, can limit oncosts.

Similarly for the product prototype, if it can be made in "sellable" quality in the kitchen, lab or pilot plan, not much extra budget is needed. If the company must hire a pilot plant facility, costs can quickly increase. The food format will also influence costs.

Factory trials can be quick, clear and immediately successful. Or they are laborious, complicated, uncertain and questionable. Depending on the cost allocation agreement with the factory, they may require substantial R&D budget. Third-party manufacturers in foods often do not charge this cost directly but add it to the final transfer price.

The ability to simulate real processing condition digitally is an increasingly more important driver to keep this cost at bay without losing the confidence that the manufacturing afterwards will be efficient.

The cost for the validation of new packaging material can be surprisingly high. If the packaging is an integral part of the processing, the lines are high speed and high throughput. The factor leadership will want to test thoroughly, before they accept a new product. The deviation of a few percentage points in line efficiency can make a huge difference in the supply chain costs. A case study in Chapter 15 discusses this in more detail. Storage and distribution trials would come on top. "*In silico*" computer simulation tools, big data and AI promise new great opportunities to do that more cost-effectively.

Similarly, certification requires substantial testing and inspection. Moving a non-certified location to certified kosher or halal is a substantial effort in product design and manufacturing. And launching a product with a nutrient content claim requires in some countries pre-market approval and disclosure of a shelf life and product tests.

Even after-launch possibilities to optimize the product and packaging post launch are possible. A new food launch can happen under time pressure, because suddenly there is a gap in the market or just a particular launch slot with the trade. There is no time to optimize the cost structure. But once launched, it is needed. Similarly, it may turn out that a food product was "overdesigned" containing costly elements that are not relevant for the buyer.

INNOVATION MEASUREMENT: FRONT- AND BACK-END INDICATORS

Every organization working on innovations wants to know how successful their new product development efforts are. Like most other processes in the company, the business requests a view on the "return on investment" to take the learnings and eventually strengthen the innovation machinery or add it.

For a company that made several new launches in the past, a simple balanced scorecard is a suitable tool to review the current innovation effectiveness. It often has four different perspectives: Financial, customer, internal process and learning and growth. For each of them so-called "key performance indicators" (KPIs) are formulated, allowing the organization to evaluate their innovation effort. Frequently used KPIs for commercial food products include those shown in Table 8.3.

TABLE 8.3 Typical key performance indicators for innovation

INDICATOR	*MEASURING SYSTEM*
NPD revenue	The total revenue that comes from new products launched in a defined period. It measures the impact of innovation on the top line and reflects market acceptance of the new product.
Return on innovation investment	The ratio of net profit from NPDs versus the cost of innovation activities. This KPI evaluates the efficiency and effectiveness of innovation spending and the profitability of new products.
Innovation pipeline strength	The number and value of new product ideas in different stages of development, for example, in an innovation funnel from ideation to launch. This KPI gives an idea about potential future revenue and growth from innovation and therefore the quality of the innovation portfolio.
Time to market	The average time it takes to develop and launch a new product, for example, from idea generation to commercialization. This KPI reflects the agility and speed of the innovation process and the ability to respond to changing customer needs and market opportunities, Critical in FMCG markets.
Market share	The percentage of a market segment that the NPD has captured after a given time. It indicates the competitive advantage and differentiation potential of new products and the market penetration and expansion.
Consumer/customer satisfaction	The degree to which they are satisfied with the quality, performance and value of new products. Good KPI to assess loyalty and the alignment of innovation with consumer/customer expectations and preferences.
Customer/consumer retention	The percentage of customers that continue buying NPD after first purchase. The KPI tracks the repeat purchase behaviour and hence loyalty of customers for NPD.
Innovation capability	Availability and quality of resources, processes and tools that support innovation activities. A measure of the available infrastructure needed for innovation.
Innovation culture	The levels measured are employee engagement, collaboration and creativity in the innovation process. This KPI helps to measure the innovation culture and the ability to generate and implement new ideas.

Many more are possible; their selection is based on the business strategy. If a company has decided to invest in new differentiating science or technology, KPIs like number of patents, the strengths of scientific collaborations or the number of recruited and trained experts or scientific publications are good measures.

Alternatively, R&D and marketing together can share KPIs about the brand awareness, by measuring how consumers recognize and recall a new product or a brand name. Or the number of product-related complaints is used as a shared target between manufacturing and R&D to minimize design and manufacturing defects.

Three aspects are important: Firstly, good KPIs are "SMART", which means they are "specific, measurable, achievable, results-focused and time-based". A general, unspecific, hardly measurable, non-validated measure does not help. If the importance and validity of the indicator is questionable, discussion happened and not performance tracking.

Secondly, one should choose only a few KPIs, since for each of them, data need to be gathered, analysed and discussed. More data do not necessarily provide more clarity; they can also create fog and trigger unproductive discussion.

Thirdly, where helpful indicators can be split into "leading" and "lagging". Lagging indicators measure output after the fact. The tool compares past performance with present. Internal business data from the past are used as a reference to improve in the future. Typical examples are revenue, profit, customer satisfaction, number of ideas or incremental turnover from past launches. They are useful to identify trends. They may, for example, show that the incremental turnover from recent product launches is going down and the cannibalization rate is going up. So they are good when looking for trends, but they reflect the past performance and results of innovation efforts and do not provide direct insights for potential problems or future opportunities.

On the contrary, "leading indicators" focus on inputs and actions necessary for success; they can be influenced, and the impact should be visible quickly. Customer feedback on product quality or employee engagement is a typical leading indicator for food innovation. They allow one to understand what is currently happening early in the process so that any corrections can be made.

Leading and lagging indicators are complementary and can be used together to manage food innovation effectively. And it is a choice: If it has priority to push ongoing innovation in the right direction, leading indicators are more suitable. If the impact of innovation on the return on investment of a company is discussed, lagging indicators are better.

Often KPIs are set in a consistent frame and not as standalone targets. A "balanced scorecard" is such a frame where the organization goals are connected. The frame is called "balanced" if it expresses a wider perspective: Beside financial and customer aspects it can cover internal business process or learning and growth (Tarver, 2024). Each perspective has a set of objectives, measures, targets and initiatives that are linked to the overall strategy. KPIs are derived from these elements and used to monitor and evaluate the progress and results of the strategy.

A newer variant are "OKRs", or "objectives and key results", a more direct link between high-level purpose (objectives) and clearly defined outcomes (key results) which are needed to achieve the objective. OKRs became popular, as they provide more context and direction for the goal achievement. OKRs are more specific and measurable than KPIs and are often more change-oriented than KPIs. A few typical examples for food R&D are the following:

- Test the new plant-based burger with 300 customers versus key the competitor and achieve superiority.
- File two patents or intellectual property registrations within 1 year.
- Reduce the average time required to bring a new concept to market ready by 15%.
- Achieve a satisfaction rating of 9 based on product quality and performance.
- Increase the success rate of research projects, measured by 70% of projects meeting predefined milestones.

Comparing the innovation effectiveness between similar entities is called benchmarking. Enterprises can benchmark themselves to identify their strengths and weaknesses. The entities just need to be comparable, and the relevant data need to be known. But getting those data within the industry is not easy, since nobody wants to make the competition smarter by giving them good ideas. But it can also be done using quantitative indicators, qualitative assessments, surveys, interviews, case studies or peer reviews. Where possible the benchmarking is done on output, like number of patents, publications, prototypes, launches and outcomes, and not on input, like people working in R&D, budget, etc. But just because data are countable, they are not relevant. A single patent about a soft drying process can be much more valuable than 20 smaller design patents on packaging formats. And two dozen range extensions do not mean more innovation productivity than one new platform product.

Good benchmarking is not a light task. The World Benchmarking Alliance (WBA) recently benchmarked the top 350 global companies active in agriculture, commodities, animal protein, processing and manufacturing, retail and food service. They used 45 different indicators in four different sectors including, social, nutrition, environment and governance (WBA & The Food Foundation, n.d.).

Management consulting firms, like McKinsey, aim to measure innovation effectiveness with simple correlations based on publicly available data (McKinsey & Company, 2018). They may simply compare two ratios: First, how the R&D spending is converting into new sales. And secondly, the ratio of gross margin to new product sales. There are fewer indicators, but done for the totality of all products and a longer rolling cycle of 3 years, representing the average life cycle of FMCG products.

REFERENCES

Aptean.com. (n.d.). *Aptean Media Viewer: Rich Insights to drive your success.* https://www.aptean.com/en-US/media-viewer?type=i&d=aptean-food-and-beverage-global-outlook-2023-en&height=full

Carroll, J. D., & Chang, J. (1970). Analysis of Individual Differences in Multidimensional Scaling via an n-Way Generalization of "Eckart-Young" Decomposition. *Psychometrika*, *35*(3), 283–319. https://doi.org/10.1007/bf02310791

Conjointly. (2024, May 2). *Idea screener - Conjointly.* Conjointly. https://conjointly.com/products/idea-screener

FoodChain ID. (2024, April 25). *Brand risk in food supply chains - FoodChain ID.* https://www.foodchainid.com/resources/current-food-supply-chain-threats-is-your-companys-brand-reputation-at-risk/

Johnson, J., & Vickers, Z. (1987). Avoiding the Centering Bias or Range Effect When Determining an Optimum Level of Sweetness in Lemonade. *Journal of Sensory Studies*, *2*(4), 283–292. https://doi.org/10.1111/j.1745-459x.1987.tb00423.x

Kim, H., Hong, J., Song, C., Shin, H., & Kim, K. (2010). Sensory Characteristics and Consumer Acceptability of Fermented Soybean Paste (Doenjang). *Journal of Food Science*, *75*(7). https://doi.org/10.1111/j.1750-3841.2010.01771.x

Klinger, J. (n.d.). *What is toll manufacturing: Foods, functional additives, & innovation.* https://www.customprocessingservices.com/blog/functional-food-toll-manufacturing

McKinsey & Company. (2018, October 8). *How to take the measure of innovation.* https://www.mckinsey.com/capabilities/strategy-and-corporate-finance/our-insights/how-to-take-the-measure-of-innovation

NielsenIQ. (2021, August 12). *Unlocking consumption: Four consumer groups driving spend - NIQ.* NIQ. https://nielseniq.com/global/en/insights/analysis/2021/unlocking-consumption-four-consumer-groups-driving-spend/

PR Newswire. (2021, October 21). *43% of consumers are loyal to at least one food and beverage company, encouraging emotive and sensory branding strategies.* https://www.prnewswire.com/news-releases/43-of-consumers-are-loyal-to-at-least-one-food-and-beverage-company-encouraging-emotive-and-sensory-branding-strategies-301404992.html

Planbeyond. (2024, February 20). *What is TURF analysis & when to use it | Market Research.* PlanBeyond. https://planbeyond.com/blog/turf-analysis

Protected, E. (2023, December 11). *Understanding toll manufacturing* - Mar-Kov. *Mar-Kov.* https://mar-kov.com/understanding-toll-manufacturing

Rossiter, S. (n.d.). *Super taster vs. non-taster: Does it matter for your health?* https://www.ifis.org/blog/supertasters#:)

Seccuro, M., Levin, B., & Seccuro, M. (2022, September 7). *In-House manufacturing – a differentiator in today's food industry.* Intrepid Investment Bankers. https://intrepidib.com/in-house-manufacturing-a-differentiator-in-todays-food-industry/

Symmetry Media. (2023, July 20). *The rise and fall of new coke: Unveiling Coca-Cola's legendary marketing blunder.* https://symmetrymedia.ca/2023/06/03/the-rise-and-fall-of-new-coke-unveiling-coca-colas-legendary-marketing-blunder/

Szabó, E., Szakos, D., Kasza, G., & Ózsvári, L. (2021). Analysis of the Target Group of Lactose-Free Functional Foods for Product Development. *Acta Alimentaria.* https://doi.org/10.1556/066.2020.00168

Tarver, E. (2024, February 24). *What is a balanced scorecard (BSC), how is it used in business?* Investopedia. https://www.investopedia.com/terms/b/balancedscorecard.asp

Thomas, J. W. (n.d.). *TURF analysis.* https://www.decisionanalyst.com/whitepapers/turfanalysis/

U.S. Food and Drug Administration. (2018, August 20). *Third-party audits and FSMA.* https://www.fda.gov/food/food-safety-modernization-act-fsma/third-party-audits-and-fsma

WBA & The Food Foundation. (n.d.). An aligned methodology for food industry benchmarking. *Food Industry Benchmarking Toolkit.* https://assets.worldbenchmarkingalliance.org/app/uploads/2021/12/TFF-WBA-Toolkit_Report-Section2.pdf

Innovation Pipeline *Managing Time and Risks* 9

STRATEGIES TO ACHIEVE GROWTH AND PROSPERITY

Food startups are often the purest "innovators"; here a single team or the whole company focusses on one single idea. For them it is critical to have "the right idea", as they often do not have the option to counterbalance the intrinsic business risk of one idea with a second independent possibility. They are "all in" entrepreneurs. If their idea is not viable, it should fail quickly, so that something new can start.

Bigger and more mature food enterprises are normally not "putting all their eggs in one basket". They have an existing business that creates revenue, and a new idea must fit with what is already on the market. Their innovation program reflects the market they are acting in. Line and range extension builds the existing business further. Rejuvenation and refreshment take care of what is already there. And even innovation and breakthrough receive their title in light of what the company already has out there.

They use stage-gated processes, linear or circular (see Chapter 7) of different flexibilities to bring ideas successfully to market. They also must decide to "do the right thing"! Putting some extra money into sales and distribution may in the short term be more profitable than NFPD. But what if a competitor sees their market and comes along with a new product and takes substantial market share?

Innovation is the lifeblood of the food business, ensuring longer-term growth. And that innovation can be the next ice cream flavour, a packaging rejuvenation, a new hybrid crop, a tool to predict consumer preference through artificial intelligence or the next whole grain bread convenience consumer pack of gluten-free, low-insulin, ancient grain cereal bread sustainably sourced by modern biotechnology.

And since many NFPDs fail, it's of critical importance for the long-term prosperity of a business to find the best mix of innovations, balancing risk and opportunity and their best balance between the optimal return and their appetite for risk.

It was probably Eric Schmidt, Google's former CEO, who applied an old, and controversial, learning theory to innovation: The "70/20/10" rule (ITONICS, 2024). He said that the innovation effort in a company was for 70% on the existing core markets, for 20% on adjacent markets and only 10% on new markets. Many FMCG companies use this as directional guidance. Nobody follows this dogmatically but putting half to two-thirds of resources behind ideas in the core of the business, like refreshment and new variants, another third to 20% behind initiatives to stretch the core and extending the range and 10% or a little more behind breakthrough and new markets is good guidance. Actually, approximately 10–15 years ago, an even more conservative 80%/18%/2% split was discussed. The rationale stemmed from empirical business data showing that companies allocating their resources accordingly show the best business performance in the long term (Nagji, 2021). Behind that conclusion is an inverse risk-return ratio. The (potentially) highest revenue comes from conquering new markets, but these also have the highest risk, while innovations in core markets often are low risk, but also with much smaller revenue.

The guidance has changed. Today the factual challenges for the food industry are bigger, but also the potential of new science and technology to create fundamental new solutions has increased. AI, plant based, precision fermentation and all types of biotics are pushing solutions into the markets. Each market leader must look at outside markets, as others are looking into theirs. Clayton Christiansen's innovation dilemma reminds them to invest more in the more unpredictable (Christensen, 2016).

DOI: 10.1201/9781032724591-9

The risk level a food enterprise can take is influenced by the market position they are in. A market leader that loses attractiveness has more need to push outwards than in a still nicely growing one. All big food companies had these moments in their history, where they started to disinvest in parts of their core to create the funds to conquer new segments.

A company lagging in an attractive market may have to shift resources from their core to a new, more aggressive initiative to attack the leadership position of others.

From the perspective of the existing commercial business, three different innovation territories can be described: Core, adjacent and transformational. Each needs a different strategy. A successful player in an established market may encounter problems to enter a "new market". Their innovation skills and ways of working have matured so much over the years of commercial success that they struggle to do things differently now that is needed to win the new markets. Past success in an old market is not the entry card for new, unknown territory.

The list of those new skills needed is long. It could be the "type of creativity". The well-known "core" and "perhaps" partly the "adjacent" markets" can be effectively analysed with known procedure skills. Being trained with all what is known about these "home markets", it will be possible to predict new variants and range extensions, increasingly via "*in silico*" work supported by AI. First attempts here are already hitting the market (Business Today, 2024).

But the entry key for "new territories" can be more intuitive, design oriented and less data driven. I am not sure if AI would have found ideas like popcorn, frozen fish fingers or salted egg in ice cream.

One can even ask, if it is most promising to ask the same R&D people that successfully support the core to identify the next outside opportunity. Outside or standalone R&D centres are an option some companies in foods use: They have their regional or local development centres with often centralised science and technology departments.

An interesting question is the best relation between both. Searching for "new growth territories/ breakthroughs" outside or in central research facilities maybe better not too much boxed by the daily product needs. It is much harder than to think "outside the box". But if it leaves those innovators too much in "ivory science tower", they may fall too much in love the problem and be too distant to create a solution for needs of the markets. That's the decision of the company and many of those prefer to build a big "window to the market" for their "new territory searchers".

The funding of such units is part of the relation. Closer to market, development units are often funded from the business revenue of the unit. The shorter-term return on investment shapes the team. The new product launches of a regional R&D team created 5% growth. So they can expect next year's budget to be a bit bigger, while launch failures could question their capabilities.

R&D teams hardly ever have the impression that they have sufficient funds. If now the central science "pathfinder" team receives its budget also as a result of that business, for example, as a 20% charge, the local R&D team may perceive that as a "tax burden", and look more critically at the contribution of the central team. If they perceived that what they have accomplished in the D2D battles in the markets disappears in "central research", without any clear output, they may be less open to new suggestions from the central team; a phenomenon known as the "not invented here syndrome". To avoid that it could be better to apply a different, not P&L-dependent funding mechanism, for example, by strategically funding them from different, top management-controlled budgets.

Breakthrough ideas also need a different management process. The innovation team cannot simply rely on the feedback of consumers/customers for something the consumer has not experienced before. A radically new environmentally friendly food packaging solution may receive an initial consumer "OK". But if the solution compromises one consumer drive, the convenience or the price, the idea may not receive a warm consumer welcome. But in 5 years from now, when the need to protect the environment is even more apparent, the solution would be absolutely "spot on". Calculating the future value of longer-term innovations is more difficult than to estimate the "net present value" of new core launches in the next 6 months. And while the whole portfolio, short, medium and long term, needs to be reviewed with different glasses, the company needs to build such a project and idea portfolio that serves them now and in the future.

For the discussion about the innovation pipeline, this often means deciding between dozens or even hundreds of potentially similar promising options. Nobody has the resources, time, skills and money to run them all. Hence, an effective selection process is needed which helps to find the "good ones", avoid the "bad ones" and understand why some are "ugly".

The review estimates the strengths of the initial idea, the potential in more markets, the level of uncertainty in timelines and resources and the potential interdependencies with other initiatives. An idea may could be brilliant in one context, but a "recipe for disaster" in a different one.

Setting up a session to "review" or "innovation pipeline fill" is again best done by a cross functional team, with representatives of different functions contributing with their specific expertise and experience. The setting of a common project goal is important. During this process the participants will feel from time to time that they leave "safe ground" or their "functional comfort zone". The process requires openness to the ideas but also confidence in their own statements. If the briefing is building a balanced portfolio of "safe bets" and "big gains", a mixture of experienced employees, new arrivals or even external people can be powerful. Experienced R&D people know the existing markets from their own experience, and they often have a solid view of what will work, why and how. The "new recruits" often do not have that experience, but they also do not see the organizational barriers that may have hindered great ideas in the past. Their thinking is fresher.

The expectations between "functions" during the discussion can create discomfort. A new variant with a tricky ingredient mix can cause stomach pain with R&D when asked to judge "*ad hoc*" the feasibility. They may envisage immediately the search for new suppliers, complex material testing, auditing processes, unpredictable shelf-life tests and a universe of prototypes. Or marketing cannot judge the size of the consumer benefit or brand fit for a technology that excites the technical team, while market research cannot say "*stante pedes*" how to test best with consumers. And when all innovators are enthusiastic, the supply chain may feel uncomfortable, as the new idea requires massive line modifications and heavy capital expenditure, while QA is afraid the production will lead to lots of products with defect, mounting waste and unacceptable amounts of unsellable products. And the legal/regulatory doubts that the claim or story is feasible and can be sufficiently supported. Even finance feels uncertainty, as future profit shows high uncertainty while the team is asking now to commit lots of money.

Often the business pressure to come up with new ideas is high, so the different functions may push each other to make firm statements about the prospect of a new idea. Too much of that will damage the necessary trust and openness to try and decide best under a remaining level of uncertainty.

I will now briefly describe a few of the basic management tools that can help the team to make the best suggestion while admitting the difficulty to reliable predict the future.

PORTFOLIO MANEGEMENT TOOLS

While a project is a time-bound activity to create a unique product or service with defined scope and resources, an innovation program is a meaningful collection of several, often very different projects.

If this portfolio is aligned along a time axis, it's a pipeline or funnel. "Pipeline" is a suitable descriptor, as it transports the purpose of innovations – bringing new energy to the business. And "funnel" conveys that between entry and exit a selection process will happen. Or in other words: Much more ideas will enter the innovation funnel than leave it.

In foods many innovations are planned for a particular time. Many launches are fixed by external circumstances like trade slots, festive moments, season or weather. If a company has the innovation capacity to launch three new variants and knows that the work on those new variants needs 30 weeks to be launched, while the launch slot is mid-April, for example after, Ramadan, they must start their project at the latest in August the year before. Work on more variants can start when the required resources are free again or new resources get on board. This approach works well in a predictable market with space for new variants, an experienced team and a stable and well-understood technology to make the products.

The level of predictability will decrease when new markets adjacent or far away are targeted. For other projects the results of consumer and technology testing are less certain. A feasibility trial may fail, or the consumer response may not be good enough. The idea maybe wrong or right but needs more time and more resources, so it becomes a bit "ugly".

But the "new product" portfolio is the totality of all activities aiming to create new things. And because single projects in foods can easily fail, the portfolio encompasses the totality of risks and opportunities and offers the possibility to act in a balanced way. A few projects with a great incremental growth potential but higher intrinsic risks need to be balanced with a series of low-level, low-risk projects with average growth perspective and high feasibility.

The simplest approach to review a portfolio is to simply list all ideas on a sheet of paper and judge each of them all against the agreed set of criteria, like market potential, development time or technical difficulty. The cross expertise then scores them based on their functional knowledge. In the next step every function scores the aspect of the idea they are experienced in: The product needs a new mixing and blending station. Manufacturing explains to the team how difficult that would be. The product may not be stable. R&D says why and what could be done to overcome the limitation.

Assume a team is briefed to develop new "immunity and natural defence" products. Their review of ideas could include the items in Table 9.1.

If needed, the different aspects can be weighted by giving scoring points on an agreed scale. This kind of numerical scoring is useful to prioritize. If no other criteria – like time to market, etc. – can be used, such scoring helps to brings a priority in a sort of forced ranking, when the list of potential ideas is longer than what the company can work on. The answers are based on individual judgment and personal views; they may need refinement later on. Sales volume, costs, resource and risk judgements are estimates. Errors and misjudgement are unavoidable, but do not endanger the outcome of the exercise. If the scoring happens transparently and consistently, the group vote will ensure that scores are not biased for specific ideas or expectations.

For such discussion graphical tools are useful, as they can guide the discussion to the most critical points. A food product portfolio can also contain projects that are not "going to market" but form a necessary stepping-stone for the commercial launches. These interdependencies need clarification.

Say a food company wants to enter the fast-growing health and wellness space and starts a workshop to build their new product portfolio. The leadership informed that they urgently need the first products need to hit the market in 6 months. The team knows how long it takes to make three new bar variants and has already started to make some low-sugar variants, rich in protein. The company can also make other product formats, for example, light meals and instant soups, and they have a reliable third-party manufacturers to make a hot drink. The technical aspects of the different formats, taste, appearance, portion size and packaging are well known and do not require immediately new technology, but for now the necessary resources are occupied. The company sees no problem to make new variants later, but to differentiate those future launches from the more and more commoditizing market of energy bars and weight management solutions, the company suggests a science program to generate credible evidence that their new products have a superior functionality versus the competition. This science program will not "hit the shelf"! It will be published and communicated at a point in time. But it must deliver a critical part of the product launch mix, a superior functionality, as without that the new variants would not differentiate in the market. All of this requires detailed planning and some longer-term studies with ambitious outcome.

The company may also see good opportunities to offer a new solution in the "hot field" of mental health, say for relaxation, stress resistance, performance, resilience or instant energy boost. In principle they can use their existing product format and use the technical synergy. But the "claim area" is relatively new – the enabling science does not yet allow for hard, differentiating claims. So for now softer, more emotional messages based on general nutrient content claims should flank the launch. Less competitive and not yet ready for a sizable science program, but a good packaging program to truly refresh the range.

TABLE 9.1 Typical screening questions for product or service ideas with the function responsible to assess

EXPERTISE AREA	*KEY QUESTION*	*ANSWER LEAD*
Consumer appeal	How well does the new idea meet the preferences, how likely are they to buy the product?	Marketing/market research
Market potential	How large and attractive is the market segment and how competitive is the idea?	Market research
Brand fit	How well does the new idea align with the existing brand identity, values, how does the product enhance or complement the brand portfolio and image?	Marketing
Product feasibility	How technically feasible is the new idea in terms of product formulation, processing, packaging, shelf life, safety and quality?	R&D
Cost and profitability	How much money (resources) are necessary to develop, produce, and distribute the idea? What is the expected return on investment and what profit margin can be expected?	All – guided by finance
Sustainability	How environmentally friendly is the new idea in terms of raw materials, energy, water, waste, emissions and packaging?	Procurement; sustainability manager
Innovation	How novel and unique is the idea? Does it tap into an existing habit or require some behavioural change? How does the product offer a distinctive value proposition and competitive advantage to consumers and the company?	Marketing and R&D
Regulatory compliance	How well does the new food idea comply with the relevant laws, regulations, standards and codes of practice in the target markets? How easy will it be to obtain the necessary approvals and certifications for the product?	Regulatory/legal
Overall risk	How risky is the idea in terms of technical, market, financial, legal, environmental or social aspects? How likely are the risks to occur?	All, for their expertise
Scalability	How easy will it be to scale up to meet the expected demand? What are the potential challenges involved in scaling up?	Supply chain/planning
Adaptability	How easy can the idea be modified to better other consumer segments or changing trends? How flexible is the product design in terms of flavour, format, size, etc.?	R&D; supply chain
Synergy	How well does this idea fit with others in the company, leveraging resources, capabilities and competencies? Are there significant synergies?	All, within their expertise
Strategic fit	How well does the project match the company's vision, mission, goals and objectives?	Leadership

And finally, since the latest pandemic more and more people search for "immunity and resistance to disease" benefits. Again, the same product formats could serve as "carriers" for those benefits, but a program has to prove that specific micronutrient and herbal extract combinations could have a superior effect. A well-designed but rather short-term program should focus on "consumer testimonials" to give the launch a scientific underpinning.

The innovation portfolio now shows different health and wellness consumer benefit areas, often bound together by a common pool of food formats and technology, but separate science programs of very different depths and lengths. Priority is given to weight and energy management products, with substantial effort, while the mental benefit area is technically immature, and the immunity territory requires a different science approach.

CONSUMER/TECHNOLOGY MATRIX

A simple, yet insightful tool to look at the innovation portfolio of a foods enterprises is the plotting of all projects in a "consumer/technology" or "C/T matrix". The graph offers the opportunity to evaluate the novelty of an idea from the market perspective and the technology needed to realize the innovation. Since even small-to-medium-size companies can have dozens of NPDs at a moment in time, the tool provides a good view about the strategic balance of the total portfolio.

The x-axis represents the consumer value perception and the y-axis represents the technology used to deliver the innovation.

In a typical set up the C (consumer) perception splits into five segments: No change, variant, improvement, new benefits and new core product. Similarly, the y-axis or T (technology) axis has four categories: Base, incremental, next generation and radical (see Figure 9.1).

The positions on the axis have been defined before, so that all people working with the matrix have a common understanding: A frequently used possibility is shown in Table 9.2.

The different areas then point out the strategic intent of the project and the key characteristics. A project aiming to generate a variant of the already existing portfolio with base technology, so well known and already in use, supports the existing brand offerings in the present business core (Table 9.3).

On this basic level the tool is used to compare the program with the business strategy and helps to reprioritize activities or guide the search for new, yet uncovered opportunities.

FMCG companies using the CT will set the different axis points in a way most meaningful for their strategic discussion. For some, a longer debate whether a new idea is a "variant" or perhaps in "improvement" is not relevant. For others this discussion may bring very valuable insights how best to go to market, without having a pedantic debate. For yet other companies, a detailed differentiation between "radical" and "next generation" wouldn't create much added value.

These classifications are not self-explanatory. Normally it requires cross functional alignment about what a classification means. A company can define the usage of blockchain technology as "breakthrough", delivering an unmet consumer need for the consumer in tracing back the whole product supply chain.

TABLE 9.2 Consumer and technology matrix areas representing the technology and project intention

CONSUMER AXIS	*PROJECT INTENTION*
No change	Meets existing needs without recognizable difference
Variant	A different way of meeting the existing consumer needs
Improvement	Meets existing need better than others
New benefit	Satisfies a known, but yet unmet need
New core	First launch of a revolutionary new product changing consumer perception
TECHNOLOGY AXIS	*TECHNOLOGY USED*
Base	Uses standard foods technology
Incremental	Standard technologies in a new way or slightly "tweaked"
Next generation	New technology with product differentiation potential
Radical	First usage of a technology in the foods sector

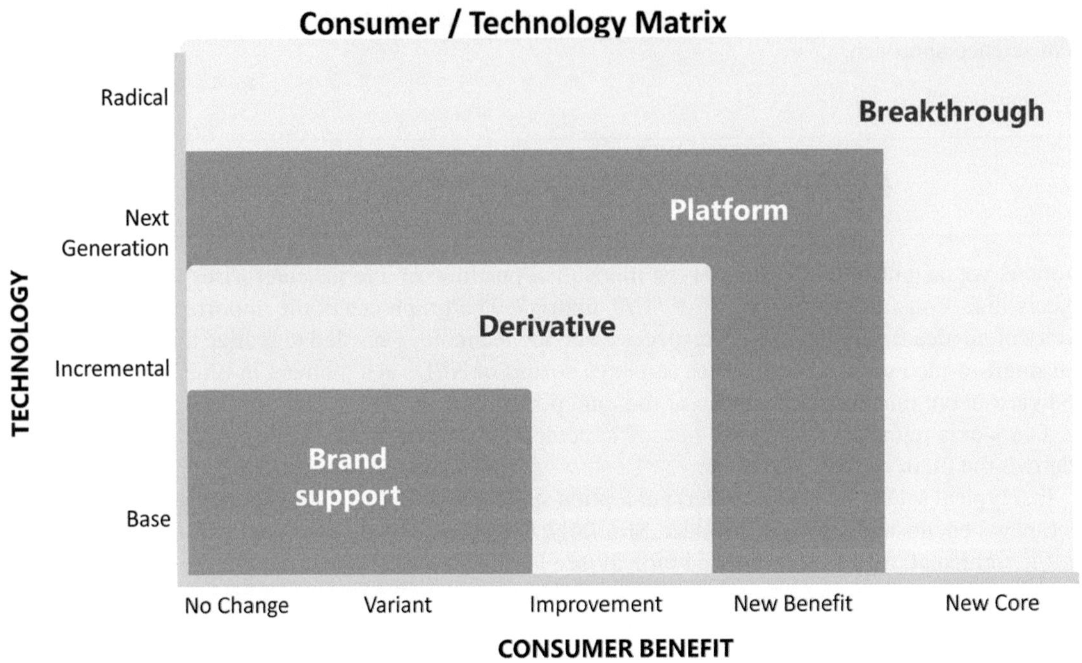

FIGURE 9.1 The different sectors in a C/T matrix.

Another brand or company may see it more as "derivative", since this company is already using this technology and subsequently provides most of that information on their website for the interested audience.

The right perspective is the company's. Incremental technology is understood as the use of standard technologies in a somewhat new way with minor tweaks of the technology. Its usage should not bring unexpected surprises.

Whether a team qualifies the launch of a four-seal standup pouch from existing manufacturing lines as an "incremental step" in packaging technology packaging or a "new consumer benefit" depends on how they judge the benefit in the market. A competitor not fully equipped with that technology but serving a market where such products already exist may judge the ideas as "next generation technology" and "variant".

Is the introduction of a "vegan" range a new platform or just a derivative? The answer sits in the concrete market circumstances and how the company looks at them. The first vegan ice cream on the market allows vegetarians to enjoy such a treat, so it is a new platform, but the introduction of "vegan jelly sweets" can be a derivative. Bigger companies tend to calibrate their matrixes differently for different

TABLE 9.3 Project positioning as the result of a balanced view about consumer benefit/novelty and technical complexity

Brand support	Perceived low consumer value and using base technology. Supports existing brands and helps to protect current market share.
Derivative	Low consumer value perception but using incremental or next-level technology. It extends an existing product range and should bring incremental market share.
Platform	A new benefit for the consumer or new core product with base or incremental technology used. Can be a new product line or a new way of making the product.
Breakthrough	New consumer benefits or new core product using new innovative technology. A new product category or a new way of manufacturing.

product categories or markets. True novelty in vegan ice cream may look different than in bouillon cubes or energy bars.

Once the team has an aligned understanding about the classification on the consumer and technology axis, the positioning of all projects in scope starts. R&D may make first proposals about the "type" of science and technology need to realize the idea. That brings the position on the technology axis. Next marketing gives their view on the character of the consumer benefits. Or, in marketing-driven companies it is the other way around: A new weight management snack option is seen as unmet and then R&D explains the characteristic of the required technology.

Once the first project is localized, others follow. The comparison with the first one will help to position the next ones. Has an idea more consumer appeal than the one before? Isn´t that technology more radical or difficult than the other ones? The discussion around all of them and from different expertise positions will lead to a fair "calibration." The classification is "an educated guess", not necessarily 100% accurate. That good enough, since "it is better to be vaguely right than precisely wrong"!

Once all projects are placed in the matrix, a total portfolio view becomes possible and shows the spread of all ongoing activities relative to their assumed consumer benefit and the science and technology needed to launch the idea to market.

That already gives a good idea of (a possible outcome is shown in Figure 9.2). how much idea or project pool fits with the business strategy. The populated matrix shows how good the idea pipeline is aligned with the business strategy.

But the position in the matrix gives immediate business direction. Imagine a project plans to apply modern biotechnology in plant-based foods. That deserves a "radical" in terms of technology! But the technology as such is in many societies still very controversial and encounters resistance from consumers. If this project now intends to use genome editing biotechnology (the rearrangement of homologues DNA without the introduction of foreign genetic material), it is critical to understand what the consumer benefit is. A better pesticide or herbicide resistance giving better yields and maybe less environmental impact could become valued by consumers differently than a superior nutrition profile.

Other features of the projects can also be included in the graph. Two options are frequently used: Either the bubble size is used to characterize the business opportunity, as expected turnover or profit, or

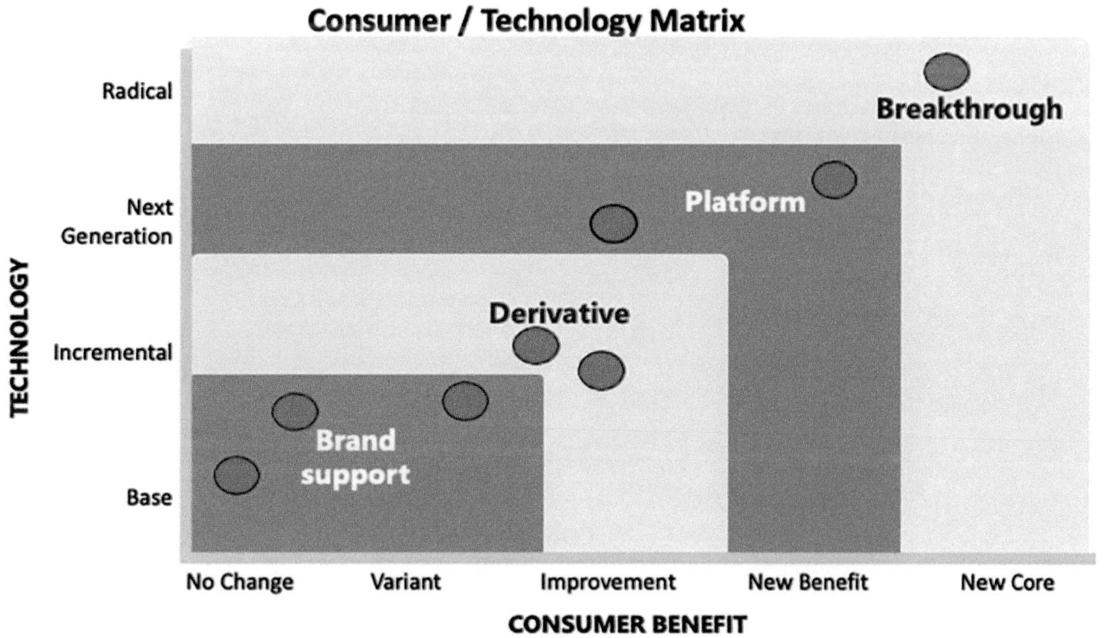

FIGURE 9.2 An example of a rather typical innovation portfolio in the food business.

the estimated resource requirements are plotted. A few "bubble classes" make good sense to bring in more discrimination between the opportunity, maybe just "small", "medium" and "big". Too much segmentation can end in "analysis paralysis", as this is still a guess to predict the business success in the future or all the work and money need to realize the idea. More project management and dedicated market research will be needed to specify that better.

It is important at this stage that all ideas are judged in comparison to each other; that is the strengths of a cross functional team discussion. A somewhat general judgement can be accepted. But R&D should not underestimate the technical effort in one project, just because they somewhat "like" the technology and the one needed in another project less.

A recurring feature in these matrices is that the more "up right" ideas show general more "gain" and more "pain". That means the resources needed are higher, but the potential extra turnover (or profit) is also bigger.

Equally important for the business can be the estimation of the effort needed to realize each project idea. For the decision about the best-fitting projects, it is often important to get an early view on the resources needed to run those projects. That can be drafted in a separate chart or in the same one, with a new colour or symbol.

For enterprises with larger R&D departments the unit is often "FTE", that is, "full time equivalents". It estimates the number of full-time working trained R&D people the project needs per time unit: 1.5 FTE per year for a project would mean one person completely in the next year and another one for half of its time.

Such estimates will help department heads, for example in R&D, to estimate the number of new projects the total R&D or a specific function can support in the planned period. They know the headcount in their department and know how many hours individuals are working, so they understand their capacity. In reality larger parts of the team will work on ongoing projects, so they can mirror their free capacity

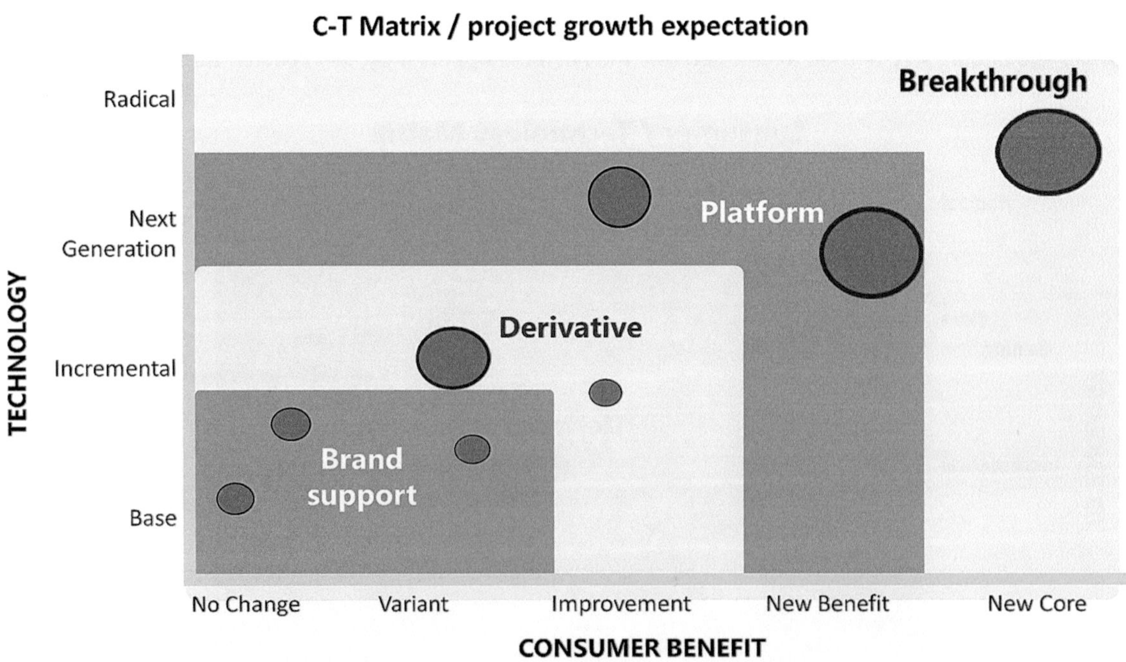

FIGURE 9.3 A C/T matrix with various projects in different strategic positions. The bubble size reflects the expected incremental turnover. The bubble size may differentiate roughly small from medium and big turnover opportunities or estimated values, in case the confidence of the cross functional team is good enough to predict those.

against the project demand. For the fully operational planning, the specific skills and time requirements are then still needed. Early in a project prototypes and new formulations are needed. Later perhaps process technologist and packaging engineers are required.

All the "out of pocket" money needed for a project, ranging from prototype making, upscaling and full consumer research, can also be captured in the "project bubble size" (see Figure 9.3).

As said, this graphical presentation indicates how well the new product ideas are aligned with the existing one, the company is already working on and the business strategy A company with a strong position in a healthy growing, profitable market segment might decide to put most new initiatives close to their core, as "brand support" or "derivative". Figure 9.4 shows the extreme of such a strategy, where all plans or focussing on the growth of the business core.

All projects aim at -known consumer needs with just a little stretch and apply existing and overall well-tested technology, so the technology shows little risks, and the projects will likely add some incremental turnover. If that is enough to match the ambitions of the business strategy (the companies' growth plan), the portfolio is fit for purpose. A portfolio like the one noted earlier signals many "low-hanging fruits", so launch opportunities without major technical needs or challenges. Typically, such a portfolio would fit a very mature, conservative market segment, like wheat flour, canned traditional soups or bread, etc. A foods company would follow such an approach, where they do not see much innovation potential or market segment, or in an area they want to leave long term (sometimes also captured under the phrase of "milking strategy"). For others, the longer-term question would be if such "conservative plans" are good enough to stay competitive in the market. There is always the risk that another company comes along and powerful attacks the existing core. So, if the same company gets aware of a new about "radical" technology, that has the potential to completely change the dynamics in the market, or the company only hears about a potent new competitor coming in from outside, they might quickly decide to put much more effort into new "platforms", or "breakthrough" solutions to keep their business healthy in the future. A situation described first in the innovator's dilemma by Clayton Christensen (Christensen, 2003). And new platforms and new technology are normally not sitting on the shelf, ready to be picked. Their developments need time and effort. That means the switch to solutions

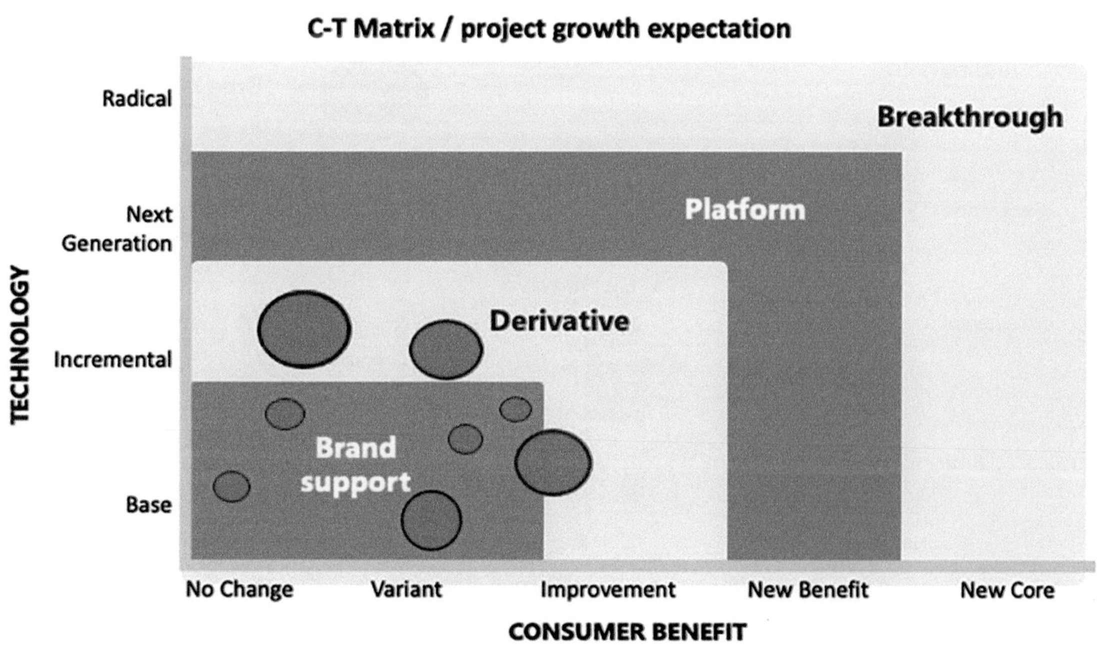

FIGURE 9.4 An unbalanced low-risk/low-chances product portfolio.

with new or untested new technology; the project needs to start well ahead of time, when the solution is needed in the market.

Should a highly successful sausage and meat patty producer just look at some lower-impact range extensions in their assortment? Or is it better to start investing into plant-based protein meat or dairy analogues to capture the current growth trend towards plant based? Are plant-based alternatives only a short hype or a niche opportunity for a few? But what happens if plant-based alternatives become a game changer, massively impacting the outlook of the existing animal protein market? Stretching into plant-based technology would require new knowledge and technology, so new ideas and projects, which are extra effort.

Unbalanced in a opposite way is when most projects are "in next" or "radical technology". Focussing most resources on technologically demanding new projects gives the chance for products with really new benefits in a couple of years. But the technology development could fail or a changing regulation impact the application possibilities. And it might take a long time and intense effort to get the enabling science and technology ready. Until then, the existing healthy core will only get less innovation attention, since major parts of the resources are bound in longer term development. Also, competition can spot the weakness, launch many new variants and steal market share and shelf space in the shops. Figure 9.5 shows such a hypothetical portfolio focussing on new platforms.

The brand core may need more attention to maintain at least the current value for the company. If the business is a startup or the company is confident, they can truly exploit the technology. They may have many filed patents, so they can be relatively sure that they are the one using the solutions first to exploit the business opportunity. And if they are also convinced that the consumer or customer will like the product made by this technology, such a polarized portfolio can make sense.

The different areas in the C/T matrix are normally not equally populated. Combinations like a really "new consumer benefit" or a "new core" product that can be easily made with "base technology" are more the exception than the rule in FMCG. Similarly, an identified "radical new technology" for a low or no incremental consumer benefit is hardly an attractive business position, unless it allows, for example, a radical step in cost reduction for a well-established business category. Such project can elevate the profit

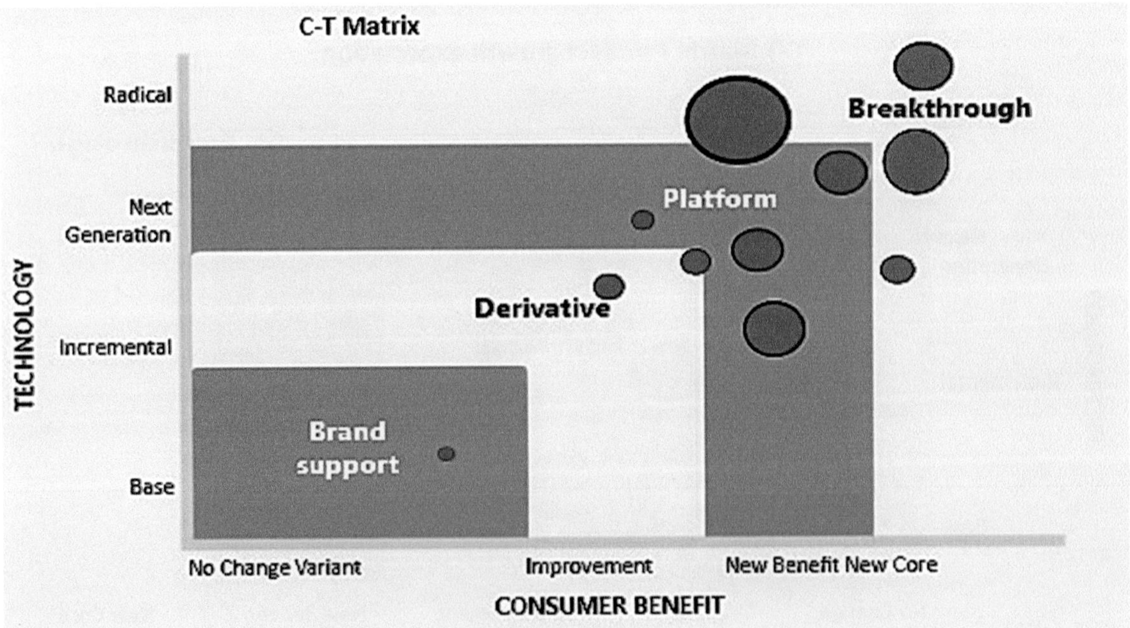

FIGURE 9.5 Ahigh-risk/high-chances project portfolio with less attention for the nurturing of the existing core.

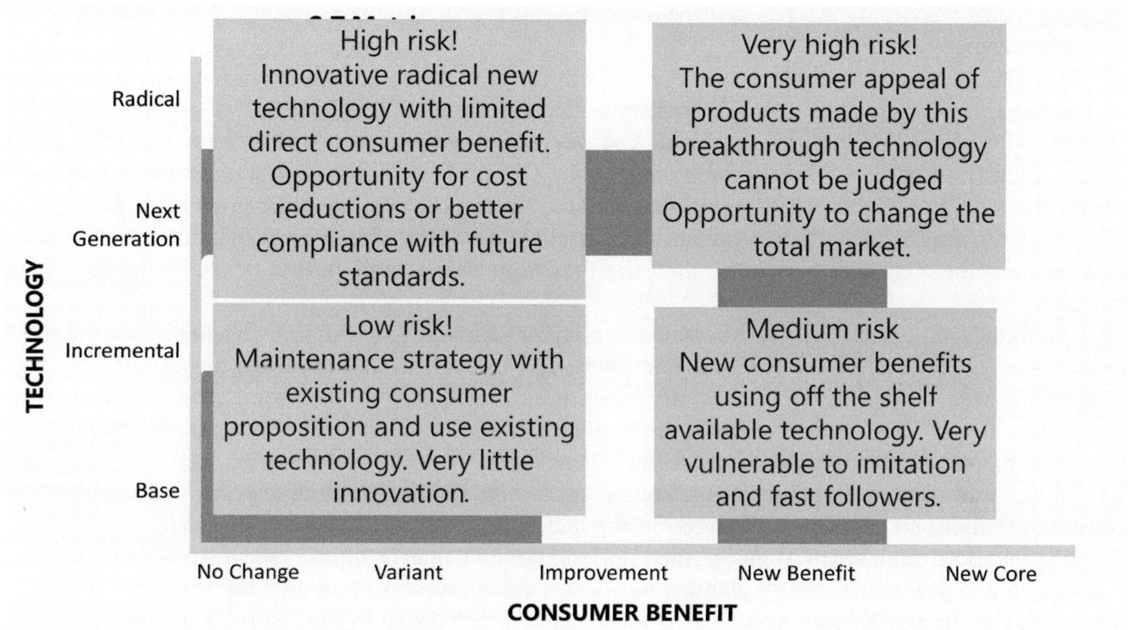

FIGURE 9.6 Overall risk assessment of the ideas in the different quadrants of the C/T matrix.

margin of the products made by the new technology. The consumer has no benefit. But this solution could then also be used to sell to sell existing products at a lower price, which could bring a lot of incremental turnover and market share. Then it's a clear consumer benefit and the team may want to revisit the positioning in the matrix. Similarly, if such a project with radical new technology helps to pre-empts a likely change in food regulation, it would make the business "future fit". An expected change is the safety assessment of some food additives could be regarded as such a situation or the development of new recyclable packaging solutions. Figure 9.6 summarized the average risk level of projects in that area, the opportunities and threats.

In summary, the underlying four quadrants in this matrix represent different opportunity types, all with advantages and side effects, which need to be managed. "New" symbols higher risk and higher chance. Some of those risks can be reasonably managed; others are not predictable. The investment into a radical, but predictable new technology to satisfy an unmet need much better sounds like a good option to ensure future growth. But hopefully the time needed to get the technology ready is not too long, and no competitor is working on it as well and perhaps already more advanced. Here, a search in patent databases can provide important news in this area.

The bigger opportunities often need more time and effort – they are more resource demanding and a couple of years away from market, and they are an opportunity, not a given promise. For bigger manufacturers of branded food products, a balanced portfolio is often the preferred option. Investing in some "big bets" provides an opportunity for a big and successful launch in the future, while some projects in "brand support" or "derivative" ensures a constant flow of new products into the market with incremental growth and profitability.

In a creative session the lower-left and top-right corners of the matrix normally fill quickly, but ideas that land in the two other corners are worth an in in-depth discussion.

A project with standard technology but the potential to deliver real new benefits or a completely new product sounds like a "must go". Big gain at low effort! A bit like a white elephant! Unless the enabling technology is proprietary for the company, it is worth checking why this idea has not hit the market yet if it can be realized so easily. And even if the company gets the "first mover advantage", how long will that last before competition follows?

It can also be useful to apply a radical new technology first on a known consumer benefit. Here, the first introduction into market tests the consumer reaction. Imagine you found a way to deliver the full saltiness with a very low sodium ingredient. You can introduce the launch with a "lot of noise" to get all the attention from markets and key opinion-formers. Or you launch first "below the radar", find out if the products really reduce sodium intake and take the time to reformulate other products before you go full steam. Or you can make food products with an edible packaging. Before praising that one, you could find out if (a) the consumer minds the shelf-life reduction and (b) will start to consume the packaging material.

One important dimension in innovation planning not explicitly visible in these graphs are the timelines. Different ideas will have different lead times. The extension of an existing product range through fortification with nutrient mix requires often only a bit of prototyping plus some advertisement work. That's done in a couple of months! But if the technology needs first tests of science, opinion-former support, new manufacturing design, a new factory, new suppliers, new products, etc., then it is likely still several years away from first launch. A company that wants to bring such an idea ASAP to market will constantly commit all necessary resources during that time.

It is therefore valuable to align the time horizon the C/T matrix should reflect. In FMCG practice, that's often 2–5 years, as the NPS planned for market entry even faster are already in execution and do not need to be discussed here. And looking more than 5 years ahead in the fast-moving consumer good industry is a speculative exercise, as many consumer trends, market circumstances and enabling technologies may have taken different directions.

For an ambitious group of innovators in a company, the C/T matrix is a great. It allows a structured process in which every expert adds their knowledge. And by calibrating the various opportunities fairly to each other, a sustainable innovation plan will emerge that can be shared with the company leadership and can be revisited from time to time.

COMPLEMENTARY SEARCH

Many ideas never make it to the market, and others are so costly or lengthy that they struggle to "earn their money back". A good innovation strategy considers that and aims to still have enough future successes in the pipeline to reach the growth plan. So, they need to put "more eggs into the basket". But more eggs mean more effort and a split in attention.

A good innovation portfolio is more than a balanced collection of independent projects. Activities within an enterprise normally do not run in isolation; they influence each other. Hence, being able to understand how project can benefit or detract value from each other can make a big difference.

Experienced innovation managers are aware of this. Starting at the same time two equally complex processing projects, where perhaps new production lines with different technology and suppliers are needed, can overstretch the technical capabilities of the company. But the development of three more variants of an existing product requires only little extra effort.

A good first step is resource planning, where the project demand is compared with the resources available. After that the search for tensions and synergies between projects start. And the gain is a longer-lasting competitive portfolio. Nobody would say "no" to this, but the big question behind this is how to do it. Researchers have started to articulate principles and formulate guidelines for the business.

This question has been investigated for mixed portfolios in foods containing new products and new processes (Hullova et al., n.d.). Here is a critical relation within a project and within the innovation portfolio. There is hardly a new project that only needs product development. And maybe except for breakthrough value improvement projects, there isn't any new process that does not require product development. Hullova and colleagues studied representative project cases from the UK food and drink industry to identify the critical capabilities (Hullova et al., 2016). They suggest starting with the identification of complexity and novelty in the projects together with the assessment of all available resources and knowledge (internally and externally), the selection of the well-fitting integration mechanism and final knowledge capture and storage. One key proposition is that there are different complementarities in such projects and between them, and they first need to be correctly identified before trying to manage them. In an earlier work the authors suggest different types of complementarities.

A "reciprocal complementarity" portrays the synchronic adoption of product and process innovation. That creates opportunities for further improvements, including resources and capabilities sharing. The result is better alignment between product design and manufacturing, with improved efficiency and cost-effectiveness between product and process innovation.

A different type of relation between projects the authors called "amensalism". That's a term taken from biology which describes a symbiotic relation where one organism is harmed and the other is unaffected. The most obvious case is when a bigger project competes with a smaller one for needed resources and hinders the progression of that project. Or a process innovation moves into a direction where the ongoing product innovation becomes obsolete, for example, through implementation of production equipment that cannot be used for a new, better, innovative product. In case of such interaction, a sequential project management is often the solution.

More typical for the food business is a different type of "amensalism". Here, old processing technology is used too long and hinders the timely implementation of newer equipment making products that better fit the consumer needs. The long fully depreciated machineries remain in use and block technology for better, more innovative products and formats.

Not all projects interact with each other. A "brand-led" innovation, where the marketing team makes a new product campaign with artwork changes and some support from nutrition, will hardly compete for resources with the installation of new dosing equipment at the same time. But the new technology should not damage the core of the campaign. If the brand wants to communicate more about the benefits of fortification, the dosing equipment in the factory should enable that.

In their considerations the authors focus on the identification of influences between process and product innovation (Hullova et al., 2019). First, individual projects must be assessed for their complexity and novelty compared to the resources available. Once that has happened the best-fitting integration mechanism should be used to manage the identified complementarity, largely based on experience acquired in the past.

If a new process innovation is complex and therefore risky, it might be better to formally run the implementation in a controlled stage-gated process, without new product testing. In-pack sterilization of liquids requires careful validation. Testing different new product types in that phase is not adding value. But if the new process idea is simpler, testing it for different solutions may give new and better product options. A weight- or volume-based dosing equipment can be tested with several mixtures without impacting the technology roll-out.

In any case, reviewing the product portfolio systematically, searching for synergies, antagonism, conflict and better sequencing, can greatly increase the pool of NPD an organization can work on.

PORTFOLIO OPTIMIZATION

The optimizing of the pool of new ideas is done in connection with the existing company portfolio. Only startups do not need to bother much here. All others must consider the consequences of the new product launches for their existing business.

The methodology to access the individual strengths and weaknesses of ideas in the innovation processes has been described at length earlier. Doing the same for a bundle of projects or ideas is different. Here, the combined value or costs of all projects must be considered to estimate if the sum of all is bringing additional value or costs or taking away some. Providing the tools that allow the best decisions is called "portfolio decision analysis" (PDA). Generally, the following practical problems are then considered:

- *Resource constraints*: There always more ideas than people, time and money to work on it.
- *Benefits*: For a single new food product, the direct commercial value is often leading. But for the total portfolio, project halo effects can bring extra value.
- *Interrelationships*: That is the complementary analysis, where 1 + 1 can be less or more than 2.
- *Portfolio fit*: The innovation has a connection with products already on the market. A planned refreshment may exclude some new ideas from progression.
- *Decision priorities*: New ideas span a wide field. Prioritizing in a group of non-similar projects can be difficult. Manufacturing may ask R&D for engineering support to improve efficiency, while R&D wants to build competitive AI capabilities in-house.

In the meantime, models have been developed to facilitate such decision-making, but a detailed discussion is beyond the scope here. For those interested, the article by Liesiö et al. forms a good starting point (Liesiö et al., 2021).

A "helicopter" view has its merits, particularly in moments where a bolder strategic decision needs to be made. Usually, a company has a product portfolio in the market and at least some ideas "in the innovation funnel". The task now is to find new ones that fit the existing portfolio and do not disturb the ongoing innovation program. Enough ideas are there; now the decision is needed which ones to progress and scale.

Success is not guaranteed, so one answer could be to keep a "a few irons in the fire" and see later, with more knowledge, which one is the right one. The risk is with this strategy is that the idea becomes resourced. Being forced to split the available R&D resources and elsewhere between several ideas can limit focus and reduce speed. In a dynamically evolving market, the prioritizing of a particular idea appears risky. But what if the competitive dynamics force the painful changes and a former leader becomes a follower?

Innovation is inherently risky; possible failure of F&B launches cannot be eliminated; it must be managed. And it is difficult to know where innovations will emerge, while searching everywhere is impractical. Here, executives must set the boundary conditions for the "opportunity" spaces. Such decision process is a combination of gut feeling, intuitive visions and careful strategic analyses.

Once the idea is captured, a governance body should monitor the expected gain, the time needed and the intrinsic risks of the totality. There's no single mix that's universally right. Many food enterprises make the mistake of overloading their innovation pipelines with relatively safe, short-term and incremental projects that have little chance of realizing their growth targets while staying within their risk parameters. Some spread themselves too thinly across too many projects instead of focussing on those with the highest potential for success and resourcing them to win. If there is $100,000 for product testing available, one can decide to do some high-level good testing with five different products or a sophisticated in-depth analysis on consumer need and target groups on two.

Shifting resource reallocation is often not a quick decision. Companies have many different product ideas in their innovation pipeline, and some of them are more mature than others. It's not easy to shift resources needed from the further developed to a less mature one.

R&D's contribution here is twofold: First, they must indicate when they start to become overstretched. A request for too many prototypes will limit the quality of the individual ones. More prototypes more quickly or fewer ones more thoroughly. Secondly, R&D has responsibility for the alternatives presented. They know their products best and can see them in context with the competitors. They must speak up when they see their work is not competitive enough and not hope that the marketing concept, the distribution strengths or advertising program will lead the idea to victory. When the advertising stops, the weakness will become visible.

CASE STUDY: COLOUR-CODED VEGETABLE SOUP

Several years back, a well-known maker of savoury products developed the idea to launch a vegetable soup range, where each soup would basically carry a natural colour code: White, green, yellow, red and orange. Ideally each soup had a nutrient profile that could be represented by the natural colours that came from broccoli, peas, cabbage, bell pepper, carrots, tomato, turnip, etc. All were low in sodium, had little fat, healthy protein and moderate calories and the claim to bring at least one portion of daily recommended vegetables per serving. On paper, such formulations could be made. But when translating these suggestions into differentiation and appealing soup dishes, the R&D team reached its limit. The prototypes were good, colourful and differentiating. The taste and aroma were good, basically in line with those of other soups in the market. A sort of "taste parity" was combined with a very exciting new "health concept", based on natural colours, portions of health veggies and good nutrient profiles. It was a great market opportunity. The clean natural colours required some attention in the crop selection, looking for more premium vegetables. And for differentiation, good herbs and spices were used. The material costs increased a bit, and the tight quality control necessary to keep the differentiation potential of all coloured soups after sterilization finally led to a price premium in the market. First consumers enthusiastically tried the new soup range, triggered by a flamboyant advertising campaign. But most of them decided afterwards not to buy them again. They now found they could have the same tasty soups from others a bit cheaper. Taste is king in foods! And at least at that time and in that country, an extra health benefit was not something the consumer was willing paying for.

But maybe it is time now to pick up the idea again: New process technologies allow full safety with much less heat, microbial safety can be guaranteed with less heat load and more colourful and tasty vegetables are available. And the consumer may be much more willing to integrate a brilliant health and natural plant-based meal solution in the dietary routine.

OTHER DECISION-MAKING TOOLS IN NPD

In the innovation pipeline management, the recurring question is which ideas to choose, where and how to invest the valuable resources, when and how to accelerate and which project to better not touch, slow down or disinvest. Such decisions require an "educated guess" about the future, the effort needed, the chances of success and the risks involved. And not taking a decision because the perspective is not clear enough is often not a good alternative, as it distracts effort and attention from alternative plans. Decision needs to be taken by the leadership team. These management tasks are often described as "decision-making under uncertainty" and basically describe the process of choosing an action when the outcomes are not known with certainty.

This is a common challenge in new product or technology development. Innovation management tools work very well for projects with a short to medium time span. In food innovation a 6- to 24-month time span can be managed. This is less suitable for longer-term projects, like the development of a new product platform, the investment into new enabling basic science or the large-scale implementation of a new technology. For those long-term plans, other approaches and tools are available to facilitate the necessary decision-making. The core elements of such approaches are similar:

- The wish to use the best scientific methods and evidence in decisions made
- Uncertainty to describe the magnitude of the problem and the corresponding benefits accurately
- A need for time-bound decision-making prior to clarification of uncertainties

- The incidence of a trade of among distinct adverse outcomes.
- The knowledge that no quick feedback will be available to see whether the desired outcome has been achieved by the decision taken

A couple of those tools used in the food sector are briefly described next.

ROBUST DECISION-MAKING

Robust decision-making (RDM) is a method that helps decision-makers deal with uncertainty and complexity in innovation processes by finding robust strategies, characterizing their vulnerabilities and evaluating trade-offs among them. RDM works by following four main steps:

- Defining the problem and the objectives.
- Generating and evaluating a set of scenarios that represent different sources and levels of uncertainty
- Identifying and comparing a set of strategies that perform well across the scenarios
- Choosing the strategy that is most robust to uncertainty, or the one that has the least regret or sensitivity

Essentially this addresses the risks of an innovation while trying to maximize the chances of success. A risk can be defined as combined knowledge about the probability and severity of a consequence. The risk assessment should find those actions that avoid unacceptable consequences.

The key features are that RDM is less about defining the single most probable one and developing the plan for that (Wikipedia Contributors, 2024). Instead, it uses the criterion of robustness to review all options and is not aiming to optimize a specific scenario. The idea is to have a project plan that is less sensitive to unexpected changes.

Therefore, RDM accepts uncertainty as unavoidable and works on "vulnerability-and-response-option analysis". For that purpose, combinations of uncertainties that will have the most impact on the choices are assessed. In a sense it takes the assumptions about the future as what they are: Assumptions! And it does not try to convert the assumption into a fact by collecting more and more information. And the decision process tries to be iterative, with feedback loops, monitoring step and decision cascades, rather than strict "go/no-go" moments.

In the foods business this decision tool shapes a longer-term science program. A relatively short-term "going to market" project is not benefitting from such analysis. It would slow it down in a situation with relatively low uncertainty.

A good example for the application of RDM applied to food science and technology is the study by Hauschild et al. where the authors explored the trade-offs and synergies between risk and sustainability assessment for different food production technologies (Hauschild et al., 2022).

They used RDM to compare alternative approaches to balance risk and sustainability assessment in situations where the objectives of both approaches differ. One striking example is the usage of recycled plastic material in food packaging. From an environmental perspective, plastic reusage should be maximized to reduce the need for new virgin plastic production and landfills with used material. But from a human safety perspective, the potential health risk of migration of contaminants into the packed food often leads to a strict limitation of the usage of recycled plastic. The authors compared different approaches and concluded that in this case the approach using sustainability assessment as a baseline and iteratively applied risk assessment elements brought the best results in terms of specificity, reliability and relevance. So RDM provided better-informed decisions about technology choices in the food sector.

Before taking important and costly decisions about which project to progress with and which one to stop, one would like to have all relevant information at hand. But unfortunately, it is not so evident when all "all information" is available! There is always some information missing that is potentially relevant. Is it better to try and get that to increase certainty or is it just more wasted time and money? The "value of information" (VOI) principles – the expected value of additional evidence to reduce uncertainty – can be used to judge here.

Let's assume a new food company wants to launch a range of new soups based on market research evidence that there is an opportunity in the market. And in order to launch in time, an urgent decision is needed about the investment into the best sterilization technology: "In line" or "in pack" sterilization. Different expensive technologies and hard to combine! Machine ordering and installation will take more than a year, and it's a heavy financial undertaking. One cannot progress both options. One route must end, so that the other one can come alive. That could be a moment for the tools provided under decision-making under uncertainty.

DYNAMIC ACTIVE PLANNING (DAP)

This is more of a "quantitative approach". Rather than forcing a full stop of one or more options, this tool enables a quick shift of necessary support and effort from one idea to the other. In that sense it is less a decision-making tool than a flexible funding tool. It helps to keep an option open until a decision can be taken without wasting too many resources. It many organizations it has replaced the static annual budget planning process. It is described as an alternative to static planning in an agile world. Fundamentally it shifts attention from predicting the longer-term future to a goal-oriented planning where strategy and flexibility are in balance.

It keeps an eye on the bigger picture to make a broad, responsible plan for the year ahead. But it also requires you to revisit that plan regularly – perhaps quarterly – to ensure it's where it needs to be right now. A dynamic plan allows employees to take some risks, to be able to pursue unforeseen opportunities, respond swiftly to threats, adopt new technology and carry out new ideas. It requires more milestones, with clear success criteria and an ability to then shift effort, support and resources quickly.

And this approach will benefit from an earlier done "complementary search". Flexibly re-allocating resources from one activity to another is only good if the consequence are transparent. Can a project still survive when cut from resources? Or can it be delayed without impacting other ideas depending on the smooth progression of the project?

CASE STUDY: SODIUM REDUCTION

The food industry has learned long ago that salt, sodium chloride or table salt, not only provides saltiness but serves as a powerful enhancer of other tastes as well, besides being a potent preservation tool. But nowadays all relevant data show that people globally are consuming too much salt. The fair share of the packed branded food makers in this intake is substantial. A technical alternative as powerful and affordable as the original is not available. Not on the ingredient market and not in science and technology. And so the international food companies have decided to put significant R&D effort into collaborative research to detect and development new sodium reduction tools, understand the dynamics of the umami taste perception and its enhancement by sodium and the options for masking the bitter taste which seems to be often associated with possible sodium replacers. The overall objectives are to create a technology for a broad range of low-sodium, health food alternatives. At the beginning of the science program, nobody can predict which of the technical approaches will succeed and which one might fail. It is also not clear if these ideas will lead to a stand-alone solution or if they may work synergistically, enhancing each other in their effect. And while in most of the existing food formats the company is offering, the sodium amount should be lowered significantly to comply with highest nutritional standards, it is not known if there could be one solution for all or a range of alternatives for different product categories. Beside so many questions of the enabling end, it cannot be predicted how the consumer, key opinion-formers and regulating authorities would react to the upcoming solutions.

All three branches of long-term program are substantial and resource-intensive and bind a significant amount of the funds and R&D skills available in the company's research base. And while it is not even clear if there could be an effective new solution, doing nothing and waiting or just cutting the amount of salt in the products is not option.

But it is obvious that choosing one fundamental option to find the solution right now, based on the best scientific data available right now, is not possible with sufficient confidence. Instead, each subprogram needs clear milestone and resource planning and the option to quickly react to new results and unexpected hurdles.

While in one arm the product developers were reaching out to various types of ingredient suppliers to articulate their need for salt (sodium) replacers, others formed a scientific partnership with academia, to set up cell assays able to perform high-throughput screening of huge ingredient banks. A different team scanned natural food stuffs for constituents that that could be isolated and cleaned up to provide a functional alternative. Meanwhile, market research and nutrition worked on study designs to understand the saltiness perception of consumers more holistically and beyond the physiological function of the salt receptors on the tongue. Computer tools were generated to define the sensorial perception of new ingredients and mixes in specific food formats to predict the saltiness for others without longer human tests.

Initially the cell and biochemical approaches looked very good and passed proof-of-principle tests. The research programs were quickly strengthened to scan more ingredients faster and develop higher yield extraction processes. The consumer research program was slowed down a bit to wait for the first truly successful hits out of the program, while the developers continued to test commercially available options in lead applications and kept on searching for new suppliers, while getting feedback their computer program.

In the next phase some hits of the cell assay high-throughput screening made material available that could be tested with sensory experts and eventually consumers. Some budgets and resources were shifted from the market research and biochemical purification part to design studies and test the material.

It took a while before it became clear that the hits from the cell line were not equally potent in delivering saltiness and beside this brought extra unwanted off notes. Some extra money was quickly made available to quickly review options to purify those ingredients in a way that the accompanying aroma notes could be split off. Proof of principle was not convincing, and the route was slowed to a lower level. The free coming resources could be used efficiently to boost the purification of umami and salt booster from natural plant-based material. New materials were brought into the program, and new purification methods were tested in the lab and pilot plant.

The new small-scale isolates were funded extra for sensorial tests and performed well. While the product developers partially shifted their focus to these new leads, others started to scope for potential upscaling partnerships. Technically the upscaling worked well, but the underlying cost of the purification started to raise concerns about the commercial feasibility. Meanwhile the discussion with external experts and regulators began to show that such ingredients could probably not be labelled "natural", an important label for the alternative.

The research program continued! New alternatives were found and tested, and key suppliers had to pick up the requirements and started to offer improved solutions. Only the dynamic active planning enabled the necessary flexibility to prioritize options, slow down others, mange surprises and change directions, keeping the program alive. In reality the total program was even much more complex and flexible and delivered suitable results. Not perfect yet in all cases and lead applications, but functional! And the program is still looking for better solutions.

On a more strategic level DAP is deployed to support the setting of a roadmap from the World Economic Forum and the UN Food and Agriculture Organization to help countries accelerate inclusive food systems innovations (World Economic Forum, 2022).

INNOVATION PROTECTION

Innovations in foods promise market success, but often require substantial up-front effort and investment. So, it is no surprise that enterprises working on innovations strive to protect the results of their efforts. For this, in principle, different options exist. Less often used in foods are copyrights, which protect authorship that have been fixed in a tangible medium. That can be on a website or artistic.

More relevant are trade secrets, covering formulas, processes or other business information that has commercial potential. To be considered a trade secret, the information must be not generally known to the public, have as the secret an economic value and reasonable steps taken by the rightful holder of the information to keep it secret. A typical mechanism is the confidentiality agreement for business partners and employees. An exact spice blend used in a famous dish or the method of preparation can be a trade secret, and Google's search algorithm or a famous chef recipe are examples. Consequently, a trade secret infringement occurs when someone improperly acquires the trade secret or improperly discloses it without consent. This cannot only happen intentionally but also through a mistake or accident.

Breaches of non-disclosure agreements, industrial espionage, theft, fraud and bribery are all illegal means of acquiring a trade secret. For the person signing a confidentiality agreement, this can also mean being sued in case a trade secret is made public through work on a unsecured private laptop in a public place (Justia, 2023). Many companies regularly train their employees, especially in R&D, about these aspects.

Other legal ways to gain intellectual property rights (IPRs) in the food business are patents and trademarks. Trade secrets are more the fall-back options. In case the chance of a patent is rather low, it can be better for an enterprise to protect their knowledge through a trade secret.

Creating intellectual property is perhaps the most viable way to convert an idea into asset. There are two routes for how this is possible, and both can be progressed simultaneously, and they can complement each other: Trademarks and patents.

TRADEMARKS

Trademarks are signs that distinguish and identify edible products of one maker from those of another one. They come as a specific design, a name, a symbol or a combination of these. They ensure quality and reputation of the food under the trademarks and help consumers to recognize the origin of the products they buy. For example, Oreo is a trademarked product name for a type of sandwich cookie. No one else can use the name Oreo to sell their cookies, or they would infringe on the trademark rights of the owner, Nabisco. Trademarks can also protect other distinctive features of food products, such as their shape, colour or packaging. For example, Toblerone has a trademark for the triangular shape of its chocolate bars, and Coca-Cola has a trademark for the contour shape of its bottles: The words, the letter type and the specific background red colour. These trademarks prevent competitors from copying the appearance of these products and confusing consumers (Bonamark, n.d.).

But such design elements do not simply protect the product; they can bring responsibility. In 2022 the American confectionery Mondelez International, owner of the Swiss Toblerone chocolate brand, decided to move the production of their iconic product out of Switzerland. The consequence was that they had to remove the iconic drawing of the "Matterhorn" mountain top from their packs, as the Swiss labelling rules do not allow the usage of a national symbol on something that is produced elsewhere (Koirala, 2023).

The application for a trademark is done by filing an application for registration with the national/regional trademark office. For multinational organizations an international protection is often wanted. Rather than filing in every local trademark office, one can use WIPO's Madrid System (WIPO, n.d.) and get coverage in up to 130 countries. The trademark fees for a new name in just one class of goods (there are actually 34) in just one country are just 1000 euros, but in case global protection should be built in several classes, this can quickly get up to 100,000 euros and more.

Without such protection, the name or logo still has basic protection under common law trademark rights, but suing a violator and claiming damage will be more difficult.

Trademark registration is often easier than patent application and is widely used in the food and beverages industry. For example, Coca-Cola has registered more than 500 patents and 20,000 trademarks worldwide, covering its products, packaging and equipment.

PATENTS

Patents offer in principle very robust protection. Eligible as food utility patents are food processes, machines, manufacture methods or the composition. A design patent protects the aesthetic appearance of the invention. And biological patents are granted for the application of biological technology, its products, genetically modified organisms (GMOs) and genetic material. International controversial surrounds if natural organisms or genes can be patented. Microorganisms, like probiotics, can be patented if they not naturally occurring. Alternatively, the method of use, here a health application of a naturally occurring microorganism, can be patented.

No doubt, patent rights are generally an important asset to protect the business. The size and depth of technology in an industrial sector is a good predictor of the patent intensity.

In the classical food and agribusiness Bayer AG seems to be the patent top holder with 1071 food and agriculture tech patents in the agriculture and forestry sector between 2002 and 2022. Most of them came from Monsanto, which is a reason why the bought the company (GlobalData, n.d.). Enterprises like BASF SE, Corteva and China National Chemical Corp follow, before the biggest food company in the world, Nestle, set its mark. The second-largest number of patents came from Deere & Co, not so surprising, as the company is an expert in agricultural machinery and equipment. Their applications fall under food and agriculture. Runners-up are BASF SE, Corteva Inc China National Chemical Corp and then Nestle SA, the world's biggest food company. Agrochemical, machinery and plant breeders have dominated the space.

The classical view is that the packaged food industry is based on mature, well-known technology. And yes, many of fundamental technologies to make food safe, shelf-life stable, nutritious and convenient have been developed many years back and enabled the rise of the food industry.

Pasteurization is a process of food preservation in foods are treated with mild heat of about 100°C (212°F). It goes back to the French microbiologist Louis Pasteur and the 1860s and his demonstration that thermal processing would deactivate unwanted microorganisms. It then took still a couple of years before the pasteurization came into wider practice. Other preservation technologies are much older! Air drying has probably been known for 15,000 years. Jacques-Arsene d'Arsonval's idea to use lyophilization as a low-temperature dehydration process was first invented in 1906, but that freeze-drying went virtually unnoticed for years and industrial freeze-drying of foods took off in the 1950s. And Peruvian Incas can argue that the way they stored their crops, like potatoes, on the mountain heights above their mountain villages at sub-zero temperatures and strong dry winds would be the first freeze-drying.

These "old" technologies are still sometimes active patent areas. And there are new examples of technologies that the industry has been built upon 100–150 years ago. And completely new ones were added in the past years. Ultra-high-temperature (UHT) processing in the 1960s, ultra-high-pressure processing (UHP), also known as high-hydrostatic-pressure processing (HHP), came into commercial use in the early 1990s. Applying these technologies to food often requires a lot of time and resources, and it needs good confidence that the payback will warrant the up-front investments needed. The benefit of the new technology versus existing ones needs to be so much better than what is existing that the consumer/customer is willing to pay a premium.

Later applied biology and chemistry opened a different patent field where new enterprises began to protect their knowledge. Impossible Foods, founded only in 2011, has today more than 200 global patent applications in the field of "alternative protein", many of them centred around "hemeproteins", one of the key challenges in the replacement of animal protein.

Other areas with heavy patent activity in the food sectors are nutraceuticals, e-commerce, food safety and transparency, personalized nutrition, restaurant digitization, digital food management, food waste reduction, robotics and or 3D food printers. Interestingly they describe the context in terms of a consumer/customer benefit and less often the technology used.

Like in all other fields a food patent must be novel, non-obvious and useful. Novelty is set by the absence of "prior art", so the invention has not been mentioned in public. That is not only patent databases and publications but also a consumer test with a prototype can create "prior art". Whether an idea is obvious or not is defined by "PHOSITA", a fictive "person having ordinary skill in the art" (Patent Experts, 2024). This description leaves space for discussion. Just adding a new ingredient, like a flavour for the first time to an existing food type, maybe ice cream, would be obvious for a developer, so would not be seen as an invention. But if then the inventor describes a surprising effect caused by that flavour addition, the chances for legal protection are better.

To obtain intellectual property, the patentable idea must be written as an application. The draft patent often contains a series of different claims, from a very narrow description to broad ones. Claiming "hemeproteins" in general will most likely not stand the "prior art search", but by adding many claims of increasingly more specificity, about the origin of the protein, how it is made, what the composition is, what it does or not does and how it is measured, the chances of getting claims granted increase significantly.

That's also why many companies do not aim to capture their right in one famous patent. There are nearly always ways to circumvent the granted patent in a way that legally allows competitors to remain active in that field. A popular strategy is the write many patents in the invention field, basically building series of fences that make it difficult for others to use the idea. Professional patent attorneys know how to do it. They would actually provide advice on the best formulated claims and may even ask the inventors to specifically perform extra experiments to support a particular claim.

Once obtained, a patent is a legal right that grants the inventor the exclusive right to make, use, sell or import the invention for a limited period.

Like trademarks, a patent application requires filing it with the patent office of the country or region where you want to protect your invention. Technical information about the invention must be disclosed to the public in a patent application. Patents may be granted in any field of technology, from an everyday kitchen tool over new plant varieties to radical new sterilization technology. The application must include a detailed description of the invention, how it is made and used, possibly with drawings or diagrams to help explain the invention. This full disclosure is required for food patents. The inventor must provide a detailed explanation of how to make the product so that anyone trained in the art could replicate it. The criteria used to decide about the eligibility of a patent application are summarized as "Wands factors" (PatentAssociate.com, 2023). They include:

How much experimentation is needed to put the invention into practice? Too extensive or unreasonable experimentation can prevent the granting.

The amount of guidance given. The inventor could try to be not too precise about the details of an invention. But not giving sufficient guidance can also be a hurdle for granting. Here "PHOSITA" is the reference. That is a legal fiction used to judge if an invention is non-obvious or has an inventive step.

Practical examples/prototypes increase the chance of granting. They show that the invention can be made. And the nature of the invention influences the enablement assessment, together with the state of "prior art", the sum of all existing knowledge in the respective field.

The examination process at the patent office will determine if an invention can be patented. Different patent offices have different guidelines for the examination, and the process can take several years.

Consultation with a patent attorney from the beginning is useful. Once an idea is considered to be patentable, the attorney helps in several ways. The prior art scan will help to see which ideas in the field have already been published.

Patent databases can be scanned with various search strings (WIPO, n.d.). A scan done in August 2022 for "taste enhancers" revealed 820 hits. Narrowing the search to "umami taste enhancer" still gave 44 hits, adding "natural" resulted then in 10 hits. Adding "salt free" delivered no hits. In case someone would have found such a molecule, the initial outlook is promising. And in case someone would have a new idea for a salt-based umami taste enhancer, the deeper look into the patents would show where novelty, and hence space for IPR, exists. That could be a particular application, or a specific amount or an additional function or benefit or a different way of making this substance, etc.

In many cases the search will help the inventor and the patent attorney to sharpen the invention and articulate the claim. It may even suggest the next series of experiments to substantiate the most powerful claims.

The duration of patent protection depends on the type of patent and the country where the patent is granted. But in general, utility patents last for 20 years from the date of filing the application. Design patents last for 14 years from the date of grant. In some cases, extensions or adjustments to the patent are possible, for example, if delays in the examination process or regulatory approval happened.

While patents are good ways to protect knowledge, it can also be laborious and resource intensive, with long lead times and sometimes an unclear perspective about the chances of success or unclear commercial value. And sometimes established food business have built up competitive knowledge through a series of small improvements. Each improvement step is too small to be patented, but in sum now there are superior manufacturing skills, enabling a company to sell their products at a lower price point than the competitors using the same "old" technology. A bit more efficiency, a slightly longer production run, a little less waste, or a better way to use rework, etc., can give competitive knowledge that cannot be legally protected. Many food companies have such knowledge that they do not want to share with others.

If now a food company with in-house processing needs more manufacturing capacity to manage temporarily increasing sales volumes, they may search for suitable third-party manufacturers, using similar processing. To get the best price from them, they could share their processing knowledge to enable a manufacturing efficiency comparable to the in-house one. Non-disclosure agreements would bind the partner to not share what they learn. But commercial contracts do not last forever. And the contractor will not forget what they learned during the cooperation. So, next time they can use what they learned to produce with the same efficiency for someone else or create their own brand.

To prepare for such a situation, where production at a third-party manufacturer is required, enterprises try to build firm understanding about their core technological skills. Patent attorneys and R&D then produce an overview about every valuable in-house knowledge they have and sort it. These confidential documents give guidance which knowledge could be shared with others and which elements must be kept inside. Such a technology strategy can limit the risk of sharing valuable competitive knowledge unintentionally. The need for outsourcing can come very suddenly. A machine breakdown or the chance to enter a new market often requires a very fast decision about the option for outsourcing. In those moments there is no time for complex analysis of potentially competitive knowledge. A result of such analysis could also be to rearrange the manufacturing in such a way that external parties can manufacture efficiently, without getting full transparency for all technical steps. The knowledge holding company could decide to supply critical part of the product composition, for example a sophisticated "premix" or "pre-prosessed intermediate", to the 3PM. In such a situation, the company has put its valuable knowledge into a "black box", where the "secrets" of producing it are not disclosed to the third-party manufacturer. And at a moment where the contract between both businesses ends somewhere in the future; the 3PM would still not be able to make the "black box" as good as the original company.

REFERENCES

Bonamark. (n.d.). *Most-famous-trademarks-and-their-history*. Retrieved June 8, 2024, from https://bonamark.com/content/most-famous-trademarks-and-their-history

Business Today. (2024). Retrieved June 7, 2024, from https://www.businesstoday.in/visualstories/technology/ai-creates-a-new-coffee-blend-in-finland-delighting-experts-with-its-unusual-recipe-and-flavour-130660-22-04-2024

Christensen, C. M. (2015). *The innovator's dilemma: When new technologies cause great firms to fail*. Harvard Business Review Press, ISBN 9781633691797.

GlobalData. (n.d.). *Europe: Top food and Agri Tech patents holders (2002–2022)*. https://www.globaldata.com/data-insights/macroeconomic/europe-top-food-and-agri-tech-patents-holders-in-the–sector-2129751/

Hauschild, M. Z., McKone, T. E., Arnbjerg-Nielsen, K., Hald, T., Nielsen, B. F., Mabit, S. E., & Fantke, P. (2022). Risk and Sustainability: Trade-Offs and Synergies for Robust Decision Making. *Environmental Sciences Europe*, *34*(1). https://doi.org/10.1186/s12302-021-00587-8

Hullova, D., Simms, C. D., Trott, P., & Laczko, P. (2019). Critical Capabilities for Effective Management of Complementarity between Product and Process Innovation: Cases from the Food and Drink Industry. *Research Policy*, *48*(1), 339–354. https://doi.org/10.1016/j.respol.2018.09.001

Hullova, D., Simms, C. D., Trott, P., & Laczko, P. (n.d.). Critical Capabilities for Effective Management of Complementarity between Product and Process Innovation: *Cases from the Food and Drink Industry*. https://bura.brunel.ac.uk/bitstream/2438/20918/4/FullText.pdf

Hullova, D., Trott, P., & Simms, C. D. (2016). Uncovering the Reciprocal Complementarity between Product and Process Innovation. *Research Policy*, *45*(5), 929–940. https://doi.org/10.1016/j.respol.2016.01.012

ITONICS. (2024, May 31). *The 70:20:10 rule of innovation to navigate the future | ITONICS*. https://www.itonics-innovation.com/. https://www.itonics-innovation.com/blog/702010-rule-of-innovation

Justia. (2023, October 15). *Trade secret infringement & potential legal defenses*. https://www.justia.com/intellectual-property/trade-secrets/infringement/

Koirala, K. (2023, April 5). *RIP Toblerone logo. Here's why they had to change it*. Mashed. https://www.mashed.com/1219802/rip-toblerone-logo-heres-why-they-had-to-change-it/).

Liesiö, J., Salo, A., Keisler, J. M., & Morton, A. (2021). Portfolio Decision Analysis: Recent Developments and Future Prospects. *European Journal of Operational Research*, *293*(3), 811–825. https://doi.org/10.1016/j.ejor.2020.12.015

Nagji, B. & Tuff, G. (2021, February 24). *Managing your innovation portfolio*. Harvard Business Review. https://hbr.org/2012/05/managing-your-innovation-portfolio

PatentAssociate.com. (2023, January 27). *35 USC 112: Patent claim ambiguity*. https://patentassociate.com/2021/03/10/patent-claim-ambiguity-35-usc-112-rejections/

Patent Experts. (2024, February 13). *How to patent a food product in 10 steps*. https://patentexperts.org/patent/how-to-get-a-patent/food-patent/

Wikipedia Contributors. (2024, April 19). *Robust decision-making*. Wikipedia. https://en.wikipedia.org/wiki/Robust_decision-making

WIPO. (n.d.). *Trademarks*. https://www.wipo.int/trademarks/en/

WIPO - Search International and National Patent Collections. (n.d.). https://patentscope.wipo.int/search/en/search.jsf

World Economic Forum. (2022, May 20). *How inclusive innovation could transform food systems - And help to end world hunger*. https://www.weforum.org/agenda/2022/03/food-systems-innovation-transformation/

Critical Design Elements of Food

10

Nowadays, many consumers buy most of their food items either as fresh food in the hygienic "modern trade", protected through packaging, or order it in clean restaurants. Food safety is then not the top concern. If the packaging looks unbroken and the best before date is far enough in the future, we trust the product quality. Only when we need to inspect our stocks at home, visit a fresh market to buy fresh locally grown unpacked goods or want to taste tempting street food during our holidays are our senses are sharpened. We start a deeper inspection to judge the safety and quality of the food item. On the open market, it is good to come early, trust the vendor, check the food item carefully and treat it properly before eating it ("cook it, peel it or forget it"!). In the traditional open "wet" markets in many countries, we can observe customers first looking into the eyes of a fish, then smell the fish, give the flesh a little poke and critically inspect the gill colour before they decide to buy it.

The capability to keep food safe for longer was a cornerstone of our civilization. Early preservation techniques like salting, drying or fermenting allowed people to settle, grow crops, harvest at the right time and preserve until the next harvest was ready.

In the industrial revolution many people left the countryside to make their living in the booming industrial centres. Here they lost their direct farm-based food supply. Enormous advances in science, technology and growing entrepreneurial activity made many food items safe through adequate pre-treatment and protective packaging. The direct food quality inspection was no longer possible, so food producers had to sign with their names for product quality and safety. This was the beginning of mass market food brands.

Today food brands mean more than this simple quality and safety guarantee. It's a story, an experience, a personality and a relationship. A brand shares values, vision and mission and reflects the culture. Today a food brand does not only feed stomachs; it feeds feelings (The Branded Agency, 2023).

In the lower-income markets in developing and emerging countries, traditionally the open markets are the main places to buy food. Here, food safety remains top of mind. A spoiled fish, a piece of meat placed too long in the tropical climate or a street vendor preparing their dish to hastily means not only disappointment, discomfort and illness but also a potential income loss plus extra expenses to see the doctor and to buy the medicine.

Since ancient times rulers aim to protect their population from unsafe food by food laws. In the United States, President Abraham Lincoln formed in 1862 the Division of Chemistry, which in 1906 became the Food and Drug Administration. The initial focus was on meat products, the slaughtering and sanitary conditions, followed by initiatives on chemicals and additives. In 1962 President John F. Kennedy declared the Consumer Bill of Rights, through which consumers got the right to be safe, informed and heard. Shortly later the Fair Packaging and Labeling Act enforced the identification of the commodity, including name, location and identity of the manufacturer, packer or distributor and qualitative and quantitative information about the food item.

Since then, the food chain has gained a lot of experience to manage known food-borne risks. But still, an estimated 48 million Americans become infected with food-borne illnesses every year (IFT, 2019). One reason is the constant appearance of new hazards, like pesticides, herbicides, foreign bodies, other food contaminants, process intermediates and synthetic products that have appeared and threaten health and wellness. Mostly these things happen unintentionally, but sometimes simple greed leads to fatal developments.

DOI: 10.1201/9781032724591-10

Some years back in China, a trimer of cyanamide, chemically C3N6H6, normally used in the plastics making was added to infant formulas. Why? A key quality criterion in milk is the protein content, and the standard analytical ways to determine protein content is to measure the amount of nitrogen through tests developed by Kjeldahl or Dumas. The aforesaid cyanamide trimer, also called "melamine", not only contains 66% nitrogen, but it's also very cheap. That brought a few people to the idea to dilute the milk with water and then add melamine to mimic a high protein content, so they could charge diluted milk for the price of a protein-rich one. The result was 300,000 ill people, more than 50,000 infants in hospitals and 6 that died (Wikipedia, n.d.).

The six individuals later accused – two of them were then sentenced to death – knew what they were doing. The obligatory "material safety data sheet" (MSDS) for melamine said in bold letters on the front page ". . . a substance of very high concern. . . . Having probable serious effects to human health. . . " ("ECHA", 2022).

The scandal ruined the reputation of the Chinese food system and destroyed the trust in the food industry for many years. I remember that for many years whenever I was traveling from Europe to Asia, friends asked me to bring infant formula powder from Europe, nothing else! Just that!

These criminal cases are accompanied by increasingly more microbial outbreaks: *Campylobacter, Clostridium botulinum*, *Escherichia coli* O157:H7, *Listeria monocytogenes*, *Salmonella*, *Staphylococcus aureus*, *Shigella*, *Toxoplasma gondii*, *Vibrio vulnificus* and many more cause food-borne illness and death in many parts of the world.

Despite impressive progress in science and technology, food safety has lost nothing of its importance. Food quality goes beyond that. While food safety is the undifferentiated fundament, food quality is a choice, including the totality of features and characteristics satisfying consumers/customers.

Safety is a non-negotiable! A very affordable common wheat bread fulfils the safety standards as good as the premium organic, whole, multigrain, seed-enriched version. The safety is given by regulation, the necessary standards must be implemented, and they are constantly monitored, while the quality standard is a choice the manufacturer has based on expectations of which quality the consumer wants at a given price or as result of comprehensive consumer research. Both standards, safety and quality, are reflected in the product, ingredient and packaging specifications to control them.

For a professional food developer, it is critical to understand the wide span from basic food safety and its management over the measures of sensorial quality up to non-perceptible health claims and promises.

R&D normally "owns" the specifications, and trained experts in quality teams ensure that the latest knowledge and best practices finds their ways to "design" the essential "safety" and "intended quality" into the food product.

Food safety and quality is a sophisticated topic that needs a closer look at the different dimensions. Therefore, I will now split the whole topic into three parts:

- The essential food quality and safety
- The responsible communication about safety and quality, which is represented by the food labelling
- The methods to measure quality and safety

While all of this belongs together and cannot be handled in isolation, the guiding principles show differences.

FOOD SAFETY AND QUALITY

Food safety refers to the conditions and practices that preserve the quality of food to prevent contamination and food-borne illnesses. That is a science-based approach that aims to ensure that the food does not contain substances that could harm our health. It's mandatory, and all governments have a variety of measures developed to control and monitor food safety. And while the principles are similar, countries sometimes show remarkable differences in the implementation. It is therefore good that in an interconnected world, the World Health Organization (WHO) made attempts to strengthen food safety from a global perspective.

Basically, all stakeholders in the E2E chain have responsibility. In the European Union the food law requirements have been largely harmonized. The basis is regulation No 178/2002 which lays down the general principles and procedures of food safety and the role of the European Food Safety Authority (EFSA) (EUR-Lex, n.d.). The primary goal is to warrant the free movement of safe food within the EU, contributing to the health and well-being of citizens.

In their "Farm to Fork Strategy" the commission captures the responsibilities of the different stakeholders along the chain:

- Farmers and producers are responsible for cultivating crops, raising livestock and adhering to the agricultural practices to safety standards. This includes the correct use of pesticides, hygiene procedures during harvesting and animal welfare.
- Food processors and manufacturers are the initial food processors and must maintain hygiene, quality control and traceability throughout their production, including safe handling, storage and transportation of ingredients.
- Food service operators are those preparing and serving the food to consumers, so for them the safety regulation includes adequate cooking temperatures and kitchen hygiene practices.
- Retailers need to ensure proper storage and display of food, the control of expiration dates and the cleanliness in their premises.
- Consumers have to handle food at home properly; that includes storage, cooking and avoiding cross-contamination and the obligation to remain informed about food recalls.
- Regulatory authorities have to set the standards, enforcc compliancc and conduct inspections.
- Food safety: The regulation emphasizes a high level of protection for human life and health. Harmonized food safety requirements across member states facilitate the smooth movement of food and feed products.

Classically, the department called "Quality Control" (QC) is charged with the responsibility to keep products safe with the planned quality. They lead the process of inspecting and testing the final manufactured food products to ensure that those meet the specifications and standards set by the manufacturer and the regulatory authorities. It takes the position that the end-to-end product-making process requires a tight control to eliminate any dangers.

In its origin, quality was an internally focussed view, looking at the materials coming into the manufacturing and the products leaving it on their way to the market. Quality control focusses on the inspecting and testing of the manufactured product to ensure that it meets the specifications and standards set by the manufacturer and the regulatory authorities. Typical common quality control issues that may arise in packaged food are contamination, spoilage, mislabelling, adulteration or defects.

Today, the mission of the quality team is wider and captured by the term "quality assurance" (QA) (ASQ, n.d.a). That describes the quality management system that ensures the fulfilment of all quality requirements internally to other functions and the leadership team and externally to consumers, customers, government agencies, regulators, certifiers and third parties. In this set-up quality control is the inspection aspect of quality management.

A different way to capture the essence of QA is the description as "the operational techniques and activities used to fulfil requirements for quality" (ASQ, n.d.b).

So, QA may involve conducting audits, training staff, monitoring suppliers and verifying records to ensure that the production process is consistent and reliable. Often the scope is spanned wider and starts for food enterprises on the farm level and reaches to consumer waste management. Sustainability, the ambition to be "good for people and the planet", is therefore nowadays part of the respective protocols, with clear outward direction to suppliers, customers and consumers.

In this, food quality is the degree to which food products meet the standards and expectations of consumers and regulators and in packaged food is a crucial factor for consumers and manufacturers alike. That affects the safety, nutrition, shelf life, taste and appearance of the food products. Consequently, the focus is to monitor and control the quality of packaged food throughout the production, distribution and storage processes. Some of the methods used to ensure product quality in packaged food include:

- Using appropriate packaging materials and techniques that protect the food from physical, chemical, and biological hazards
- Applying good manufacturing practices (GMP) and hazard analysis and critical control points (HACCP) systems that prevent or reduce the risk of contamination and spoilage
- Conducting regular inspections and tests on the raw materials, ingredients and finished products to verify their compliance with the specifications and standards
- Implementing traceability and recall systems that enable the identification and removal of defective or unsafe products from the market
- Providing clear and correct information on the labels and packaging of the food products that inform the consumers about the ingredients, nutrition facts, allergens, expiry dates and storage instructions

Often quality departments in food enterprises are part of the manufacturing and that for a couple of good reasons. Being close to the manufacturing allows the continuous monitoring and testing of raw materials, production processes and finished products and a tighter control of potential hazards in the production process and the compliance with all regulations and standards. This reduces the risk of legal issues and ensured continuous effort to proactively prevent defects, keep records and report of progress and issues.

This protects consumers and upholds the manufacturer's brand integrity. So at least part of the team is based where the manufacturing is happening, in the factory, where they monitor in a fixed schedule the incoming materials and the final food products based on the agreed specifications and reference samples.

Sometimes, the quality teams are part of the R&D organization. Then they are part of the product and process design team and can steer safety and quality aspects more directly into the design of food product and process.

A third option is to keep QA fully separate to maximize the level of independence for the team ensuring quality and safety. As part of the supply chain, they could be under pressure of production targets and efficiencies, with responsibility to maximize production output and interpret the quality standards a bit more liberally. And in R&D they could be perceived as innovation skid or "more holy than the pope" in their attempt to maximize safety and minimize uncertainty.

Bigger companies need more dedicated quality and safety resources. Here, operational QA teams are collocated with the production facilities but guided by central teams in the local head office and advised and directed by an independent global expert team, which collocate with the central research, where they apply the latest techniques, deep scientific expertise and an evidence-based approach to mandatory toxicological safety clearance for people and the planet (Unilever, 2023).

Contemporary quality assurance in the food business aims to increase customer confidence and satisfaction by ensuring that food products meet specific quality and safety standards, for which a few key concepts are essential:

- The quality control as such must monitor and test raw materials, ingredients and finished products to ensure they meet the set criteria for taste, appearance, texture, nutritional content and safety. Quality control also involves complying with regulatory standards and conducting sensory evaluation.
- Standard operating procedures (SOPs) are the documents that describe specific procedures, protocols and best practices for each stage of the food production process. SOPs ensure consistency and quality in sanitation, hygiene, handling of raw materials, processing methods, equipment calibration and packaging.

- GMP provides a higher frame to maintain high standards in food production facilities. GMP covers areas such as facility design, equipment maintenance, personnel training, sanitation, pest control and documentation.
- Hazard management is the implementation of a system to identify and control potential hazards in the production process. For instance, HACCP is a system that involves conducting a thorough hazard analysis, setting critical control points, establishing control measures, monitoring procedures and maintaining detailed records.
- Supplier quality assurance to establishing criteria and procedures to evaluate and select reliable suppliers, including regular audits and inspections to ensure they meet quality and safety standards. Additionally, it involves maintaining open lines of communication with suppliers to address any quality-related issues promptly.
- Training and education programs for employees to enhance their understanding of quality assurance principles and practices. Well-trained employees are essential for maintaining quality standards and identifying potential risks.
- Documentation and record keeping of all quality-related activities, such as test results, inspection reports, supplier certifications, training records and customer feedback. This documentation serves as evidence of compliance and provides a reference for continuous improvement.
- Continuous improvement, especially a culture of continuous improvement where quality assurance processes are regularly reviewed and evaluated. Feedback from customers and stakeholders to identify areas for improvement, implementing corrective actions and measuring the effectiveness of the actions is also in scope.

Several management systems are used to execute the underlying processes effectively.

GOOD MANUFACTURING PRACTICE

This is the set of standards that ensures food safety and quality, protects consumers from adulterated or misbranded food products and supports the development of international food standards. GMPs contain guidelines about personnel hygiene, plant sanitation, equipment maintenance, product labelling and record-keeping. They are necessary to ensure the safety and suitability of food at all stages of the food chain. Compliance with a GMPs regimen improves product quality, lowers the risk of contamination and thereby increases consumer confidence. Or said differently, compliance with GMPs helps to reduce the risk of regulatory action, such as product recalls or legal penalties, which can not only be costly but also damaging to a manufacturer's reputation.

Investment into GMP improves efficiency in the operation, as it can help to cut waste and often improves process control, and through that better productivity.

Fundamentally, GMP covers the food production end to end (E2E), from raw materials used to the finished product up to the consumer usage. The Food and Drug Administration regulates GMPs in the United States, and they've published a list of standard GMPs for food manufacturers. This list covers the main GMPs that food businesses should know.

The main categories are captured in the "5 P's": People, premises and equipment, products and primary materials, processes/paperwork, and procedures:

- *Personnel*: Food safety depends on the people. So, all food handlers must be properly trained in food safety and good manufacturing practices and need to be aware of all potential hazards that can occur during food production.
- *Equipment*: All equipment used in food manufacturing must be suitable for the intended use, should be easy to clean and help minimize risks, so that for example the food-contact surfaces of the equipment are free of contamination.

- *Suitable Facility Location*: The facility should be located in a suitable area, free from contamination and away from sources of pollution.
- *Sanitation and Hygiene*: Regular cleaning and disinfection of surfaces, equipment and utensils are mandatory. This includes personal hygiene.
- *Pest Control*: All pests (rodents, insects, birds, etc.) need to be kept out of the food production and storage areas.
- *Raw Materials*: All material used for production must be of a specified quality and must be stored, handled and used in a way that doesn't contaminate the finished product. The ingredients must meet the required specifications for safety and quality.
- *Storage and Transportation*: Storage and transportation conditions must maintain safety and quality, with full protection against physical, chemical and biological hazards, including temperatures and contamination prevention.
- *Labelling*: All products need a name, identity, net weight or measure and the name and address of the manufacturer or distributor. The packaging must include a list of ingredients and any allergens that are present in the product.
- *Quality Management*: This is the guiding principle that should be followed throughout the food manufacturing process, ensuring that all food products meet the required quality standards. These standards include testing raw materials, monitoring production processes and conducting final inspections.
- *Validation and Verification*: On a regular basis all raw materials and processes are monitored, inspections are performed and the records of all food safety activities are stored.
- *Documentation and Record-Keeping*: All food safety procedures must be properly documented, keeping records of raw materials, production processes, final inspections and other food safety activities. Documentation helps ensure that all food safety procedures are being followed and allows for quick corrective action in case needed.
- *Inspection and Auditing*: Facilities have to undergo regular inspections by government agencies. These inspections help ensure that the facility is complying with all food safety regulations. In addition, third-party audits are often conducted to verify compliance with customer certification standards.

The size of a GMP manual for foods can vary depending on the specific requirements and regulations of the country and the actual situation. A GMP manual for a hospital canteen would focus on other aspects than a wheat mill or a slaughterhouse. It needs to cover all aspects of food production, from the raw materials used to the finished product, and should include detailed procedures and guidelines for ensuring food safety and quality. The size of the manual will depend on the complexity of the food production process and the level of detail required to ensure compliance with GMP standards.

HACCP

GMP and HACCP are both systems intended to ensure the safety of food, chemicals and pharmaceuticals. GMP sets the basic principles of food safety so that products meet legal prerequisites for safety and quality. HACCP adds requirements for risk mitigation around the products and ingredients and brings in the requirements for traceability.

The HACCP system follows a couple of principles:

- Hazard analysis to identify potential hazards and measures to control them
- Definition of critical control points (CCPs), where the hazards can be controlled
- Setting limits for each CCP to ensure that the hazard is controlled

- Monitoring those CCPs to ensure that they are under control
- Specify corrective action when CCP is not under control
- Verify that the HACCP system is working effectively
- Maintain records of the HACCP system

To act properly on food hazards, they are usually split into four major classes: Biological, chemical, physical and allergenic hazards.

- Biological hazards are all types of microorganisms such as bacteria, viruses, parasites or fungi. They are the main causes of morbidity and mortality globally, especially in less developed markets. Some common examples are *Salmonella*, Campylobacter, and enterohaemorrhagic *E. coli*.
- Chemical hazards are harmful substances that can be "naturally" found in food or are unintentionally added on the way from farm to fork. Typical examples are toxins, heavy metals, pesticides and cleaning agents.
- Physical hazards are foreign objects in the food products, like hair, dirt, stones and insects. Material specifications contain details about the limits for typical contaminants.
- Allergenic hazards mean the presence of allergens in food products, including peanuts, tree nuts, milk, eggs, wheat, soy, fish and shellfish.

All potential hazards are managed at the CCPs, that is the moment in the production process where a hazard can be prevented, eliminated or reduced to an acceptable level.

Control points do not need to be very sophisticated; they just need to be effective! Thermal processing, vulgo: "cooking" – is the most common method of removing biological hazards. Controlled by time and temperature a proper thermal treatment kills microbes. But it can also negatively influence the product quality. Overcooking makes the product safe but often damages quality. Hence, more accurately one speaks about the "temperature time regime" where the microbiological pathogens are certainly killed or reduced to a safe level.

A typical example of thermal processing with a critical control point is the cooking of meat. The details are global best practice. To properly heat chicken, the thermal cooking CCP requests the poultry meat to at heated to minimum 74°C (or 165°F), which leads to a 5-log reduction for the typical pathogen *Salmonella*. In other processes higher temperatures can be applied. In the roasting of cocoa roasting a 10-minute roasting at 150°C leads to the same reduction. Other methods for inactivating *Salmonella* include irradiation and chemical treatment. Irradiation involves exposing food to ionizing radiation, which can effectively kill or inactivate bacteria, including *Salmonella*. Technically food irradiation is a superior sterilization method. It's safe, effective and often with minimal impact on the fresh quality attributes. Still, it's often only allowed for special food categories, like dried herbs and spices, it must be labelled and some countries the technology is broadly banned. Alternatives are chemical treatment with chlorine or peracetic acid to kill or inactivate harmful microorganisms.

Low temperature is another powerful mechanism to control the growth of many microbiological pathogens. Particularly for storage, a chiller or freezer at 40°F or freezer at 0°F (4°C and –18°C, respectively) is required.

Temperature is not only controlling microbiological activities but also enzymatic degradation. Under very low temperatures, enzyme activities such as browning, oxidation and protein breakdown stop. Typically, vegetable and fruit produce require cool temperatures to keep them fresh during, for example, during transport.

For chemical hazard testing cooking can also be the suitable method to reduce some chemical contaminants in foods. A study in 2002 showed that adequately cooking beef patties reduces the level of polychlorinated dibenzodioxin, a common contaminant considered toxic at high levels (Vu et al., 2022).

The opposite can be true as well. Then thermal conditions must be controlled because the excessive processing can generate contaminants including aromatic hydrocarbons and N-nitroso compounds such as acrylamide. Such contaminants have proven to be carcinogens in animal trials and cannot be removed after cooking. Too much heat can therefore not only reduce the nutritional quality but also stimulate the forming of potentially toxic substances (Iammarino et al., 2024). And safe limits here can often only be confirmed through elaborated chemical testing. Several methods for measuring N-nitroso compounds in food and other products are available, ranging from gas chromatography-mass spectrometry (GC-MS), to detect and quantify N-nitroso compounds in various matrixes, over headspace-GC-MS and direct injection GC to liquid chromatography-mass spectrometry (LC-MS).

Using such a method as CCP has its difficulties! In-line measures are very difficult and even offline testing in one's own lab may can lead to a long response time to run the process adequately.

For other products the addition of additives may be the method of choice. An example of this is in the manufacture of juices in PET bottles. Juices are normally acidic having a low pH that prevents bacteria growth, but not moulds. Therefore, prewashing the PET bottles with sodium benzoate solution is an appropriate control method for acidic drinks. But this additive is regulated in some countries as it can cause allergic reactions to vulnerable consumers.

Similar for packaging, the food packaging materials must also be inspected. Food packaging nowadays is highly functional and often a pretty complex composition of plastic and metal. In case of cans, its iron and the canning of food under acidic conditions can cause corrosion when the wrong type of can is used. Generally, packaging materials can become contaminated through the chemicals from the food processing equipment or through environmental contamination.

The physical hazard testing in foods can luckily be simpler. Optically pre-inspection in the QA lab often identifies foreign bodies like sand, stones, hair or insect parts sufficiently. Various types of detectors can help to increase the throughput.

But physical contaminants can signal much more than just the potential to injure consumers. They are also leads for biological hazards. Rodent hairs or insect in a food material is not only unsanitary; it carries a substantial risk of pathogenic microorganisms, making it a food safety hazard.

CCPs need to span the whole end-to-end process. A supplier of ingredients can confirm the safe quality of their products through a "certificate of analysis" (COA), where they confirm that the delivered charge of material is "in specification"

The other end of the value chain with storage and transport must be incorporated as well. During the storage of raw materials and transport of products, they are somehow susceptible to damages such as punctures, cross-contamination, rodent bites and others. Most issues arise because of poor sanitary practices as well as improper storage. Once a finished product is compromised, there is not much you can do since there are no other succeeding steps aside from distribution.

Authorities like the FDA or Codex offer practical help for the food industry to create CCP decision trees and to identify the CCPs in the process. A elementary general flow diagram can be applied. A very recent one was published in February 2023 by the Codex Alimentarius Committee on Food Hygiene (Evans-Lara, 2023). The latest one is shown in Figure 10.1.

Several consulting companies are available to help with the task. An example is "FoodDocs", a 2017 established company providing an AI-powered food safety management system. Their software includes a HACCP builder helping to create a food safety system (FoodDocs, 2023). On their website the company brings alive the necessary "thinking process" by the example of industrial raw chicken breast marinating and roasting. On their website they discuss the difference between marinating and roasting processing steps and their suitability as a CCP.

After the identification of the control points comes the control of those. This is done by setting critical limits, normally by defining minimum and maximum values at which you can consider that the hazard is within acceptable levels.

The important next question is: What happens, if those critical control points standards are not met? Basically, this means that the food safety hazard it was designed to control is not effectively working. In the HACCP procedures, a type of "corrective action" must be described to bring back control ensuring that the food produced is safe for consumption. This action needs to be effective under normal working conditions and it needs to be measurable. The specific corrective action will depend on the nature of the CCP and the hazard it is intended to control. If the CCP is related to cooking temperature, the corrective action might be to continue cooking until the established temperature is reached. If the CCP is related

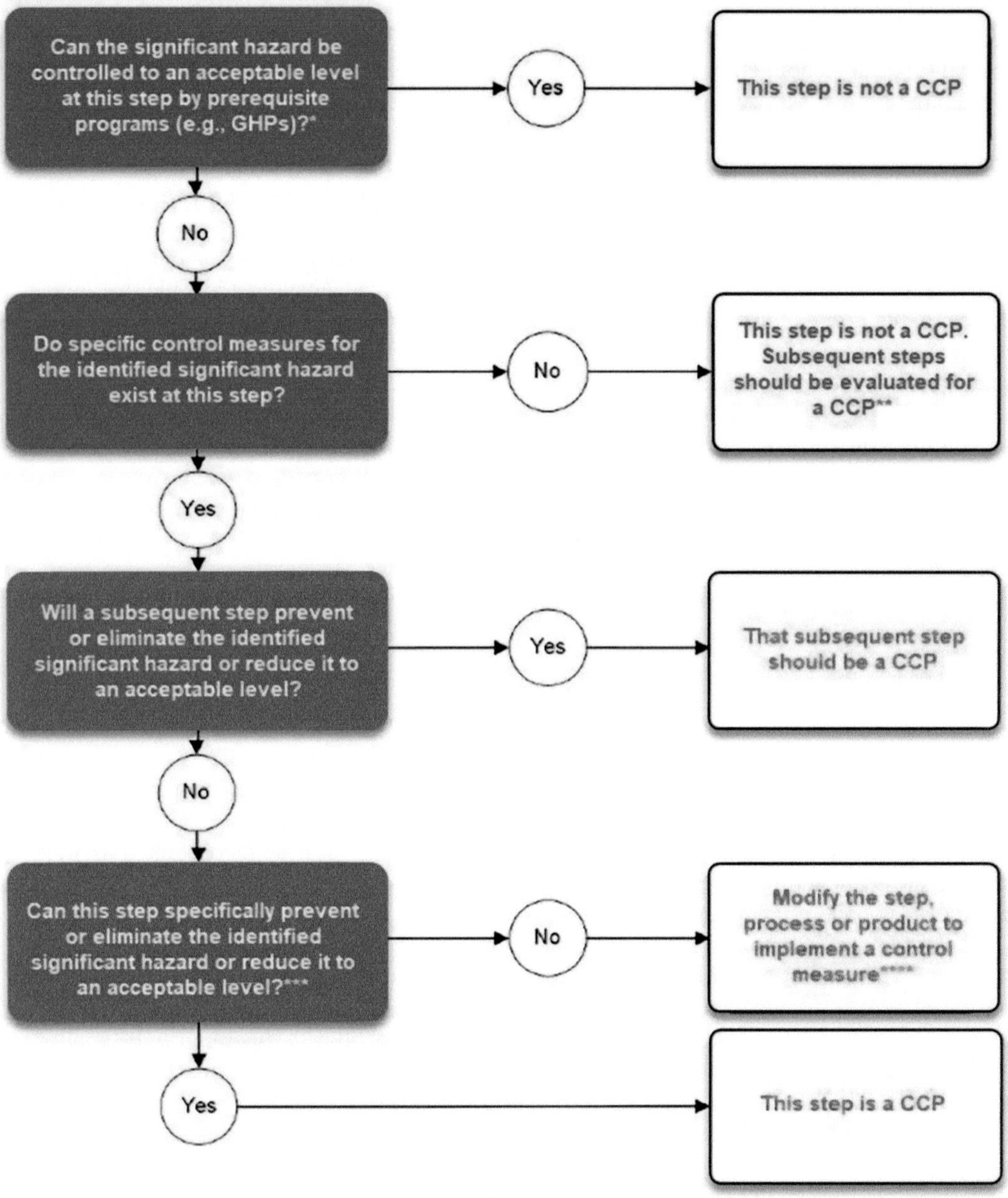

FIGURE 10.1 The CCP decision tree. (From Evans-Lara, 2023; Formatted by HACCP Mentor.)

* the significance of the hazard, so the likelihood of occurrence and the severity are considered. A key question is whether it could be sufficiently controlled by prerequisite programs such as GHPs. ** If a CCP is not identified, process or product have to be modified to implement a control measure. ***A control measure might work in combination with a measure at another step controlling the same hazard. Then, both steps are CCPs. ****and return to the beginning of the decision tree.

to ingredients or the overall product formulation, the corrective action might be to redesign the recipe to meet the established criteria.

The corrective action itself needs to be monitored, verified and documented. The documentation serves as records to auditing bodies and serves as proof that the HACCP food safety plan is effective.

CASE STUDY: CHILLED READY MEALS

This case is modified from (Wallace et al., 2010). A food company in Northwestern Europe recognized the opportunity to launch a range of ambient ethnic ready meal products under their own brand. The products should be in microwaveable trays and ready to eat in 3–5 minutes.

A cross functional team from engineering, manufacturing QA, R&D and procurement held a workshop to set an E2E HACCP workshop to identify critical control points and their management, including all microbiological, physical and chemical hazards, including allergens and compounds that cause intolerant reaction. The head of production received earlier formal HACCP training and was therefore assigned as HACCP leader.

The scope of the study was set from raw and packaging material, the material storage, the manufacturing process, the subsequent storage of the finished goods up to distribution and delivery at the retailors warehouse. Maximum shelf life was set as 10 days, under prescribed storage and usage conditions and was validated through production trials and microbial tests. The meals were heated to pasteurization temperatures, then quickly chilled, and stored and quickly distributed chilled.

It was agreed to review the protocol at least once a year and updated as required. A full HACCP study would be carried out prior to every new product launch in case of any changes in the product manufacturing or in case new raw material or pack materials would be introduced in the factory. After product safety assessment authorization and sign sign-off would be made.

The two-step high/low significance assessment procedure was used to detect the significant hazards from the list of all potential hazards for each step in the sourcing and supply chain. This assessment estimated the likelihood of occurrence and severity of such incidence. Hazards considered "high/high" both for likelihood and severity were considered significant. For the risk classification the Codex decision tree with its four key questions was applied:

- Can the hazard be controlled to an acceptable level at this step by prerequisite programs (e.g., GHPs)?
- Do specific control measures for the identified significant hazard exist at this step?
- Will a subsequent step prevent or eliminate the or reduce it to an acceptable level?
- Can this step specifically prevent or eliminate the identified significant hazard or reduce it to an acceptable level?

(Evans-Lara, 2023)

First the team made an overall flow chart starting from the material specification, the material receipt at the factory gates to the manufacturing and distribution.

HACCP PLAN

That scheme shown here only represents to top line flow charts; more detailed charts were made to describe all the activities happening in the handling of the incoming raw material, including:

- The unpacking of the different material coming fresh, frozen, chilled or dried
- The check against the buying specifications, including extra visual and analytical inspections, laboratory testing and certificate check.
- The storage of the different materials fresh frozen, chilled or dry
- Any pure pre-mixing of different components prior to processing

Similar for the packaging materials, the check against agreed specification and the handling and storage with the main split between food contact material and outer packaging (secondary and third).

The next level captured the manufacturing process, the modification of the raw materials, mainly mixing cooking and cooling and the filling of the tray:

- *One arm*: Cooking rice/pasta according to recipe, decanter and cool down (e.g., blast chill)
- *Second arm*: The cooking of sauces according to recipe, then either cool or pumped to filler
- The fusion of pasta/rice and the sauces
- The hot fill into the CPET trays
- Adding of hot topping/garnish
- Tray closure with lid
- Rapid cooling (blast, etc.)

Afte that the further packaging and the storage of the finished good was defined:

- Foreign body detection, for example, metal
- Sleeve wrapping.
- Data coding (time, production line)
- Out case packing
- Label attachment
- Palettization
- Storage (temperature, time, e.g., 0–5°C, max 2 days)
- Loading and distribution (temperature time, 0–5°C)
- Delivery at customer warehouse

The different modules were then subjected to the hazard analysis, where each step was analysed for any potential hazard and judged by the likelihood and severity of a hazard to happen. If both likelihood and severity were estimated as high, the hazard was defined as "significant". Consequently, the control measures were further specified. For the different cooking steps, that is done by written and operator-signed cooking instructions. The details of the cooking are further specified by measurable data limits, the definition of the necessary equipment and the outline for the responsibilities for execution and documentation. The result is summarized in Table 10.1.

The responsibility of R&D in this HACCP started with the definition of the material specifications for all incoming goods. These specifications were signed by the supplier and included the limits for each hazard causing a contaminant, plus the definition of the way the required product quality is confirmed, including analytical measure to be used, sample size or content of the certificate of analysis delivered by the supplier.

TABLE 10.1 Details for an effective control at a thermal CCP

STEP: COOKING	*HAZARD: PATHOGENIC MICROORGANISMS*
Control measure	All cooked ingredients with min temperature and min time
Critical limit	Min 72°C in ingredient core
Monitoring	Temp measure with calibrated probes
Monitoring responsibility	Line operator at cooking station
Corrective action	Heating until 72°C is achieved
Correction responsibility	Shift leader
Record keeping	Electronic cooking archive

QA would test all incoming material in line with the agreed inspection plan and release the material only into production after they passed validation. If the supplier is trusted and certified, they can specify all relevant parameter through a CoA, which can replace the full in-house inspection. Beside the material specifications, R&D defines here the processing conditions for the mixing, cooking and cooling steps.

CASE STUDY: HOT FILL LIQUID SOUP

A company in India planned to launch a new liquid vegetable soup range. Needed was a HACCP plan for vegetable soup that is hot filled into plastic bottles. The process still had to be developed, but the company had a preference for hot fill polypropylene bottles, capped with plastic lids that have foil liners. The key task was to find those steps where a lack of control could cause, allow or contribute to a microbiological hazard for the consumer in the final product.

The most logical control point was the cooking temperature, but to an extent this was a balancing act. Higher temperature would of course be safer, but too extensive heat would bear the danger of overcooking the soup, which would reduce the original flavour and aroma and added "cook notes", not wanted in this type of vegetable soup.

To manage that the company established an additional cooling step straight after their hot fill. The guidance was a heating to minimum 74°C and fill at same temperature. For acidified foods with a pH lower than 4, the recommendation had defined heat processing conditions that would ensure a 5-log reduction of all critical pathogens, here *E. coli* O157:H7, *Salmonella enterica* and *L. monocytogenes*. The "holding time" at the recommended temperature was added to cover all "cold spot" in the product container.

Cooking temperature was set as a critical step and the filling temperature was defined as an additional control point. The recommended regimen then was hot fill at minimum 74°C for at least 40 seconds. Additional guidance was to pre-warm the containers, close the bottle and invert the sealed pack to ensure the pasteurization of the unavoidable container headspace. The practical guidance added a safety margin and therefore recommended minimum 82°C for typical products like tomato-based pasta sauce with a pH below 4.0.

After the finishing of the recipe formulation, one soup turned out to have a pH of about 4.4, which classified the formulation as low acidic. Reformulation to lower pH was not regarded suitable, as the consumer test showed much better buying intention for the higher pH formulation. The soup also did not contain any compounds known to be inhibitors of microbial survival or growth, but instead had amino acids, sugars and other compounds that could aid in the survival of bacterial pathogens. The other vegetables, like peppers, garlic and cabbage were known to contain antimicrobial compounds. But as their amount would fluctuate, this could not be further considered.

Together with an external lab the company performed a challenge test to measure time/temperature necessary for a needed 5log reduction of the target microbes, plus the five times standard error estimated as an arbitrary safety factor for recommended thermal processes.

Finally, it was agreed to the conclusion that not the cooking temperature, but the filling temperature and time is the critical control point.

Since the bottle was made from polypropylene, guaranteed max filling temperature was 74°C for the maintenance of the packaging material integrity. These limits were sufficient heating of the filled bottle to kill any pathogens attached to the final packaging. And the final inversion step, combined with 5 minutes holding time ensured that headspace and cap were sufficiently heated.

BRITISH RETAIL CONSORTIUM

Yet another globally recognized certification program is the British Retail Consortium, also known as BRC, Standard for Food used by over 17,000 suppliers in 90 countries. It's also recognized for best practice in food safety, quality and responsibility (Foston, 2023). The BRC Global Standard for Food Safety sets out the requirements for food manufacturers to meet to comply with legal obligations and consumer protection regulations. It covers all stages of food processing as well where food is handled, from packaging fresh foods to slaughterhouses and all stages of further processing, again building HACCP principles and GMP.

The key chapters require senior management commitment and continuous improvement, include a broad HACCP plan with a food safety and quality management system, site standards, personnel and product and process control.

Progressively, the traceability of all material used traceability and the risk of food fraud is critical for companies and their reputation. The latest version of the BRC therefore includes those chapters, together with more elaborated auditing processes and training on vulnerability assessment methods, so companies can take proactive measures to safeguard their products from counterfeit or fraudulent activities.

Beside the simple need to continuously improve a company's food safety and quality, hence complying with both national and international food safety regulations, such certificates enhance the company's profitability through better reputation, In many cases it's also a prerequisite allowing food makers to do business with retailers, manufacturers, ingredients companies and food service. In a globalized world, certified enterprises are the basis for global partnerships. And for all selling under brands to other business or to consumers, it's an essential element of brand protection.

ISO 22000 AND FSSC 22000

Many food company websites refer to ISO 22000. That is the most popular voluntary international standard for food safety management systems. ISO (International Organization for Standardization) is an independent, non-governmental organization setting the standards based on Codex Alimentarius guidelines. It can be used by any organization regardless of its size or position in the food chain. The standard maps what a company must prove as food safety hazards. Companies following this standard often are certified, and the resulting document is required to do business with others in country or overseas.

ISO 22000 combines the core elements from the older ISO 9001 and HACCP to ensure the robust ways to food safety in the global supply chain. Beside the Quality elements (ISO 9001) it typically contains Environmental (ISO 14001) and Occupational Health & Safety (ISO 45001) parts. These are even more suitable as a guide for food manufacturers and processors.

FSSC 22000 of the "Food Safety System Certification" fully builds on ISO 22000 but is more sector specific and contains additional requirements, such as food defence, food fraud prevention and allergen management (FICSI, 2023).

SPECIFICATIONS

In most food companies R&D is the "owner" of the specifications. These documents cover all materials and processes and contain the detailed descriptions of the quality, quantity and safety requirements for all materials used and made in the manufacturing hierarchy.

In the food industry, these specifications are often presented in a hierarchy, a structured set of documents that outline the detailed descriptions and criteria for products, raw materials, processes and packaging. These specifications are crucial for ensuring food safety, quality, legality and authenticity throughout the supply chain. A simple breakdown would specify:

- *Raw Material Specifications*: They outline the requirements that raw materials must meet before they are used in food production. They include information on the nature and characteristics of the materials and define acceptable quality standards. They part of the risk assessment and ensure consistency in the quality of the raw materials used.
- *Packaging Specifications*: The equivalent for packaging material used ensuring compliance with food safety regulations. The describe materials, dimensions, design and labelling of the food container, as well as its compatibility with the product and its environmental standards. They cover all packaging material from primary packaging (in contact with the foodstuff) to the transport cartons and container.
- *Process Specifications*: Specify all steps involved in the manufacturing process, including the details of the handling of materials, production methods and quality control measures.
- *Product Specifications*: Describe the composition, quality, performance and regulatory compliance of a food product. This includes the physical, chemical and microbiological characteristics of the food, such as its ingredients, nutritional value, shelf life, sensory attributes and safety standards They serve as a contract between the manufacturer and the customer and ensure that the product meets the agreed standards.

The hierarchy is often a little bit more complex. A typical specification may have more layers in between, like premixes of raw specific raw material composition, their processing treatment, like a mixing or preheating, The assembly of different pre-treated premixes with other ingredients in a primary processing step and eventually subsequent ones.

The product specification captures the single picked item. That is not necessarily the "stock-keeping unit" (SKU), the item sold. SKUs can contain various numbers of single unity, often identical ones, but also mixed ones.

The primary packaging can also contain different elements, a wrap and a foil, a tub and a lid and a label. That requires in each case a specification. For the distribution several options may exist, for example, different palletizing schemes, to which some shrink foil and pallet label is added.

But let's here focus for the quality and safety discussion on the major ones. The packaging specification defines the physical and functional characteristics of the packaging materials and components used to protect, preserve and transport the food product. Key elements included are:

- Dimensions, weight and shape of the packaging unit
- The materials, layers and coatings of the packaging components used
- The functional properties, like barrier properties, mechanical strength and thermal stability of the packaging materials
- The printing, labelling and sealing methods of the packaging unit
- Nowadays as well the environmental impact, recyclability and biodegradability of the packaging materials
- All regulatory compliance, safety and quality standards of the packaging unit

The next covers the other incoming material: The raw materials. A typical raw material specification describes not only the product but sets minimums for value-adding constituents, like minerals and vitamin content, aroma-giving volatile substances, etc. For unwanted materials, like contaminants and foreign bodies, maximum values are included. And preferably for all values specified, a commonly accepted analysis method is specified (i.e. ISO).

Some specification demand also details the production method of the ingredient or list methods that are excluded. Ethylene oxide (ETO) treatment has been banned under European legislation for herbs and spices, and most users do not want irradiation; even so it's legally permitted, when declared. A short summary of the key elements in a raw material specification are listed here:

- Name of the material.
- A description of the material, including biological, chemical and physical characteristics.
- Limits for microbiology per lead organism, maximum values for all possible types of contaminants, like herbicides, pesticides, chemicals, heavy metals or other foreign material.
- Detailed composition of the material, including additives and processing aids.
- Country of origin.
- Method of production.
- Packaging formats or unit of measure.
- Delivery methods.
- A description of the labelling, lot ID and coding for traceability.
- Storage conditions and shelf life.
- Preparation and/or handling before use.
- Acceptance and rejection criteria.
- Requirements for certificates of analysis for high-risk materials or vulnerable materials.
- Special requirements such as allergen information, organic status, GMO status, fair trade and ethical sourcing policies.
- Information about compliance with statutory and regulatory requirements, where relevant.
- A requirement for suppliers to notify of any authenticity issues with the product.
- A requirement for suppliers to notify of any changes to the product.
- Formal agreement between the supplier and purchaser.
- Document control features, such as author, date and page numbers.
- The approval status, revision history, and attachments of the document. The purpose of the specification is to ensure that the raw material meets the legal and customer standards, and to facilitate the traceability and compliance of the food products (Food Fraud Advisors, 2019).

The sources of all the details in the raw and packaging specifications are often the supplier's data sheets, the attached certificates of analysis, complemented by companies own analysis or laboratory reports. Often, they are part of the contract with the supplier, together with other necessary documents and certificates, and they form the fundamental building blocks for the finished product and the product specification. Depending on the type of product and the process conditions, the food product may not be made in one continuous process, but from preformed batches.

Such premixes or batches form the next level. These specifications contain the ingredients they are made of and all resulting changes from that processing step. Various dry materials may have been diluted in water, or a liquid solution has been heated up, or the step has resulted in some material loss (waste). The pre-mix specification needs to reflect the physicochemical changes. Generally, if this semi-finished product is stored between initial making and final food assembly, it gets a specification, which is more than the sum of the components used to make it.

From pre-mixes, other raw materials and combinations the final product is made. Again, this specification provides detailed information about the ingredients, nutritional values, allergens, shelf life, storage conditions, and other relevant aspects of a food product. It builds the product ingredient label.

Basically, the product specification reflects composition of the materials used. The composition can be in modern formulation tools be calculated by "roll up" calculation. The ingredient lists all material that went into the making of the product. The nutritional content is the addition of the nutrients of all raw materials, etc.

But this simple sum needs to be corrected by all influencing factors all processing steps had. The product may have been heated up and hence lost the water that evaporated. Heat-sensitive nutrients like vitamins have been reduced in the same process. Some processing aids may have been added during manufacturing that need to be declared on the label. Or air has been added and that changes weight and volume density calculations. An obvious example is ice cream, where approximately 500–600 g of ice cream mix then results in 1 L of ice cream.

A typical process specification for packaged foods is a document that defines the quality and safety requirements of the food product, as well as the methods and equipment used to produce, package, store and distribute it. The process specification often contains information such as:

- The ingredients and their sources, quantities, specifications and allergens. For example, a process specification for the making of a bakery product may list the type and origin of the flour, sugar, eggs, milk, etc., as well as their weight, moisture content, purity, etc., and any potential allergens such as gluten, lactose, nuts, etc.
- Detailed processing steps and parameters, such as temperature, time, pressure, pH, etc. For example, the process specification may describe the mixing, baking, cooling and slicing of the cake, as well as the optimal temperature, time and speed for each step, and the target pH and water activity of the final product. R&D sets the max here in terms of detailed machine setting parameters to ensure production quality and safety.
- The packaging materials and their specifications, such as size, shape, colour, seal, etc. This can include the thickness of the plastic film, cardboard box and label used to wrap the pallet.
- The labelling and coding requirements, such as product name, date of manufacture, expiration date, batch number, etc.
- The storage and transportation conditions, typically temperature, humidity, light exposure, etc.
- The testing and verification methods and criteria, such as microbiological, chemical, physical and sensory tests. This can include sampling plans and frequency for testing for microbial contaminations, chemical residues, physical defects and sensory attributes, like appearance, aroma, flavour and texture.
- The corrective actions and preventive measures in case of non-conformity or deviation, including the type of test, the mandatory procedure and when elements are out of specification the measures to prevent recurrence (e.g., review and improve the process control).

All specifications can be linked logically to describe a food product from its ingredients, the formulation, the different packaging levels, the label and the distribution process. They are the skeleton of the product and its assembly and form the basis for all the quality assessments from supplier to market.

SAFE BY DESIGN

Safe by design fundamentally describes a pre-market approach addressing safety issues already during the product development in early phases of the innovation circle. This can even start in the crop breeding phase through the usage of safe organisms, prediction tools, metabolomics and bioinformation in modern biotechnology (Van Der Berg et al., 2020).

Safe by design is used in foods in the concept, lab, scale up, product storge and distribution phase. In the formulation phase, substances are intentionally added to food to maintain the safety and good quality of the final product. Most of these shelf life–extending ingredients are regulated as additives and need pre-market approval. Often preservation is not their only functionality; they can improve other functions like taste, emulsification, thickening or flavour enhancement.

Using specific substances as preservatives tools is not new science. Simple materials like salt, sugar, sulphur dioxide or low pH have been used for centuries in food making. But in the past 100 years the industry has created many powerful actives. They act as acidity regulators, simple pH control agents, or to change or maintain pH. The aim is to prevent microbial spoilage. They can be organic or mineral acids, bases, neutralizing agents or buffering agents. Typical agents are the sodium salt of acids, like sorbic acid, acetic acid, benzoic acid and propionic acid. Acidity regulators can be recognized in Europe by their E number, such as E260 (acetic acid), or simply listed as "food acid". Generally, food is regarded safe when the pH is below 4.2, a value well below the point where *Clostridium botulinum* spores do not grow any longer. Preservatives can be split into antimicrobial preservatives and antioxidant preservatives. Some have both functions. The application is often guided by the food or beverage that should become safer: Antimicrobial preservatives like sulphur compounds inhibit the growth of bacteria, for example, in wine or dried fruits. Salts like sorbic acid help with the preservation of potato products, cheese and jam. And benzoic acid is often used as antibacterials and antifungals in cucumbers, low sugar jams, jellies, dressings or condiments.

Additives can be natural or synthetic (see Chapter 11), and their "industrial" origin often raises suspicion. Hence, since a couple of years research is on the way to identify natural preservatives to enhance food safety (Teshome et al., 2022). Many plant extracts, especially from herbs and spices, have this potential. But the cost of those and the often more difficult application still limit their usage.

Often their "protective power" is much lower than the one of chemical additives, Therefore, applying only one of them may give insufficient protection. But in combination or in addition with other protective mechanisms, safe food can be designed.

SAFER AND BETTER OVER TIME

Food quality means simply meeting the expectations and preferences of consumers, such as appearance, texture, flavour, labelling, packaging, ingredients and more. In most cases, the product quality of food items relates to the maintenance and full preservation of the "original" goodness. Basically the "fresh food items" is the gold standard, and by preserving and storing them some decrease in a few quality attributes is accepted. Only a few farmed goods improve their quality like taste, aroma texture or nutrient profile during initial storage. Bananas may be such a case, as they are getting sweeter and softer over time.

Proper storage can radically improve initially processed food. Cheese would come to mind, bringing up more complex and intense taste as it ages, combined with a firmer texture. Honey crystallizes over time, enhancing its sweetness and creaminess. And the whole group of fermented foods such as yogurt, kimchi, sauerkraut and kombucha may increase their probiotic content and increase pleasant acidity during storage, which can benefit the digestive system and the immune system. Not to forget all-time favourites like wine and beer. But here it's the controlled fermentation and what follows that bring out the intended quality.

Traditional condiments are often the result of long-established, traditional fermentation processes. Vinegar, kombucha, soy, fish, sriracha or oyster sauce are iconic ingredients in the local kitchens. This local knowledge has been passed from generation to generation without full understanding of the underlying microbial processes. The tight control about the material, process and storage conditions ensured mostly a safe and delicate product. The rise of brand, packed food created the opportunity to market such products nationwide and throughout the whole year. To get a safe and affordable industrial product, food science and R&D had to deeply understand the ongoing physical, chemical and biological reactions and redesign and simplify the vulnerable local processes.

"Kecap manis", or sweet soy sauce, is one of the most popular condiments in Indonesia. It has been made for centuries by fermentation of soybeans with palm sugar, water, salt and spices, sometimes star anise, cinnamon and clove. The traditional production starts with a controlled fermentation of soybeans with *Aspergillus oryzae* at 25–30°C for 2–3 days to make a soy sauce, "koji". The koji is subsequently mixed with a specific amount of saline, creating a soy sauce mash, which is called "moromi", in which a mould is enzymatically digesting proteins and starches from the substrate over several weeks or months.

The tasks of product developers, process engineers and biochemists are to understand this traditional recipe, identify the leading microbes, understand their physiological pathways and reactions and the generation of the actives to scale the traditional manual process to industrial scale, without losing the often very complex key sensorial compounds and attributes. In a later stage experiments can start to optimize the "yield" in the process, by varying substrate, leading strain or growth conditions, like pH, salt, water activity, etc. (Zhou et al., 2019). It can require a lot of testing, trialling, analysing, separation, recombination and tasting to split the value-giving characterizing components from the other.

But for the rest, the maintenance of the original quality parameters after production is the key point of attention. Packaged food quality is initially defined by the type of food and the packaging used to protect it. Then the external conditions like storage and distribution directly impacts the speed of quality loss and the shelf life. The affecting physical, chemical or biological processes alter appearance, texture, flavour, aroma or nutritional value of the food. Some of these processes include oxidation, moisture loss or gain, microbial growth, chemical and enzymatic reactions and nutrient degradation. Packaged food quality can be maintained by using appropriate packaging methods, such as vacuum sealing, modified atmosphere packaging or active packaging. These methods can reduce the exposure of the food to oxygen, moisture, light or other agents that can cause quality deterioration.

CONSUMER-RELEVANT QUALITY STANDARDS

The quality team does not only act within the company and "downstream" towards the material supplier side, but as well "upstream", to the market, the retail and customer environment and the consumer. The easiest way to fulfil that task is to visit the places where the products are sold, the retail stores of the modern trade, the cash and carry markets or the open markets in developing and emerging markets.

The simple idea is to get a realistic impression of how the products are appearing at the point of sale and how they stand up versus competing products. In many markets consumer do not contact the company care line to comment about the product. With this feedback loop disappearing, store checks are becoming even more important. And they can show challenges that happen to the product on the way from factory to the point of sale, the distribution chain.

Consumer-relevant quality standards (CRQS) in food are important to ensure customer satisfaction, loyalty, and trust and legal compliance. Under CRQS are those criteria condensed that determine the acceptability of food products by the consumers. For foods these standards often include sensory attributes, such as appearance, colour, flavour, texture and aroma, as well as nutritional, safety and ethical aspects of product and packaging. This "quality monitoring" can happen on three levels: "On pack", "in pack" and "in use".

These criteria go far beyond safety standards. They reflect the quality ambition of a brand and should show all the product design principles implemented by R&D. They are captured in the "finished good specifications". The quality function should also have reference samples so that they could easily confirm that the samples are "in specification". Additionally, samples are visual documentation is available showing possible product and packaging quality defects, deviations from the designed standard, still safe, but with an inferior quality aspect. This can for example describe inferior print and seal quality dimensions of the packaging or minor physical product deviations.

All the possible defects can then also get a classification by a "traffic light" system. Here, fully quality compliant samples are "green", and those having minor mistakes are "yellow" and inacceptable defects are signed "red". The results of the sampling are captured and fed back to the manufacturing, where new corrective actions can be agreed upon if needed.

Like other "store checks" such trade visits are important for food R&D as well. By seeing their own products at the point of sale, aside their competitors and alternative offerings, new insights are coming up. The packaging delivers the first impression. Is it undamaged, well closed, shows the intended print quality, displays all relevant written information and does it have the impression on shelf that it was expected to have? Any visible damage or quality deterioration can signal problems in manufacturing or a particular aspect of the distribution. Or it signals a "design defect" – here the product was correctly made and handled but shows inferior quality. This often requires an improvement on specification level.

The results of the investigation are normally first shared with the supply chain and distribution network to determine the root cause. Damaged primary packaging (the packaging in which the food product is embedded) may indicate that the secondary packaging (the transport packaging) or the palletizing scheme is not fit for purpose. Normally the packaging R&D will have specified all relevant parameters for all packaging and transport layers through stress simulation tests, like simple drop tests or full transport trials. But sometimes even subtle changes on the way to market can have drastic effect on product quality. The retailer may have changed the storage system, and this now puts more stress on the transport packaging. Or a product is delivered on a new route, a very bumpy road or by ship, and the constant agitation has destabilized the emulsion.

A comparison of reference samples in the factory and products sold in the market gives a good idea about the cause of a quality problem. If the team identifies a statistically relevant number of "weak seals" in a pouch, it could be time to inspect the sealing jaws in the filling line, control the sealing temperature or check the adhesiveness of the packaging layers.

But when it becomes evident that the product cannot withstand the known environmental stresses in the regular distribution, R&D can consider a design defect. Maybe the product packaging is simply no longer fit for purpose. A corrective action in the design is needed. This sometimes happens, when a cost saving or VIP workshop has found a savings opportunity through a packaging strengths reduction, reducing the layers by a few microns, or cutting the carton strengths in the transport packaging. The tests and simulations done before the relaunch of the product did not indicate any stability change, but now it seems that it's not sufficient. The specifications describe the measurable parameter and include the tolerances that can be accepted based on the quality ambition of the brand. A packaging seal is close, but not exactly parallel, the tear strip does not work in 3 out of 10 cases. Is this acceptable or does it need an immediate corrective action? The specification should be clear how long a product is "in specification" and a sellable good and when not any longer.

Defects can also appear only in one channel, particularly when distribution and market conditions at point of sale are different. In many developing and emerging countries, the modern trade has well-controlled chilled distribution and modern warehouses. This "stresses" a food product much less than the long and often bumpy way through heat and humidity after a wholesaler is bringing it to the remote markets up country after having repacked the palettes.

Then the company now needs to decide if they make their SKU fit for all channels and circumstances, taking the worst condition as reference, and putting more money into the necessary product improvements. Or if they split the product into two SKUs, where the less protected distribution gets a new specification for stronger packaging. The latter may minimize direct on costs but increases the complexity in the company from the material handling over the manufacturing up the data and system management.

HURDLE TECHNOLOGY

And while permanent usage of additives is standard or "old school" food safety design, newer approaches like the "hurdle technology" have started to gain more ground. Hurdle technology is an integrated approach for basic food preservation methods to create safe, stable, yet nutritious foods. As such, it is tried in food preparations and has been used successfully in many countries for mild but effective preservation of foods. Particularly for products with ambient shelf life and hot and humid conditions, alternative routes to stabilize products without chemicals are interesting.

The idea combines the conscious use of mixture of preservation methods, which could be of physical, chemical and biological origin. It requires a deep understanding of all the factors governing food preservation in the food items where it is used. Typical elements are pH, temperature, water activity, preservatives and competing microorganisms in raw materials, ingredients and finished food products. Nowadays the approaches are often clustered into formulation intrinsic factors like pH, the famous preservatives, extrinsic factors like storage temperature and packaging atmosphere and the processing factors (heating, drying, fermentation).

Extra consumer attractiveness comes from the avoidance of classical chemical preservatives. It's the application of technology to preserve maximum quality of food while keeping full food safety. It's the "art" of scientifically integrating individually mild preservation factors to a combination bringing enhanced product quality at full safety. From a science perspective, a "strategic combination of hurdles leads to metabolic exhaustion and disruption of cellular homeostasis, thereby controlling microbial populations in a product" (Gragg & Brashears, 2021).

Beside the classical methods the modern non-thermal technologies are moving into focus. Examples are pulse electric field, UV, ionization, ultrasound, high pressure processing and atmospheric cold plasma (Bigi et al., 2023).

Current applications are often combination of two or three of those factors, like usage of natural antimicrobials like vanillin and low-temperature storage in combination with short UV-C or the addition of sodium benzoate/potassium sorbate to blanched fruit. Often, traditionally effective tools like heat are still essential but can be performed at lower methods and thereby better-preserved beneficial temperature labile substances like vitamins and antioxidants. And the simple combination of 13 g of salt + 180 g of glycerol per kg flour has been shown in South Africa to improve the shelf life of the bread from 1 day to 1 week (Abdullahi & Dandago, 2021).

Particularly for intermediate moisture foods (IMF) or semi-dried foods these approaches are followed, as they may open new possibilities for products like fresh food, but with a substantial longer shelf life. In many tropical and subtropical countries, the distribution of fresh or semi-dried items cannot happen in temperature-controlled distribution chains, so that industrial products, safe by design at ambient temperature, are very welcome to supply food products also in rural and remote areas.

The concept of safe and sustainable by design is intended to accelerate the widespread market uptake of new and alternative chemical products and technologies that deliver greater consumer confidence in their safety, environmental and societal benefits and advance the transition towards a circular economy and climate-neutral society. This involves working together with policymakers, civil society, academia and others to achieve stronger consensus on criteria for chemicals that are safe and sustainable by design. These criteria must address the three pillars of sustainability – environmental, social and economic factors – and take a life cycle approach.

The product formulation is another layer to keep product safe during shelf life and after opening. Regardless of how the product is built, some of its ingredients play a critical role in maintaining food safety. Safe product formulation contains a lot of critical control examples especially when it comes to preservatives, both natural and artificial. One of the common CCPs related to formulation is the addition of an acidulant which can be vinegar, citric acid which comes from lemon or other acidic fruits, and lactic acid.

TRACEBILITY

Food traceability means the ability to track and trace the movement of a food product and its ingredients throughout all stages of the supply chain, from the consumer back to the source and from the source to the consumer (U.S. Food and Drug Administration, 2024a). It includes the unambiguous linkage of all steps in the supply chain from harvest over primary processing, food product assembly and packing to distribution. For all food products and their ingredients, it is the ability to access all information related to food.

Authorities like the FDA have established a "traceability list", which records all food items for which additional traceability records are needed. The motivation is to enable faster identification and quick removal of potentially contaminated food from the market. Food items have been risk-ranked based on:

- The frequency of outbreaks and occurrences of illnesses, the severity of illness
- Likelihood of contamination
- The potential for pathogen growth, with consideration of shelf life
- Manufacturing process contamination probability and industry-wide intervention
- Consumption rate and amount consumed
- Cost of illness (U.S. Food and Drug Administration, 2024b)

Thus, items like cheese, eggs, fruits, vegetables and fish are included. Overall, it is estimated the up to 30% of the total food consumed in the United States will be affected by this new regulation. The regulation will require all parties in the food chain to be able to provide within "24 hours portable end-to-end electronic records. during a food-borne outbreak or food recall investigation" (Unnikrishnan et al., 2024). The compliance date for all this additional information is January 20, 2026. Other countries and regions, like the EU, have similar plans.

No doubt, these regulations will provide good tools to manage food safety in wide open markets more effectively. More transparency in the practices applied in the chain can also help to strengthen environmental standards, allow more effective supply demand planning, harmonized supplier qualifications on a higher level and perhaps even allow for price premium for the first movers. But the massive transformation in the food chain will also add significant costs, which can overwhelm smaller- or lower-margin entities in the chain.

Fortunately, technology can help to bring the needed transparency to life. Sensors, machine learning, blockchain and AI technology (see Chapter 5) can be used to reliably check all claims made, ranging from safety, over origin, authenticity to environmental and social and work standards. (OpenSC, n.d.). More and more companies in the food chain are applying these tools.

In Europe, one of the eye-opening events, urging much better traceability in increasingly complex supply chains, was the 2013 "horsemeat scandal" in the UK and other European markets. It surfaced when a routine control of a frozen food range by authorities in Ireland detected first "unclear", later in retest then, "equine" and "pig" DNA in beef-labelled meals like lasagne and Bolognese or burger patties. After complex investigation, it turned out that one source of the fraud was a French company importing meat from Romania providing that to another well-known French frozen meal maker, producing in Luxembourg, who sold his products to many international retailers in Europe (Taylor, 2023). As a result, consumer trust in UK fell by nearly 25%, and all affected retailers had to invest each millions of euros in the coming years to improve the traceability of their meat sourcing.

FOOD FRAUD

The international food supply chain is still getting longer and more complex. While there is on one side a trend to more locally produced food, there is as well an interest in new and exotic food items or local items outside their seasonality. Consumers are looking for more genuine, traditional and ethnic food items. Both the internationalization of the food industry and the will of consumer to pay premium for "high reputation" food offers opportunities for some to earn more money.

Food fraud is understood as the intentional deception of consumers or businesses about the quality, authenticity, or origin of food products mostly for economic gain. The Food Authenticity Network (FAN), has estimated the annual cost of food fraud globally with $49 billion (Food Authenticity, 2024). The 2020 report also identified 1193 food fraud incidents in 2019, affecting 137 different food categories. Some of the most common types of food fraud include adulteration of honey, olive oil, spices, fish, meat and dairy products.

It can take various forms, such as adulteration, substitution, dilution, counterfeiting, misrepresentation or theft. And of course, food fraud can pose serious risks to public health, food safety, consumer trust and fair trade. A few striking examples of food fraud are:

- Adding sugar to honey to increase its volume and sweetness
- Selling regular beef as Wagyu beef, which is a premium product from Japan
- Injecting shrimp with gel to make them look larger and weigh more
- Mixing horse meat with beef for minced meat for lasagne
- Adding melamine to milk to increase its apparent protein content
- Labelling products as organic when they are not
- Applications of unlabelled or non-permitted manufacturing processes

To fight the increasing number of food frauds, the authorities release reports with suspicious case. In Europe the European Commission publishes a monthly newsletter listing the suspicious cases in the member states (European Union, 2024). For January 2024, the report listed 277 cases, nearly half of them affecting fruits and vegetables. In about 80% of the suspected cases an implicit claim violation was flagged; the rest were assumed to be product or documentation tampering.

Product altering often includes the non-compliant usage of food additives, the addition of cheap ingredients or the substitution of value-giving ingredients or the usage of unapproved processes. Misleading denominations and false health and nutrition claims are noted in many food categories, and the usage of non-approved food items or the detection of residues of not-permitted pesticides.

The consequences for the health of the consumer can be minor, like labelling an aged cheese from Spain "Grana Padano" or potentially dangerous (non-permitted food items, residues), like the contamination of a food item with undeclared allergens. A dairy drink containing soy protein residues or vice versa, a fish contaminated with shellfish. The consequences can be life threatening.

Not delivering on pack guaranteed food constituents, for example, nutrients, deprives consumers of the nutritional value or quality they expect from the food products they have purchased.

False claims are made for a food product claimed to be natural, bio-organic, vegan organic, etc. These ones could potentially go unnoticed by the consumer for a while, but once detected could pose a huge blow to the reputation of the brand. Similarly, selling products as halal and kosher while they contain some minor pork-based ingredients like gelatine is, of course, fraudulent. The detection of such case happens often during custom or market controls; only few are flagged by the media, consumers, whistle-blowers or companies themselves.

With such risks for the consumer and the reputation damage for food brands and companies, effective risk mitigation is critical. From the food company perspective, this requires good supply chain transparency, trust in traders and suppliers and also regular audits and strict adherence to HACCP principles.

The global "food authenticity network" offers prevention tools, fraud reports, mitigation guidelines and methods for authenticity testing and serves as an expertise centre for such analysis, ranging from specialize academic groups to international life sciences companies, like Eurofins (Food Authenticity, n.d.). Especially the DNA test to identify species is constantly becoming better and more differentiating. Since 2019 trained experts can, for example, distinguish buffalo milk mozzarella from cow milk mozzarella (Chapple, 2019).

The motivation for food fraud is the possibility for illegal economic gain. The concept of "food defence" takes the problem broader and includes all activities to cause harm, including terroristic attacks. And food fraud is also different from unintentional contamination, which is caused by negligence, error or accident.

QUALITY TESTING IN FOOD

As explained earlier in Chapter 9, the basis for an objective quality testing is given by the specification. Often, wherever there is a specification a quality check is performed, expressing the jointly agreed quality standards and the non-discussable safety standards.

When is material is delivered to the doorstep of a food factory, the quality department will receive the material first and perform necessary quality checks. It is part of the relationship between the supplier and the manufacturer to agree which parameters are actively controlled, how often the controlled and how. With a new, less known but certified supplier, the manufacturing side may decide to test more parameters and more often. And with a well-known long-standing supplier the control frequency can be reduced or to a minimum extended and be replaced by "compliance proof" through the supplier. A batch-wise CoA confirms that the material is completely "in spec" and ready for use.

Generally, these material investigations include the method selected and the preparation of a sample, the assay as such and the calculation and interpretation of the results obtained. The choice of analysis method is usually based on the objective of the analysis, characteristics of the method itself and the food matrix involved.

The validation of the method is important, as is the use of standard reference materials. For routine production tests and uncomplicated items, a robust and less sensitive test may be more practical than a sophisticated, highly accurate test used to support legal claims, for example, on nutrient content. Often, a particular measurement in a specific food matrix requires very special tests. A seemingly simple viscosity or texture analysis in a starch-bound liquid or a dough needs good validation in the target material.

The next pages provide introductory information on the topics summarized earlier that lays the foundation before covering various specific methods of food analysis in other chapters.

Often, the required attributes may include analytically more complex methods to determine a particular flavour profile, a volatile ingredient or an aroma, an ingredient content or a viscosity parameter.

There are several globally accepted sources of information for food analytics. FCC is one of them; the acronym stands for "Food Chemicals Codex" and represents a collection of internationally recognized standards for the purity and identity of food ingredients. The FCC standards are reviewed and approved by independent experts and are cited by regulatory bodies around the world (Creydt & Fischer, 2018).

Besides the FCC, there are other standards for food ingredients that may be used in different regions or for different purposes. The FDA also publishes food ingredient chemistry reference documents that provide purity specifications and methods for detecting ingredients in food. Additionally, some countries or organizations may have their own standards for food ingredients, such as the European Union, Codex Alimentarius or AOAC International.

The Association of Official Analytical Collaboration, (AOAC) develops and validates standards, methods and technologies for food and agriculture analysis (AOAC International, n.d.). AOAC collaborates with government, industry and academia to ensure the safety and integrity of food ingredients and products around the world.

The balance between the analytical tools used and the human judgement requires often careful consideration. Where possible, objective analytical methods are preferred, but they can require immense effort, training, infrastructure build and costs. Therefore often, sensorial standards in food items, especially taste and aroma, are just specified as "typical", which means "human judgement". A trained expert can do that reliably, based on consensus between supplier and customer, what the "typical" taste or aroma is. This is a simple and cost-effective test that works very well for many food items. But other food items like herbs, spices, meat or dairy alternatives show local characteristics that need to be aligned, so that a trained human panel is sufficient to confirm the suitability in production.

Sometimes the detailed chemical analysis of aroma or aroma components is useful Something like an artificial vanilla flavouring should "typically" contain vanillin, ethyl vanillin, guaiacol, vanillin mandelic acid, eugenol and piperonal. The necessary RT-HPLC assay is based on analytical-grade substances, sophisticated columns, accurate calibration and trained staff. Nature knows more than 300 different types of vanilla, and the differentiation of those flavour and aroma profiles can become more than a demanding PhD thesis. The actual or seasonal environmental conditions can heavily influence the sensorial properties. Different crops are not identical. Supplier and user should agree in common testing/testing sessions the range of what is "typical" and what "not any longer", jointly identify reference samples. They can also agree preferably an "in application" testing, so that the component is tested in the final product.

A company ordering this flavour could set up their own analytical lab to control the profile from time to time. Or they trust the CoA the supplier provides from a very controlled, closed chemical synthesis process.

In particular aromatic botanicals show an immense variety, and they are inconsistently defined. Take basil (*Ocimum basilicum*) as an example here: The fragrant herb that belongs to the mint family. Under this name up to 150 species or cultivars are known, with hugely different aromatic profiles. People in Europe normally expect the "sweet" or "Genovese" or "Italian" basil with large, shiny, green leaves and a sweet, spicy flavour reminiscent of mint and clove. They would not easily accept "purple basil" with its dark red leaves and a slightly stronger and more peppery note. Lemon type basil is a hybrid between sweet basil and American basil, with light green leaves and a fresh, citrusy aroma and flavour. Lime basil is another hybrid with narrow, bright green leaves and a zesty lime scent and taste and is less sweet. But that is not all aroma direction basil can take: Cinnamon basil is a variety of sweet basil, but with dark green leaves, purple veins and stems. Very popular in Asian cuisine is another, Thai basil, with a strong, licoricey-like flavour. And a Thai would always differentiate between this one and the holy basil, with green or purple leaves and more peppery and a clove-like flavour and slightly bitter taste.

There are more variants, but the point is made, I guess! A new supplier, a new growth area or an existing supplier, different environmental conditions or the opportunity for a fantastic contract on the spot market can shift the sensorial properties of this signature ingredient in such a way that it is no longer "typical "and "in spec".

In the following part, different possibilities to measure and analyse relevant food safety and quality parameters will be discussed. As always, all of them have advantages and limitations. Therefore, it is important to be as clear as possible for which aspect a test is needed. And combinations of two or more methods often generate a more detailed view. For safety measures biological, physical and chemical procedures are applied. The times where a "praegustator" had to taste the food before the rulers or celebrities could start to eat are long gone!

Today, there are many options to determine sensory food attributes (Rodrigues et al., 2024). The best choice must be made in the actual business context, often a balance between speed, accuracy and costs. Food quality testing with human individuals is widely used. Once properly "calibrated" these people can judge the relevant quality dimension in a human holistic impression. These trained experts cannot reliably quantify the amount or number of aromatic food component, nor do they have a common "food attribute language", but they can judge if the agreed standard is met. If food is a piece of music, they judge the performance of the orchestra, not the exact contribution of a single instrument or player. Factory QA departments often use trained panels for the routine check of their production process.

SIMPLE DIFFERENCE TESTING

There are several types of product difference tests in the food industry, each with its own advantages and limitations. Product difference tests are used to determine if foods differ in certain aspects such as odour, taste or texture. These tests are done frequently and are used for the more routine tasks in food product development, like production quality testing, value improvement programs (VIPs) or calamity management. The three most often used ones are the "triangle test", "duo-trio test "and "paired comparison test".

TRIANGLE TEST

A triangle test is one of the easier sensory evaluation methods and focusses on the identification of sensory differences between two products. It is often used in food product development to test the impact of changes in ingredients, processing or packaging on the overall sensory properties of the product, like calamities and VIPs (see Chapter 3). In a triangle test, each assessor is presented with three samples, two of which are identical and one of which is different. The assessor must identify the odd sample and describe the difference perceived (Campden BRI, n.d.). The number of correct answers and comments are then statistically analysed to verify whether the difference is significant or not.

The number of assessors needed for a triangle test depends on the level of significance, the power and the expected proportion of correct responses for the test. The level of significance is the probability of rejecting the null hypothesis when it is true, and the power is the probability of rejecting the null hypothesis when it is false. The expected proportion of correct responses is the proportion of assessors who would correctly identify the odd sample if there were a difference between the products. You can use statistical tables or formulas to calculate the minimum number of assessors required for a given level of significance, power and expected proportion of correct responses. For example, according to ISO 4120:2021, if you want to use a level of significance of 0.05, a power of 0.90 and an expected proportion of correct responses of 0.50, one will need at least 24 assessors for a triangle test.

DUO-TRIO TEST

A duo-trio test is a sensory method that is used to determine whether there is a perceptible difference between two food products. It is a type of difference test that is often used in product development and improvement. In a duo-trio test, the testers receive three coded samples, with one of them identified as the reference. They are then asked to indicate which of the other two samples matches the reference. They are not asked about any specific attribute differences.

This test is fast and simple: First, two food products that are suspected to have a sensory difference are prepared and coded with random three-digit numbers. Then reference sample is randomly selected from one of the two products and given to the testers along with its code. Then the testers are also given the two other samples, one from each product, without revealing their codes or identities. The testers are instructed to taste the reference sample first and then the other two samples, and to indicate which of the two samples matches the reference sample in terms of overall sensory impression. They normally are not asked to identify or describe the difference between the products, only to select the matching sample.

PAIRED PRODUCT TEST

A paired product test is a type of sensory evaluation method that involves presenting two samples of a food product to a panel and asking them to compare them based on a specific attribute, such as taste, texture, appearance or aroma. The purpose of this test is to determine whether there is a significant difference between the samples in terms of preference or acceptance. Paired product tests are used for various applications, such as product development, quality control, shelf-life testing or marketing research.

All three test designs are inexpensive, can be set up quickly set up, are easy to perform and do not require too much training from the assessors. These tests, like others, need to be controlled for the influence of external factors, such as the order of presentation, the number of samples, the environmental conditions or the tester's mood, sensorial capabilities and motivation. This can be controlled by randomizing the presentation order of presentation and usage of a balanced designs to avoid order effects and biases.

The biggest limitations are that they do not reflect the actual consumer behaviour or preference; they just detect differences, which doesn't say much about consumer liking or disliking of products. Similarly, they do not account for the variability among testers, such as their sensory acuity, experience or expectations. These tests are therefore very suitable for quality control in production or "value improvement projects", where the key objective is to "match current". Implicitly, all these tests assume that the consumer likes the original and now changes are needed. That's why this difference testing can deliver quickly valuable insights. The task is to spot a difference between samples. Same means success.

In practice, appropriate sample sizes and statistical methods should give sufficient power and accuracy. And the good control of the environmental conditions during testing further increases the validity. The training of the participants better also covers the necessity to avoid any conversation or "opinion exchange" during the testing. Once all results are collected, the valuable opinion exchange can start. This discipline is sometimes difficult to maintain. When tasting the product samples, a leader lead may be tempted to convey early message about sensorial parameters. Or in factory QA tasting the responsible production lead may be tempted to convince the panel that all samples are "full in specification". Similarly in a calamity activity, when a material is "out of stock", the procurement representative desperately trying to find replacement material would be very relieved to find his replacement sample indistinguishable. And in VIP, procurement and finance would only be too happy to see the cheapest alternative as the most promising solution, while on the other hand, the marketing colleagues of a brand is in tendency critical about any changes to products of his brand.

When it comes to truly new food product development, more sophisticated product evaluations are needed to develop the standard to be matched, to be successful in the market. Trained human panels are needed than to translate the elements of a sensorial impression, particularly for taste and smell, into actionable product development briefs.

HUMAN PANEL

Technically the easiest sensory evaluation is by a human panel. In many companies these are the people that have experience with the product category, are familiar with the necessary quality and are active in the new product development. If the development department is located at the manufacturing factory, a mix of experienced people from R&D, QA and production often take the task. In this situation that can be the same team that is controlling the running production by sample testing. In cases, where the R&D department is located outside the factory, maybe because it's a centre developing product for various sourcing units in a globally, the job is often done by QA alone.

It's advisable to test and perhaps specifically train the internal panellists. Not everybody has the same sensorial capabilities. Some people are so called "super tasters" for a particular taste dimension; others can better memorize a taste or aroma. The composition of the panel doesn't need to be balanced, democratic or hierarchical! The R&D leader or the factory managers can be part of the team if they qualify and not because they are the ones that will be held accountable for the quality later.

Training a sensory panel involves selecting and training individuals to evaluate the sensory properties of food products. The process follows the typical scheme:

- *Selection*: Panellists are chosen based on their ability to perceive and describe sensory attributes, as well as their availability and willingness to participate in the training and the routine testing sessions. It can be beneficial to include suitable people form the sales department, but if those individuals are often skipping the testing, because they are busy "in the market", the potential benefit of having people in the sales organization that do know the product they have to sell can turn into an impractical difficulty.

- *Screening*: Panellists are screened for their sensory acuity, using tests such as odour and taste recognition, and their ability to differentiate between similar stimuli. Often, the target taste or odour is presented in various concentrations, so that a threshold level can be identified.
- *Training*: The panellists will be trained to use standardized sensory evaluation methods and terminology and to evaluate products objectively and consistently. Particularly if more complex sensorial parameters need to be verbalized, it's crucial that the panellist share a commonly understandable vocabulary. Mouthfeel may not be texture, sour is not astringent and meaty not umami! In cases where the test only comprises a silent numerical scoring of the individuals without important verbal exchange, this may be less critical. For R&D in product development, the possibility to articulate the sensorial impressions very well is very important, as this discussion guides incremental improvement steps.
- *Calibration*: The different panellists need to be "calibrated", so they are using the same criteria and standards when evaluating products. Particularly, when numerical scoring is done, the setting of common standards is essential. On a scale from 1 to 5 the panellist should be able to place a given saltiness on a comparable level.
- *Monitoring*: Chosen panellists are monitored over time to ensure that they continue to evaluate products consistently and accurately. Things can change, and sensitivity can go up or down. A simple viral infection can have lasting impact. That's been known since the COVID pandemic and the frequently reported loss of smell.

The specific methods used for training a sensory panel can vary depending on the type of product being evaluated and the goals of the sensory evaluation program. Some common training methods include descriptive analysis, where panellists are trained to identify and describe specific sensory attributes of a product, and difference testing, where panellists are trained to detect differences between products.

PANEL SELECTION

A taste panel can be recruited from within the company or from outside – that could be students from universities, respondents from a media invitation or consumers recruited by an agency specializing in these tasks. Some food companies use their own employees for such panels. That's probably the simplest solution, but bears a bias risk, since employees recognize their "own" products and judge them subjectively. Results of such a test with an "internal panel" also often cannot be used for "activation", advertising or claim support. The regulating authorities would not accept on-pack or media statements about a product superiority if the panel is from the company.

Here it helps to discuss early in the project what kind of "story telling" or "claim making" is planned to support the product launch. In case the launch activation includes the idea to create evidence for "superior taste", "preferred by x% consumers", etc., the independence and quality of the judging must be ensured. For the routine tasks like shelf-life test, production sampling or competitor monitoring, an internally trained panel is an excellent solution.

The candidate screening starts with the questionnaire and tasting exercises. Standard protocols recommend to screen at least two times as many candidates as finally needed (DLG, n.d.). The selection criteria are a combination of physiological and personality-related attributes, including overall health status; allergies; availability, personality; verbal creativity; concentration; motivation; smoker; dietary habits; education; sensitivity; specific anosmia (loss of smell); previous experience; and medication.

The screening starts with simple exercises to assess the sensory acuity of the individuals for the five basic tastes: Sweet, sour, bitter, salty and umami. For this purpose, five clear aqueous solutions containing a dilution of tartaric acid, quinine sulphate, sucrose, sodium chloride and MSG are used for that. This exercise drives the final selection of individuals for the panel.

People's abilities to sense basic tastes and olfactory signals differ hugely. The taste receptors in the oral cavity are dedicated protein structures detecting chemical substances and transmit this as neuronal signals to the brain. But in contrast to, for example, pain receptors, all taste receptors are up- and down-regulated, so that their sensitivity can change depending on various factors, such as signal intensity diet, inflammation or disease. People reducing their salt intake will after a while have a similar saltiness perception than before on higher salt intake levels. Not all could sense them, particularly at lower concentrations. A few others are so called "supertasters." A genetic variation affects their taste receptors and allows then to detect taste at very low levels. Often, these "supertasters" have more receptor structures to detect bitter compounds.

Having supertasters in the team is useful. They also react during concentration, where "normal" tasters would sense any difference. The "fortification" of food with iron, in many countries a much-wanted nutritional enrichment, can bring "metallic-bitterness". Having a supertaster here can prevent a product launch with iron levels that some consumers can perceive (see also the case studies in Chapter 16).

SENSORIAL TESTS

In all stages of a new food product development sensorial inspection and tasting are essential. Beginning with the material inspection, production sample evaluation and shelf-life testing, ingredients, semi-finished goods, like pre-mixes or finished products need to be controlled for quality. These are routine processes performed by experienced and preferably trained people in the company. Essentially, everybody in the company could be part of that test-and-taste panel. The idea is to identify the difference between the reference sample and the actual sample.

Precondition for the members is that they have experience with the reference standards, the final product or key ingredients, so that they have a reference in mind to which they can compare the actual sample. Normally, the different routine sensorial tests have a standing panel performing the job routinely.

Production samples are taken after a fixed schedule, maybe three times a day, one per shift and line, etc., and are checked by member of the production team and people from the quality function. These people are based in the factory or sourcing unit so, organizing once-a-day sample testing is no major planning effort. If R&D is also based in the factory, they are often part of the team. In many food companies R&D is the "owner" of all specifications, including the finished good specification. They are therefore the ones understanding the product design best, and they have agreed to the specification, the tolerances and the minor and major defects in standardized protocols during handover of the product to production.

In other situations like the innovation process, R&D and marketing may perform the testing sessions together. Both functions are key in the new product development – they have seen the prototypes, agreed to the key attributes in product development briefs and organize and observe the flanking consumer research. If not too complicated from an organizational perspective, it's good to include factory members. They then have an early option to discuss and understand the key attributes of the product and consider opportunities and treats in the upscaling to the commercial production.

From time to time products or semi-finished products like pre-mixes are bought from third-party manufacturers. Such commercial relationships are often handled by the procurement function in a company. And in such situations, it is helpful to have them as also in the testing, so they are directly involved in the evaluation and can react better towards the supplier. For the evaluation of bought in raw materials, the combination of material experts from R&D and procurement, sometimes called "material expert teams", is the preferred option.

Most important is that the specific team can perform the task on a routine basis. In many bigger food companies not all functional departments are at the same location. Head offices are in the city centre or a well-connected business district, while the manufacturing unit is on the outskirts or on the countryside, where material logistics and infrastructure are more suitable. Sometimes R&D departments are not even in the same region or country, or the product is made in a country different from the one where it is going to market.

Pragmatic solutions are needed here. Panellists should have sensorial experience with the product tested and should be able to pick out differences. Preferably, they have been checked and trained for their sensorial abilities, in particular taste and aroma, before the testing session. And they can follow a repeatable scorable procedure. Especially in the routine tastings, like the shelf-life tests, participants should have standard forms, digital or on paper, where they can score a sensorial impression like sweetness, lemon or crunchiness on a given scale, with 1–5, 0–10 or something different, but always the same. A too sophisticated scale does not increase accuracy, as panellists may not have the ability to score confidently minor differences in a sensorial impression on a 10-point scale. But it is important to pre-define how long a product is "in specification" and can be sold and when that is no longer the case.

In all those cases trained human professionals are the judges. Sometimes they are "just" trained on the job, as they have known the product for a long time and know how it is specified; this can be experienced members of the factory team. Or panellists have been screened, tested and trained, so they able to articulate their impression in a more descriptive way.

For a formal training, a couple of methods have been described that will be briefly described now.

If a reliable validation of product samples is needed, one may need 100 or more subjects. The product testing as such should happen in a neutral location. Some food companies have built up their own research facilities with individual sensory booths, where the data are directly recorded on a computerized data collection system The sample presentation protocol describes how the samples are presented, how long they are tasted and how long the break between the samples should last to prevent assessor fatigue and allow sensory receptors to recover.

Descriptive sensory analyses differ from other sensory testing methods in that it seeks to profile a product on all its perceived sensory characteristics. In this chapter, the process of implementing a descriptive sensory program will be reviewed, with some discussion of new approaches and applications. Variations of descriptive sensory analysis will also be considered, including The Flavour Profile Method™, Texture Profile Method™, Quantitative Descriptive Analysis™, Quantitative Flavour Profiling, Spectrum™ method and Free-Choice Profiling.

SENSORY ANALYSIS

Sensory science is a younger discipline and tries to understand how people perceive and respond to foods and beverages in a variety of settings. Its applied science used to build competitive advantage in industrial food production by using accurate and reliable measurement of sensory characteristics to design successful products. That's done by observational and experimental studies to come to improved sensory and consumer methods to create better.

Having good access to competitive sensory science in food is a critical success factor for food product development. Attributes like taste, aroma, aftertaste, satiation and fullness, but also packaging and branding need to be optimized on their own and in combination. Among the sensorial impressions, smell is the most powerful stimuli triggering memories almost instantaneously. The sensation travels through the nose, passes by the amygdala, where emotions are created and comes to the hippocampus, where the memories are stored, before they reach "higher-order structures" (Levitan, 2022). Or in other words: Smell is noticed by memory and emotion before its consciously registered.

For packaged food, that is a clear limitation. If the packaging should fulfil one of its key functions to protect the food from external challenges, it cannot pass food smell through the protective barriers. Sometimes attempts are made to odourize the packaging. But that can easily become a two-edged sword, as consumers can understand that as cheating and manipulating, when they consciously get aware of that.

Another big challenge in today's packaged product world is the reduction of key components like sugar, fat or salt. Healthier alternatives often fail to deliver the full trigger package of the original. And as these ingredients often form a major part of the formulation, modifying one of them influences many other aspects in the food matrix.

Good sensory analysis can help to identify and articulate the lead triggers of delight; it could be a particular flavour, taste, aroma, a special texture or mouthfeel or just a cracking noise.

FLAVOUR LEXICON

The aroma of food is undoubtedly one of the important sensory impressions and directly influences consumer choice and acceptance. The combination of taste and smell triggered by hundreds of molecules and the verbal exchange about these impression requires a common language, where the personal perception become aligned and described with common vocabulary. The application of a "common flavour language" makes it possible to describe the sensory qualities of foods much more precisely – an indispensable help for food professionals to develop, evaluate and improve their products. And researchers can use it to study the physiological, psychological and cultural factors that influence flavour perception and preference.

The start is a flavour lexicon – basically, a set of words to describe the flavour. In the past the so-called "ASTM lexicon" was the starting point (ASTM, n.d.). ASTM, known as "American Society for Testing and Materials", develops international standards and publishes them for voluntary use as a consensus. ASTM helps sensory professionals to describe the aroma, flavour, texture and appearance of products in a consistent, scientific and non-subjective way. The lexicon includes product examples for each term to illustrate its meaning and usage.

Most work on the vocabulary list is done by trained panellists as both qualitative and quantitative approaches. Reference lexica have been previously developed, but often become modified for the specific task or food product category. The most suitable "vocabulary" is shaped by a product type, sample, test location, culture and language. It is not carved in stone, but dynamic and constantly evolving, optimized for the specific purpose. The language should be differentiating in critical attribute areas, while elsewhere a detailed differentiation of the vocabulary is not adding value.

Wine tasters often describe colour, clarity, aroma, taste and mouthfeel of wine. They then use terms like dry, fruity, oaky, tannic and acidic. For fruity notes they quickly go deeper and speak about berry notes, citrus, apple, peach, apricot, fig, raisin, etc., then add floral notes like rose, jasmine, lavender and violet. For red wine, darker, more spicy notes are relevant, so the tasters may describe pepper, clove, cinnamon, nutmeg and ginger. And to express the new notes coming from wine aging in oak, terms like vanilla, coconut and caramel are used. These dimensions are often less relevant for white wine, while here herbal characterizations like grass, mint, eucalyptus, sage and thyme are useful for differentiation. But both can show earthy notes like mushroom, truffle, wet stone, chalk and graphite.

And a panel evaluating chocolate will use a part of the wine vocabulary like spicy, nutty, sweet, caramel, vanilla, coconut, etc. But they would also need words to express but smoky, minty coffee, salted, dark, milk, white and ruby marbled impressions.

So different product groups need different languages, and enterprises must make efforts to find capable individuals, allow them to build a common language and keep them trained and accessible.

And new products require new training; the panel must "recalibrate" their vocabulary with that a lively sensorial description for the new product brief to support product development.

The flavour profile method and the sensorial attribute screening are two techniques that are widely used to evaluate the sensory characteristics of food products.

DESCRIPTIVE SENSORY ANALYSIS

The "descriptive sensory analysis" aims to generate quantitative data describing the similarities and differences among a set of products. The flavour language is the tool to profile all perceivable sensory attributes in food. Several variants exist today like the Profile Method™, Texture Profile Method™, Quantitative Descriptive Analysis™, Quantitative Flavour Profiling, Spectrum™ method and Free-Choice Profiling (Murray et al., 2001).

FLAVOUR PROFILE METHOD

One of the oldest sensorial profiling protocols is the "Flavour Profile Method" introduced by Arthur D. Little after the Second World War (Lawless, 2012). Based on the acquired flavour language, the individual rates quality and intensity of different food attributes such as appearance, aroma, taste, texture and aftertaste, with the help of a descriptive scale. A scale for sweetness could range from 0 (no sweetness) to 10 (very sweet). The collected data can be applied to plot a graph or a spider web chart to compare sensorial profile of similar products. The flavour profile method is useful in various stages of new food product development, starting from consumer preference testing product development, quality control, to shelf-life testing.

Descriptive analysis by a panel of trained judges or assessors is probably a powerful tool to compare sensory properties of food products relatively objectively.

SENSORIAL ATTRIBUTE SCREENING

Something different is the sensorial attribute screening. It's often used for recruitment-trained panels and tests the general sensory capabilities of candidates. It's also used with untrained consumers who indicate whether they can detect a specific attribute in a product, such as vanilla, chocolate or mint. For the development of food products, it is important to be able to describe impressions and express views about often subtle and complex sensorial attributes. For a superior product development, basic impressions like sweet, salty, fruity, bitter, sour, etc., is often insufficient.

The use of trained descriptive panels is a cost- and time-intensive task. It makes good sense for companies holding a major chair in a well-established product category. A trained panel is much better suited to articulated unmet fine nuanced opportunities in an already segmented market. The effort is not always needed to have success in new food product development.

Newer profiling methods directly use partly trained or untrained consumers to get sensory insights:

- Methods like Free-Choice Profiling (FCP) start with semi-trained panels and use open-ended questions. The consumers have not a common language, so they freely describe their sensorial impression. The task later on is to cluster those articulations and translate them into a actionable product brief.

- Polarized Sensory Positioning (PSP) normally uses untrained, naive consumers and exposes them to polarizing impressions (very strong vs. very weak).
- With Flash Profiling, individuals use their own vocabulary, which makes the method fast, but limits exactness, especially in more complex sensorial descriptions, like flavour (Kemp et al., 2018).

While these faster tests have their limitations in the details of the description of taste, flavour and texture, they often find application in general product characterization, competitor scans or more routine quality assurance, like shelf-life testing. Even a panel not trained in a common flavour or aroma language can learn quickly to adjust and align their vocabulary to each other. That means when the team can decide for themselves, like in shelf-life tests, the screening can be very reliable and effective. A difficulty will only come up if they need to express their sensorial impression to an outside person or group.

Sometimes it is necessary to start consumer research for new product development with the identification of the fundamental sensory descriptors. Sensorial attribute screening is a method of evaluating the sensory properties of a product or a material based on a predefined set of criteria. It is often used in product development, quality control or consumer research to identify the most relevant sensory attributes that differentiate products or materials. For example, sensorial attribute screening can be used to compare different types of chocolate based on their appearance, aroma, flavour, texture and aftertaste. Sensorial attribute screening can be performed by trained panels or by consumers, depending on the objective and the context of the study. The process of sensorial attribute screening involves selecting a representative sample of products or materials, defining the criteria for evaluation, designing a sensory test, collecting and analysing the data and interpreting the results.

The test can be done with trained or naïve consumers and consists of the basic steps and starts with the collection of representative samples and criteria definition. The test as such can be descriptive analysis, hedonic scaling, preference ranking or check-all-that-apply. Data are collected and statistically analysed to identify the differences and similarities among the products based on their sensory attributes. For the interpretation of the results, graphical tools, like spider plots and bar charts, are often used.

WIDER SENSORY SCIENCE

Sensory sciences are not only applied to use trained and untrained human senses to characterize the sensorial attributes of food. In the tests described earlier, the research aims to describe the sensorial attributes "objectively". That's why they apply robust statistics. The descriptive sensory evaluation provides the sensory profile of the product sensory plus a qualitative measurement of each attribute's intensity. Newer sensory techniques search for insights how these impressions of the five senses are perceived by the testers. That offers the opportunity to go beyond the characterization of a product and come to an understanding how sensory signals influence elements consumer liking and how they interact (Institute of Food Science and Technology, 2021). These are affective testing methods.

Quantitative descriptive analysis (QDA) often uses trained panels to optimize the different sensorial attributes and also in relation to each other (Marques et al., 2022). This understanding of the "higher integration" of different sensory attributes helps successful new food product development. It tries to understand the correlation between single sensory characteristics, like taste, aroma, texture and appearance from the signal uptake over storing (encoding), up to recalling them when making decisions to understand the purchase intent of consumers. And even though much effort is done to "objectivize" this individual output, they remain subjective measures and can fail to capture the full set of all sensory impressions for food products The data generated here are more "subjective". These tests with humans, whether trained ("QDA") or not ("FCP"), find their limits in the ability of the participants to fully articulate their impressions.

To compensate for that, efforts have started to use instrumental analysis to mimic human senses and make the output more objective. These approaches go beyond electronic senses like electronic noses or tongues and include biometric measurements, like facial expressions, eye tracking, heart rate and the use of virtual reality (Rodrigues et al., 2024).

Eating and drinking is perhaps the most multisensory experience humans can have. Understanding better how all the different sensorial impressions build the holistic experience is another way to successfully develop new food products.

When we consume food, our senses collaborate to create a holistic experience. The appearance of food influences our expectations and enjoyment, smell triggers flavour perception and mouthfeel (texture) can affect satisfaction. And the saying "we eat with our eyes" just captures the effect a visual cue has on our taste expectation. A particular colour raises taste expectations. Interestingly these expectations are also culturally influenced. In an experiment with a clear blue drink, testers in Taiwan expected more mint taste, while testers in the UK more expected raspberry (FoodUnfolded, n.d.). And the colour can even override aroma impression. Colour seems to be most important sensory cue when in setting aroma expectations. That is maybe one reason why parts of the food industry are reluctant to replace the strong and vibrant artificial colours with natural colourings. Weaker-coloured food and beverages are often perceived less tasty.

The optical cue can be so strong that an orange-coloured drink is perceived as orange flavour, even though it's cherry-flavoured (Spence, 2019). Basically, all sense influences the taste and flavour expectation, and an improved understanding of the results of multisensorial integration is valuable input for food development. Sometimes the colour is even used to disconfirm the aroma expectations. Black-inked pasta or counteractively coloured dishes in molecular gastronomy will probably not increase liking but creates specific experiences the consumer remembers better. They may therefore encode and retrieve relevant products better – an important aspect in the design of food packaging and wider marketing psychology.

The research in multisensorial experiences for foods is not limited to the effect of the human senses. Today, we have solid evidence has been created on the influence of extrinsic sensory factors, like packaging, cutlery, background music, light, temperature or health claims on food choice (Wang et al., 2019).

For a holistic food product design, it can be helpful to understand the dynamics between all these multisensorial experiences: Which sensory inputs lead to a unified experience (yellow and lemon aroma), which are more dominant, building hierarchy (yellow colour and cherry aroma) and which ones influence each other (green colour and environmentally friendly).

Often simpler and more objective are machine and technology-based attribute testing methods, but here the accurate measurement is not linked to better consumer liking or increased sales.

ANALYTICAL TEST OPTION LIMITS

The technical food sensory analysis deals with the characterization and evaluation of the sensory properties of food and food products. They can complement and replace human testing systems.

Basically physical, chemical or biological testing aims at maximum sensitivity, trying to detect substances at as low as possible limits. Sometimes it said that the final goal is to measure concentrations down to zero. But since all detection systems are based on the sensitivity of instrument-based signals and their sensitivity, measures can never reach "true zero". Or: Absence of evidence is not evidence of absence!

But for the detection of all those substances that trigger the human senses, the sensitivity of technical equipment is no limitation. E-noses nowadays have 100 or more different sensors and can detect thousands of volatile components. The human nose seems able to detect up to a million scents (Bushdid et al., 2020). But closing that "differentiating" gap is not the key challenge for machine-bound solutions. The biological "concept" of flavour works on the integration of taste, smell, oral, optical and maybe even audio inputs. Its shaped by cognitive descriptions, the active representation of words we learned and then often it modulated by attention, past experiences and emotions or simply the level of hunger (Rolls, 2016). The attempts to objectively and more precisely detect flavours can lead to excellent detection of numerous volatile molecules, many of them in very low quantities, but even the full capture of a combination of sensor signals may not represent the flavour impression in someone's brain. Still these equipment's have found usage spaces in which they reliably detect sensorial leads.

ELECTRONIC NOSE AND TONGUE

Closest to the idea to complement human sensorial analysis by analytical equipment are probably the electronic nose (e-nose) and electronic tongue (e-tongue) (Wang & Chen, 2023). An e-nose is a gas sensor array with much higher sensitivity. The e-tongue is the technical relative but has chemical sensor arrays instead. The working principle of a modern versions is based on the delivery of the sample and the exposure to appropriate sensors.

The material of a sensors must react differently to different compounds in the sample. That can be metal oxide semiconductors, conducting polymers, quartz crystal microbalance or piezoelectric materials. Once the compound is detected, the sensors produce electrical signals that can be converted in digital values, and those values are then sent to a computer for further analysis, for example, through pattern recognition or comparison with a database content for identification.

We discussed the fundamental differences in the signal detection in a human nose earlier. Beside this an electronic nose is more expensive than a panel test, as it requires sophisticated hardware and software, continuous maintenance and calibration and laborious sample preparation. The sensors are often sensitive to temperature, humidity, pressure, gas velocity and vapor concentration, and therefore an electronic nose is not able to adapt to changing environmental conditions, such as temperature, humidity and background noise. For a panel, these factors can be easily adjusted leading to more consistent results.

An e-nose can detect specific volatile molecules precisely but will not capture the complexity and subtleness of human perception. In that sense a panel test can provide more nuanced and subjective feedback, as well as emotional (NIH, 2020) and cultural responses (Tan & Xu, 2020). If the wanted or unwanted aroma is the result of a more complex molecule mix, rather than a dominating component or simple mix, the analytical measures cannot predict the human sensorial impression.

For wider application, firmer standardizations are important, mainly for the standardization of the sample preparation, the specific sensor array and the data storage online libraries build from those standardized assays, like online libraries.

An electronic tongue attempts to mimic the taste perception of human taste buds. Again, a challenging competition: While we speak about the five different tastes distributed in up to 10000 taste buds on the tongue, the molecular structures and receptors show more variations. And the taste perception is also the result of the integration of more sensorial factors like visual, olfactory, sapictive (the perception of tastes) and mechanical and influenced by factors like culture, individual preferences and context.

Their sensory units are characterized by cross-selective sensory array, a signal receiving and sending device and a pattern recognition device at the end. The e-tongue lacks the specificity of the taste buds. It recognizes many different organic and inorganic compounds, based on a stored signal spectrum-profile from each sensor, so that substances in the liquid solution are recognized by their unique fingerprint.

Their main usages in the food and beverage industry is the safety and quality control in food and beverage industries, where consistent taste profiles and the absence of detrimental substances matter. In applications where the presence of a few molecules defines the food quality, this technology has its clear advantages. Th e-nose applications therefore are often in the food quality maintenance, In the meat industry, an electronic nose is used to detect meat spoilage or contamination by measuring the volatile organic compounds emitted by the meat, such as carbon dioxide, ammonia or ethylene. These gases are produced by the microbial growth and enzymatic reactions that occur in meat as it ages and rots. Future work on new sensor development could widen applications internationally (Wasilewski et al., 2020).

Beside the DNA analysis, the e-nose and e-tongue can be used for food authenticity assessment, for example, to find food adulterations (Peris & Escuder-Gilabert, 2016).

OTHER ANALYTICAL TOOLS TO DEVELOP AND TEST FOOD

Beside the standard analytical equipment like in a food developer's laboratory – water activity meter, moisture balance, pH meter, refractometer, colorimeter or a texture analyser, to measure like hardness, chewiness and crispness – some more sophisticated analytical tools are used in food analysis.

At the top-line level, they are separated by the physical-chemical principles applied. Chromatography means separating components out of a mixture based on their different affinities for a stationary phase and a mobile phase. That allows scientists to analyse composition, purity and quality of food ingredients – those wanted as well as those unwanted. Popular variants are discussed in Table 10.2.

More sophisticated chromatography combines different principles and thereby increases the resolution.

A different physical approach for material investigation are the spectroscopic methods. They are widely used in foods. The interaction of emitted electromagnetic radiation with the food sample allows one to identify and quantify constituents of food. Spectroscopy helps to provide information about chemical

TABLE 10.2 Analytical tools based on chromatography

CHROMATOGRAPHY TYPE	*PRINCIPLE*	*TYPICAL APPLICATION*
Ion-exchange chromatography (IEC)	Separates based on electrical charge	Vitamin C content or spoilage Indicators in food and beverages, such as pyruvic acid and lactose in milk
High-performance liquid chromatography (HPLC)	Separates on the basis of polarity	Used to detect artificial sweeteners, colorants and preservatives
Gas chromatography (GC)	Separation principle is volatility	Determination of proteins, vitamins, amino acids, fats and flavour compounds
Liquid chromatography-mass spectrometry (LC-MS)	Combines HPLC and mass spectrometry to identify and quantify molecules based on their mass-to-charge ratio	Used to detect contaminants, such as pesticides, hormones and antibiotics
Size-exclusion chromatography (SEC)	The separation in based on size	Used to analyse molecular weight distribution of polymers, such as starch, cellulose and pectin
Affinity chromatography (AC)	Uses the relative level of retention of substances through interaction with an immobilized matrix	Used to isolate specific biomolecules from food samples, such as enzymes, antibodies and hormones
Thin-layer chromatography (TLC)	Uses different absorption rates	Used to screen for compounds like sugars, amino acids, lipids and alkaloids

TABLE 10.3 Analytical tools based on spectroscopy

SPECTROSCOPY TYPE	*PRINCIPLE*	*APPLICATION*
NMR (nuclear magnetic resonance)	Uses radio waves to identify the structure of components through influence of the spin state, without prior purification	NMR methodology can explore various aspects of foods, including fraud detection, safety, authenticity, traceability, geographical and botanical origin, farming methods, food processing, maturation, and ageing.
Raman spectroscopy	Uses a laser beam to induce vibrations in the molecules of the food sample, which then release light in different frequencies	Used to measure food quality parameters, such as moisture, fat, protein, sugar, starch and fibre content
IR (infrared)	Stimulates vibrations in the molecules in the food sample and is used to identify the functional groups and chemical bonds	Organic molecules, such as aldehydes, ketones, esters, ethers, and amides or food contaminants
UV (ultraviolet) spectroscopy	Detects the electronic transitions in the molecules of the food sample	Used to search for compounds having conjugated double bonds or aromatic rings, such as carotenoids, flavonoids and anthocyanins
Visible spectroscopy	Uses normal light to cause such electronic transitions and is used to find colour pigments	Measures chlorophylls, carotenoids, anthocyanins and caramel; used to determine freshness and shelf-life
Fluorescence spectroscopy	Uses high-energy light and allows detection of fluorescent structures	Works with, for example, vitamins, amino acids, proteins, lipids and enzymes
X-ray–based methods	Creates spectra of different intensities and shows higher internal structures and composition of the food sample at different scales	Used to analyse density, porosity, texture, crystallinity or presence of foreign objects

composition, physical characteristics and sensory attributes, up to authenticity, traceability, quality and safety. Commonly used ones are described in Table 10.3.

But these methods need the application of statistical and chemometric approaches, such as pre-processing, exploratory analysis, variable selection, regression, classification and data fusion to capture spectral features, reduce noise and dimensionality and enhance data visualization and may help to build predictive models or integrate data from multiple sources. Combination or sequential use of two of those methods can be used for even more sophisticated tasks.

Increasingly important is the detection of genetically modified organisms (GMOs). Traditionally the application in foods is the genetic modification of crops to increase their robustness, creating better resistance to pests, common crops diseases or other environmental stress, like salt or draught. In a sense the aim is to have more crop, either improving yields on existing fields or by creating possibilities to create farmland on unproductive land.

Polymerase chain reaction (PCR) is the technique to detect and quantify GMO in foods. By using specific primers that are binding to the "new gene" or a regulating sequence of such genes, the DNA becomes amplified in a controlled cycle of reaction by the polymerase. If done carefully, the methods allow not only the detection of very small amounts of such foreign material but also the quantification of the suspected material (Marmiroli et al., 2008).

Instead of isolating, purifying and amplifying the DNA in a food sample, one can aim to detect the protein that is produced from the GMO gene. The enzyme-linked immunosorbent assay (ELISA) uses specific antibodies that binding to such protein copied from the GMO gene. ELISA testing is simpler and cheaper than PCR testing, but it may have lower sensitivity and specificity.

Depending on the food and components an enterprise is interested in, a wide spectrum of highly sensitive equipment is available, but it does require a substantial investment and a continuously high level of maintenance, a group of highly qualified analytical specialist staff and a critical mass in the lab. For a food company, that investment only makes sense when sophisticated analytical tests are critical on a routine basis. Many food companies have therefore started to give their analytical tasks they have to external service providers.

Today, there are many reliable and accredited analytical labs that can perform those tests with accuracy and precision. The biggest and best-known ones are:

1. Eurofins Scientific is a global leader in food testing and has more than 62,000 employees and 900 laboratories in 61 countries. They provide a comprehensive range of food testing services, covering microbiology, chemistry, allergens, contaminants, GMOs, nutrition, authenticity and more. Beyond that Eurofins also provides consulting, auditing and certification services for the food industry (SGS, n.d.a).
2. SGS, the world's leading inspection, verification, testing and certification company, owns over 2500 laboratories and offices in 140 countries. They also provide a wide range of food testing services, including microbiology, chemistry, pesticides, residues, toxins, heavy metals, additives, allergens, GMOs, nutrition and more. SGS also offers inspection, auditing, certification and training services for the food sector. They also provide retail store checks (SGS, n.d.b).
3. Intertek is a global provider of quality and safety solutions for the food industry, with over 1000 laboratories and offices in 100 countries. Intertek offers a variety of food testing services, such as microbiology, chemistry, contaminants, allergens, GMOs, nutrition, shelf life, sensory analysis and more. Intertek also provides inspection, auditing, certification and consulting services for the food industry (Intertek, n.d.).
4. ALS, another globally active provider of testing and analytical services for the food and beverage industry, has over 350 laboratories in 65 countries. ALS offers a very similar range of food testing services, such as microbiology, chemistry, contaminants, allergens, GMOs, nutrition labels, organic certification and more. ALS also provides inspection, auditing and consulting services for the food sector (ALS, n.d.).
5. Mérieux NutriSciences has over 100 laboratories in 26 countries. Mérieux NutriSciences focus is on food, offering all necessary food testing services, such as microbiology, chemistry, contaminants, allergens, GMOs, nutrition, authenticity and more. Like some of the others they also offer inspection, auditing, certification or training services for the food sector. For food companies not having their own R&D team or stretching into new product groups, the company is providing product development (Merieux Nutrisciences, 2024).

Using the service of those certified experts gives extra credibility for the analytical results. If the core of a new product launch is anchored in the presence and amount of a beneficial ingredient, like a vitamin or mineral, or the absence of something, like GMO, the independently performed analysis by an accredited lab is sometimes mandatory, but in all cases adds credibility.

SHELF-LIFE TESTING

Testing of a product shelf life is one of most important tests done for food. It simply defines the length of time that a food product may be stored under known conditions without becoming unfit for use, consumption or sale. A shelf-life study assesses the food product quality and hygiene and gives food manufacturers and confidence of their product's shelf life. The test will reduce the risk of product recalls and maintain brand integrity and reputation, keeping the brand and consumers safe. The testing gives insights into the root causes of reduced shelf life and offers opportunities to improve products or processes to extend shelf life and ultimately profitability. And as an appropriate setting of the shelf life is so important for consumer safety and satisfaction, it is regulated.

On the global level, the "food code" in Codex Alimentarius provides a collection of international standards, guidelines and codes of practice to protect the health of consumers and ensure fair practices in the food trade. Organizations like the ISO, an independent, non-governmental organization, have undertaken the task to shape and publish standards related to food safety. The outcome of those expert consultations are specifications for products, services and systems to ensure safety, quality and efficiency.

The standardization builds an important cornerstone of the international trade, as it gives a framework with common language and content for all businesses and organizations globally, but it does not specify methods and tests that should be used to determine shelf life. This is left to the discretion of the organization.

It is the responsibility of food manufacturers to determine the shelf life of their products, so that they can provide safe and good-quality food. The shelf life of food is the time between the production or packaging of the product and the time when it becomes unacceptable under certain environmental conditions. That sounds simple and straightforward, but it often requires a tight and sophisticated material management in the background.

An average packed food product can contain between 10 and 25 different ingredients, all of them having a specific shelf life. Some of them may only have a few days' shelf life, others 2 years or more. And some may already be in stock for 80% of their shelf life, while others have just been brought by the supplier. Some material may be stored in good moisture barrier and low oxygen packaging at room temperature, chilled or freezing conditions.

The declared shelf life is normally set before the launch; therefore, it's important for the company management to understand the dynamics and details of the material flow and storage. The ingredients used in a prototype, first factory trial or first production need to reflect the typical age and quality used later.

The primary factors influencing shelf are exposure to light, heat, moisture, transmission of gases, mechanical stresses and contamination by microorganisms.

There is a variety of methods to determine the shelf life of foods, including sensory tests in which trained people uses the five senses to inspect the appearance, taste, colour and smell of the product. This subjective human test helps to define the sensorial quality and is done after the tests for microbial alteration and some other physicochemical tests have confirmed the food product safety. Primarily the shelf life is determined by the type of ingredients, the applied manufacturing process, the type of packaging used and the storage conditions of the final product.

Thus, the shelf-life of a product can be defined as a group of microbiological, chemical and sensory tests performed on food products that determine at what point in time is the product is no longer safe to consume and when the product has lost its quality markers. The shelf life is also a guide for the consumer about the period that food can be kept before it starts to deteriorate, assuming that the stated storage conditions have been followed. Storage conditions heavily influence the shelf-life of the product and can either be refrigerated, frozen or ambient depending on the product.

The probably most reliable shelf-life method are real-time studies, so simply storing the product under conditions like those that it will be subjected to in market and monitor its changes at regular time intervals. This method creates a very accurate estimation of the time it takes for a product to deteriorate. That's suitable for food products with short shelf life, but not a good option for products that are designed to be stable for 9 months, 1 year or even more.

Other commonly used methods can be grouped into five different groups which will be briefly presented (TUV NORD, 2021).

THE RULE OF 10

Q10 is not a term specifically related to food safety and not a shelf-life prediction tool. It is a term used in various fields, including biology, chemistry and pharmacology, to describe the temperature coefficient of a reaction or process. It's a rule in biology used to describe the change in the rate of a biological process, such as enzyme activity or metabolic rate, that occurs with a 10°C change in temperature. The basis is the paramount effect of temperature on food related reactions. In food safety, Q10 may be used to describe the effect of temperature on the growth rate of microorganisms or the rate of various chemical reactions that affect food quality. Simple online tools allow the calculation of the product specific reaction rate at a 10°C increase (WESTPAK, 2023).

ARRHENIUS MODEL

The Arrhenius equation was first proposed by Swedish chemist Svante Arrhenius in 1889. It also deals with temperature dependence of reaction rate and calculates energy of activation. In food science it is used to model the rate of chemical reactions that affect food quality, such as lipid oxidation, protein denaturation and non-enzymatic browning. By measuring the rate of these reactions at different temperatures, the Arrhenius model helps to estimate the shelf life of a product.

Like any model, the reliability depends on the quality of the data used for calibration and the underlying assumptions made. When used appropriately, the Arrhenius model can provide reliable predictions of the shelf life of food products. For "chemically" simple products, like frozen fish, it provides accurate predictions (Wu et al., 2014). The accuracy goes down in more multicomponent systems or under more extreme conditions.

In practice, food developers would not rely completely on the results of models and calculations. They would set samples away for microbial testing and evaluate mainly product quality through shelf-life tests.

SENSORIAL SHELF-LIFE TEST

With the final prototype (design lock), the question of the period during which the food keeps the intended sensorial properties is asked. The "sensorial instruments" here are experienced, and sometimes trained individuals sensing the characteristics of foods during storage are used. It is good if they are familiar with the product in the test and understand the composition. That helps them to pay attention to expected quality changes in taste, colour, texture, aroma, etc. It is better if they are more sensitive than the target group of consumers. That will ensure that the shelf-life limit is set for a moment where the consumer would not detect it.

Initial tests are the paired-comparison test, the duo-trio test, and the simple triangle test. The paired-comparison difference test simply judges two samples beside each other on the difference in a particular attribute. A disadvantage of the paired-comparison test is that attribute in questions needs to be stated. That's not easy for criteria "off-flavour", a typical factor in shelf-life test. A duo-trio compares two shelf-life samples with one standard sample on all relevant attributes. In a triangle test, panellists are asked to identify differences between products by comparing three samples: two identical and one.

Such tests do not provide any information about the type of changes; that's why more frequently sensory descriptive tests are used. Here the relevant attributes like colour, flavour and texture are established and scaled, for example, from 1 to 5. The reference sample is set a 5 for all parameters and then the shelf-life samples are tested against the reference. In an additional free text field they add comments to explain their vote. The team or the company policy has pre-agreed up to which deviation a product is still good for sale. Common statistical tests, like ANOVA, are used to evaluate the significance of the results.

The prototype used initially in a shelf-life testing should be replaced as soon as possible with the first production samples. Only the latter truly represent the commercial product in the market. The same formulation in a pilot plant or first factory trial does ensure identity. Small changes in the processing or packing or the use of different raw material batches can impact the shelf life.

CHALLENGE TEST

This method is suitable to predict the microbial safety of a product and is used to assess the ability of a food product to support the growth of all types of microorganisms. The test involves inoculating the food product with a known amount of the target microorganism and then monitoring its growth over time under controlled storage conditions. The results of the test can be used to predict the shelf life of the product and to determine whether any additional preservation measures are needed to ensure its safety and quality. Challenge tests are commonly used in the development of new food products and in the evaluation of changes to existing products, such as changes to formulation or packaging consists of experimentally introducing pathogens or microorganisms into the food during the production process, thus exposing the product to real conditions it will undergo in real conditions.

In typical challenge tests strains of *L. monocytogenes*, *C. botulinum* or specific yeast and mould strains are used to infect food product to evaluate whether these strains multiply in the tested product under the declared storage temperature. The choice of the challenger organism depends on the product to be tested, e.g. *Listeria* for dairy products, *C. botulinum* for sausage and *Aspergillus* in cold sauces and beverages.

SURVIVAL METHOD

This is the least often used method and not a standalone test for shelf life. It's a study based on the opinion of the consumer about the physical characteristics of the product. The study can help to understand the attitude of consumers towards a product with different production dates to get an idea if they would consume it. This can be combined with stricter test on the product safety to can show the relationship between the shelf life and the perceived quality of the product.

PREDICTIVE MICROBIOLOGY

The fundamental idea here is to reliably predict the behaviour of microorganisms in foods under foreseeable conditions. This means mathematical models describing the behaviour of microorganisms under established conditions. It's a type of "predictive microbiology".

This methodology studies the different microbial responses of foods to varying environmental conditions, based on mathematical and statistical models, in order to predict the behaviour of the microorganisms in the product. These models consider factors such as temperature, pH, water activity and the presence of preservatives to predict the rate of microbial growth and the time it takes for the microbial population to reach a certain level. The basic concept has been known for many years, but due to the exceptional improvements achieved in computer software modelling and simulation tool advances, the method has regained attention (Pérez-Rodríguez & Valero, 2012). The different types of predictive models allow predicting growth, inactivation and probability of growth of bacteria in foods under a variety of environmental conditions and can include additional confounding factors like the interaction with other microorganisms. A selection of modelling tools can be found on websites of relevant authorities, like UC Davis, and are free to use (UC Food Safety, 2024).

The most advanced prediction tools start with so-called "primary-level models", which describe the change in microbial numbers over time, and combine those with "secondary-level models", which show the microbial growth changes under the influence of more environmental factors, such as pH, temperature or the water activity (a/w), which is a major factor influencing microbial growth. A popular first level tool is "GInaFiT", a freeware add-in for Microsoft Excel established to bridge the gap between developers creating predictive modelling approaches and end users in the food industry (Geeraerd et al., 2005). "Tertiary-level models" use that information to generate graphs, predictions and comparisons (Eissa, 2012).

But microbes are not the only factor heavily impacting shelf life. Flavour, aroma and colour changes due to oxidation, irradiation, lipolysis and heat, induced by light, metal ions or excessive heat during processing or storage dramatically change the product quality. Visible deteriorations are increased viscosity, gelation, sedimentation and colour change. Hydrolysis, lipolysis and proteolysis can cause changes in flavour, texture and vitamin content and bring up bitter taste.

Biochemical reactions can lead to lipid oxidation, which can result in loss of essential fatty acids and increase of rancidity, non-enzymatic and enzymatic browning reactions affecting colour, taste and aroma and nutrients and can even end in the formation of toxicologically suspect compounds, like acrylamide. More physical changes include separation, gelation, sedimentation, etc.

ACCELERATED SHELF-LIFE TEST (ASLT)

Many ambient stable food products can have shelf life between 1 and 2 years. The option to wait for a full shelf-life test after the first production sample has been made would delay a launch so much that the market opportunity might have gone. The accelerated shelf-life test offers a reliable alternative.

In an accelerated stability test, the samples are stored at elevated stress conditions, often higher temperature, oxygen pressure humidity and light exposure. Degradation at the recommended storage conditions can be predicted using known relationships between the acceleration factor and the degradation rate. Temperature is often the main acceleration factor due to the fundamental relation with the speed of chemical reactions causing degradation, as characterized by the Arrhenius equation.

The study at exaggerated temperature and humidity conditions over a predefined duration allows predictions on the shelf life of the product at the planned market and distribution conditions and provides a good idea about the expiration date. The exact conditions used for ASLT should be chosen based on product stability characteristics, typical climatic conditions during distribution and retail storage. Chocolate developers will set the ASLT for dark chocolate at a higher temperature than for milk chocolate, as the fat melts earlier here (Subramaniam, 2009).

Company experience with the product format, a good understanding of the real storage and distribution conditions or the use of Arrhenius-type calculation then allows a reliable prediction of the real shelf life. The key benefit of ASLT is speed to market in cases where the product has a long shelf life.

However, ASLT is normally accompanied by regular shelf-life tests running over the full shelf life. By comparing the quality attributes of the different time points in both tests, correlations can be optimized to further improve the future shelf-life testing. And the regular samples can be used as reference samples in case consumer complaints appear post launch. The comparison with the stored samples is instrumental in finding the root cause of a quality deterioration, which can happen long after the product has left the factory warehouse.

In the product development phase products are often already stored in their final packaging. That can test the key functions of the packaging to protect, and it may give early signs of interaction between the product and the packaging under harsh conditions.

Panel testing during shelf-life test can also be used to evaluate fortification stability of food products with sensitive nutrients. A well-trained panel is at least as sensitive as other expensive analytical test, for example, the stability of omega-3 and omega-6 fatty acids (Sobel, 2014).

In the future, AI will probably play a more important role in food shelf-life prediction (Chhetri, 2023). Many of the current assessment tools, especially in microbiology, are based on mathematical algorithms and understanding of the effects of external factors. Incorporating robust data about food constituent stability, like nutrients, the effect of various environmental factors and protection schemes and the relevant interactions between them, opens the door for a reliable shelf-life prediction "*in silico*", saving food enterprises substantial time and efforts, once fully accepted by the authorities.

CASE STUDY: FAT LOWERING

A company in Asia active in the mayonnaise and dressings segment had developed a superior solution to produce a fat-based dressing with less oil but the same sensorial attributes, like taste, smell and mouthfeel than the market-leading full-fat variant with 74% vegetable oil. The project was very attractive as it promised material cost savings on one side due to the reduced amount of the expensive oil. And it was on trend, with a consumer need for "more regret-free indulgence", so same great taste, but fewer calories.

The legally valid "standards of identity" for mayonnaise demanded a minimum of weight based 65% vegetable oil. Technically, even lower-fat products would have been possible, but the company decided to keep the standard to be able to call it mayonnaise and not confuse the loyal users.

A typical mayonnaise is made from oil, egg yolks, vinegar or lemon juice plus seasonings. With the help of a dedicated software tools the R&D team was able to quickly create a new formulation where the gap from the reduced oil amount was closed with new ingredients, keeping the typical taste and aroma at a little lower pH, while being safe. The prototypes were successfully tested with consumers, so the products of the factory trial went into ambient shelf-life test. As the company was offering their mayonnaise in a large variety of formats ranging from single-use small 15-g sachets to 500-mL Doy packs in a big glass container, it was decided to put the glass container into full shelf life testing, as this "look \[through" transparent format was seen as most sensitive in terms of the critical emulsion stability.

A few months after successful launch into the market, the company's care line started to receive the first complaints from loyal consumers about a "strange" off-taste. While there is always a certain level of "noise" after a new product launch, here the reports about a "weak bitter" taste, often coming from loyal users of the brand, was worrying. The company, all trained experts in R&D, quality and supply chain collected samples back from the market and compared them with their reference samples in production. By testing samples at different ages, from various locations and in all different formats, they hoped to trace back the origin of the consumer reaction. Initially, results were confusing! In some tests, a few samples indeed had an unusual note, while others, with the same production date and same formulation, had not. A wider investigation into the distribution network and shelf-life conditions did not also provide an explanation. Only after tasting a statistically significant number of samples through the trained panel did a pattern appear: A small portion of the early launch volumes produced in single-serve laminated pouches had that taste. the glass sample did not show it, and even the bigger volume pouches, made from the same of the same PE-Alu-PE laminate, were inconspicuous.

And a bit later the hypothesis emerged that the very slight lowering of the pH influenced the small pouch integrity, eventually causing a bit of delamination on the inner pouch structure, so that "bitter" super tasters would detect a taste difference. note. This effect only happened in a situation where the ratio between packaging surface and product volume was high.

This was not a safety concern, but was a nonacceptable quality issue, potentially damaging product and brand reputation. Hence, the product was quickly adapted, and the old small-volume samples were replaced.

REFERENCES

Abdullahi, N., & Dandago, M. A. (2021). Hurdle Technology: Principles and Recent Applications in Foods. *Indonesian Food and Nutrition Progress, 17*(1), 6. https://doi.org/10.22146/ifnp.52552

ALS. (n.d.). *Food safety & quality testing services.* https://www.alsglobal.com/en/food-safety-and-quality-testing

AOAC International. (n.d.). *About AOAC International.* Retrieved June 13, 2024, from https://www.aoac.org/about-aoac-international/

ASQ. (n.d.a). *Quality-resources/quality-assurance-vs-control.* Retrieved June 12, 2024, from https://asq.org/quality-resources/quality-assurance-vs-control#:~:text=Quality%20control%20can%20be%20defined%20as%20%22part%20of,is%20more%20the%20inspection%20aspect%20of%20quality

ASQ. (n.d.b). *Quality-resources/quality-glossary.* https://asq.org/quality-resources/quality-glossary/q

ASTM. (n.d.). *Sensory evaluation standards - Standards products - Standards & Publications - Products & services.* https://www.astm.org/products-services/standards-and-publications/standards/sensory-evaluation-standards.html

Bigi, F., Maurizzi, E., Quartieri, A., De Leo, R., Gullo, M., & Pulvirenti, A. (2023). Non-Thermal Techniques and the "Hurdle" Approach: How Is Food Technology Evolving? *Trends in Food Science & Technology, 132*, 11–39. https://doi.org/10.1016/j.tifs.2022.12.015

Bushdid, C., Magnasco, M. O., Vosshall, L. B., & Keller, A. (2020). Humans can Discriminate More than 1 Trillion Olfactory Stimuli. *Science.* https://doi.org/10.1126/science.1249168

Campden BRI. (n.d.). *Sensory triangle testing, discrimination test at Campden BRI.* https://www.campdenbri.co.uk/videos/sensory-triangle-testing.php#:~:text=The%20triangle%20test%20is%20principally%20used%20to%20determine,had%20an%20impact%20on%20it%27s%20overall%20sensory%20properties

Chapple, A. (2019). *New test to combat buffalo mozzarella fraud uncovers mislabelled products.* Quadram Institute. https://quadram.ac.uk/buffalo-mozzarella-test/

Chhetri, K. B. (2023). Applications of Artificial Intelligence and Machine Learning in Food Quality Control and Safety Assessment. *Food Engineering Reviews, 16*(1), 1–21. https://doi.org/10.1007/s12393-023-09363-1

Creydt, M., & Fischer, M. (2018). Omics Approaches for Food Authentication. *Electrophoresis, 39*(13), 1569–1581. https://doi.org/10.1002/elps.201800004

DLG. (n.d.). *Practice guide for sensory panel training Part 1: Requirements in accordance with DIN EN ISO 8586:2014-05.* https://www.dlg.org/en/food/topics/dlg-expert-reports/sensory-technology/practice-guide-for-sensory-panel-training-part-1

ECHA. (2022). Member state committee support document for identification of melamine. In *SVHC Support Document - Melamine.* https://echa.europa.eu/documents/10162/49f9ee64-f18d-778a-3052-1a292a3f32a0

Eissa, A. H. A. (2012). *Trends in vital food and control engineering.* InTechOpen. https://doi.org/10.5772/2354

EUR-Lex. (n.d.). Regulation - 178/2002 - EN - EUR-LEX. https://eur-lex.europa.eu/legal-content/EN/TXT/?uri=CELEX:32002R0178

Evans-Lara, A. (2023, May 17). *Revised CCP decision Tree adopted by Codex.* HACCP Mentor. https://haccpmentor.com/revised-ccp-decision-tree-adopted-by-codex/

FICSI. (2023, October 5). ISO 22000 vs. FSSC 22000: Understanding the Key Differences in Food Safety Standards - FICSI. *Ficsi Blog - Ficsi Blog.* https://www.ficsi.in/blog/iso-22000-vs-fssc-22000-understanding-the-key-differences-in-food-safety-standards/

Food Authenticity. (2024). *The-UK-food-authenticity-programme.* https://www.foodauthenticity.global/the-uk-food-authenticity-programme

Food Authenticity. (n.d.). *Locate methods for authenticity testing.* https://www.foodauthenticity.global/methods

FoodDocs. (2023). *Food-safety-compliance-software.* https://www.fooddocs.com/food-safety-compliance-software

Food Fraud Advisors. (2019, November 25). *Raw material specifications.* https://foodfraudadvisors.com/template-purchasing-specification/#:~:text=).

FoodUnfolded. (n.d.). *How does colour affect taste and the way we eat?* https://www.foodunfolded.com/article/how-does-colour-affect-the-way-we-eat

Foston, B. (2023, June 13). */what-is-brcgs-global-food-safety-standard-explained.* https://www.ideagen.com. https://www.ideagen.com/thought-leadership/blog/what-is-brcgs-global-food-safety-standard-explained

Geeraerd, A., Valdramidis, V., & Van Impe, J. (2005). GInaFiT, a Freeware Tool to Assess Non-Log-Linear Microbial Survivor Curves. *International Journal of Food Microbiology, 102*(1), 95–105. https://doi.org/10.1016/j.ijfoodmicro.2004.11.038

Gragg, S. E., & Brashears, M. M. (2021). Hurdle Technology https://api.semanticscholar.org/; CorpusID:107899902.

Iammarino, M., Marino, R., Nardelli, V., Ingegno, M., & Albenzio, M. (2024). Red Meat Heating Processes, Toxic Compounds Production and Nutritional Parameters Changes: What About Risk–Benefit? *Foods*, *13*(3), 445. https://doi.org/10.3390/foods13030445

IFT. (2019, September). *A historical look at food safety*. Retrieved June 12, 2024, from https://www.ift.org/news-and-publications/blog/2019/september/a-historical-look-at-food-safety

Institute of Food Science and Technology. (2021, February 25). "What is sensory science?" - Think Piece. https://www.ifst.org/membership/networks-and-communities/special-interest-groups/sensory-science-group/what-sensory

Intertek. (n.d.). https://www.intertek.com/

Kemp, S. E., Hort, J., & Hollowood, T. (2018). *Descriptive analysis in sensory evaluation*. John Wiley & Sons Ltd. https://doi.org/10.1002/9781118991657

Lawless, H. T. (2012). *Flavor profile method*. Food Science Text Series (pp. 79–84). https://doi.org/10.1007/978-1-4614-5713-8_9

Levitan, A. S. (2022, April 28). *The strongest sense – Sydney Levitan*. https://sites.psu.edu/sml6791/2022/04/28/the-strongest-sense/

Merieux Nutrisciences. (2024, June 7). *Home - Global | Merieux Nutrisciences*. https://www.merieuxnutrisciences.com/

Marmiroli, N., Maestri, E., Gullì, M., Malcevschi, A., Peano, C., Bordoni, R., & De Bellis, G. (2008). Methods for Detection of GMOs in Food and Feed. *Analytical and Bioanalytical Chemistry/Analytical & Bioanalytical Chemistry*, *392*(3), 369–384. https://doi.org/10.1007/s00216-008-2303-6

Marques, C., Correia, E., Dinis, L., & Vilela, A. (2022). An Overview of Sensory Characterization Techniques: from Classical Descriptive Analysis to the Emergence of Novel Profiling Methods. *Foods*, *11*(3), 255. https://doi.org/10.3390/foods11030255

Murray, J., Delahunty, C., & Baxter, I. (2001). Descriptive Sensory Analysis: Past, Present and Future. *Food Research International*, *34*(6), 461–471. https://doi.org/10.1016/s0963-9969(01)00070-9

National Institutes of Health (NIH). (2020, May 7). *Humans can identify more than 1 trillion smells*. https://www.nih.gov/news-events/nih-research-matters/humans-can-identify-more-1-trillion-smells

OpenSC. (n.d.). *OpenSC - Home*. https://opensc.org/

Pérez-Rodríguez, F., & Valero, A. (2012). *Predictive microbiology in foods* (pp. 1–10). Springer. https://doi.org/10.1007/978-1-4614-5520-2_1

Peris, M., & Escuder-Gilabert, L. (2016). Electronic Noses and Tongues to Assess Food Authenticity and Adulteration. *Trends in Food Science & Technology*, *58*, 40–54. https://doi.org/10.1016/j.tifs.2016.10.014

Rodrigues, S. S. Q., Dias, L. G., & Teixeira, A. (2024). Emerging Methods for the Evaluation of Sensory Quality of Food: Technology at Service. *Current Food Science and Technology Reports*, *2*(1), 77–90. https://doi.org/10.1007/s43555-024-00019-7

Rolls, E. T. (2016). Flavor: Brain processing. In *Flavor* (pp. 143–160). Elsevier. https://doi.org/10.1016/b978-0-08-100295-7.00007-4

Sobel, R. (2014). *Microencapsulation in the food industry: A practical implementation Guide*. Academic Press.

Spence, C. (2019). On the Relationship(s) Between Color and Taste/Flavor. *Experimental Psychology*, *66*(2), 99–111. https://doi.org/10.1027/1618-3169/a000439

Subramaniam, P. (2009). *Shelf-life prediction and testing*. (pp. 233–254). Elsevier. https://doi.org/10.1533/9781845696436.2.233

Tan, J., & Xu, J. (2020). Applications of Electronic Nose (e-Nose) and Electronic Tongue (e-Tongue) in Food Quality-Related Properties Determination: A Review. *Artificial Intelligence in Agriculture*, *4*, 104–115. https://doi.org/10.1016/j.aiia.2020.06.003

Taylor, S. (2023, September 25). What did we learn from the Horsemeat Scandal and should we still be worried? *The Hub | High Speed Training*. https://www.highspeedtraining.co.uk/hub/horsemeat-scandal-facts-and-effects/

Teshome, E., Forsido, S. F., Rupasinghe, H. P. V., & Keyata, E. O. (2022). Potentials of Natural Preservatives to Enhance Food Safety and Shelf Life: A Review. *The Scientific World Journal*, *2022*, 1–11. https://doi.org/10.1155/2022/9901018

The Branded Agency. (2023, October 8). *Feeding feelings: How the food industry masters the art of emotional branding*. https://www.brandedagency.com/blog/feeding-feelings-how-the-food-industry-masters-the-art-of-emotional-branding

TUV NORD. (2021, July 19). *Predicting product shelf life using stability studies*. https://www.tuv-nord.com/in/en/blog/blog-details/article/predicting-product-shelf-life-using-stability-studies

UC Food Safety. (2024, March 27). *Pathogen modeling programs*. https://ucfoodsafety.ucdavis.edu/processing-distribution/building-food-safety-plan/pathogen-modeling-programs

Unilever. (2023, November 10). *Leading safety and environmental sustainability sciences.* https://www.unilever.com/planet-and-society/safety-and-environment/leading-safety-and-environmental-sustainability-sciences/

Unnikrishnan, S., Von Koeller, E., Mutz, M., Fovargue, H., & Yong, K. (2024). *The Time Is Ripe for Food Traceability.* BCG Global. https://www.bcg.com/publications/2021/food-traceability

U.S. Food and Drug Administration. (2024a). *Tracking and tracing of food.* https://www.fda.gov/food/new-era-smarter-food-safety/tracking-and-tracing-food

U.S. Food and Drug Administration. (2024b, March 20). *Food traceability List.* https://www.fda.gov/food/food-safety-modernization-act-fsma/food-traceability-list

Van Der Berg, J. P., Kleter, G. A., Battaglia, E., Bouwman, L. M. S., & Kok, E. J. (2020). Application of the Safe-by-Design Concept in Crop Breeding Innovation. *International Journal of Environmental Research and Public Health, 17*(17), 6420. https://doi.org/10.3390/ijerph17176420

Vu, G., Zhou, H., & McClements, D. J. (2022). Impact of Cooking Method on Properties of Beef and Plant-Based Burgers: Appearance, Texture, Thermal Properties, and Shrinkage. *Journal of Agriculture and Food Research, 9,* 100355. https://doi.org/10.1016/j.jafr.2022.100355

Wallace, C. A., Sperber, W. H., & Mortimore, S. E. (2010). Case Study 2: Manufacturing - Prepared Meals. Wiley Online Library, 271–291. https://doi.org/10.1002/9781444328653.oth2

Wang, M., & Chen, Y. (2023). Electronic Nose and Its Application in the Food Industry: A Review. *European Food Research & Technology, 250*(1), 21–67. https://doi.org/10.1007/s00217-023-04381-z

Wang, Q. J., Mielby, L. A., Junge, J. Y., Bertelsen, A. S., Kidmose, U., Spence, C., & Byrne, D. V. (2019). The Role of Intrinsic and Extrinsic Sensory Factors in Sweetness Perception of Food and Beverages: A Review. *Foods, 8*(6), 211. https://doi.org/10.3390/foods8060211

Wasilewski, T., Kamysz, W., & Gębicki, J. (2020). Bioelectronic tongue: Current Status and Perspectives. *Biosensors & Bioelectronics/Biosensors & Bioelectronics (Online), 150,* 111923. https://doi.org/10.1016/j.bios.2019.111923

WESTPAK. (2023, August 9). *Accelerated aging calculator [AAT Calculator] | Westpak.* Westpak. https://www.westpak.com/resources/calculator/accelerated-aging-time-calculator

Wikipedia. (n.d.). *Melamine-in-milk.* Retrieved June 12, 2024, from https://en.wikipedia.org/wiki/2008_Chinese_milk_scandal, https://www.sciencebase.com/science-blog/melamine-in-milk.html.

Wu, H., Wang, Z., Luo, Y., Hong, H., & Shen, H. (2014). Quality Changes and Establishment of Predictive Models for Bighead Carp (*Aristichthys nobilis*) Fillets During Frozen Storage. *Food and Bioprocess Technology, 7*(12), 3381–3389. https://doi.org/10.1007/s11947-014-1340-8

Zhou, W., Sun-Waterhouse, D., Xiong, J., Cui, C., Wang, W., & Dong, K. (2019). Desired Soy Sauce Characteristics and Autolysis of *Aspergillus oryzae* Induced by Low Temperature Conditions During Initial Moromi Fermentation. *Journal of Food Science and Technology/Journal of Food Science and Technology, 56*(6), 2888–2898. https://doi.org/10.1007/s13197-019-03742-5

Food Labelling and Regulation

11

Labelling is the act of providing information about a food product to consumers through labels, tags, marks or other descriptive matter. Food labelling is intended to protect consumer health, inform consumer choice and promote fair trade practices. Some food information is mandatory by law, such as the name, ingredients, nutritional value, origin, expiry date and handling instructions of the food product. Other food information is voluntary, and the manufacturer is free to inform about items such as suitability for vegetarians or vegans, sustainability or gluten-free claims. The "Codex Alimentarius" is a collection of food safety and product quality standards issued by the Food and Agriculture Organization (FAO) and the World Health Organization (WHO) and serves as a benchmark for all national or territorial food labelling regulations. These standards are intended to protect consumer health globally and ensure fair practices in the food trade (Food and Agriculture Organization, n.d.). The core elements of these regulations cover food labelling, food additives, pesticide residues, contaminants, animal feed, antimicrobial resistance, biotechnology, nutrition and labelling and health claims.

BACK-OF-PACK LABELLING

While the front of the packaging material is the side where the brand can attract buyers through the creative expression of its identity, the back side is the space for objective and mandatory information. While older research indicates that 59%–66% of consumers say that they study the backside information when looking at a new product, experts believe the true percentage to be much lower (Cosland, 2022). Typical mandatory elements are the ingredient list, nutritional information including additives and allergens, product net quantity, mandatory symbols about dietary standards on recycle types, possible accreditation labels and usage instruction, manufacturer and production details. Sometimes the retail price is also included.

The most important elements will be discussed in some more detail now.

Ingredient List

The purpose of the ingredient list is to transparently inform the consumer. It allows people to make informed choices based on their dietary preferences, allergies or other dietary restrictions. Therefore, several aspects of content and form are strictly regulated. The mandatory part starts with the following:

- Name of the food
- The list of all ingredients, normally in descending order of weight, with legal names
- Allergen information, in a highlighted format
- The data of manufacture
- Country of origin
- Name and address of the food manufacturer

DOI: 10.1201/9781032724591-11

- Net quantity
- Special storage conditions and/or conditions of use
- Instructions for use, if needed.

The font size of the ingredient list text is also often specified. Where the single pack is too small, the secondary multipack must display the information. A single 8-, 9- or 10-g bouillon cube cannot display the ingredient list, so here the information must be given on the multipack carton or sachet, unless the cube is sold as single, like is often done in wet markets in developing and emerging market (D&E) countries. In this case, it will need a secondary pack, where the information can be provided in readable form.

The "Quantitative Ingredients Declaration" (QUID) informs the buyer of packaged foods about the percentage, weight or volume of relevant ingredients. The declaration of the amount is mandatory where the ingredients are name giving for the product, shown in picture or graphic form on the pack or essential to characterize the food, as expected by the consumer. A "Spaghetti Bolognese" product in western markets, therefore, has to list the amount of beef in the product. This informs consumers about the relative proportions of different components in the food. And at the same time a product whose packaging states "100% chicken breast" may contain only 75% chicken, the rest is coating (Blythe, 2023).

Many countries allow exceptions in the QUID. The most prominent are added water, concentrated ingredients, volatile products and herbs and spices. The label also allows the manufacturers some flexibility in the change of the composition for minor ingredients. If an oil is below 2% of total weight at the point of manufacturing, the statement "vegetable oil/fat" can be used, if in brackets the types are mentioned (palm, coconut).

Compound ingredients are those that are made themselves from different products. Listing for example "crème fraiche" is not sufficient, but must include the primary constituents like cream, skimmed milk, or preservative potassium sorbate.

Food additives and flavourings must be included and identified by their category. For example, a colouring must be listed by its specific name or number (in Europe, e.g., curcumin or E100).

In summary, the ingredient list is tightly regulated and helps consumers to understand the characteristics. It gives manufacturers different options to declare the product composition, but does not always add transparency. The sometimes very detailed quantities provided may actually go beyond the abilities of the industry to dose them exactly on a routine basis and with natural ingredients and there may be fluctuating compositions.

E Numbers

The safety assessment and approval of E numbers in Europe is the responsibility of the European Food Safety Authority (EFSA), an independent agency of the European Union that provides scientific advice and communicates on existing and emerging risks associated with the food chain. They conduct safety assessments of substances intended as food additives. Each substance deemed safe to use is given a code, called an E number. An old industry joke says, the "E" stands for "edible". Officially it just signifies "Europe". In other parts of the world the same numbers are used just without the "E". So, E300 in Europe stands for "ascorbic acid", or "vitamin C". In Australia it would be "300" (ascorbic acid)".

The E number list includes molecules that are made chemically but are identical to natural ones, such as the abovementioned vitamin C or E300, but also pure artificial substances, like E102, also known as Tartrazine.

The substances are clustered by ranges, and the ranges link to the International Numbering System for Food Additives (INS) at Codex (CODEX General Standard for Food Additives (GSFA) Online Database, n.d.), bundled into functional classes and purpose.

In the past decades the trust in E numbers has deteriorated. Consumers perceive these substances as chemicals that do not belong in food, and they doubt their harmlessness. The fact, that high-level expert groups have thoroughly tested the substances is often not known or seen as not reassuring.

Generally, although preservatives have increased the safety of food items, not all additives are deemed equally safe by all experts. As a result, the allowed usage of some substances differs between countries.

The restrictions of particular E numbers vary between countries, often influenced by the discussion of a size of the safety margin added after all toxicity tests.

Toxicity tests aim to determine the degree to which a substance negatively impacts the normal biological functions of an organism in relation to the concentration of the substance, the exposure time and exposure route and substance concentration. Toxicity tests are often performed in cell, tissue or animal model systems, and the translation of those results into advice for humans is difficult.

"The dose makes the poison" is a well-known saying attributed to Paracelsus, a Swiss physician and alchemist who is considered to be the father of toxicology. In high enough doses most food ingredients can become toxic. Therefore, many substances ranging from caffeine to MSG and vitamin C have received a safety or acceptable daily intake (ADI) level.

Hence, for the experts it's often not a real surprise that test results lead to very different conclusions. Amaranth (E 123), for example, is used as a colouring agent in foods. Europe and Australia permit the use in foods up to a concentration of 30 mg/kg. But in the United States it is banned due to concerns it can cause cancer, based on a Russian study performed in 1971 in rats. Later, the design of the study was heavily criticized, and the FDA initiated its own studies with very little evidence that amaranth would be harmful. But through the results of the Russian study the public was concerned, and as there also was no real industrial pressure, the FDA decided to leave the ban (Wikipedia Contributors, 2024a–c).

Setting safety limits in foods is a difficult task with a good portion of uncertainty. So, no wonder that experts' opinions differ. But, if even the specialists cannot agree on the safety of a substance, how should the average consumer get their trust in the scientific opinion? And in food safety science, like in all other sciences, knowledge progresses and new insights, maybe on long-term effects, come only later, after longer observation or with new methods and different approaches.

In Europe, six substances in the additive section E100–E199 became famous as the so-called Southampton Six. They were approved as food colourings, but in 2007 a UK study found a link between two mixtures of food colouring and an increased level of hyperactivity in children. The EFSA evaluated the study and concluded that there was only limited evidence that the two mixtures of additives would have a small effect on the activity and attention of some children. With that uncertainty the authorities felt they must use the "precautionary principle" and ruled that products with those colours would need to carry the statement "*may have an adverse effect on the activity and attention of children*" (foodnavigator.com, 2016). The EFSA kept on testing and did not find evidence for such effect and later even increased the ADI for two of the colouring agents.

But food manufacturers, especially those of child food and beverages, realized the effects of this debate in the media and immediately started to stop using these colourings, initiating massive reformulation waves, for example in ice cream. Since the alternatives were less vibrant, a replacement was not without risk. Vibrant colours signal freshness, taste and aroma. Reducing them often means reducing likeability. And those alternatives were more expensive, but charging the consumer for that was not a viable option, while the shareholders were not pleased with the margin dilution.

Another example here is E621, or monosodium glutamate (MSG). MSG is a very popular food additive enhancing the flavour of food, particularly the meatiness. But it has been the subject of controversy and criticism for many years. Media are filled with stories about headaches, flushing, sweating, dizziness, brain damage, overweight and obesity. The FDA, EFSA and the Joint FAO/WHO Expert Committee on Food Additives (JECFA) looked again and again at the safety of MSG, without finding evidence for a harmful effect.

So, in this case, and maybe a few others, it may not be the safety aspect that troubles consumers. It could also be the feeling of being the victim of another "chemical trick" of the food industry to pretend a food quality that is not there. Cooking causes massive losses of vitamin C, which can be compensated through addition of ascorbic acid, or E300. A dressing or sauce looks more appetizing with a little red colour, like E129 or Allura red! And meat bouillon has a stronger well-rounded meaty note with a bit of MSG, E621, added. Additives bring an essential function to food items and can compensate for some quality losses in processing. And they can be added at very low cost, so an attractive price can be kept.

Nutrient Content

Another mandatory element of food packaging is the nutrition facts label. The purpose of the nutrition content table is to provide consumers with information about the nutritional content of the food they eat. In most countries the nutrition declaration must provide the content per 100 g or 100 mL of

- Energy, expressed in kilojoules (kJ) and kilocalories (kcal)
- Fat and saturated fat
- Carbohydrate and sugar
- Protein
- Salt

Voluntarily these data can also be given for:

- Mono-unsaturated and polyunsaturated fatty acids
- Polyols (sugar alcohols or bulk sweeteners)
- Starch
- Fibre

All vitamins and minerals permitted by law can also be declared. The amount can be additionally given per portion or per consumption unit of the product (serving). Details vary between countries. The regulation also prescribes details of the type and size of the printing.

The manufacturer is responsible for the correctness of the data provided. In some countries, the manufacturer can take data from nutritional tables and databases like the one provided by the U.S. Department of Agriculture (USDA). But more often they must perform laboratory analysis of their product, using standard methods for determining the nutritional content of food.

Some tolerances between the declared and factual level is acceptable, since it is often not possible for foods to contain the exact nutrient levels labelled, due to natural variations and variations from production and during storage. But clearly, the nutrient content of foods should not deviate substantially from the labelled values. That would mislead consumers and be considered fraud.

The specific tolerances allowed vary by country and region. The European Union, for example, has given detailed guidance on tolerances for the control of compliance ranging from relatively wide tolerances for naturally present nutrients to much more narrowed ones for technically enriched or fortified products (Abby Hannah, 2024). That makes sense, as the intentional addition of nutrients in the process (fortification) can be much better controlled than the natural nutrient formation under changing environmental conditions.

Especially companies offering product with added nutrients in hot and tropical climates often must pay good attention to the quantification of the nutrient loss during storage and distribution. In many of those developing markets, up to 50% of the trade is still via simply booths in traditional or wet open markets. Here products can be exposed to harsh conditions for a long time, which can dramatically lower the micronutrient content. For the declaration on the label, companies often have to "overdose" the respective nutrient to ensure the nutrient content at the end of shelf live.

FRONT-OF-PACK NUTRIENT LABEL (FOPNL)

While the back of pack provides sufficient detailed information about the product and producer, concerns have been raised that this "back-of-pack" information is too complex for the rushed shopper. As a solution, easily recognizable and understandable "front-of-pack nutrient labels" are suggested. These can be simple icons or designs with detailed nutritional information, a traffic light system, stars or numerical scores.

The discussion around these labels is complex and touches various dimensions. But all suggestions are in a sense judgemental, signalling food items as "good" or "not good". In detail the controversial aspects are:

- *Effectiveness*: At this stage it is not clear how well the FOPNL supports consumers to make healthier food choices. Data are mixed! Some studies report an increased consumer understanding about nutritional quality, while others find only marginal impact on actual consumer behaviour (Braesco & Drewnowski, 2023).
- *Mandatory or voluntary*: Producers of "less healthy" food see their business endangered, when their offerings are mandatorily and prominently marked as "not healthy". These food makers see their products unfairly targeted and stigmatized. A voluntary label would give producers the choice to highlight the nutritional aspect.
- *Industry impact*: For FOPNL supporters, the longer-term goal is to force the food businesses to reformulate their products into healthier versions. It is not clear if this would truly lead to a nutritional enhancement or just a tactical "health-washing" where the nutritional goodness is overpromised.
- *Standardization*: The lack of a global standard has created many different versions of the FOPNL. That's a challenge for international food businesses and can be confusing for consumers in different countries (UNICEF & Pazos, n.d.).

Most resistance is articulated for the type of "warning labels", where today critical nutrients like fat, calories, sugar or salt are flagged in vibrant red colour front of pack, when specific limits are exceeded. The mandatory FOPNL in Chile is a representative. On the opposite side of the spectrum is the simple "Healthier Choice" logo used in Singapore, Brunei and Malaysia, which informs about the healthier choice. In such systems the manufacturer applies at the local health promotion board for the use of the logo, and if the product meets a set of nutritional criteria, which was set out by the Department of Health, it can be used on the pack as a promotion tool.

Understandably, an institution like the United Nations (UN) urges for a globally harmonized design to avoid consumer confusion. But beside the necessity for an international alignment between countries and industry, evidence would be needed regarding which FOPNL design would best serve the purpose. Figure 11.1 shows the logo the Malaysian Government has chosen to indicate healthy food choices.

FIGURE 11.1 Malaysia Healthier Choice Logo. (From Hong, 2017.)

Nutri-Score

A recent example of a judgemental nutrition label is the Nutri-Score. This five-colour nutrition label is based on a rating system in which food categories are scored on nutritional constituents like energy density, simple sugars, saturated fats, salt, fruit and vegetables, fibre and protein (see in Figure 11.2) All these components give points depending on their amount present in 100-g points depending on the level present, while the first four elements result in negative scores. The score then is basically the sum of up to ten points for "baddies"" like energy density, sugar, salt, fat and up to five "goodies". The lower the score, the better the nutritional value of the product.

Currently, the Nutri-Score system is the basis in a few EU countries. And it faces both support and criticism. Science is used to either strengthen the position of either supports or critics. A recent discussion about the validity and efficacy of the scoring system has now even led to scientific reviews where the correlation between the affiliation of the authors and the study results are investigated (Peters & Verhagen, 2024). Concerns have been raised that the system may favour some processed foods with low nutritional value, while devaluing some natural products. Consumers need to understand that just choosing "good" score products does not automatically lead to a healthy diet. The central tension remains: On the background of an undoubtedly massive wave of diet-related diseases, food and health experts are looking for ways to motivate consumers to choose more "healthy" food items to get to a healthier diet. Others, particularly those producing food products, simply see the danger that the food and beverage (F&B) category they are operating in is simply "punished" with a label that can cut their sales. They find many reasons why it is difficult to try and build a healthy diet through a quite simple algorithm that only counts a few constituents on a 100-g basis. Meanwhile the Nutri-Score systems seem to find increasing acceptance in the population and trade in Europe (Oqali & Santé Publique France, 2021).

The NOVA food classification system, developed by Brazilian nutrition researcher Carlos Monteiro at the University of São Paulo, Brazil (Monteiro, 2009), goes a step further. The scheme has no logo yet, but the idea is to classify products mainly by the nature, extent, and purpose of the industrial food processing. The leading idea is that industrial processing makes food unhealthy, or at least non-nutritious. The force behind this is the profit-driven industry, which always strives for lowest-cost ingredients and highest consumption. So, they make products convenient and appealing, hyperpalatable it is sometimes called, and are therefore responsible for the rise of diet-related non-communicable diseases and obesity. I discuss this in detail in Chapter 16, since this frame can influence the direction of future food product development.

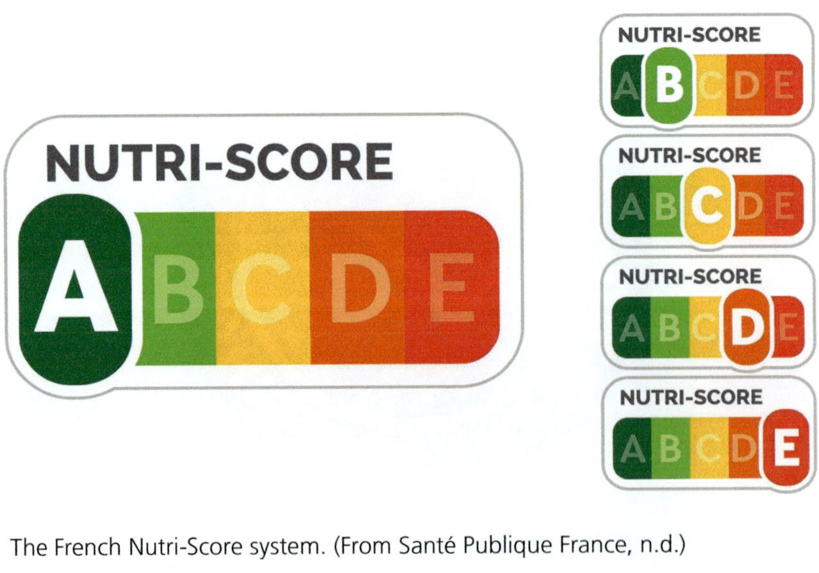

FIGURE 11.2 The French Nutri-Score system. (From Santé Publique France, n.d.)

A freely accessible database like "Open Food Facts" lists packaged food products with three different scores: Nutri-Score, Nova-Score and Eco-Score (World.openfoodfacts.org, n.d.), covering the scored products from a nutritional, processing and environmental angle.

Health Star Rating (HSR)

The "Health Star Rating" in Australia and New Zealand is another version of a front-of-pack labelling system that has been developed in collaboration with all stakeholders, governments, food industries, public health experts and consumer groups (Australian Government Department of Health, n.d.-a and n.d.-b). It was launched in June 2014 and includes most packaged food and beverage categories present in the market. It is voluntary and judgemental, classifying products by their Health Star scoring. The "scoring algorithm" gives points for the presence of "qualifying" nutrients (like protein, fibre, and sometimes calcium, fruit and vegetables) and deducts points for disqualifying nutrients (overall energy, sodium, total sugar and saturated fat). The calculation gives the final Health Star Rating with ½ to 5 stars.

The food producers that want to use the Health Star Rating load the required nutrient data into the online available calculator to score their products and calculate the rating (Australian Government Department of Health, n.d.-b). To start, the producer must place the product into one of the pre-designed categories:

- Non-dairy beverages, jellies and water-based ice confections
- Milk and dairy
- All other foods
- Dairy foods with ≥75% dairy
- Oils and spreads,
- Cheese (with calcium content >320 mg/100).

This initial classification helps to score product more "like for like", since it is not meaningful to compare a pizza with a yoghurt. The main categories of beverages, general food and oils and fats each have a subsequent score for dairy, which allows a meaningful calcium scoring, as this micronutrient seems to be of superior importance here.

The outcome of the rating can be displayed front of pack with a given design (see Figure 11.3).

Products with an HSR ≥ 3.5 stars can be promoted as healthy. Some specialized categories (like baby food) are excluded, but for packaged food and beverages in the market, locally made and imported, the system can be used.

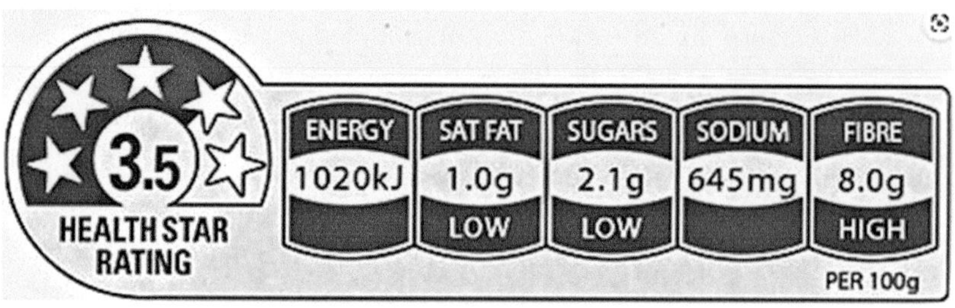

FIGURE 11.3 Example of the front-of-pack Health Star Rating.

CASE STUDY: HEALTH STAR RATING

A few scoring cases are provided as help for companies that want to score their products. The example here is for a Pizza Supreme (modified from: Australian Government Department of Health, n.d.-c).

PRODUCT	*PIZZA SUPREME*
Nutrition content (per 100 g)	Energy: 1125 kJ Protein: 13.9 g Saturated fat: 5.2 g Total sugars: 0.9 g Sodium: 743 mg Dietary fibre: 3.3 g
Category determination	Not a dairy food Not a non-dairy beverage Not an oils and spreads food It is a general food product, Category 2 (*means, calcium score is not relevant*)
Is policy affected?	No, it's not in an exclude category (*like baby food)*
Product form	As sold *(compared, e.g., to powder, where dilution needs to be included)*
HSR baseline points	By using the appropriate scoring table for "Foods" Category 2 (per 100 g)

Baseline Points	**Energy (KJ)** *Per 100g or 100ml*	**Saturated Fat** *Per 100g or 100ml*	**Total Sugar (g)** *Per 100g or 100ml*	**Sodium (mg)** *Per 100g or 100ml*
0	≤335	≤1.0	**≤5.0**	≤90
1	>335	>1.0	>5.0	>90
2	>670	>2.0	>8.9	>180
3	**>1005**	>3.0	>12.8	>270
4	>1340	>4.0	>16.8	>360
5	>1675	**>5.0**	>20.7	>450
6	>2010	>6.0	>24.6	>540
7	>2345	>7.0	>28.5	>630
8	>2680	>8.0	>32.4	**>720**
9	>3015	>9.0	>36.3	>810

The total HSR baseline points = (5) + (4) + (9) + (0) = **16**

HSR Fruit and Vegetable (F&V) points

For the fruit and vegetable score the amount of concentrated and non-concentrated vegetables are relevant:

Product has:

4% concentrated + 23% non-concentrated vegetables, and 100% – 27% = 73% non-veg ingredients

Formula to be used:

$$\frac{(\%\text{ non-concentrated F\&V}) + (2 \times \%\text{ concentrated F\&V})}{(\%\text{ non conc.}) + (2 \times \%\text{ conc.}) + (\%\text{ non-F\&V ingredients})}$$

Results in: 23 + (2 × 4) / (23 + (2 × 4)) + 73 = 31/104 × 100 = 30% F&V

As this is a mixture of concentrated and non-concentrated F&V, the scores for non-contracted F&V applies:

Points	Column 1 % concentrated fruits or vegetables	Column 2 % non-concentrated fvnl
0	<25	≤40
1	≥25	>40
2	≥43	>60
3	≥52	>67
4	≥63	>75
5*	≥67	>80

Leading to HSR Vegetable points: 0

HSR Protein (P) points

Rule: A product scoring 13 or more points on the HSR baseline can only score for protein if they score at least 5 V points.
The pizza scored 16 HSR baseline points and zero V points.
Therefore, no score for protein is permitted and **P points = 0.**

HSR Fibre (F) points

This is a scoring table with dependencies from other factors: 3.3 g

Points	Protein (g) *per 100 g or 100 mL*	Dietary fibre (g) *per 100 g or 100 mL*
0	≤1.6	≤0.9
1	>1.6	>0.9
2	≥3.2	>1.9
3	>4.8	>2.8
4	>6.4	>3.7
5	>8.0	>4.7
6	>9.6	>5.4
7	>11.6	>6.3
8	>13.9	>7.3

3.3 g fibre gives F points = 3

Final HSR score

Formula to be applied:
Final HSR score =
HSR baseline points – (V points) – (P points) – (F points)
(16) – (0) – (0) – (3) = 13

Final HS rating

Table to be used for non-dairy foods:

Category 2 products must score	
≤ −11	5 stars
−10 to −7	4½ stars
−6 to −2	4 stars
−1 to 2	3½ stars
3 to 6	3 stars
7 to 11	2½ stars
12 to 15	2 stars
16 to 20	1½ stars
21 to 24	1 star
≥ 25	½ star

13 points qualify for 2 star rating (no health messages allowed).

Limitations of the Health Star Rating

The Health Care Rating case study shows the difficulties/limitations a "health rating system" for single products is facing. A food item cannot simply reflect pro rata daily dietary requirements; hence, linking all to the caloric content can be misleading. Major food categories have different profiles: Oil and fat do not provide protein or fibre and are not relevant for calcium intake.

Soups can provide a good portion of vegetables, legumes and lentils. But they may have a lot of salt. The question on how to bring the pre-treatment of key ingredients like fruits and vegetables into the equation is not an easy one.

Five years after an official review was done to address points of concern that have been raised. Recommendation was made to change "the way the HSR is calculated for non-dairy beverages, based on adjusted sugars, energy and FVNL points and to better discern water (and drinks similar in nutritional profile to water) from high energy drinks" (http://healthstarrating.gov.au/internet/healthstarrating/publishing.nsf/Content/formal-review-of-the-system-after-five-years).

Scoring products "as prepared", after addition of other food components, remains another difficulty. People do not consume milk powder, but how much is added and if it is skimmed or full fat milk, can influence the score. And a bouillon cube needs dilution with water and maybe the addition of some vegetables and meat (lean or fat), which can lead to different outcomes. Here it is suspected that food producers would put the most favourable preparation method on the pack and use the score to promote their products.

The voluntary nature of the system is perhaps the biggest criticism. Companies can score their products, and use the logo front of pack, for those items with a favourable score.

The official Health Star website announced that about 10,300 products use the Health Star on pack, which is less than half of the products on the market. Others calculated that only 27% of eligible products got a score (Crino et al., 2018).

FUTURE PROFILING SYSTEMS

A discussion has been ongoing for many years about a more impactful consumer-relevant nutrition labelling. A recent review about the status of such labels in four Southeast Asian countries, Malaysia, Thailand, Philippines and Vietnam, openly criticized the slow roll out and implied that the food industry would try everything to avoid a mandatory FOP, nutrition labelling and scoring (Pettigrew et al., 2022). Often the starting point is the factual rise of non-communicable diseases and an assumption that better, clearer on-pack information would be a powerful tool to prevent that. Packaged food is often flagged as non-nutritious and unhealthy and manufactured in a way to make the food super palatable and seduce the consumers to eat too much. (See also chapter 17 about "ultra-processed food".) Under this argumentation an FOPNL gives the consumers salient, easily understandable nutrition information on food products, which lets them make healthier food choices.

The second, more indirect effect would be the extra pressure on the food manufacturers to reformulate their products to make them healthier. First studies suggest a link between diets, scoring low in nutrition profiling systems and an increase in body mass index (Carruba et al., 2021).

The food industry is accused of trying everything to delay, dilute and derail the introduction of such judgemental labelling, the list of tactics they are accused of using contains:

- Reshaping the evidence base through favourable own studies, in-house research selective funding of supportive research. This makes them a part of the solution, rather than the problem, and that earns them a seat at the policy-making table.
- A different strategy is to warn of premature judgement, for example, by the articulation that the aetiology of obesity is so complex that individual products cannot be held accountable for the effect.
- And if their science influencing is not effective enough, they may switch and start to influence governmental groups like trade and industry to slow the process down. And if even this does not help, industry may start to threaten governments with litigations or the closing down of factories and services.

It's hard to deny that the overweight and obesity wave is the result of taking in more calories than what is burned. And nutrition labels should inform consumers understandably about the nutritional facts of packaged foods and the role of that to the overall diet. More than 120 United Nations (UN) member states already mandate nutrient declaration, including fat, salt, energy, and sugar, and 44 member states have it already.

What is still missing is the evidence that FOPNL has an impact on the food choices people are making. One can ask if it is necessary to have that evidence, before trying it out. Hence, while the WHO admitted that there is no real evidence, they also urged to "... *implement comprehensive policy approaches to create healthy food environments including nutrition labelling policies.*" (WHO, 2021). And a recent multi-country study on FOP labels in *Nature Journal* showed improvements in both understanding of nutritional quality and to some extent in food choices (Pettigrew et al., 2022).

So, FOPNL labelling is a step worth trying. But how this is done best requires further investigations. Some experts believe that "non-interpretive systems", like Guideline Daily Amounts (GDAs), which provide "g of nutrients" and "% of daily allowance", are not understood by consumers and therefore an "interpretive" system is needed. The main argument against such a judging system is the oversimplification of food product quality and the fairness of one product score in a product in the diet: And stigmatizing specific products will affect consumer perceptions and brand reputation, so the companies affected will try a lot to protect their business.

A voluntary scoring system is more acceptable to the food industry. Here, companies can "test" their products, and a "good" outcome can be used to differentiate from competitors and convey positive health messages. A mandatory scheme will put many product categories in the spotlight of being "not good". This is not really motivation to "improve" the nutritional profile within the boundaries of the specific food category. Therefore, a different way to use the score could gain momentum rather than only using these systems to "score" existing products.

As an alternative way encouraging manufacturers to actively reformulation those products of their portfolio that have a nutritional weakness or improvement option, was recently suggested (Leroy et al., 2021).

This approach starts with a classification of food items into product categories, where similar products can be "fairly compared". The key nutrients relevant to specific food categories that are driving nutritional quality are selected, say sugar, saturated fat, sodium, fibre and protein. On the basis of a reliable, public and popular database, for example, the "Global New Products Database" (GNPD), standards can be set for the "best of class" products. The database compiles nutrient profiles of all new launches of branded packaged foods and beverages in 62 countries worldwide (Mintel, 2023). That then sets in a new profiling system with stretching, but feasible, nutrient targets for reformulation of food items in that category. By using settings like "low in" for disqualifying nutrients, like salt or sugar, and "source of" for qualifying nutrients, like protein and fibre, an algorithm can cluster the products in the database according to their category of relevant nutrient. By setting a 50% threshold for the relevant nutrients in a category as average and a top 15% threshold as ambitious, a realistic reformulation can be set.

Product makers can see the average value in their food category, and the "best 15%" with respect to the chosen limit, like, for example, sodium and saturated fatty acid (SAFA) content for pizza. Since the best 15% are all existing products on the market, a principal feasibility should be given; even so, taste preferences differ between countries.

This would create a form of rolling, continuous-improvement target. And that could encourage the reformulation of commercial top sellers; those products with high commercial importance volumes and high potential move the needle in improving public health.

The idea of rolling reduction targets for "nutrients of concern" as such are not new. Some international companies have already started years back to set nutrition profiles for their key product categories, combined with the promise to move the nutrition profile towards the set targets (Unilever, 2024). But a common, publicly available database would also allow many more players in the food sector to adopt realistic reduction targets, without the necessity of massive baseline data collection work. Those enterprises would get credible, yet realistic, guidance for the redesign of their products.

And while the scientific discussion continues, authorities increasingly consider using product composition data as a basis for taxation initiatives or the tighter regulation of marketing campaigns.

NUTRITION AND HEALTH CLAIM REGULATION

The idea that food is contributing to health and well-being is not new! "Let food be thy medicine and medicine be thy food!" is the famous quote attributed to the Greek physician Hippocrates, the father of medicine. He did not consider food and medicine as the same thing, but he recognized that different foods could have different effects on different people.

Documented is the quote in the 1926 book *The Key to Rational Dietetics*, of the French-American business man Otto Carque, who was promoting a balanced diet with natural, raw food (Wikipedia Contributors, 2024b). At that time a niche esoteric movement! Fifty years later, in a decade of increasing social and environmental movements such as vegetarianism, organic farming and alternative medicine, the quote got traction, and the ancient Greek wisdom resonated well with people believing in natural and holistic ways of healing and living.

This trend on health food broadened, and soon the first food companies started to design specialty foods with extra health benefits. With the appearance of this new "functional" food type, new regulations were needed to manage the "storytelling".

Nowadays, health food and claim regulations are rooted in the principles of the Codex Commission (CAC). The aim is to protect consumer health and set fair practices in food trade. The basic objective standards for a declarable nutrient are how much is needed per day (the reference value) and under which conditions a nutrient can be quantified. This allows "nutrient content claims", including "source of", "high", "low", "free" or "Nutrient comparative claims" signalling "reduced", "less than", "fewer", "increased" and "more than".

The next level is "nutrient function claims". They explain the physiological role of the nutrient in the body. A typical statement shows the form: "*Nutrient X contributes to (physiological role of nutrient X) in the body*".

Beyond that, "health claims" territory starts: Here, relations between a food or an ingredient and a health aspect are stated, suggested or implied.

The most difficult claim level, but also the most attractive one, is the claims that state the reduction of disease risk. Here the claim suggests a clear connection between the consumption of a food and the reduced risk of developing a disease or unfavourable health condition. That's the "Let food be your medicine", but now with scientific proof of a significant alteration of a major risk factor for a disease or health-related condition.

These Codex guidelines served as the basis for national or regional (EU) health claim regulations. The guidelines differ from each other, which is understandable since the nutrition status, local diets and eating habits differ between cultures. Three frameworks dominate: Japan, United States, and Europe.

Japan

Japan was the first country to set up a suitable framework in the mid-80s. It is not a surprise that this movement started there, since Japanese people are world known for their longevity, living in a country and culture with strong beliefs in the health benefits of traditional food, green tea, miso, natto, seaweed and fermented foods. And the sober facts were that the Japanese society had to find new ways to manage the health care cost of a quickly aging population. The government defined a new product category; they called it "Food for Specific Health Uses (FOSHU)", for "*food containing an ingredient with functions for health and claiming physiological effects on the human body.*" (Ministry of Health, Labour and Welfare, n.d.). Under patronage of the government, a national exploration project took up the work in 1993. And when the *Nature* magazine first reported about the initiative, it coined the term "functional food", under the headline that "Japan explores the boundary between food and medicine" (Swinbanks & O'Brien, 1993). The initially very tight regulation was relaxed a bit in 2015, now allowing new claims associated with fatigue, eyes, memory, stress, sleep, joints, blood flow, muscles and body mass index, combined with more flexibility in the clinical protocols and the required data. This opened a route, so that now more than 1000 products have received a claim since the beginning of the FOSHU regulation (Statista, 2023).

United States

The United States followed in the '90s with their first health claim act, but without providing any formal definition of the term "functional" (Martirosyan & Singh, 2015). They split the whole territory into three parts: "nutrient content claims", "structure/function claims" and "health claims".

"Nutrient content claims" inform about the level or range of the nutrient, like "low fat" or "contains 100 calories".

The "structure/function claims" indicate the effect of a nutrient/ingredient on the structure or function of the organism ("calcium builds strong bones").

A "health claim" goes a step further and describes the relation between a food or component and a disease that is a risk for the population. This moves the product into pharma territory; hence, the difference between food and medicine needed clarification. To be regarded as food, the product must have an appropriate nutritional profile, with at least 10% of the daily value for one or more of six key nutrients.

In the "health claim" group, two types of claims exist: The "Qualified Health Claims" and the "Authorized Health Claims".

"Qualified health claims" are those supported by "some" scientific evidence. To inform consumers properly, the "some" in the health claim is classified by the "level of evidence" that must be stated with the claim. This can range from good, "some scientific evidence", to relatively weak, "little scientific evidence".

It's a fair approach, since this approach allows communication about a possible health effect, even at times when the scientific evidence is only emerging. It is often difficult to establish a positive effect for the average population. This is surely more difficult than for drugs, where the individual is diagnosed, the health condition is clinically established, and the "cure" is dosed individually. But for the interested average consumer, not trained in "the art of science", it may be too tough to reflect on the academic strengths of scientific significance.

The final and most competitive group of claims are the "authorized health claims". They require "significant scientific agreement" and are the only ones approved by the FDA through pre-marketing approval. What the "Significant Scientific Agreement" is remains a constant and open expert debate. But a company aiming to go that route would need to show data from double-blind, randomized and controlled clinical intervention trials. Experience shows that this is not achieved by a few sharp clinical trials, but rather a program, with dozens of such interventions (Díaz et al., 2020).

Up to now the FDA has approved 18 of such health claims and the review process alone sometimes takes several years. Hence, food companies willing to take that route need a long-term plan, time, stamina, resilience sufficient budget and a fallback plan.

Europe

While the process in the United States already is not the simplest, the European Union took 10 more years before the member states could agree on the regulation on nutritional and health claims. A leading thought was as well to protect consumers from false and misleading claims and use science to judge that. In many other markets, functional food claims are based on scientific evidence, but in Europe nutrition and health claims need pre-market entry authorisation. Even trademarks, brand names that suggest that foods have beneficial characteristics, should be accompanied with authorised claims.

Since the start, 30 nutrition claims and 265 health claims have been authorized. That means any food business can now use a nutrition claim if the product fulfils the specific conditions (Collins & Verhagen, 2022).

Nutrient and energy claims can be made as content claims, like "source of" or as comparative claim like "*x* less than *y*".

Nonspecific health communication, like "healthy for you" must be accompanied by a health claim, which captures function claims with generally accepted scientific evidence or based on newly developed evidence (Food Safety, n.d.d).

Other articles regulate "reduction of disease risk" or "child health claims".

Anybody wanting to make a health claim must provide the supporting evidence to the EFSA for evaluation. This is a tough process. From over 45,000 submissions made under article 13.1, only 2331 made it into the health claim register and only 262 of those were authorized (European Commission, n.d.b). Today about 40 of such claims have been approved.

Compared to other national regulations, the process has not become simpler or more predictable, yet. In theory, EFSA promises a scientific opinion within 5 months after complete data submission, but before the formal document is accepted for review a lot of time can pass by. French food group Danone can tell a story about the difficulty to get a health claim approved by EFSA (European Food Safety Authority, 2010).

While the various markets still have somewhat different interpretations of what functional foods are, a preliminary consensus circles around formulations like: "*Similar in appearance to conventional food, consumed as part of the usual diet, with demonstrated physiological benefits, and/or to reduce the risk of chronic disease beyond basic nutritional functions*" (Chao & Health Canada, 2000). Or "*foods or dietary components that may provide a health benefit beyond basic nutrition*" as the International Food Information Council (IFIC) described it. The Europeans took the effort to nail down what a functional food is even more watertight: "*… if it is satisfactorily demonstrated to affect beneficially one or more target functions in the body, beyond adequate nutritional effects, in a way that is relevant to either an improved state of health and well-being and/or reduction of risk of disease*" (Henry, 2010).

Relevant for food innovators in EU markets is the adjacent "Novel Food Regulation", in force since 2018. This framework figures the conditions under which a new and innovative food can be brought to the EU market. And "Novel" here is any food that had not been consumed to a significant degree by humans in the European Union before 15 May 1997.

Compared to Japan, where a deep traditional belief in the health power of food exists, the regulators in Europe have chosen a much more protective and conservative approach. Every food item that cannot be proved to have been consumed in Europe before that date falls under that regulation. This includes, for example, vitamin K, extracts from existing Antarctic krill, phospholipid rich oils from *Euphausia superba,* up to a range of agricultural products from third-world countries, like for example, chia seeds, noni fruit juice or wakame. Also included are food products coming from new production processes like UV-treated milk, bread, mushrooms or yeast (Food Safety, n.d.c).

No wonder the European Union in the meantime realized that the application process they put in place needed simplification. But it is still a laborious and sometimes painful exercise, starting with an application with detailed information about the production process, the evidence for a history of safe use in Europe, and full nutritional and toxicological information. This data set the EFSA uses for their risk assessment, which is followed by the legislative process, that finally concludes whether this new food can be approved for European consumers.

The EU member states have shaped this regulation to limit new technologies and especially new biological material from entering the market. In most European countries at least half of the population was sceptical about new biological techniques in food, especially if it was sensed as a "genetic modification".

The result is a framework that does not really support innovations. It gives much more weight to safety than to novelty (European Commission, n.d.a).

An example is wakame! Since the 8th century, this is a popular dried seaweed in east Asia. Originally loved for its salty, but light sweet umami taste, it turned out to have a unique nutritional profile with high amounts of omega 3 fatty acids – one of the best "nutrients-to-energy ratios" known – plus it has good levels of vitamins A, C, E, B1, B2, B6, and B12, iron, magnesium, potassium, boron, silica, selenium, chromium and, of course, iodine. A "superfood" profile! And on top of that, wakame contains fucoxanthin, a chlorophyll pigment that carries the promise of burning fat tissue and hence is a perfect companion for those watching their weight. But it needed years-long Novel Food authorization before it could finally enter the menu in Asian cuisine restaurants in Europe. In the universe of "new food products", wakame is a fast-growing cultural curiosity for those wanting to explore new sensorial impressions.

The sheer number of new scientific insights about food sources or processes provides nowadays many opportunities to build a healthier and more sustainable "future fit food chain". Here, the regulation should evolve further, to ensure consumers protection without blocking solution providers.

Health and wellness are the leading trends in foods, and foods truly need regulation to handle the essential safety aspects of consumption, but also a process is needed to allow companies to make attractive health promises to consumers. Today, that road to success is often too long and bumpy.

Evidence from *in vitro*, animal and observational studies is only the starter for well-designed human studies, and perhaps a couple of them. In contrast to all sensorial food attributes, nutritional and health benefits cannot be perceived immediately. A positive effect often needs time, and the buying consumer needs confidence that the extra investment into such premium food is paying back and supporting a better and healthier or longer life. And when indeed more people invest extra into healthy eating, that can improve public health at lower health costs for the society, which in the end was the original intention of the first functional food regulation.

Elsewhere in the world, national nutrient and health claim legislation also shows differences, often caused by different sets of available data about nutritional intake of the population, their diets and the view of the accessors on the evidence provided. A few years back, Prof. E-Siong Tee from Malaysia compared the various claims allowed in Asia and found 77 allowed nutrient function claims in China, 56 in Thailand, but only 17 in Japan and not a single one in Brunei (E-Siong Tee et al., 2021).

ALLERGEN MANAGEMENT

Allergens are characterized as substances that trigger an abnormally vigorous immune system response. A relatively harmless challenge to the body causes the immune system to fight. The consequences of these overreactions can be fatal. Symptoms of food allergies include hives, swelling, itching, anaphylaxis and dizziness. Food allergies are common. In fact, it is estimated that about 10% of the population are affected and these percentages are rising (Fong et al., 2022). Although it's possible for any food to cause an allergy, most food allergies are caused by just eight foods. According to the Food and Drug Administration (FDA), eight food items cause 90% of food allergic reactions in the United States. These are cow's milk, eggs, tree nuts, peanuts, shellfish, wheat, soybeans, and fish. Allergies to mustard, celery, lupine, sesame and molluscs also can pose a problem. Although up to 60% of allergies occur in childhood, some symptoms of allergies disappear in adulthood. Food allergies can show up at any time in our lives, even during older adulthood.

Often, in discussions the terms food allergy, food intolerance, and food sensitivity are used interchangeably, but scientists and medical experts do distinguish between them.

The allergic reaction is the most severe problem, and it happens as an exaggerated response by the body's immune system against a seemingly harmless substance in the food. The classic example is the potentially life-threatening difficulty to breath combined with a drop in blood pressure as the result of exposure to peanuts or seafood.

Celiac disease is frequently reported as another prominent example of an allergic reaction, where gluten is the allergen. It affects about 1% of the Western population. And the consumption of gluten indeed initiates a complex inflammatory reaction that can make people with celiac disease very sick. Current clinical data suggest that in celiac disease, cells of the small intestine are damaged by an immune reaction that is triggered by gluten. The intestinal mucosa is damaged, and the small intestine loses the ability to absorb nutrients. Symptoms are diarrhoea, flatulence, abdominal pain, joint pain, weight loss, fatigue, anaemia symptoms, even osteoporosis, a higher risk of developing some tumours, impaired blood coagulation and malnutrition.

Some clinical experts do not define celiac as a true allergy, basically, because, in contrast to all other allergies, the immune response is not mediated by immunoglobulin E.

Avoiding gluten is the only solution for this problem. Gluten is found in a large variety of grains, including wheat, rye, barley, semolina, bulgur and farina. Many processed foods also contain gluten. People with celiac disease must also be careful about cross-contamination, when a gluten-free food meets a gluten-containing food.

The symptoms of food intolerance include gas, bloating, diarrhoea, constipation, cramping and nausea. Symptoms of a food allergy include hives, swelling, itching, anaphylaxis and dizziness. Food intolerance is the difficulty to digest certain foods. Up to 25% of the population may be affected (NSW Food Authority, n.d.).

The best-known example is lactose intolerance. The reason for this is a shortage or absence of the enzyme that can split lactose, the main milk sugar. The intestinal lactase activity peaks at birth and reduces progressively with weaning. Only in some adults the enzyme remains sufficiently active. Globally, about 35% of the adult population has lactase persistence, with a large variability across different geographic regions, depending on dietary culture, with a persistence of up to 90% in the Nordic populations versus 10% in Southeast Asia.

Like most mammals, newborn babies can digest milk sugar from breast milk without difficulties. But towards the end of the weaning the expression levels of the lactase gene decreases, when the babies diet shifts to non-dairy sources.

Where then is this remarkable difference between populations in lactose sensitivity coming from? A widely accepted explanation for this striking difference is the "gene–culture coevolution hypothesis". The positive selection of the lactase persistence phenotype comes from the observation that pastoralist populations often show high levels of lactase persistence (Wikipedia Contributors, 2024c). Horse-breeding steppe people acquired the ability to digest milk sugar. And when these people started to move westward some 4000–6000 years back, regular consumption of milk in West European started. As a rich source of nutrients, milk gave those populations an evolutionary advantage. And this advantage was bigger than the original discomfort coming from the inability to digest milk sugar.

For example, in East Asia, historical sources also attest that the Chinese did not consume milk, whereas the nomads who lived on the borders did. This reflects modern distributions of intolerance. China is notably a region of poor tolerance, whereas in Mongolia and the Asian steppes, milk and dairy products are a main nutrition source.

Those affected by lactose intolerance have more lactose sitting in the digestive tract, which can cause stomach bloating, inflammation and diarrhoea. Compared to a real allergy, lactose intolerance is less dangerous. That is because in the small intestine indigestible food moves into the colon, where bacteria start to break it down. The result is an influx of hydrogen gas and water into the colon, which causes the well-known symptoms, such as cramping, diarrhoea, bloating and flatulence (Harvard Health, 2023).

Yet another unwanted effect after eating certain foods are symptoms not related to food intolerances or food allergies, or celiac disease. These are called food sensitivities. Though there is controversy around what exactly happens in the body of someone with a food sensitivity, it appears that exposure to specific foods may create an immune reaction that generates a multitude of symptoms. The symptoms are not life-threatening, but they can be quite disruptive and include joint pain, stomach pain, fatigue, rashes and brain fog. Gluten is probably the best-known trigger of food sensitivities. Avoiding dairy products is a surefire way to avoid symptoms; some, like milk, tend to produce more severe symptoms than others, like yogurt and cheese.

The best tool to identify a food sensitivity is careful observation. By removing different foods from the diet for a couple of weeks, and then one by one reintroducing them, its often possible to identify the food item causing symptoms.

A recent review, looking at published data, evaluated the potential of pro- and prebiotics to help with digestion of milk products in lactose-intolerant people. Five studies out of a total 830 articles found in the literature made it through the quality cut. And in those, two strains, *Limosilactobacillus reuteri* DSM 17938 and *Lactobacillus acidophilus* DDS-1, showed promising results, together with the prebiotic GOS (RP-G28). These appeared to be efficient in reducing symptoms after the ingestion of milk products (Giugliano et al., 2023).

During the last century, more and more people globally are self-reporting lactose intolerance and starting to avoid dairy products. So, it is no wonder that the global dairy alternatives market size in 2023 was already valued at nearly30 billion U.S. dollars and estimated to have an average annual growth of nearly 14% (Fortune Business Insights, n.d.). It seems driven by food lifestyle changes especially in the developed western markets.

But an undogmatic look at commercial plant-based beverages reveals that the majority of products are nutritionally not fully comparable to animal milk, having lower protein content and amino acid scores, less essential amino acids and less micronutrients such as calcium, vitamin D, potassium, magnesium and riboflavin. Until today, there is no evidence that the compositional difference between the plant-based milk drinks based on soy, oat or rice is impacting the quality of the overall diet. But some authors recommend further studies on the influence of the processing of the plant-based alternatives.

Due to the massive impact allergen intake can have on human health, the allergen regulation for food sets detailed rules to protect consumers suffering from food allergies or intolerances. The food business, industry, traders, restaurants and other operators like street food vendors must provide accurate and clear information on the presence of substances that can cause allergic reactions.

Regulations all over the world list the substances that cause allergies or intolerances and that must be indicated on the labels of prepackaged and non-prepackaged foods. The Annex specifies the names of these substances or products, such as cereals containing gluten, or crustaceans, eggs, fish, peanuts, soybeans, milk, nuts, celery, mustard, sesame seeds, sulphur dioxide and sulphites, lupin and molluscs. Furthermore, companies must prevent or limit the cross-contamination of allergens in their equipment, conveyances and containers, and/or use precautionary allergen-labelling statements (for example, "may contain") indicating that the absence of allergens in the products cannot be guaranteed.

For a company to claim gluten-free on the label, the product needs to have less than 20 parts per million (ppm) of gluten, confirmed by enzyme-linked immunosorbent assay (ELISA) or liquid chromatography-mass spectrometry (LC-MS/MS).

And similar to food allergies, immune-mediated food hypersensitivity has emerged as a major public and personal health burden in the last centuries. According to WHO, the prevalence of food allergies has increased in both developed and developing countries over the past few decades, especially among children. The exact causes of this increase are not fully understood, but some possible factors include genetic susceptibility, environmental exposure, dietary changes, hygiene hypothesis and microbiome alterations (Food-Safety-Food-Allergies in the Allergen Management, n.d.).

So clearly, with globalisation of the food business a harmonized global approach by all actors in the food sector is critical. Good hygiene practices (GHPs), and HACCP systems in manufacturing, retail and food service have become global standards as allergen management practices.

The control of the critical control points here must acknowledge that processes like heating, and high-pressure processing often do not inactivate the allergenic component, which is often a protein. And even special treatment for protein activation like enzymatic degradation or acid hydrolysis may destroy allergenic proteins.

Beside the major eight allergens (crustaceans, eggs, fish, milk, peanuts, soybeans, tree nuts and cereals containing gluten), other food allergens such as sesame seeds, buckwheat, celery, mustard, molluscs and lupin are listed in some countries.

The allergen management approach needs to be a truly global end-to-end (E2E) one (Home | CODEXALIMENTARIUS FAO-WHO, n.d.). It starts already with harvest, storage and transportation of the primary agronomical good, the effective cleaning of containers and transport vehicles and safe physical separation or storage of commodities with different allergen profiles and employee.

The branded food manufacturer must show correction of label errors, full control over cross contamination, effective cleaning/sanitation, care with rework handling and a proper production sequencing and robust and transparent change in management procedures to handle formulation, ingredients, process and supplier changes.

Food manufacturers frequently handle multiple allergens, often on the same processing line or shared equipment. Most often they do not have the opportunity to run allergens on dedicated lines. However, through efficient production planning, it is often possible to minimize allergen contamination, for example, by running allergen-containing products at the end directly before the next full line cleaning. This needs to include the control of cross contamination between different lines in a production hall or factory. Such cross over points need to be identified in the hazard analysis and then controlled.

An alternative is the redesign of the manufacturing process, by defining a processing design, in which the allergen is added as late as possible in the process, so that the whole "upstream" remains unaffected.

With the allergens defined and procedures and protocols to control them in place, the important practical question is, When are they present and when are they not? Or as the expert would call it: What are the established threshold levels in foods for the priority allergens?

Again, the FAO/WHO took on the task and in 2021 assembled an expert committee, comprising scientists, regulators, physicians, clinicians, and risk managers from academia, government, and the food industry to set those limits. Looking at different approaches, the committee concluded that the principle of establishing a "Benchmark Dose" and the "Probabilistic Hazard Assessment" would be most suitable (DSpace, n.d.).

A benchmark dose is a start value from which a reference dose can be built. A dose–response assessment! The other difficult concept is the "probabilistic risk assessment", a systematic and comprehensive methodology to evaluate risks associated in complex technological settings (like a nuclear power plant). And like in all risk assessments, two main angles are critical here: 1. The magnitude (severity) of the result of the risk, and 2. The likelihood (probability) of occurrence of that risk consequence.

In this context: For the general population the probability of being lactose intolerant is significantly higher than the probability of suffering anaphylactic shock from milk.

The result are values for reference doses given as milligrams of protein from the allergenic source. Depending on the allergen, this value differs for many between 1–5 mg, whereby milk and sesame do not yet have one.

Having a value now demands a quantitative measure! For most allergens, quantitative ELISA kits are most used. They confirm the absence of an allergen. Other methods such as polymerase chain reaction and liquid chromatography coupled to mass spectrometry (LCMS) may be even more sensitive but are simply not suitable for routine analysis in the food industry. With a hazard analysis for allergens, threshold levels for those substances and sufficiently sensitive, robust test methods, a sufficient control of allergens in the food chain is possible.

But sometimes the absence of an allergen in a food product cannot be guaranteed, or the inclusion of the allergen may even be intended, as it is a value given part of the formulation. In these cases, clear information is key! And this is important not only in the retail food industry, but in food service as well, where food items are brought to customers, outlets, restaurants, canteens, hospitals, etc. The line of thinking here is that consumers are aware of their own allergies or insensitivities. In contrast to many others, these people are reading the back-of-pack information to understand the food composition and make sure nothing they are sensitive to is in the package.

So, food items can contain known allergens, but they must be emphasized clearly in the list of ingredients, for example, by using a different font, letter size or background colour, so everybody can see it easily. Restaurants clearly state it on the menu.

From this discussion, it should be clear that responsible allergen management in the food chain causes substantial effort in all angles of the business. And there may even be situations where a cross contamination cannot be fully excluded. For those cases, a "precautionary allergen labelling (PAL)", sometimes called advisory labelling, is possible. This refers to voluntary labelling to indicate that one or more allergens could be unintentionally, but unavoidably, present in a product and thus pose a risk to susceptible consumers. The statement is by no means an alternative to GMP or HACCP, but allows for the possibility to market products that despite all care in the manufacturing could still contain allergens.

This option triggers a not untypical discussion between company functions. Example: For a range extension, an allergen-containing variant should be added to the portfolio. The handling and management of that allergen in a manufacturing process shared with other products causes extra investment into equipment, regular assays, new management procedures, a different production scheme and extensive cleaning. The discussion between marketing, supply chain and R&D would centre around the question: What would be less profitable, the new complex manufacturing regimen and the cost its adding or the potential loss of consumers/customers due to the precautionary labelling?

However, precautionary allergen labelling should only be used when, following a thorough risk assessment, a genuine risk of allergen cross-contact within the supply chain is identified that cannot be removed through careful risk management action. Excessive use of precautionary allergen labels can unnecessarily limit consumer choice and devalue the warning for consumers with a food allergy or intolerance.

LIFESTYLE, DIETARY RESTRICTIONS

In many cultures religious sentiments play a crucial role in the food selection, with a direct implication on consumers based on the faith they follow. Some of them are very old! Hinduism started about 4000 years ago in today's India, and nowadays there are approximately 1.2 billion Hindus worldwide. Like other religions Hinduism has dietary restrictions, but no central authority for their dietary laws. It is more the principles of the Hindu faith asking to avoid some foods. Central is the "Ahimsa" concept, which demands non-violence and compassion towards all beings. Today, 83% of Hindus have meat restrictions; 44% say they are vegetarian. And amongst the others, there are many that abstain from eating non-vegetarian food during special days or festivals such as Janmashtami (Blazina, 2024).

The country with the most Hindus is India. Here the primary law for the regulation of food products is the Food Safety and Standards Act (FSS), which is constantly updated about the vegan food regulation, so that people have clarity about their food intake (FSSAI, n.d.). A specific Hindi food logo does not exist.

Similarly, also for the about 2.6 billion Christians on the planet, dietary guidelines still exist. Details vary by denomination. Their holy book, the "New Testament", demands to "*abstain from food sacrificed to idols, from blood, from meat of strangled animals*". Most Christians today follow food restrictions during periods of fasting or as part of a specific season or once a week. A particular food regulation or pack logo is not in place either.

But that is different for other main religions! For the more than 2.0 billion Muslims in the world and the about 14–15 million Jews, specific food regulations are very alive, and they get dietary guidance through on-pack logos.

Globally not all Muslims eat halal food and not all Jews eat kosher food, as some may follow different interpretations or practices of their faith. Moreover, some non-Muslims and non-Jews may also eat halal or kosher food for various reasons, such as health, ethics or curiosity. Therefore, the actual number of people who eat kosher and halal food can be higher or lower than the estimated number of adherents of each religion.

Halal

According to Statista, the global halal market was valued at 2 trillion U.S. dollars in 2021 and is expected to reach 2.6 trillion U.S. dollars in 2025 (Statista, 2024a). "Halal" is an Arabic word that means "permitted or lawful". It is used to refer to any behaviour or object that is allowed under Islamic law. Halal especially relates to the Islamic dietary laws and particularly to meat that is processed and prepared according to these requirements.

The most common example of "haram" (non-halal) food is pork. While pork is the only meat that categorically may not be consumed by Muslims since their holy book, the "Quran", forbids it, other foods that are not in a state of purity are also generally considered "haram".

Islam generally considers every food as halal unless it is explicitly forbidden in a hadith or the Quran. Specifically, halal foods are:

1. Made, produced, processed and stored using machinery and equipment that have been cleaned according to Islamic law (sharia).
2. Free from any component that Muslims are prohibited to eat according to Islamic law.

Religious sentiments play a crucial role in product selection, with a direct implication on consumers based on the faith they follow, and a legitimate labelling has become an inevitable tool for consumers for their product choice.

Several food frauds in the past years have led to an accelerated effort to create sufficiently sensitive analytical methods that are able to detect pork parts, like gelatine. Nowadays, species-specific target DNA amplification, polymerase chain reaction (PCR) or monoclonal antibodies are in use (Premanandh & Salem, 2017).

In a couple of Southeast Asian countries, the respective governments have established institutions dedicated to halal certification, following the international standard ISO/IEC Guide 65:1996, which specifies the general requirements for a third-party operation. Fundamentally, it sets the comprehensive requirements for the food-manufacturing and food-servicing chain from processing to handling, distribution, storage, display, serving, packaging and labelling according to Sharia law. Hygiene, sanitation and food safety are also part of it.

But there are varying standards from one country to another. For example, a difference is made between Malaysia halal and Indonesia halal. In both countries the "halal type" is governmentally legislated, voluntarily, but they use locally different certification bodies, JAKIM, MUI, and BPJPH (Nuraini & Sucipto, 2021).

While the basic definition for halal and haram is very clear, there a different view in Islamic schools of what would make a product/process halal. Basic changes in the physical and chemical properties of a substance (physiochemical alteration), either natural or manmade, are accepted as halal by some, while other religious authorities only accept natural changes. That affects not only products like gelatine, but also food flavourings and enzymes of animal origin. The details about the stunning of animals remains also as a controversy. All the different interpretations make it complex and time consuming for companies to trade halal products over borders or for Islamic tourists to choose their food abroad. Products need to be certified, production places and lines need to be inspected and new alternative materials or supplier must be audited, all by different authorities and different schemes.

For food industry and consumers in various countries, the establishment of a single standard would surely make things easier. Combined with a robust tracing system and constant monitoring, this would enable a faster and cheaper market introduction of new halal products from other markets, giving the industry better planning tools and predictability in the certification process. And the unification of standard and analytical tools would help to limit the risks for food fraud. For the time being, international actors often take the Malaysian JAKIM standard as the benchmarking scheme, since this one covers many others.

Kosher

The Jewish religion requires food to be "kosher". It therefore must conform to the regulations of kashrut, the Jewish dietary law. Kosher food shares some aspects with halal food, such as the prohibition of pork, blood and certain types of animals. But it also has its own rules, concerning, for example, the separation of meat and dairy products and the restriction of certain cuts of beef. Here, dairy products must not be mixed or eaten together. This means that separate utensils, dishes, and cooking equipment must be used for meat and dairy, and a waiting period must be observed between eating them.

Meat must be slaughtered in a way that is considered humane and painless, called "shechitah". Certain parts of the animal must be removed before the meat can be eaten.

Fruits, vegetables, grains, and other plant-based foods are kosher, unless they are contaminated by insects or non-kosher substances. Some produce grown in the Land of Israel may have additional requirements, such as tithing or leaving the land fallow every seven years.

Kosher food must also be certified by a reliable authority that ensures the standards of Jewish law. A symbol or label called "hechsher" indicates that a product or establishment is kosher. Like for halal, there are different types of hechshers for different levels of kosher supervision reflecting the different

FIGURE 11.4 A selection of halal labels used for food.

considerations of local Jewish communities in different countries. The Kosher Alliance is one of the world's largest and most widely recognized kosher certification organization. The process as such is similar to the halal procedures. A trained religious expert, the "Rabbinic field representative" (RFR), visits the production site and investigates products, lines, equipment, ingredients, supplier and other business partners, and may even want to understand the ownership structure of the company.

On both diets, a religious person must recite prayer before the slaughter. On a kosher diet, the animal must be slaughtered by a Jew who says a blessing before and after the slaughter.

For both dietary guidelines, various labels exist for processed products, based on the certification process the organisation applies. A selection is given in Figure 11.4.

The interest in both dietary guidelines is growing. Not only because the respective populations are growing and there is an increase in tourism, but also because the stringent food and processing rules appeal nowadays to more consumers, even those not following any religious dietary restrictions. For them the tight regulations give assurance that the products are clean and safe and the food producer has adhered to ethical standards in the animal care and meat treatment. These very old regulations show what we nowadays call transparency, traceability and animal welfare. And the ongoing monitoring of hygiene and cleaning processes provides reassurance to the consumer/customer that the regulation is not only "live on paper".

OTHER ON-PACK CLAIMS

As described before, many on-pack claims are meant to inform consumers about their nutritional, health, societal or environmental benefits and they must comply with well-defined and tight regulations and standards. Beside the mandatory information, many food products have additional information providing more, often less tangible, data.

During the last 150 years, food in general has become safer through the wide application of new processing and preservation techniques. Generally, this has led to products with extended shelf life, reduced waste, and often also a better nutritional profile. But details of the technologies, the usage and background were often not shared with the consumers or they were not explained in an understandable way. The industry preferred a dialogue between professionals to agree on and define the technical standards and how they could be translated into consumer language. The consumer then was expected to trust the outcome of the expert conversation. But the distance between food manufacturer, regulators and consumers grew. And big food scandals like dioxin, bovine spongiform encephalopathy (BSE) or melamine-contaminated milk and horsemeat left suspicion, and a feeling of a conspiracy between industry and regulators.

Therefore today, consumers look more sceptically at the information provided on the pack and elsewhere about the production methods and components of the food products they buy. And clearly, some production methods are perceived as less "natural", like conventional versus organic farming, and food components like artificial additives are quickly classified as "unhealthy" and "unfamiliar".

But how natural is an industrially processed food product or can it be? Where are the benefits of food processing? And what are the disadvantages and value distractors of such industrial processing?

"Natural food" is somewhat the gold standard, the innocent state of purity and wholesomeness. So important to consumers, yet frustrating for the product makers due to the lack of any formal definition of "natural". While there is no definition of natural for packaged food, science focussed on the perception of naturalness and how the consumer rates it in various products. Science and regulation cannot provide sharp criteria to classify packaged food as "natural" or "clean label". Much research has been done to capture the "consumer understanding", as this provides a different approach for claim regulation: The expectation of the consumer! Results showed, for example, that perceived naturalness of the colouring agents and preservatives can vary substantially within and between product categories, while stabilizers/thickening agents were generally seen as less natural. But only as long as they are described by their function, not when expressed by their source. Socio-demographic factors, consumers' age and their knowledge on food ingredients also influence the perception (Maruyama et al., 2021).

There is not that single "killer criterion" for naturalness perception for packaged food. The claim regulations are then based on the consumer's perception of what natural or clean label means and how food manufacturers market their products. That's also why clean label can vary depending on the product, the consumer, and the manufacturer. Typical examples of clean label foods are fresh fruits and vegetables, whole grains, nuts and seeds, eggs, dairy products, and meat and poultry. However, some packaged foods can also be considered clean label, depending on the product, the consumer and the manufacturer. As some consumers accept foods containing colouring foods as alternative to colourants and yeast extract as natural alternatives to flavour enhancers, these products may be communicated as clean label.

Various studies with consumers in various countries on this very important topic of the industry constantly reveal different cues playing roles and influencing each other:

- The impression of the ingredient list as such.
- The identity of the ingredient, how natural they sound (fruit pulp, yoghurt, …).
- Naming options of ingredients on pack ("Sodium Bicarbonate" or "Baking Soda").
- The classification of ingredients as natural or artificial.
- How familiar the ingredients are (wheat flour vs. sorghum flour).
- How familiar the chemical names are.
- The lengths of the ingredient list, shorter ones with 10–15 ingredients are perceived as more natural than >15.
- The perceived pre-processing, physical versus chemical treatment.
- Original source of additive: Plant based, flower, fruit versus animal based, insect based.
- The association of the product with health. A candy or snack is not seen as nutritious.
- The actual combination of the artificial additives, like flavours and colourings.

Artificial colourants came into focus when observational research linked them to attention deficit disorder and attention deficit hyperactivity disorder in children (Arnold et al., 2012).

ORGANIC/BIO/LOCAL

Organic, bio, and local food is an attractive and continuously growing market! Already worth 18 billion U.S. dollars globally in 2000, this segment of bio has reached about 132.74 billion U.S. dollars in 2021 and is expected to grow by double digits into the next decade at least (Statista, 2024b). The major factors driving the demand for organic food include the growing awareness of the health benefits, environmental sustainability, and ethical consumption and food product quality (Basha et al., 2015).

At its core, organic labels promise food products made without the help of synthetic chemicals, such as pesticides, fertilizers, hormones, antibiotics and genetically modified organisms.

Whether there is a difference between bio and organic food in regulation often depends on the country where the question is asked (where the food is produced or sold). In the United States, organic food is regulated by the USDA National Organic Program. It sets the standards for organic production, certification and labelling. Bio food is not a regulated term and can be used interchangeably with organic or natural food, but without a legal guarantee.

"Local food" or "locally produced" are claims living in a somewhat overlapping mental space with "bio" or "organic" for consumers. Most of them rightly do expect as a criterion a legal limitation in terms of transport distance or same region, etc., as something that connects "local" with a reduced environmental impact and lower carbon footprint. But from there the perception often moves further into core organic territory, like no fertilisers and non-GMO; something factually not included in the "local" claims. And both are then sharing an expectation of being "more nutritious" and "better tasting" (Campbell et al., 2014). These positive associations are activated in marketing campaigns to sustain a price premium.

In the European Union, bio and organic are basically interchangeable and describe food that complies with the strict rules of the EU organic regulation. This regulation defines production, labelling, control and import rules for plant and animal-based products, as well as for processed food and feed.

European bio food regulation is anchored in the principles of organic production, aiming to preserve natural resources, ensure animal welfare and produce high-quality food without harming the environment or human health. The regulation covers not only food but also other agricultural products, cork, salt, essential oils, cotton and wool. Core rules are to avoid genetically modified organisms, rely on green energies, maintain and improve soil fertility, avoid soil erosion, implement crop rotation, transform organic waste into resources, respect high animal welfare standards and promote rural development.

Most food offerings in the trade are not pure, they are composites, made with a variety of several ingredients. Regulation therefore had to establish a limit: Only products with at least 95% by weight of organic ingredients can be labelled with the EU logo. And the list of ingredients shall indicate which ingredients are organic.

The assurance comes for the consumer in the European Union via distinctive "marks" that guarantee that a product complies with the EU organic standards and has been verified by authorized bodies. The logo also indicates the origin of the product and the code of the control body.

NATURALNESS AND AUTHENTICITY

Another very appealing proposition in the packaged food market is "naturalness". It is often used in food marketing and labelling to appeal to consumers preferring food products that are minimally processed, free of artificial additives, and/or derived from organic or biodynamic farming. Some experts anchor it in the clean label territory, while others articulate the differences between both concepts.

This expert discussion already indicated the "*vagueness and indistinctiveness of the concept, enhanced by a simultaneous lack of regulatory definition of the term.*" (Battacchi et al., 2020). The difficulty to establish a clear or consistent definition of what naturalness means in the context of packed food is reasonable. A purist view could be that anything that has been packed or (minimally processed) should not be called natural. Others would point out that there is a clear difference between a food product that has not been modified, carefully and minimally processed and without anything added. The legal guideline that exists are some general principles that require food claims to be truthful, not misleading. They, of course, apply here as well. But naturalness in foods is not regulated in a systematic or comprehensive way, and consumers have difficulties in verifying the authenticity and quality of natural foods.

Naturalness in packaged food is consumer perception. And different consumers have different expectations and perceptions of what naturalness then means. And the food manufacturers use different tools and criteria to fulfil the expectations. One option is to qualify the products as organic, preferably minimally processed and locally sourced. Others may put their hook to the expected benefits like to taste, freshness, or the overall appearance of the product. But highlighting the traditional roots of a recipe, or on the contrary communicating the usage of innovative new technologies, for a better nutritional profile, or a more environmentally friendly sourcing or processing, is a used entry point.

Most studies done around food and naturalness found inconsistent understanding of naturalness. But the interpretation of naturalness is also inconsistent. The claim is sometimes interpreted as organic or locally grown, healthier, but inconsistent from country to country (Murley & Chambers, 2019). This 2019 study done with consumers in the United States, United Kingdom and Australia demonstrated that consumers look first of all at the total product primarily to form a view about naturalness: The origin of the ingredient, like plant based, plays a role, and under the artificial additives, artificial flavours seem more acceptable than artificial colours, maybe an echo of the work done on the Birmingham 6 (E number colourings, discussed earlier in this chapter). Males, Millennials and educated participants have higher naturalness scores than other participants in their respective demographics.

The broadly agreed consensus says that highly processed usage of GMO, the usage of artificial, non-natural ingredients or processing aids would prohibit the use of "naturalness" (U.S. Food and Drug Administration, 2018).

Recently, key stakeholders made attempts to frame the concept better. The International Organization for Standardization (ISO) published in 2017 the "technical specification (TS) 19,657" as "Definitions and technical criteria for food ingredients to be considered as natural" (ISO/TS 19657:2017, n.d.). This document was a first answer to the growing consumer interest in natural food products, in the absence of a clear and consistent definition of the term "natural" in the food industry. The definition suggested the document is based on the following principles: The ingredient should be derived from a natural source, such as a plant, an animal, a microorganism or a mineral; the ingredient should not undergo any chemical modification or processing that alters its essential nature or properties; and the ingredient should not contain any artificial additives, such as preservatives, colours, flavours or sweeteners.

This document is not legally binding, but it is already instrumental as a basis for the business-to-business (B2B) discussion between the food ingredient industry and consumer goods manufacturers on this important topic.

"Authentic" is another attractive claim or statement appearing on food products! Authenticity refers more to the degree to which something is true to its original or essential reference. A great recipe for Turkish "baklava", Mexican 'takos" or Indian "butter chicken" gets extra value by being perceived as authentic and very tasty, but not by being nutritious or natural. For authentic Japanese "sushi", Vietnamese 'Pho soup", German "sauerkraut" or Korean "kimchi", the vicinity to natural may be much smaller. But both may rather be complementary aspects of good-quality food and great culinary heritage.

CLEAN LABEL

Clean label is a term that refers to food products that have mostly natural and familiar ingredients and aim for minimal processing. Clean label as such is not often found on pack. Additionally, the label often contains clear information about how ingredients are sourced and how products are manufactured.

As such, it's more an umbrella term, capturing all the products that use "less", "pure" and "without" claims. The "claim territory" is not firmly defined, and there is no specific regulation for it. But there are some general rules and guidelines that apply to food labelling and claims. Examples are:

- EU Regulation No. 1169/2011 is about the provision of food information to consumers and states that food labels must be clear, accurate and easy to understand (Food Safety, n.d.b). They must not mislead consumers about the characteristics, effects, or origin of the food. They must also provide information on the ingredients, allergens, nutritional values and expiry dates of the food.
- EU Regulation No. 1924/2006 captures nutrition and health claims made on foods and sets out the conditions for making claims about the nutritional or health benefits of foods (Food Safety, n.d.a). Such claims must be based on scientific evidence and authorized by the European Commission.

On a factual level. clean label foods are not automatically healthier, more nutritious or safer than other foods. Even so, they appeal to consumers closer to their natural state. The making of the famous premium French raw milk cheese, made from unpasteurized milk, can fulfil all the criteria of the clean label, but it bears a high risk for microbial infections if the producer does not follow strict sanitation protocols and good manufacturing practices (GMP). The American FDA regards the risk of this clean label product so high, that they decided to forbid the sales of this masterpiece of French cuisines. (See also chapter 16.)

Today the value and logic of such claims is often difficult to understand. "Virgin olive oil" is not 100% natural, but "extra virgin" is! Both are made from unrefined oil, but one has a few more natural defects and some more free acids. And both are not cold pressed nowadays but purified in a decanter (Eckelkamp, 2023).

Since science and technology cannot fully define the claim conditions, attempts are made to capture the essence of the consumer understanding of clean label claims (Asioli et al., 2017). The result is, simply spoken, not simple! A lot of different factors, product, situation or personality related, influence the perception. Health often seems to stand in the background driving concerns about modern technology processing, together with scepticism towards some ingredients and a lack of trust.

It is helpful that a few related claims like "natural", "organic", "free from" and "no added" are better fixed and have specific requirements in the EU.

- "Natural" can only be used for foods that have not been processed or only in a way that does not alter their essential characteristics. It can also be used for certain ingredients derived from natural sources, such as natural flavours or colours. However, it cannot be used for foods that contain additives, artificial sweeteners, flavour enhancers, or that come from genetically modified organisms.
- "Organic" as a claim is limited to foods that comply with the EU Regulation No. 834/2007 on organic production and labelling of organic products. This regulation sets out the standards and rules for organic farming, processing, certification and labelling. Organic foods must be produced without synthetic pesticides, fertilisers, antibiotics, hormones or GMOs. They must also bear the EU organic logo and indicate the code number of the control body and the place of farming.
- "Free from" can only be used for foods that do not contain a certain ingredient or substance that is normally present or intentionally added in similar foods. For example, "free from gluten" can be used for foods that contain less than 20 mg/kg of gluten, which is the threshold for people with coeliac disease. However, it not allowed for foods that naturally do not contain gluten, such as rice or potatoes.
- "No added" can only be used for foods that do not contain any added ingredients or substances that are normally used for technological or sensory purposes. For example, "no added sugar" can be used for foods that contain only naturally occurring sugars, such as fruits or milk. It cannot be used for foods that contain sweeteners or other substances that enhance the sweetness.

The whole clean label territory is constantly moving. Regulators try to articulate the essence better; consumers are more interested to understand the details of the processing and sourcing, and environmental expectations have joined the health expectation. For food manufacturers that want to convey a "clean product", it might be more important to meet the consumer expectation than just comply with "regulator" requirements. Ingredients like high fructose corn syrup, modified starches or refined fat and oil are legally not excluded, but for many consumers these ingredients do not qualify for such a claim. Using them in combination with such a claim can heavily impact the product or brand.

THE CERTIFICATION FLOOD

A certification mark is a graphical on-pack presentation showing consumers that a product has met certain standards. It is an effective and convenient way to reassure consumers about important –often not visible – properties of the food products. "Eco labels", for example, indicate a particular environmental performance of a product, verified by an independent organization. But today there are too many of those. A recent counting by Big Room Inc. resulted in already 456 different labels only in the Eco range (Ecolabels, n.d.).

This label overload undermines the initial purpose. Consumers become confused and lose trust and normally have neither the time nor leisure to become familiar with the details between various certification criteria. So, any innovation team that wants to effectively inform about an intangible characteristic of their food product, like Eco, organic or bio, needs to discuss the "strength of the vehicle". Two aspects are leading here: first, of course the reputation of the underlying verification scheme. Some of the most popular ones are: "Organic" (EU or USDA), "Fairtrade", "Rainforest Alliance" or "Marine Stewardship Council (MSC)". The organisations focus on different aspects and have different standards. The NPF team has an option to decide which aspect of the product they may want to highlight: Organic agriculture, fair wages and trade, sustainable sourcing or care for marine life or other parts of the environment.

That's a decision on content! A different one is the decision based on design. "Semiotics" is the study of signs and symbols and how they communicate meaning. Contemporary research in this area for food labels looks at the neural efficiency of information processing from the sensory registration into short-term memory and long-term memory. With more and more labels, symbols, packs in the shop, information front and back of pack, different designs, sizes and colours, it is becoming more important to improve the strengths of the signs to become anchored in the brain with the intended message. A recent study at Jönköping University provides a good starting basis (Schollweck & Heidelberger, 2019).

All those popular on-pack claims are somewhat related. Some areas are more clearly regulated; some are more directly linked to health or environmental concerns. So, the claim a development team is aiming for and how it should be communicated requires a good understanding of the target population.

Why are there so many different "on-pack" claims? It´s because there are several scientific theories about the decision-making processes of consumers when they are shopping. Applying the findings of those psychological concepts can be a competitive advantage. I will briefly introduce here two of them: The "regulatory focus theory" and "dual processing theory".

Regulatory Focus Theory

People buy a product because they identified a need for it and in the moment they articulated that need. It becomes a goal, which leads to an action, here the buying of a food item. The regulatory focus theory suggests that people's actions can be separated into either a "prevention" or "promotion focus" (Scholer et al., 2019). A consumer might be motivated to get something, for example some nutrients or more abstract health and well-being. Or the consumer might be motivated to avoid something, for example, a food-related risk of a disease. Taking this into consideration can help to "land a better on-pack message".

And consumers show bias in their food choices. A negative bias can be a single, negatively perceived ingredient: For example, an artificial flavouring or a single E number can already "infect" the overall assessment, while in a positive bias a single positive ingredient can lead to an overall positive assessment of the food. Whether the choice for a "natural" or "organic" claim is now a promotion or avoidance reaction is still debated. But a preference for "clean label" seems more motivated by avoidance of the typical attributes of packaged food.

For the product sales, it is more promising if the message fits the consumer's mood state. But there is no simple toolbox telling which claims would address which mode state.

So, hard and binary rules are difficult to establish, but a few elements important in the "clean label territory" are important:

- The ingredient list should be short, simple, and feature minimally processed ingredients.
- It should not contain more than five ingredients or ingredients that are difficult to pronounce.
- It should not include names that sound like chemicals or E numbers.
- It has to be free from chemicals (or any ingredient that is difficult to pronounce).
- Clean labelled products should be natural, organic, and/or free from additives/preservatives.

Not every product category has the same degree of freedom to make claims in that territory.

A candy will have more difficulties to be seen as natural, while the label on a dairy snack may not even have to say it explicitly! For a more natural category, some extra formulation effort and limited oncosts can be a promising approach. A category like ice cream, although not a natural product format, may not be perceived as the output of a "chemical kitchen", even though it often contains, stabilizers, emulsifiers, colourants and flavours.

Dual Processing Theory

Sending the right messages and doing it the right way can make the difference between a successful product and a commercial flop. A psychological framework first proposed by William James in the late 19th century and since than developed further by various psychologists provides good guidance here (Kahneman, 2011).

This theory sees two ways of thinking and reasoning, often simply called system 1 and system 2. System 1 represents the peripheral, fast, intuitive, emotional and subconscious responses. It relies on heuristics, associations and past experiences to make judgements. System 2 is the central, slow, deliberate, rational and conscious thinking process. It uses logic, concepts, models and evidence to weigh the options and outcomes. And the dual process theory sheds light on how our minds can blend automatic intuition with deliberate reasoning, shaping our perceptions, decisions and actions.

There are moments where the conclusion from both systems can be in conflict or interfere with each other, which can lead to biases and errors in our decisions. A classic example is the "lunch decision": System 1 would suggest picking something that looks and smells very appetizing or familiar, and not paying attention to the nutritional value or the price asked. That could be a super indulgent, super premium ice cream or a portion of fresh golden french fries. System 2 would compare the available options in the light of ingredients used, the calories and the price, and produce a preference for a salad or an apple.

In today's world, the consumer is often bombarded with information and impressions during shopping. And with a busy daily life, people are not willing to spend time and energy in a deep rational process: The buying decision happens in a low-involvement situation (Knox & Walker, 2003). In such situations the brain can tend to use the more peripheral, "short cut" routes for decision making. That's why eye-catching symbols and logos are important and why many experts demand a vibrant "front-of-pack labelling".

If the food shopper is now under time pressure, in a hectic environment or with young children and becomes aware of a simple, trust-building front-of-pack (FOP) signal, visual, logos or certificates, the fast system 1 processes the impression and stimulates an instant shopping decision.

But in another situation, perhaps at home in front of the computer or in the kitchen, consumers take a deeper look at the back of the pack (BOP) and prefer more detailed rational information about nutrition facts, sourcing routes or claims. Here the set of facts on-pack would feed that decision-making tool.

It's not easy to predict when each system will take a lead in decision making during shopping. So, labels on packaged food products not only provide legal compliance but also provide meaningful cues for both systems. The back of pack gives the detailed facts about ingredients, nutrition and environment to feed system 2, and an eye-catching message front of pack can convince the hasty shopper to put the item into the basket. Here may also be the core of the dispute about "judgemental front-of-pack nutrition labelling (FOPNL)". Food makers do not want intuitive messages signalling the "fast system" to reconsider the purchase decision.

FUTURE FOOD REGULATION

Food regulation refers to the laws and administrative rules that control end to end all aspects of the food supply chain aiming to protect and maintain the health and interest of consumers. Attempts to ensure consumers' living standards can include a more abstract view of economic well-being, and countries may want to protect food items from freely entering the local markets. Where the science is inconclusive, the precautionary principle protects the consumers.

The novel food regulation in the European Union is a typical case. Novel food is defined as food that had not been consumed to a significant degree by humans in the European Union before 15 May 1997. That can be newly developed, innovative food, food produced with new technologies or simply food that has been traditionally eaten outside of the European Union. The whole application and approval process is complicated, time consuming and expensive and more of an obstacle for new innovative products and processes than a help.

In the past, the weaknesses were realized, but not criticized by consumers and opinion formers. Today, the necessity makes food production more sustainable and has created new food sources like cultured meat, edible insects, fungi or algae, production side streams and by-products, to answer these needs. But those solutions frequently fall under novel food regulation (Lachenmeier & Walch, 2023). So now the protective character of the process must shift to become an innovation enabler.

Even for one of the most demonized aspects in western food markets, genetic modification, signals of change are appearing. In the past, the technology was sold to the consumer without tangible personal advantages. A more robust crop variety, a tomato that stays fresh longer or a microorganism that ferments food better was more seen as "another trick of the food industry to increase profit" than the solution of an urgent need. But today smart technologies using biological principles, like Crispr/Cas, provide the opportunity to improve all farm goods in a faster, much more predictable, reliable and controllable way than all the old classical mutagenesis and breeding programs. The discussion about the approval procedure for "homologues gene editing", the modification of an organism's own genetic material without foreign DNA, has already reached parliament level in the first countries in Europe (S&P Global, n.d.).

While more efficiency and less rejection in the regulating systems offer new food product and process opportunities, other areas may receive attention and more restriction. Food and beverage manufacturers better prepare for more plans for what are called "HFSS" products, food products considered too "high in fat, sugar and salt". Future limitations can span from location to advertising and communications bans to sales limitations, taxation and warning labels. The anti-tobacco movement has successfully developed an effective toolbox to reduce tobacco consumption. HFSS food consumption cannot be compared to smoking, since calories, fat, sugar, salt, etc., are essential nutrients. But increasingly more people and opinion leaders spot those ingredients as a key public risk for non-communicable diseases.

Food and nutrition labelling may also see a stronger push towards simple classifications. Expecting people to analyse a BOP listing of nutrients, content, the share in the total product or a serving or a 100 g versus the daily recommended intake of an average person, may be an intellectually rewarding exercise, but it is not an effective help for the busy shopper.

The different nutrition scores discussed earlier in this chapter are not carved in stone. Their effectiveness is monitored. And they become more popular. Since the introduction of "Nutri-Score" in 2017 in France, it has been rolled out all over Europe and right now is the leading system. And while it still can be unfair to some food items, the scoring algorithm has been fine-tuned constantly. Valuable ingredients like omega fatty acids, some vitamins or secondary metabolites do not score, healthy food like walnuts and olive oil only score "C", products using sweeteners could score "A" or "B". Further correction in the scoring algorithm will lead to a fairer judgement. Even big retailers and discounters have started to put the score front of pack on their own brands (Lidl.de, n.d.). So it might be time for the food manufacturers to decide if they voluntarily roll the score out over wider product portfolios, including those categories scoring less good, or wait to see the pressure for a mandatory scoring system further building up.

REFERENCES

Abby Hannah, M. (2024, May 13). Foodlabelmaker.com. Tolerances-nutrition-labeling. https://foodlabelmaker.com/regulatory-hub/ec/tolerances-nutrition-labeling/

Arnold, L. E., Lofthouse, N., & Hurt, E. (2012). Artificial Food Colors and Attention-Deficit/Hyperactivity Symptoms: Conclusions to Dye for. *Neurotherapeutics*, *9*(3), 599–609. https://doi.org/10.1007/s13311-012-0133-x

Asioli, D., Aschemann-Witzel, J., Caputo, V., Vecchio, R., Annunziata, A., Næs, T., & Varela, P. (2017). Making Sense of the "clean Label" Trends: A Review of Consumer Food Choice Behavior and Discussion of Industry Implications. *Food Research International*, *99*, 58–71. https://doi.org/10.1016/j.foodres.2017.07.022

Australian Government Department of Health. (n.d.-a). *Health star rating - About health star ratings.* http://www.healthstarrating.gov.au/internet/healthstarrating/publishing.nsf/Content/About-health-stars

Australian Government Department of Health. (n.d.-b). *Health star rating - Guidance for industry.* http://healthstarrating.gov.au/internet/healthstarrating/publishing.nsf/Content/guide-for-industry

Australian Government Department of Health. (n.d.-c). *Health star rating - Guidance for industry.* http://healthstarrating.gov.au/internet/healthstarrating/publishing.nsf/Content/guide-for-industry

Basha, M. B., Mason, C., Shamsudin, M. F., Hussain, H. I., & Salem, M. A. (2015). Consumers Attitude Towards Organic Food. *Procedia Economics and Finance*, *31*, 444–452. https://doi.org/10.1016/s2212-5671(15)01219-8

Battacchi, D., Verkerk, R., Pellegrini, N., Fogliano, V., & Steenbekkers, B. (2020). The State of the Art of Food ingredients' Naturalness Evaluation: A Review of Proposed Approaches and Their Relation With Consumer Trends. *Trends in Food Science & Technology*, *106*, 434–444. https://doi.org/10.1016/j.tifs.2020.10.013

Blazina, C. (2024, April 14). *Eight-in-Ten Indians Limit Meat in Their Diets, and Four-in-Ten Consider Themselves Vegetarian.* Pew Research Center. https://www.pewresearch.org/short-reads/2021/07/08/eight-in-ten-indians-limit-meat-in-their-diets-and-four-in-ten-consider-themselves-vegetarian/

Blythe, T. (2023, February 10). *How to create an ingredients list. The Hub | High Speed Training.* https://www.highspeedtraining.co.uk/hub/how-to-create-an-ingredients-list/

Braesco, V., & Drewnowski, A. (2023). Are Front-of-Pack Nutrition Labels Influencing Food Choices and Purchases, Diet Quality, and Modeled Health Outcomes? A Narrative Review of Four Systems. *Nutrients*, *15*(1), 205. https://doi.org/10.3390/nu15010205

Campbell, B. L., Khachatryan, H., Behe, B. K., Dennis, J., & Hall, C. (2014). U.S. and Canadian Consumer Perception of Local and Organic Terminology. *International Food and Agribusiness Management Review*, *17*(2), 21–40. https://doi.org/10.22004/ag.econ.167903

Carruba, M. O., Caretto, A., De Lorenzo, A., Fatati, G., Ghiselli, A., Lucchin, L., Maffeis, C., Malavazos, A., Malfi, G., Riva, E., Ruocco, C., Santini, F., Silano, M., Valerio, A., Vania, A., & Nisoli, E. (2021). Front-of-Pack (FOP) Labelling Systems to Improve the Quality of Nutrition Information to Prevent Obesity: NutrInform Battery vs Nutri-Score. *Eating and Weight Disorders*, *27*(5), 1575–1584. https://doi.org/10.1007/s40519-021-01316-z

Chao, E. & Health Canada. (2000). *Standards of evidence for evaluating foods with health claims: A proposed framework*. In Consultation Document. https://www.canada.ca/content/dam/hc-sc/migration/hc-sc/fn-an/alt_formats/hpfb-dgpsa/pdf/label-etiquet/consultation_doc-eng.pdf

CODEX General Standard for Food Additives (GSFA) online Database. (n.d.). https://www.fao.org/fao-who-codex-alimentarius. Retrieved June 15, 2024, from https://www.fao.org/fao-who-codexalimentarius/codex-texts/dbs/gsfa/en/

Collins, N., & Verhagen, H. (2022, September 3). *Nutrition and health claims in the European Union in 2022*. RAPS. https://www.raps.org/news-and-articles/news-articles/2022/9/nutrition-and-health-claims-in-the-european-union#citation

Cosland, J. (2022, July 19). Research-suggests-few-consumers-actually-read-package-labels. https://www.supermarketperimeter.com. https://www.supermarketperimeter.com/articles/8428-research-suggests-few-consumers-actually-read-package-labels

Crino, M., Sacks, G., Dunford, E., Trieu, K., Webster, J., Vandevijvere, S., Swinburn, B., Wu, J. Y., & Neal, B. (2018). Measuring the Healthiness of the Packaged Food Supply in Australia. *Nutrients, 10*(6), 702. https://doi.org/10.3390/nu10060702

Díaz, L. D., Fernández-Ruiz, V., & Cámara, M. (2020). An International Regulatory Review of Food Health-Related Claims in Functional Food Products Labeling. *Journal of Functional Foods*, *68*, 103896. https://doi.org/10.1016/j.jff.2020.103896

DSpace. (n.d.). https://www.fao.org/3/cb6388en/cb6388en.pdf

Eckelkamp, S. (2023, March 3). *How is Olive Oil Made? | OliveOil.com. OliveOil.com*. https://www.oliveoil.com/how-is-olive-oil-made/#:~:text=Extra%20virgin%20and%20virgin%20olive%20oil%20undergoes%20very,virgin%20is%20a%20grade%20of%20virgin%20olive%20oil

Ecolabels. (n.d.). ecolabelindex.com. Retrieved June 15, 2024, from https://www.ecolabelindex.com/ecolabels/

E-Siong Tee, E.-S., Wong, J., & Chan, P. (2021, October 30). *Functional-food-monograph*. ilsisea-region.org. https://ilsisea-region.org/wp-content/uploads/sites/21/2022/02/ILSI-SEA-Region-Functional-Food-M https://ilsisea-region.org/publication/functional-foods-monograph/

European Commission. (n.d.a). *Food and Feed Information Portal Database Version 5.0*. Retrieved June 15, 2024, from https://ec.europa.eu/food/food-feed-portal/screen/novel-food-catalogue/search

European Commission. (n.d.b). https://ec.europa.eu/food/food-feed-portal/screen/health-claims/eu-register

European Food Safety Authority. (2010, December 8). Lactobacillus casei DN-114 001plus yoghurt symbiosis and reduction of the presence of Clostridium difficile toxins in the gut. https://www.efsa.europa.eu/en/efsajournal/pub/1903

Food Safety. (n.d.a). *Evaluation of the Regulation on nutrition and health claims*. https://food.ec.europa.eu/safety/labelling-and-nutrition/nutrition-and-health-claims/evaluation-regulation-nutrition-and-health-claims_en

Food and Agriculture Organization. (n.d.). *Codex Alimentarius | Food safety and quality | Food and Agriculture Organization of the United Nations*. https://www.fao.org/food-safety/food-control-systems/policy-and-legal-frameworks/codex-alimentarius/en/

Food Safety. (n.d.b). *Food information to consumers - Legislation*. https://food.ec.europa.eu/safety/labelling-and-nutrition/food-information-consumers-legislation_en

Food Safety. (n.d.c). *Novel food*. https://food.ec.europa.eu/safety/novel-food_en

Food Safety. (n.d.d). Nutrition and health claims. https://food.ec.europa.eu/safety/labelling-and-nutrition/nutrition-and-health-claims_en

Fong, A. T., Ahlstedt, S., Golding, M. A., & Protudjer, J. L. P. (2022). The Economic Burden of Food Allergy: What We Know and What We Need to Learn. *Current Treatment Options in Allergy*, *9*(3), 169–186. https://doi.org/10.1007/s40521-022-00306-5

foodnavigator.com. (2016, February 22). From medieval burnings to modern manufacturing: A brief history of food colours in Europe. https://www.foodnavigator.com/Article/2016/02/22/A-history-of-food-colours-in-Europe-modern-manufacturing?page=7

Fortune Business Insights. (n.d.). *Industry-reports/dairy-alternatives-market*. https://www.fortunebusinessinsights.com/industry-reports/dairy-alternatives-market-100221

FSSAI. (n.d.). https://www.fssai.gov.in/cms/food-safety-and-standards-regulations.php

Giugliano, R., Musolino, N., Ciccotelli, V., Ferraris, C., Savio, V., Vivaldi, B., Ercolini, C., Bianchi, D. M., & Decastelli, L. (2023). Soy, Rice and Oat Drinks: Investigating Chemical and Biological Safety in Plant-Based Milk Alternatives. *Nutrients*, *15*(10), 2258. https://doi.org/10.3390/nu15102258

HarvardHealth. (2023, April 7). Lactose intolerance. https://www.health.harvard.edu/a_to_z/lactose-intolerance-a-to-z

Henry, C. J. (2010). Functional Foods. *European Journal of Clinical Nutrition*, *64*(7), 657–659. https://doi.org/10.1038/ejcn.2010.101

Home | CODEXALIMENTARIUS FAO-WHO. (n.d.). https://www.fao.org/fao-who-codexalimentarius/sh

Hong, T. H. (2017, May 5). 100Plus awarded Healthier Choice logo in Malaysia. Mini Me Insights. https://www.minimeinsights.com/2017/04/24/100plus-gets-healthier-choice-credential-in-malaysia/

ISO/TS 19657:2017. (n.d.). ISO. https://www.iso.org/standard/65717.html

Kahneman, D. (2011). *Thinking, fast and slow.* Penguin UK.

Knox, S., & Walker, D. (2003). Empirical Developments in the Measurement of Involvement, Brand Loyalty and Their Relationship in Grocery Markets. *Journal of Strategic Marketing, 11*(4), 271–286. https://doi.org/10.1080/0965254032000159072

Kaufland. (n.d.). *Nutri-score.* https://unternehmen.kaufland.de/nachhaltigkeit/unsere-nachhaltigkeitsmassnahmen/nutri-score.html

Lachenmeier, D. W., & Walch, S. G. (2023). The novel food regulation: a major obstacle to sustainability in the coffee industry. Proceedings of International Coffee Convention 2023. https://doi.org/10.3390/icc2023-14840

Leroy, F., Rytz, A., Drewnowski, A., Tassy, M., Orengo, A., Charles, V. R., & Green, H. (2021). A New Method to Monitor the Nutritional Quality of Packaged Foods in the Global Food Supply in Order to Provide Feasible Targets for Reformulation. *Nutrients, 13*(2), 576. https://doi.org/10.3390/nu13020576

Lidl.de. (n.d.). *Nutri-Score - Lidl Deutschland.* https://unternehmen.lidl.de/verantwortung/gut-fuer-die-menschen/gesundheit-foerdern/handlungsfelder/bewusste-ernaehrung/nutri-score

Martirosyan, D. M., & Singh, J. (2015). A New Definition of Functional Food by FFC: What Makes a New Definition Unique? *Functional Foods in Health and Disease/Journal of Functional Foods in Health & Disease, 5*(6), 209. https://doi.org/10.31989/ffhd.v5i6.183

Maruyama, S., Streletskaya, N. A., & Lim, J. (2021). Clean Label: Why This Ingredient but Not That One? *Food Quality and Preference*, 87, 104062. https://doi.org/10.1016/j.foodqual.2020.104062

Ministry of Health, Labour and Welfare. (n.d.). *Ministry of health, labour and welfare: Food with health claims, food for special dietary uses, and nutrition labeling.* https://www.mhlw.go.jp/english/topics/foodsafety/fhc/02.html#:~:text=In%20order%20to%20sell%20a%20food%20as%20FOSHU%2C,the%20claim%20must%20be%20approved%20by%20the%20MHLW

Mintel. (2023, August 22). Mintel Data Integration | Mintel. https://www.mintel.com/products/data-integration/

Monteiro, C. A. (2009). Nutrition and Health. The Issue Is Not Food, nor Nutrients, so Much as Processing. *Public Health Nutrition, 12*(5), 729–731. https://doi.org/10.1017/s1368980009005291

Murley, T., & Chambers, E. (2019). The Influence of Colorants, Flavorants and Product Identity on Perceptions of Naturalness. *Foods, 8*(8), 317. https://doi.org/10.3390/foods8080317

NSW Food Authority. (n.d.). *Allergy and intolerance.* https://www.foodauthority.nsw.gov.au/consumer/life-events-and-food/allergy-and-intolerance

Nuraini, S., & Sucipto, N. (2021). Comparison Halal Food Regulation and Practices to Support Halal Tourism in Asia: A Review. *IOP Conference Series. Earth and Environmental Science, 733*(1), 012044. https://doi.org/10.1088/1755-1315/733/1/012044

Obesity Evidence Hub. (n.d.). *Health star rating system: Proposed improvements | Obesity Evidence Hub.* https://www.obesityevidencehub.org.au/collections/prevention/health-star-rating-system-proposed-improvements#:~:text=The%20key%20recommendations%20of%20the%20five-year%20review%20of,resources%204%204.%20Removing%20the%20energy%20icon%20option

Oqali & Santé Publique France. (2021). *Nutri-score assessment report after three-year of Nutri-score implementation.* https://sante.gouv.fr/IMG/pdf/nutri-score_follow-up_report_3_years_26juillet2021.pdf

Peters, S., & Verhagen, H. (2024). Publication Bias and Nutri-Score: A Complete Literature Review of the Substantiation of the Effectiveness of the Front-of-Pack Logo Nutri-Score. PharmaNutrition, 100380. https://doi.org/10.1016/j.phanu.2024.100380

Pettigrew, S., Coyle, D., McKenzie, B., Vu, D., Lim, S. C., Berasi, K., Poowanasatien, A., Suya, I., & Kowal, P. (2022). A Review of Front-of-Pack Nutrition Labelling in Southeast Asia: Industry Interference, Lessons Learned, and Future Directions. *˜the œLancet Regional Health. Southeast Asia, 3*, 100017. https://doi.org/10.1016/j.lansea.2022.05.006

Pettigrew, S., Jongenelis, M. I., Hercberg, S., & Julia, C. (2022). Front-of-Pack Nutrition Labels: An Equitable Public Health Intervention. *European Journal of Clinical Nutrition, 77*(1), 135–137. https://doi.org/10.1038/s41430-022-01205-3

Premanandh, J., & Salem, S. B. (2017). Progress and Challenges Associated With Halal Authentication of Consumer Packaged Goods. *Journal of the Science of Food and Agriculture/Journal of the Science of Food and Agriculture, 97*(14), 4672–4678. https://doi.org/10.1002/jsfa.8481

Santé Publique France. (n.d.). *Nutri-score.* https://www.santepubliquefrance.fr/determinants-de-sante/nutrition-et-activite-physique/articles/nutri-score

Scholer, A. A., Cornwell, J. F. M., & Higgins, E. T. (2019). Regulatory focus theory and research, *O* (pp. 46–66). https://doi.org/10.1093/oxfordhb/9780190666453.013.4

Schollweck, M., & Heidelberger, A. (2019). *Understanding Certification Marks: A qualitative study on the influence of semiotics on consumers information processing of grocery certification marks.* Oxford University press. https://psycnet.apa.org/record/2019-54239-004

S&P Global. (n.d.). *Significant-gene-editing-policy-changes-in-Europe.* https://www.spglobal.com/commodityinsights/en/ci/research-analysis/significant-gene-editing-policy-changes-in-europe.html#:~:text=In%202018%2C%20the%20European%20Court%20of%20Justice%20%28ECJ%29,result%20of%20wider%20impacts%20of%20the%20Russia%2FUkraine%20

Statista. (2023, January 3). *Number of products with FOSHU label in Japan FY 2011–2020.* https://www.statista.com/statistics/1196260/japan-number-foshu-products/

Statista. (2024a, February 29). *Global Muslim market size in 2021 and 2025.* https://www.statista.com/statistics/796103/global-muslim-market-size/

Statista. (2024b, May 22). Worldwide sales of organic foods 2022 | Statista. https://www.statista.com/statistics/273090/worldwide-sales-of-organic-foods-since-1999.

Swinbanks, D., & O'Brien, J. (1993). Japan Explores the Boundary between Food and Medicine. *Nature, 364*(6434), 180. https://doi.org/10.1038/364180a0

UNICEF, & Pazos, U. (n.d.). Policy brief: Front-of-pack nutrition labelling of foods and beverages. https://www.unicef.org/media/116686/file/Front-of-Pack%20Nutrition%20Labelling%20%28FOPNL%29.pdf

Unilever. (2024, May 16). Setting standards for the future of food. https://www.unilever.com/news/news-search/2021/setting-standards-for-the-future-of-food

U.S. Food and Drug Administration. (2018, October 22). *Use of the term natural on food labeling.* https://www.fda.gov/food/food-labeling-nutrition/use-term-natural-food-labeling

WHO. (2021, September 27). *Nutrition labelling.* https://www.who.int/initiatives/food-systems-for-health/nutrition-labelling

Wikipedia Contributors. (2024a, March 19). *Amaranth (dye).* Wikipedia. https://en.wikipedia.org/wiki/Amaranth_%28dye%29

Wikipedia Contributors. (2024b, June 1). *Otto Carque.* Wikipedia. https://en.wikipedia.org/wiki/Otto_Carque

Wikipedia Contributors. (2024c, June 5). *Lactase persistence.* Wikipedia. https://en.wikipedia.org/wiki/Lactase_persistence

World.openfoodfacts.org/?sort_by=popularity. (n.d.). world.openfoodfacts.org. Retrieved June 15, 2024, from https://world.openfoodfacts.org/?sort_by=popularity

Product Development in the Ingredient Sector 12

INTRODUCTION TO THE SECTOR

The industrial food ingredient (FI) sector is a big and versatile part of the food industry. In contrast to the foods service and branded foods business, these companies normally do not sell directly to the consumer, but to those serving them. The food ingredient industry is characterized by its service to commercial customers, the food and beverage industry or the food service sector! That's called "business to business" (B2B). It traditionally provides specialty and bulk ingredients for the preparation, manufacturing or handling of food. Specialty food ingredients have technological and/or functional benefits and are essential in providing today's consumer with a wide range of tasty, safe, healthy, affordable, qualitative and sustainably produced food. Bulk food ingredients are commodities that are traded in large quantities and are used as basic raw materials or as inputs for other food products. Their key customers are the packaged food makers or the food service sector. Consumers in shops and restaurants are not the primary target group.

Today this market is versatile, with a variety of different players, including small specialized privately owned companies, acting as growers, processors or traders, and big multinational enterprises with a broad portfolio of branded offerings.

The commercial success of these companies depends on the satisfaction of their customers. And the customers have a reason why they collaborate with the FI industry. They often request "functionality", so their solutions are truly functional ingredients, having an essential application in the consumer goods. That's why these ingredients are added.

The generic term "additives" just says that these substances are added to food to preserve or enhance quality, appearance, taste or stability. To do that, these ingredients perform specific industry-wanted functions. Antioxidants prevent oxidation and rancidity of fats and oils, while emulsifiers improve the texture and consistency of products having oil and water phases, like mayonnaise and ice cream. Flavourings are substances that are added to food to impart or modify its aroma or taste.

Processing aids are also added to the foodstuff. But they may or may not end up in the finished product, and so they do not need to be disclosed to the consumer as an ingredient on the label. They are either removed or reduced to insignificant levels before the food is market ready. Typical examples of processing aids are enzymes, catalysts, solvents, anti-foaming agents, filtering materials, and sanitizers, such as filtering aids, clarification substances, adsorbers and lubricants. Processing aids do not affect any intrinsic characteristics and may be present in negligible residues in the finished product. But they, of course, must be safe.

Most of these foods provide an intentionally wanted food functionality, like colour, flavour, aroma or shelf life. Consumers do not explicitly ask for that functionality. Some bring back important attributes that have been reduced or lost during production like a flavour or colour. Others bring new benefits, like shelf-life stability or nutrient fortification (Food Ingredient Facts, 2023). But even adding vitamins and

DOI: 10.1201/9781032724591-12

minerals can be perceived by the consumer to be more of a chemical industrial design than food quality improvement.

THE DIFFERENT FUNCTIONAL SECTORS

Most of the ingredients provided are classified and regulated as food additives. In their totality, these are substances of natural or synthetic origin that are added to foods to serve a certain technological or sensory function, for example, restoring colour, countering food perishability and bacterial degradation, and bringing colour or flavour to foods.

FLAVOURS

Flavour houses are companies that specialize in creating, producing and supplying flavours for various food and beverage products. Traditionally, flavour houses work closely with the food industry to develop customized flavours that suit the needs and preferences of consumers.

Food flavours are the sensory impressions of food or other substances and are determined primarily by the chemical senses of taste and smell. The food manufacturers need them either to create a specific aroma or to restore and enhance the aroma of ingredients that have lost their smell during the manufacturing, for example, during drying.

Food flavours can be natural or artificial, depending on the source and the method of production. Natural flavours are derived from natural sources, such as fruits, vegetables, spices, herbs, flowers or animal products. Natural flavours are extracted using physical, enzymatic or microbiological processes, such as distillation, expression, maceration, fermentation or hydrolysis. Natural flavours are usually more complex and variable than artificial flavours and may contain hundreds of different chemical compounds. Natural flavours are generally more preferred by consumers as they signal authentic and wholesome food products.

Artificial flavours are created by combining synthetic chemicals that mimic the taste and smell of natural flavours. Artificial flavours are usually simpler and more consistent than natural flavours and may contain only a few chemical compounds. Artificial flavours are often used because they are cheaper, more stable and more versatile than natural flavours. Artificial flavours can also create novel and unique sensory experiences that are not found in nature.

One of the main differences between artificial and natural flavours is the degree of similarity to the original source. Artificial flavours may not capture all the nuances and subtleties of natural flavours and may have a more uniform and standardized taste. Natural flavours may have more variation and diversity, depending on the quality and origin of the raw materials, and may have a more rich and realistic taste. Another difference is the safety and regulation of artificial and natural flavours. Artificial flavours are subject to strict testing and approval by food authorities and must be labelled as such on food products. Natural flavours are generally considered safe and do not require approval or labelling, but they may also contain allergens or contaminants that are not present in artificial flavours.

Food flavours are made by flavourists, who are experts in the chemistry and biology of taste and smell. Flavourists use various tools and techniques to analyse, create and test food flavours, such as gas chromatography, mass spectrometry, sensory evaluation and computer modelling. Flavourists work closely with food scientists, chefs and marketers to develop food flavours that meet the needs and preferences of consumers.

The various "flavour houses" have somewhat different portfolios. Some are more experts on savoury, sweet or floral notes or are more experienced in particular applications, like beverages. But basically, all offer a wide range of aromatic compounds. Some examples of famous flavour houses are:

- *Givaudan*: The world's largest flavour house, headquartered in Switzerland, with operations in over 140 countries. Givaudan creates flavours for categories such as snacks, beverages, dairy, confectionery and savoury.
- *dsm-firmenich*: The world's largest privately owned flavour house, based in Switzerland, with a presence in 100 markets. dSM-firmenich offers flavours for segments such as sweet goods, beverages, dairy, savoury and health.
- *International Flavors & Fragrances (IFF):* A leading global flavour house, headquartered in the United States, with facilities in 44 countries. IFF provides flavours for sectors such as beverages, dairy, bakery, confectionery and savoury.

Others are Sensient, Hasegawa, MANE and Takasago (The 9 Best Food Flavoring Manufacturers for 2021, n.d.). Some flavour houses have existed for more than 100 years; a few others are relatively new. Increasingly, other food ingredient companies join forces with other companies in the traditional flavour industry to build synergies and provide broader solutions for their key customers, the fast-moving consumer goods (FMCG) food makers (https://www.kerry.com/taste-charts/en/2024/adoption).

COLOURANTS AND PIGMENTS

Vision is the dominating sense in humans. Visual cues influence taste and flavour perception (Motoki et al., 2023). When we choose food, we first look at it. The colour signals us what to expect in terms of taste, nutrition, safety and freshness. To get an impression about the power of vision, some winemakers offer "tasting in the dark" – where people often cannot differentiate between white and red wine even. With the beginning of the 19th century, food preparation, preservation and processing began to move from scratch cooking in the kitchen outwards to the industrial world of food manufacturing. Many preservation technologies, like heating, destroy or modify natural fragile food pigments. Preserved food is safer and more storable, but sometimes is also less attractive. So, food producers were looking for anything that was able to bring back the old vibrant colours of the raw material (Coultate & Blackburn, 2018).

Today, the global food colorants market is worth $5 billion and is expected to double its size in the next 8–10 years (Market.us, 2023). The beverage segment accounts for the largest share of the food colourant market by application, followed by the dairy and frozen products segment. This growth is driven by the increasing demand for processed (convenience) foods, especially in emerging economies, combined with a constantly growing consumer preference in natural and clean label products.

The synthetic colour segment right now still dominates the market in terms of revenue generation, due to its low cost, wide availability and stability. But the natural colour segment is growing faster, due to the health benefits and organic properties of natural pigments.

North America and Europe are the major markets in terms of revenue generation, accounting for over 55% of the industry share. Both regions have stringent regulations regarding the use and labelling of food colours, as well as a high demand for natural and organic products. The fastest-growing market is Asia Pacific, due to the rapid urbanization, rising disposable income and changing lifestyle of consumers in countries like China and India.

Food colourants are regulated by different authorities around the world to ensure their safety and quality. In the United States, food colourants are subject to certification or exemption by the Food and Drug Administration (FDA). In Canada, food colourants are subject to maximum limits of concentration by Health Canada. In the European Union, food colourants are assigned E numbers by the European Food Safety Authority (EFSA).

The European Union has established a comprehensive set of regulations for food additives, which include food colorants. Currently, 39 colours are authorized as colour additives for use in foods in the European Union.

Here are a few examples of the natural food colourants that are permitted for use in food:

- *Curcumin*: A yellow-orange pigment extracted from turmeric root, also known as E100. It has anti-inflammatory and antioxidant properties and may protect against some diseases.
- *Riboflavin*: A yellow-green pigment also known as vitamin B2 or E101. As a vitamin, it has an essential function in energy metabolism and cellular function.
- *Carotenoids*: A group of yellow-orange-red pigments that are also provitamin A. They are also known as E160. They are found in fruits and vegetables such as carrots, tomatoes, peppers and pumpkins. They have antioxidant and anti-cancer effects and may improve eye health and immune function.
- *Anthocyanins*: A flavonoid group consisting of red-purple-blue pigments. They are also known as E163. They are found in fruits and vegetables such as berries, grapes, red cabbage and eggplant. They have antioxidant and anti-inflammatory effects and may protect against cardiovascular diseases and diabetes.
- *Chlorophyll*: A green pigment with natural functionality in plant photosynthesis. It is also known as E140. It is found in green plants such as spinach, kale and parsley. It has antimicrobial and deodorizing effects and may help detoxify the body.
- *Betanin*: A red-violet pigment extracted from beetroot. It is also known as E162. It has antioxidant and anti-inflammatory effects and may lower blood pressure and improve liver function.
- *Carmine*: A natural red dye from the cochineal insect, belonging to the scale insect family, that is widely used in foods. It is also known as E120.However, carmine can also cause allergic reactions in some people who are sensitive to insect proteins. The symptoms are the typical spectrum of an allergic reaction, and the allergenicity is confirmed by serum specific IgE tests or oral challenge tests.

Some food colourants are of mineral origin: A group of white-grey-black-brown pigments that are derived from minerals such as calcium carbonate, iron oxide, titanium dioxide and carbon black. In Europe, they carry the E numbers E170–E175. They have no nutritional or health benefits but are used to adjust the colour of food products.

- *Tartrazine*: A synthetic yellow dye that is used in candies, soft drinks, cereals and sauces. It is also known as E102 or FD&C Yellow 5.
- *Indigotine*: A synthetic blue dye that is used in ice cream, candy, liqueurs and baked goods. It is also known as E132 or FD&C Blue 2.
- *Anthocyanins*: These are natural pigments that give red-purple colour to foods like berries, grapes, red cabbage and wine. They are also antioxidants that may have health benefits.

European consumers associate natural products with superior quality and more readily recognize and accept colours derived from well-known sources such as beetroot, grapes, cabbage and paprika.

For those using food colourants to make products for the international markets, like the United States and Europe, it is important to note that some colours are permitted in one market, but not the other. And these colourants can contain animal-source components, which excludes the usage in vegan or sometimes kosher/halal product categories.

INNOVATIONS IN THE COLOURANT INDUSTRY

The food colourant market focusses on natural food colours and colouring to replace synthetic ones. For most colours, natural alternatives exist today, but they are often less stable, vibrant and more expensive (Novais et al., 2022).

Some use microbial pigments as natural food colours. Microbial pigments, derived from algae, fungi and bacteria, offer important advantages such as high stability, diversity, sustainability and safety. Microbial pigments can produce a wide range of colours, such as red, yellow, orange, green, blue and purple, and have applications in beverages, processed food, bakery, confectionery, dairy, meat, poultry and seafood products. Some examples of microbial pigments are astaxanthin from *Haematococcus pluvialis* algae, carotenoids from *Blakeslea trispora* fungus and monascus pigments from *Monascus purpureus* yeast.

Another approach is to replace the colourants by colouring foods. Colouring foods are foods with colouring properties obtained from natural sources such as fruits, vegetables, herbs and spices. They are not considered as food additives, as they are not chemically modified or isolated from the source material. Colouring foods are preferred by consumers who seek clean label products with natural ingredients. Typical colours are red, pink, purple, yellow, orange and green, and typical applications are in beverage, confectionery, dairy, ice cream, bakery and savoury products. Popular colouring foods are beetroot juice concentrate, purple sweet potato concentrate, safflower extract and spirulina concentrate.

Natural food colours often have some important limitations such as low solubility, sensitivity to light, heat, and pH and limited shelf life. The adoption of new technologies and methods like encapsulation, emulsification, nanotechnology, spray drying, freeze drying and microfluidization help to improve the performance and stability of existing natural food colours. These innovations indicate the dynamic and growing demand for natural food colours.

KEY PLAYERS IN THE MARKET

Colourants are not only important in the food industry, but also in many other industrial sectors. Therefore, the challenges and opportunities for each colourant maker differ. Consequently, the different manufacturers formed different organisations.

The "International Organization of Colour Manufacturers" (IACM) is the only association representing the complete colour additives industry, including both natural and synthetics (International Association of Colour Manufacturers, 2021). An underpinning of their work is the safe usage of colourants for people and the environment.

The "Ecological and Toxicological Association of Dyes and Organic Pigments Manufacturers" (ETAD) has its focus on synthetic organic colour manufacturers (ETAD, n.d.). They also promote high standards of safety and environmental topics related to their processes. Food companies are a smaller fraction in here.

And the "Natural Food Colours Organisation" (NATCOL) represents the interests of natural food colour manufacturers (NATCOL, 2024). As the name says, their mission is more focussed in terms of product and application. Safe use, maximum levels and various types of regulatory debates form the core of their activities.

The key players in the food colorants market include companies like Sensient Technologies Corporation, Chr. Hansen Holding A/S, Döhler Group DSM N.V., BASF SE, GNT Group B.V., DDW, Kalsec Inc., Naturex S.A. and FMC Corporation (now part of Givandon). These companies compete on product innovations, prices, quality, distribution networks and geographic presence. They also invest in research and development to develop or improve new natural colorants.

There is also a difference between synthetic and artificial colorants. Synthetic colorants are made in an industrial setting with the intention to improve a colourant used in foodstuff. Artificial colour, first of all, means that a colorant is added to a food product not naturally having that colour. If "artificial coulours" is listed on the ingredient label of a food product, it does not necessarily mean that it is synthetic.

Some health risks are associated with food colourants. A few studies have suggested that artificial food dyes may cause allergic reactions, hyperactivity in children, cancer and DNA damage (Berkeley Public Health & Berkeley Public Health, 2023). However, the evidence is not conclusive, and more research is needed to confirm the safety of food dyes.

EMULSIFIERS ANDHYDROCOLLOIDS

An emulsion is a heterogeneous suspension in which microscopic droplets of an immiscible liquid, for example, a watery phase, is dispersed in another unmixable liquid, like oil. This mixing of both phases requires physical force, like vigorous stirring, to create the dispersion. Two types of emulsions exist, the oil-in-water (o/w) emulsion with fat droplets in a water continuous phase. Examples are milk or mayonnaise. The other way around, it's a water-in-oil (w/o) emulsion with fat as the continuous phase. Examples are margarine and vinaigrettes. In the last couple of years, there have been multilayer emulsions in the market, like water-oil-water and oil-water-oil. A water-oil-water emulsion describes a water continuous phase and an oil-dispersed phase, but the oil also contains another water phase in its core, whereas an oil-water-oil emulsion describes an oil continuous phase with a water-dispersed phase containing an oil core. These are often called double emulsions (o/w/o or w/o/w). These multilayer emulsions are created by processing the liquids to produce microscopic particle sizes, with the benefit of improved stability.

Once created, emulsions are inherently thermodynamically unstable and readily revert to their separate phases. Without such "active" ingredients emulsions start to cream, sediment, flocculate, coalesce or separate. Creaming happens when the light droplets move to the surface. If the droplets have higher density than the surrounding medium, they fall to the bottom, causing sedimentation. And if the droplets start to aggregate, flocculation happens. The physical end of those processes is the full separation into the two indispensable phases (Cox et al., 2020).

To keep water and fat phases stable, emulsifiers have hydrophilic and hydrophobic moieties and locate themselves at the surface of the watery and oily phase. This reduces the interfacial tension between the unmixable phases and stabilizes the emulsion. A common classification of emulsifiers differentiates them as low-molecular-weight emulsifiers, amphiphilic biopolymers and solid or colloidal particles.

In the low-molecular-weight emulsifiers, the hydrophilic part is charged or non-ionic, the hydrophilic end consists of mono- or diglycerides, sucrose or sorbitan esters, polysorbates, or natural phospholipids, like in lecithin. Some of them are acid resistant; others initiate fat crystallization, for example, in a bouillon cube.

Amphiphilic biopolymers often consist of proteins, like casein and whey, and polysaccharides. The protein ends are amphiphilic and have emulsifying and foaming properties. And the polysaccharides, like different gums and pectins, bring thickening agents to stabilize emulsions by increasing viscosity, as that limits the separation of emulsion droplets.

Amphiphilic biopolymers are natural or synthetic macromolecules that have both hydrophilic and hydrophobic parts. They can be used in food applications to create edible films and coatings that can protect food from moisture, oxygen, microbial contamination and oxidation. Some examples of amphiphilic biopolymers are proteins, polysaccharides and aliphatic polyesters. Amphiphilic biopolymers can also be used to create edible films. The applications span into environmental packaging, as they are often biodegradable, compostable, renewable and have lower carbon.

The last group are Pickering emulsifiers, which form a protective film on the surface. They tend to have higher stability, lower cost and lower environmental impact. They are made from various materials, such as silica, clay, cellulose, starch or other food-grade by-products. Applications also cover fat reduction, encapsulation of functional ingredients and use as biosensors (Klojdová & Stathopoulos, 2022).

Besides their emulsification properties, many of these molecules have additional functionality in foods. For example, carboxymethylcellulose (E466) can act as a stabilizer, thickener, firming agent, gelling agent, coating agent, glazing agent or humectant. These functional ingredients work synergistically in various food matrixes. That's why often more than just one emulsifying additive is included in the formulation.

A search done by Cox et al. in Mintel GNPD indicated for the European Union a usage of lecithin in 14% of all food products, guar gum in 6% of foods, xanthan gum in 5%, and celluloses, like carboxymethylcellulose, in 2% of foods.

Like all food additives, emulsifiers undergo safety assessments by relevant bodies such as the Joint FAO/WHO Expert Committee on Food Additives (JECFA) (international), EFSA (EU), FDA (United States) and Food Standards Australia New Zealand (FSANZ) (Australia and New Zealand). Expert groups calculate acceptable daily intakes (ADIs), expressed as milligrams per kilogram of body weight per day. Overall, food additives do not show much evidence of adverse effects. As a result, lecithin (E322), mono- and diglycerides of fatty acids (E471), xanthan gum (E415) and guar gum (E412) have no ADI specified.

Right now, some large-scale population-based cohort studies suggest potential health implications in the intestine environment for some frequently used emulsifiers (Salame et al., 2024). The microbiota dysbiosis might stimulate pro-inflammatory signalling and thereby potentially predispose consumers to diseases such as hypertension, obesity, diabetes and other cardiometabolic disorders. The potential health risks of substances like carrageenan, polysorbate 80, and carboxymethylcellulose have long been in focus.

And because emulsifiers are chemically such a diverse class, there are uncertainties about their health effects because there are no solid data about their dietary intake and there are no biomarkers identified that can be followed clinically. In the past years, emulsifiers have found new application as plant-based egg replacers or baking ingredient to modulate the postprandial glycaemic response (Alshammari et al., 2021).

Another class of functional actives are the hydrocolloids like guar, locust bean, gum Arabic, tragacanth, agar, carrageenan, alginate, dextran and xanthan. The global hydrocolloids market size was estimated at 11.23 billion U.S. dollars in 2022 and is forecasted to grown on average 6.0% in the next couple of years (Grand View Research, n.d.). About 60% of the industrial usage is in food and beverages; much of the rest in is in pharma and personal care products and cosmetics.

Hydrocolloids can absorb or entrap large quantities of water. This water-holding capacity of the functional carbohydrates is used in many different food formats to extend shelf life and to modify the viscosity and texture of food products like ice cream, salad dressings, gravies, processed meats and beverages. Such modifications are known to enhance the sensory properties of food, and hence hydrocolloids are used as the key food additives to carry out specific functions.

Most hydrocolloids impart thickness and stickiness to aqueous dispersions and form gels. Their functional spectrum is described as thickening, gelling and stabilizing. So textural properties such as brittleness or elasticity, spreadability or longevity, and creaminess or chewiness are created by the type of hydrocolloid used. Furthermore, sensory properties such as mouthfeel, opacity and taste also depend on the hydrocolloid employed. Some are used alone, while others are used in synergistic combinations.

The ability to regulate the rheological properties of numerous food products, the often-easy dissolvability in cold and hot water, and a broad stability in a wide pH range, plus the synergistic connections with other hydrocolloid products like guar gum and locust bean gum, gives a great potential in a wide variety of food products.

For gel formation, the way each ingredient reacts to water is important, including its solubility and how it performs when exposed to varying temperatures. As water is immobilized, gels take on unique characteristics depending on the type of hydrocolloid used.

Even though they are often used together in food applications, hydrocolloids and emulsifiers are not the same. They are additives with different functions and properties. Hydrocolloids are substances that form gels or thicken solutions when added to water, while emulsifiers are substances that help to stabilize mixtures of oil and water.

The market for food emulsifiers was estimated at $3.6 billion in 2023 and to grow at about 5.3% annually over the next years (Food Emulsifiers Market, 2023). Together with the about $6.6 billion market for food hydrocolloids, the perspective of this food ingredient segment remains positive, because the accelerated growth of plant-based meat and dairy needs these functional actives to make sensorially convincing formulations. And the threats are also similar to those of other food ingredient segments, health concerns and tighter food regulation, which makes the regulatory approval for new products long and expensive. Clean label solutions are in high demand by the finished goods food industry.

The key players in both markets are similar. DuPont sells emulsifiers, such as lecithin, mono- and diglycerides, polysorbates and esters, under the Danisco brand. Cargill, a U.S.-based company, is one of the largest suppliers of lecithin as a natural emulsifier coming from soybeans, sunflower seeds or rapeseed, their key crops. Nowadays they also offer citric acid esters, sorbitan esters and sucrose esters. In the last couple of years, Cargill has invested heavily in the commercialisation of soluble, clean label fibres.

DuPont provides customized service for specific applications and customer needs, like the Kerry Group, which focusses on bakery, dairy, confectionery, meat and beverage products. Other companies are dsm-firmenich from the Netherlands, Kerry from Ireland, ADM and Ingredion from the United States, or W Hydrocolloids from the Philippines.

The leading producer of synthetic emulsifiers is Germany-based BASF. This company, as well as others like Paalsgaard, are constantly working on functionally further improved emulsifiers, particularly those that can help to stabilize not only simple two-phase emulsions, but more complex structures. An example is polyglycerol polyricinoleate (PGPR), which can reduce the amount of cocoa butter needed in chocolate production, enhance the volume and texture of baked goods making them softer and moister and stabilize very low fat emulsions. More "sophisticated" emulsions, like o/w/o and w/o/w double emulsions, can offer sensorially superior products at very low fat content, for example, in mayonnaise or salad dressings and spreads, with low caloric content, but excellent mouthfeel. To bring such complex matrixes to a good stability throughout the product life cycle, very efficient homogenization with ultrasonication or high-pressure homogenisation and highly effective emulsifiers are needed to stabilize droplets with sizes between 1–10 μm.

Food technologists and researchers are working on the use of composites to create stable emulsions for various food types having good colloidal stability and resistance to oxidative and hydrolytic spoilage.

VITAMINS AND MINERALS

Micronutrients are another sizable and distinct segment in the food ingredient industry. The market for agricultural use and food and beverage fortification was in 2022 estimated at $4.6 billion, with an expected strong average growth of 8.6%. The main driver is health and well-being, with focus on energy and weight management, bone and joint health, heart disease, cancer prevention, immunity and general health.

Currently more than 140 countries have food fortification programs, many of them are mandatory for large-scale commodities like sugar, salt, oil, wheat and rice (Olson et al., 2021). Voluntary food fortification is possible in nearly all countries globally. The nutrients covered are folate, niacin, riboflavin, thiamine, vitamin A, vitamin B6, vitamin B12, vitamin D, vitamin E, fluoride, iodine, iron, selenium and zinc.

The large-scale fortification of staples aims to close nutritional deficiencies in the wider population, or well-defined subgroups. Iron, zinc, iodine, vitamins A and E and folate are typically used here. For less-severe micronutrient shortages, governments often install voluntary fortification programs, where food manufacturers can decide to enrich their products with micronutrients. A good monitoring of all fortification programs is necessary to avoid an unacceptably high intake of single nutrients beyond their safety limit. But generally, food fortification is seen as a very useful tool to improve the health of a population. Most micronutrients have an "upper intake level", and where not enough adequate data are available, safe intake levels are often specified instead.

And while food items are generally a good carrier for extra nutrients, a few aspects require consideration. These include the question of possible sensorial impact from the fortificant, the nutrient's stability and the promoting or inhibiting effects other constituents of the food matrix have. The suppliers offer pretreated micronutrients to increase bioavailability and stability by using encapsulation, emulsification and other techniques. Key companies are often from the pharma or chemical sectors, like Abbott Laboratories, Pfizer, Novartis Roche and Sanofi, but also includes companies like dsm-firmenich or ADM.

Beside the restoration of nutritional value, some vitamins are also used as colouring agents or antioxidants.

PRESERVATIVES AND ANTIOXIDANTS

Preservatives are another group of food additives, since they are intentionally added to food products to delay spoilage, maintain freshness and extend shelf life. Roughly, they can be split into antimicrobial agents and antioxidants. The purpose is to prevent physical and chemical damage (antioxidants) or biological spoilage. Some actives can fulfil both functions, like for example sulphite (Eufic, n.d.).

This part of the food ingredient market is also still growing, although at a somewhat lower rate than other segments. In 2023, the industrial market for preservatives reached about $3.2 billion with an expected compound annual growth rate (CAGR) of 3.8% for the coming 5 years (Food Preservatives Market Insights, n.d.). Today, about two-thirds of the market is coming from synthetic alternatives, but the momentum is shifting to the natural solutions, a trend generally seen in the processed food market, as that is where preservatives are predominantly used.

Antioxidants (AOXs) are often named in this group of preservatives, as their intended use is as well the extension of the shelf life of food through their mode of actions, which is the protection of oxidation. In that role, they reduce the oxidation of sensitive food constituents, thereby delaying rancidity, limiting the loss of nutrients and preventing changes in colour, flavour and aroma.

But primarily the classification is chemically. Fundamentally, these substances can release electrons to neutralize free radicals, unstable atoms adversely affecting vital biomolecules such as nucleic acids, lipids and proteins (Harvard Health, 2019). Hence, providing the body with sufficient AOXd can neutralize the detrimental effects of free radicals, and that's what gave the AOX in contrast to other preservatives a better image. Call it preservative, and it sounds like one of those chemical tricks the food industry uses to extend shelf life of their products, helping their sales and not primarily the consumer. But antioxidants helps the consumer to keep the body status in balance and maintain good health.

Another important differentiator is the classification as natural or synthetic. Natural preservatives are derived from natural sources such as plants, animals or minerals. The oldest known examples are salt and sugar, but also vinegar, citric acids and several herbal and spice extracts like rosemary act as preservatives. Especially herbs and spices or their extracts find increasing application due to their natural origin and perceived "foodiness".

Today, many different preservatives find their application in various food and beverage products. Basically all have application specifies products in which they perform good. A selection is summarized in Table 12.1

The usage of natural preservatives in food products allows a labelling as: "natural" and sometimes "organic", a competitive labelling option, often preferred by consumers. From a technical perspective,

TABLE 12.1 Typical examples of preservations and their main application

PRESERVATIVE/AOX	*NATURAL/SYNTHETIC*	*MAIN USAGE/ PURPOSE*
Citric acid	Natural	Preservative and flavour enhancer in food and beverages
Tartaric acid	Natural	Taste enhancer in oral medications, also leavening agent in baking, and general antioxidant
Malic acid	Natural	Food and beverage preservative supporting sour taste
Fumaric acid	Synthetic	Multifunctional for flavouring and coagulant, food preservative, and as an acid in leavening
Lactic Acid	Natural	Flavouring agent adding sourness taste, and an antimicrobial agent for its hydrophobic characteristic
Vitamin C	Natural	Water-soluble antioxidant, helps in tissue growth, development and repair
Starch	Natural	Used in brewing and as thickening agents in baked goods and confections
Nitrates	Both	Used in curing meats and as a preservative, naturally present in plants
Nitrites	Both	Used in curing meats and as a preservative and found naturally in plants
Sodium benzoate	Synthetic	Used as a preservative in foods and beverages
Potassium sorbate	Synthetic	Used as a preservative in foods, drinks and personal care products
Calcium propionate	Synthetic	Used as a preservative in bread and other baked goods
Sodium nitrite	Synthetic	Used for curing meats
Butylated hydroxyanisole (BHA)	Synthetic	Used as an antioxidant in foods, cosmetics and pharmaceuticals
Butylated hydroxytoluene (BHT)	Synthetic	Used as an antioxidant in foods, cosmetics and pharmaceuticals
Sulphur dioxide and sulphites	Natural	Antioxidants that inhibit the growth of bacteria and fungi; used in wines, beer, fruit juices and vegetable oils
Tocopherols	Natural/synthetic	Various forms of vitamin E and key fat-soluble antioxidants preventing oxidation; used in oils, cereals, nuts, seeds and animal feed

natural preservatives often function over a reasonable pH range and can also be used to add taste and aroma to products.

However, natural preservatives have functional limitations, often a lower efficacy and a more narrower activity spectrum than synthetic preservatives. Hence, often higher concentrations are needed, and as they come from natural sources, the batches can have variable potency and consistency, which makes the application more difficult. A wrong dosage can lead to undesirable results, affecting the colour and texture of the food or even causing sensitization or allergic reactions in some people.

These limitations are the reason why the industry developed synthetic preservatives. In this chemical design approach, the technical weaknesses are eliminated and allow the generation of broad-spectrum actives against bacteria and fungi, matching very different food product properties and applications. These substances are the result of fully controlled chemical processes, and that ensures batch consistency, stable efficiency, quality and also safety, as they have been thoroughly tested before being released to market. The "compatibility" with other ingredients in food products has been pre-tested, which therefore ensures minimal interference with other actives like fragrance, colour and taste. And, not to forget, they are often available at lower cost (in use).

Currently, they have a negative reputation in the media and minds of many consumers linked to health risks or their environmental impacts. And the evidence that at least an excess of some of those synthetic preservatives can contribute to severe health problem is adding up (Sambu et al., 2022).

Additionally, the environmental concerns are substantial, as many of these molecules are petroleum-based and basically non-biodegradable. That is one reason why antioxidant-rich traditional and natural herb and spice extracts are increasingly popular (Embuscado, 2015).

The preserving solutions generally need to be made for purpose, fitting the particular application. Solutions for meat, poultry and seafood, making up more than 30% of the total market, are different from applications in bakery (more than 20%) or beverages (about 20%).

Traditionally, the market is characterized by many synthetic preservative manufacturers, like BASF, Eastman Chemical Company or Celanese. However, due to the rising demand for organic preservatives, the market has seen new entrants, including global manufacturers include ADM, Royal dsm-firmenich, Galactic S.A., Kemin Industries, SINOPEC, Givaudan, International Flavors & Fragrances (IFF) and Tate & Lyle.

For the food product developer, the use of natural alternatives offers the opportunities to find clean label preservation solutions. But this often requires a deeper understanding of the product that should be preserved, the reaction of the preservative(s) in that product and the specific preservation function. Such a product development needs more time and effort to define the right source or mix and the right dose, ensuring good shelf-life protection of the food product under development, without impacting other wanted attributes of the food.

HERBS AND SPICES

Conventionally, herbs and spices are not grouped into the FI sector. Due to their origin, they often captured in the fruit and vegetable segment. They mainly provide unique sensorial attributes like taste, aroma, spiciness and mouthfeel. But today they provide wider functionality, like safety, preservation, shelf life, antioxidation and colour. They become powerful, natural alternatives for some of the typical FI industry ingredients (Embuscado, 2015). In the past, the herbs and spice companies regularly sold to the FI industry, which purified and extracted the actives out of the primary processed material to produce their "functional" offerings for the branded goods manufacturers and food service providers.

The increasing demand is fuelled by the consumer need for more natural, authentic, yet exotic and exciting products. Herbs and spices are trusted and well-known cooking ingredients in many kitchens of this world. They are not perceived as industrial, processed food.

They come from different parts of various plants that are grown for their aromatic, pungent or otherwise desirable substances. They provide unique flavours and aroma and often also natural colour. The leafy parts of these plants are often called herbs, while the seeds, roots, fruits, stems, flowers and bark are considered spices. But some popular flavourings like bay leaf and curry leaf are also considered as spices. Spices often provide the powerful basic flavour, while herbs are used for their fresh, clean flowery notes.

In an overall trend towards more natural, authentic, minimally processed and plant-based food, herbs and spices can bring immense culinary variety, have a long history of usage in food, are used in religious ceremonies, used in medicine for centuries and often have an exotic flair. They often also have a high perceived value. Many people in the western world remember the history of sea trading in the Middle Ages and the immense value exotic spices had, when many were worth their weight in with gold.

In 2023, the global market was estimated to be $171 billion and to continue to keep on growing 3%–4% every year for the next decade (Herb and Spice Market, 2024). A few main reasons driving this growth are:

- Herbs and spices are known to support physiological functions and have medical benefits, like antiviral and antibacterial properties. They are trusted natural food preservatives in the kitchen, and indeed science nowadays confirms that many of them have antimicrobial properties (Martínez-Graciá et al., 2015). Their essential oils can inhibit the growth of bacteria and fungi, and many have antioxidative, thus preserving, capacities. Their potential healing and protective powers make them not only a healthy addition in eating, but sometimes an alterna tive to medicine and pharma.
- Primarily, they are aroma powerhouses boosting flavour in any foodstuff. Just a little bit turns boring foodstuff into an exciting and surprising experience. Recipe developers try out various herbs and spices to create new extraordinary flavour experiences.
- With increasing tourism and travel and the media feeding curiosity in all types of ethnic cuisines, people are more curious about the uniqueness and diversity that comes from the herbs and spices used. The food service industry, with all the different restaurants, eateries, cafes, food stalls, food trucks, open markets and cooking courses, has fuelled the trend and constantly creates new dishes or variations of local favourites.
- At the same time, herbs and spices are often local and authentic. The local environment shapes their profile and gives them a unique appeal; they escape standardization.
- Spices and herbs are also well known in consumers' kitchens. They are a typical kitchen component and therefore a fully accepted food ingredient. Their pre-treatment is minimal, often just dried, so they do not suffer from a "processing image".

Amongst the globally most-known spices are pepper, ginger, cinnamon, cumin, turmeric, cardamom, coriander and cloves. Popular herbs include oregano, parsley, mint, basil, rosemary and others.

Traded formats for spices are mostly dry, less often fresh and frozen. Their often oil-based terpene, aldehyde or phenolic compounds can be dried well and become concentrated in this step. If needed in very concentrated form, the oils and resins can be extracted from those plants as oleoresins.

Food applications of herbs and spices offer great innovation potential, as there are so many new and existing ways to combine them, something known as spice pairing. Signature dishes from basically all countries and kitchens get their identity from their specific herb and spice mix. Examples are berbere (Africa), cajun spice (Louisiana), chili (Mexico, Thailand), ras el hanout (North Africa), garam masala (Indian subcontinent), herbes de Provence (France), Chinese five spice (Asia), adobo (Latin America and the Caribbean), dukkah (Egypt), khmeli suneli (Georgia) and many others (Han, 2023).

Additionally, producers and distributors regularly bring new varieties to the market. The unique sweet and sour taste of tamarind, the earthy vanilla reminding colourful "ube root" from the Philippines, new masala or harissa mixes from Asia and Africa or green vanilla-like pandan leaves from Thailand or Indonesia are well-discussed new favourites.

The biggest spice exporter is India, and some of the biggest companies like Everest, MDH or Rajesh Spices are located there. McCormick from the United States is the global leader, but other important ones are based in Europe, Latin America and other Asian countries (Nooranne, 2021).

And the producers of these plants today use various technical solutions to cultivate, process and distribute them. Precision agriculture techniques, such as remote sensing and GPS-guided machinery, improve crop yields while reducing environmental impact. Origin and authenticity can be further strengthened through the usage of modern blockchain technology, providing transparency for the long way from farm to fork.

The vulnerable side of herbs and even more so for spices are often their safety and quality. They are susceptible to contamination from pesticides, heavy metals and microbial pathogens and adulteration, where they mix with inferior or harmful substances (Food Safety, n.d.). With their often-high sales value, it makes good sense for the companies to create a tight net of stringent quality control measures, applying technologies like spectroscopy and chromatography to detect contaminants and adulterants in spices.

In the past, herbs and spice suppliers were smaller and often very specialized companies and well separated from the big multinational food ingredient suppliers like the flavour houses. But with the strong demand for more natural and authentic solutions, both segments are coming much closer. The economic perspective looks promising, but future business here often needs substantial investment in science and technology to fulfil market expectations and regulatory requirements. As a result, much consolidation and integrations have been happening. Companies like Olam, Sensient, Kerry, dsm-firmenich, and IFF are buying the spice business and use their superior technical abilities to develop natural alternatives for the ingredients in their portfolios.

CHANGING INDUSTRY

In the past, dedicated key account managers of ingredient companies kept close contact with their key customers, ensuring a high service level and detecting challenges and opportunities early. They knew the product compositions, understood the technical details of the food manufacturing, were up to date with the changing demands and quickly brought in their own experts when it comes to details. And they were not only reactive, waiting for the customers to express their wishes. They used the needs and wants they had to provide new opportunities for the food makers, without being explicitly asked for it. Flavour houses prepared showcases for new market trends in the food business of the customer, picking up trends in gastronomy early.

Examples are new combinations of sweet and spicy, sour, bitter or savoury flavours to create complex and unexpected sensations. Chili-infused honey, dark chocolate and chili pepper almonds, spicy bakery, mango and habanero salsa, and lemon and rosemary cake.

And they constantly improved and specialized their key ingredient portfolio. Companies making savoury meals or snacks or cooking aids are not looking for “chicken flavour”. For them to differentiate and fulfil consumer demands, white meat, skin, breast, roasted, boiled, fried, grilled, hen, old hen, Asian and American flavours are more suitable.

One of the most dynamic aspects of the food industry are the ever-changing flavour trends. Nowadays flavour represents more than just a good smell or aroma. New functional properties are needed in plant-based space.

If a company decides to develop a sensorially beef-matching plant-based patty alternative, it knows that the exchange of the animal protein and fat against plant-based ones has a tremendous impact on the composition. The new mixture needs different aromas and tastants, starches, fibres, colourants, emulsifiers, stabilizers and gelling agents to prepare a competitive alternative. It is laborious and difficult to try individual solutions from various ingredient suppliers before a sensorially convincing and commercially viable product is developed. New plant-based alternatives need many different functional ingredients to match the sensorial properties of animal-based food. But they must be clean label and natural in the future, to resolve the tension of “chemical” substances in a more healthy and environmentally friendly food. With this, the food ingredient industry can much better serve specific consumer segments, such as those with vegan, gluten-free or keto diets.

Instead of a collection of different ingredients, it’s more effective to get a premix, where key functionalities, like colour, taste, smell and texture, are already adjusted and the development work can focus on the right combination of plant proteins and fats and oils.

Certain fruit and vegetable flavours, combined with bright colour, convey health food. And rather than leaving it there, a mix containing some nutrients and secondary metabolites can be used to create products with health benefits. For example, turmeric, ginger, berries and green tea can provide anti-inflammatory properties or benefits.

For the ingredient industry. the market is changing. Through acquisitions of smaller specialists, the bigger ingredient companies get more vertical integration and more control of their key materials and conversion processes, similar to what they have in their classical chemical businesses. This ensures reliable quality, traceability and sustainability of those ingredients.

Natural ingredient solutions are no longer an option, but increasingly a necessity. And a successful player in this industry must master the latest science and technology to ensure their solutions for the industry match all requirements in functionality, safety, quality and sustainability.

So, the "functional solution" providers in the FI industry are disappearing. Where 15 or 20 years back these companies were specialized in flavours (and fragrances), or nutrients, colourants, texturizers, taste enhancers, emulsifiers and sweeteners, they now aim to provide broader solutions to their customers. Until now, the "specialist" often provided a matching solution for a specific functionality, like the emulsifier, colourant, antioxidant, flavour or taste enhancer and R&D in the branded food company was busy combining all actives for all necessary functionalities in the food product they are developing. This was a complicated and time-consuming effort, where every change of one food ingredient had to be tested for its potential impact on others. The specialists brought their leading expertise, often a strong national presence and long term trust.

The globalization of the whole industry heated up the competition and pushed consolidation in the different sectors. Partners are now global, and opportunities and challenges are bigger and more complex and require different skills and resources.

Providing single solution for colour, taste, stability or health is no longer the preferred option. Branded food manufacturers prefer integrated solutions with several key functionalities and easy application in their food products. By now, many ingredient suppliers have learned the potential interactions between actives in given food matrix, so they can offer the manufacturers a tested "plug and play" solution at a better price.

A series of bigger mergers and acquisitions (M&A) enhance the capabilities of the ingredient suppliers, expand their reach and create great synergies. Mergers and acquisitions can give excellent access to unique sources, competitive knowledge and patented technology. At the same time, developing new solutions becomes more research intensive and the necessary investment into new science and technology can go beyond the financial capabilities of some companies. But where larger parts of the technical infrastructure, labs, pilot plants, testing and tasting facilities, large-scale manufacturing, storage and distribution can be shared, impressive synergy and cost savings potential appears. Some of the notable M&A deals are listed next. (For a wider overview, see Food Dive, 2021.)

- IFF, a leading flavour and fragrance house, completed a $26.2 billion merger with DuPont Nutrition & Biosciences business to create a global leader in taste, scent, nutrition, biosciences and health ingredients.
- Kerry, 35 years back, was a small private milk company in Ireland. It first bought Quest Food Ingredients, then Cargill Flavour Systems, Red Arrow and a few others. In 2021, it acquired clean-label preservatives maker Niacet, announcing their ambition to lead in the fast-growing market for clean label food protection solutions.
- Givaudan, a world known flavour and fragrance house, strengthenedits natural colours business with the purchase of DDW, a U.S.-based natural colour company for food and beverage applications.
- Ingredion acquired KaTech, a German provider of advanced texture and stabilization systems for food and beverage products.
- Kemin acquired Proteus Industries Inc., a U.S.-based maker of clean label antimicrobial solutions for meat and poultry products.

- Symrise, a Germany-based flavour and fragrance company, acquired Sensient's fragrance business to include aroma chemicals and fragrances for personal care products and to strengthen its position in natural ingredients.
- DSM, or Dutch State Mining, shifted in the last years from an oil and mining company to a multinational corporation active in the fields of health and nutrition. Recently, they confirmed a merger with Firmenich, a Swiss flavour and fragrance maker.

The new companies provide increasingly integrated functional solutions; their sales teams can offer more products and increase their support level for valuable customers. It's now possible for them to understand the needs of their customers more broadly and spend more time on providing broader functional solutions. What maybe in the past was clearly separated R&D work at the ingredient supplier and the food maker now shifts to more of a co-development and innovation partnership. Product development can happen more quickly. The food producer can use the superior technical skills of the ingredient supplier, who understand the needs of their customer and the consumer market behind them much better. Consumers request better food products, more natural, more healthy and environmentally friendly. This requires access to new capabilities. And this can go far beyond new certificates or access to analytical skills that ingredient suppliers routinely have. It can include support by their own clinical program access, regulatory support, key opinion former (KOF) network support, effective lobbying and access to new academia and media channels.

CONTEMPORARY CUSTOMER SERVICE

The "new" ingredient supplier wants to be an end-to-end (E2E) testing partner for the makers of branded consumer products or food service. They promise co-development, customized solutions based on their own technical and market insights on top of superior products and operations, highest quality standards and scientific support.

Food additives are tightly regulated, and they carry the image of harmful substances for human health. In a sense, these ingredients are materials of concern. They are understood as essential components to make food safe and sensorially attractive, but they are more accepted as necessary than truly wanted.

The roots of some of the big food ingredients enterprises often are classical heavy industry sectors, DSM, now dsm-firmenich, was the abbreviation for "Dutch State mining" or BASF; in food commodities, like ADM and Cargill; or in flavour and fragrances, like IFF, Symrise and Givaudan, with their leading chemical science.

What these companies share is the management of large-scale operations and a belief in the benefits of new science and technology. In contrast to many food brand owners, they actively search for "technology push". When their research shows potentially exciting new technology, they do not hesitate to invest heavily into further exploration. The later benefit for the customer and the consumer, needed of course to make the investment profitable at the end, has to be somehow visible, but does not need to be confirmed by "market pull", before the opportunity will be taken.

With this stronger focus on science and technology, a better understanding of the requirements to mature those ideas and perhaps more trust in the impact of innovations and that the initial investment will pay back later, this part of the food industry is spending much more resources on their R&D programs, sometime up to 10%. "Bright science into brighter living" is, for example, DSM's proud motto (DSM Integrated Annual Report 2021, n.d.).

And of course, while part of the R&D of these companies is busy trying to sell all the solutions their "applied science and development" centres have in stock, other major parts of R&D focus on new science and technology and breakthrough work. They are more used to making major investment decisions based on a 5- or even 10-year horizon, very different to the typical business case of branded food manufacturers.

To serve the ideas on both sides of the business well, they manage the projects differently. Ready-to-use ""off-the-shelf" solutions, based on existing technology, are quickly implemented with the customers in fast and flexible "agile" project management processes (see Chapter 7), while the progress in big new technology projects, with heavy investments run well controlled through a rigid management process, like to stage gated processes, discussed in Chapter 7, to control the business risks.

So, the research and development intensity starts shift ingfrom the big multinational branded goods makers to the food ingredient suppliers, willing to invest heavily upfront into enabling science and technology to commercialize competitive solutions.

Short term, one big opportunity is to move the consumer perception about additives about in a positive direction, where the safety testing is understood and trusted, while providing new natural, nutritious, authentic and clean alternatives, with the well-known functional efficiency at a competitive price.

The publications produced by the Natural Food Colours Association (NATCOL) to move colouring foods that "have not been obtained through selective extraction" out of the additive space can be understood as part of that strategy.

CASE STUDY: EGGS FOR MAYONNAISE

Branded food makers are not only searching for the supplier they need; sometimes they provide help for new companies to start in the industry. A leading maker of branded mayonnaise in Southeast Asia decided to outsource their in-house egg processing. Their flagship product, premium mayonnaise, is made with oil, vinegar, some lemon juice, whole eggs and egg yolk. In earlier times, the eggs were broken in house and immediately conserved in a spice mix and then used up quickly. When new safety guidelines demanded a pasteurization step for the broken eggs, the situation changed. Due to spatial limitations of the factory, the extra processing turned out to require a major investment and rebuild of the factory. The decision was made to use the remaining space in the factory for other manufacturing, while trying to outsource the egg breaking.

The trusted supplier of the whole eggs was asked to invest in the necessary industrial egg breaking, splitting and pasteurization equipment in return for a guaranteed longer-term commercial contract.

Technically, mayonnaise is an oil-in-water (o/w) emulsion with a. highly concentrated internal phase. As food emulsions are thermodynamically unstable with a tendency towards coalescence, flocculation and creaming, a good mayonnaise can have a shelf life of over 1 year with no change in the organoleptic acceptance. The key reason for this impressive long-term transport and storage stability is the high viscosity generated by the close packing of droplets.

The lecithin in the egg yolk is responsible for that tight packing. And the amount of egg yolk (lecithin) seems correlated with the stability of the emulsion. Whole egg is as well part of a good formulation, as it serves as a thickener. The balance between the egg yolk and the egg white is important in the formulation to achieve the desired organoleptic results and the superior product stability. Any contamination with egg white proteins (albumins) is assumed to cause instability, by bridging single well-dispersed droplets into bigger aggregates, destabilizing the emulsion and leading to flocculation (Ariizumi et al., 2017).

For the mayonnaise-making company, a stable country supply of egg yolk was an additional watch out point. The international egg market was traditionally already a volatile business, but with the international alignment on the standards of animal welfare for cage-free eggs, sourcing became a massive challenge. Helping to build a local supplier while simplifying the in-house manufacturing processes was therefore a logical step to create a sustainable business.

The collaboration required a close exchange of detailed technical knowledge, full alignment on the ingredient specification, agreement on analytical tests and their validation and a long-term contract for the supplier to balance the capital investment needed to supply the egg yolk.

Throughout the contract period, that egg supplier started to realize that it couldn't find a commercially attractive market for the egg white, the by-product of the egg yolk production. Just disposing of it as animal feed was seen as non-sustainable, and so both companies agreed in a collaboration to optimize the egg yolk and whole egg composition in the final product, without impacting shelf life and organoleptic attributes. Egg white proteins (EWPs), in addition to their nutritional value, possess excellent functional properties that facilitate their use in the food industry. The international contacts of the mayonnaise producer were used to find applications better using the major functional properties of EWPs such as gelling, foaming, and emulsifying used in puddings, desserts, cakes, bread, meringues and cookies (Razi et al., 2023).

This case is an example of a deeper and more strategic collaboration between a brand goods maker and an ingredient supplier, where the supplier, with the help of the industrial customer, built up additional technical experience to master new food processing steps.

REFERENCES

Alshammari, N. A., Taylor, M. A., Stevenson, R., Gouseti, O., Alyami, J., Muttakin, S., Bakalis, S., Lovegrove, A., Aithal, G. P., & Marciani, L. (2021). Effect of Intake of Food Hydrocolloids of Bacterial Origin on the Glycemic Response in Humans: Systematic Review and Narrative Synthesis. *Nutrients, 13*(7), 2407. https://doi.org/10.3390/nu13072407

Ariizumi, M., Kubo, M., Handa, A., Hayakawa, T., Matsumiya, K., & Matsumura, Y. (2017). Influence of Processing Factors on the Stability of Model Mayonnaise with Whole Egg During Long-Term Storage. *Bioscience, Biotechnology, and Biochemistry, 81*(4), 803–811. https://doi.org/10.1080/09168451.2017.1281725

Berkeley Public Health & Berkeley Public Health. (2023, September 28). *New report shows artificial food coloring causes hyperactivity in some kids.* https://publichealth.berkeley.edu/news-media/research-highlights/new-report-shows-artificial-food-coloring-causes-hyperactivity-in-some-kids

Coultate, T., & Blackburn, R. S. (2018). Food Colorants: Their Past, Present and Future. *Coloration Technology, 134*(3), 165–186. https://doi.org/10.1111/cote.12334

Cox, S., Sandall, A., Smith, L., Rossi, M., & Whelan, K. (2020). Food Additive Emulsifiers: A Review of Their Role in Foods, Legislation and Classifications, Presence in Food Supply, Dietary Exposure, and Safety Assessment. *Nutrition Reviews, 79*(6), 726–741. https://doi.org/10.1093/nutrit/nuaa038

DSM Integrated Annual Report 2021. (n.d.). *Our focus in Innovation - DSM Integrated Annual Report 2021.* https://annualreport.dsm.com/ar2021/report-by-the-managing-board/strategy/our-focus-in-innovation.html

Embuscado, M. E. (2015). *Herbs and spices as antioxidants for food preservation* (pp. 251–283). Elsevier. https://doi.org/10.1016/b978-1-78242-089-7.00011-7

ETAD. (n.d.). Who we are? https://etad.com/who-we-are/

Eufic. (n.d.). *What are preservatives and what are common examples used in food?.* https://www.eufic.org/en/whats-in-food/article/what-are-preservatives-and-what-are-common-examples-used-in-food

Food Dive. (2021, December 23). *M&A deals that shaped the ingredients space in 2021.* https://www.fooddive.com/news/ma-deals-that-shaped-the-ingredients-space-in-2021/611776/

Food Emulsifiers Market. (2023, August 2). https://www.futuremarketinsights.com/reports/food-emulsifiers-market

Food Ingredient Facts. (2023, June 15). *Sources of food ingredients - Food ingredient facts.* https://www.foodingredientfacts.org/facts-on-food-ingredients/sources-of-food-ingredients

Food Preservatives Market Insights. (n.d.). https://www.mordorintelligence.com/industry-reports/food-preservatives-market

Food Safety. (n.d.). *Herbs and spices (2019–2021).* https://food.ec.europa.eu/safety/eu-agri-food-fraud-network/eu-coordinated-actions/herbs-and-spices-2019-2021_en

Grand View Research. (n.d.). Hydrocolloids Market Size, Share & Trends Analysis Report by product (Gelatin, xanthan gum, carrageenan, alginates, pectin, guar gum, carboxy methyl cellulose), by function, by application, by region, and segment Forecasts, 2023–2030. https://www.grandviewresearch.com/industry-analysis/hydrocolloids-market

Han, E. (2023, July 25). *25 spice mixes from around the world.* https://www.thekitchn.com/. https://www.thekitchn.com/spice-mixes-189368

Harvard Health. (2019, January 31). *Understanding antioxidants.* https://www.health.harvard.edu/staying-healthy/understanding-antioxidants

Herb and Spice Market. (2024, March 18). https://www.futuremarketinsights.com/reports/herbs-and-spices-market

International Association of Color Manufacturers. (2021, August 10). *What are colors? - International Association of Color Manufacturers.* https://iacmcolor.org/what-are-colors/

Klojdová, I., & Stathopoulos, C. (2022). The Potential Application of Pickering Multiple Emulsions in Food. *Foods, 11*(11), 1558. https://doi.org/10.3390/foods11111558

Market.us. (2023, November 6). *Food colorants market size, share | CAGR of 9.3%.* https://market.us/report/food-colorants-market/#:~:text=Key%20Takeaways%201%20Market%20Size%20and%20Growth%3A%20In,into%20natural%20and%20artificial%20colors.%20...%20Weitere%20Elemente

Martínez-Graciá, C., González-Bermúdez, C. A., Cabellero-Valcárcel, A. M., Santaella-Pascual, M., & Frontela-Saseta, C. (2015). Use of Herbs and Spices for Food Preservation: Advantages and Limitations. *Current Opinion in Food Science, 6*, 38–43. https://doi.org/10.1016/j.cofs.2015.11.011

Motoki, K., Spence, C., & Velasco, C. (2023). When Visual Cues Influence taste/flavour Perception: A Systematic Review. *Food Quality and Preference, 111*, 104996. https://doi.org/10.1016/j.foodqual.2023.104996

NATCOL. (2024, May 19). *NATCOL | Natural Food colors Association.* https://natcol.org/

Nooranne, S. (2021, May 19). 5 largest spice companies in the world. Insider Monkey. https://www.insidermonkey.com/blog/5-largest-spice-companies-in-the-world-942272/4/

Novais, C., Molina, A. K., Abreu, R. M. V., Santo-Buelga, C., Ferreira, I. C. F. R., Pereira, C., & Barros, L. (2022). Natural Food Colorants and Preservatives: a Review, a Demand, and a Challenge. *Journal of Agricultural and Food Chemistry, 70*(9), 2789–2805. https://doi.org/10.1021/acs.jafc.1c07533

Olson, R., Gavin-Smith, B., Ferraboschi, C., & Kraemer, K. (2021). Food Fortification: The Advantages, Disadvantages and Lessons from Sight and Life Programs. *Nutrients, 13*(4), 1118. https://doi.org/10.3390/nu13041118

Razi, S. M., Fahim, H., Amirabadi, S., & Rashidinejad, A. (2023). An Overview of the Functional Properties of Egg White Proteins and Their Application in the Food Industry. *Food Hydrocolloids, 135*, 108183. https://doi.org/10.1016/j.foodhyd.2022.108183

Salame, C., Javaux, G., Sellem, L., Viennois, E., De Edelenyi, F. S., Agaësse, C., De Sa, A., Huybrechts, I., Pierre, F., Coumoul, X., Julia, C., Kesse-Guyot, E., Allès, B., Fezeu, L. K., Hercberg, S., Deschasaux-Tanguy, M., Cosson, E., Tatulashvili, S., Chassaing, B., & Touvier, M. (2024). Food Additive Emulsifiers and the Risk of Type 2 Diabetes: Analysis of Data from the NutriNet-Santé Prospective Cohort Study. *The Lancet. Diabetes & Endocrinology, 12*(5), 339–349. https://doi.org/10.1016/s2213-8587(24)00086-x

Sambu, S., Hemaram, U., Murugan, R., & Alsofi, A. A. (2022). Toxicological and Teratogenic Effect of Various Food Additives: An Updated Review. *BioMed Research International, 2022*, 1–11. https://doi.org/10.1155/2022/6829409

The 9 Best food Flavoring Manufacturers for 2021. (n.d.). https://www.innovaflavors.com/blog/the-9-best-food-flavoring-manufacturers-for-2021

The Food Service and Solution Sector 13

The food service sector incorporates all businesses, institutions and companies providing home meal preparation. Under this heading, a broad range of enterprises are captured. Typical services are:

- *Full-service restaurants*: Where customers are served food while seated
- *Fast-food restaurants*: Providing quick service, often a standardized and more limited menu, limited seating and takeout service
- *Cafés and bars*: Where mainly drinks are served, but often also with a limited food service
- *Catering services*: Providing food service to remote sites and venues, typically for all types of events
- *Ghost kitchens*: Professional food preparation and cooking facilities preparing delivery-only meals
- *Institutional services*: For hospitals, convalescent homes, social services, prisons, rehabilitation centres and institutionalized patients
- *Military service*
- *School meal programs*: In public schools, colleges and universities
- *Travel meal services*: Provided during travel by air, bus, boat or train
- *Relief feeding*: During famines and other natural disasters, and in refugee camps
- *Company restaurants and cafeterias*
- *Food trucks and street food*

The main commonality of services in this versatile sector is the fact that they serve food to customers. Some offer premium sophisticated cuisine, others typical local top dishes. In quick serve restaurants (QSR), preparation time and standardization is a key factor. Some only deliver to other locations. Many are in independent commercial enterprises, but in other services, the meal provision is part of the overall service and not separately charged. A few need highly trained culinary leaders; others require professionals with a broad view about food and nutrition or people without food education.

With a value of approximately \$3.5 trillion, the food service sector in many countries has a massive impact on their national gross domestic product (GDP) (Food Service Industry, 2023). But it is very difficult to capture all market participants, and the growth perspective for the separate segments differ hugely (Food Service Industry, 2023). The market share of each type varies by region and country, depending on the consumer preferences, income levels and cultural factors. Global changes in lifestyle and economic prosperity let many people reduce their own cooking and are expected to fuel further growth in this sector, which is forecasted to grow with a compound annual growth rate (CAGR) between 5%–10% in the next decade.

In the past few years, the sector has seen dramatic changes. With the pandemic, the dine-in part of the sector came to a standstill, resulting in massive restaurant closures, while online, takeout, home delivery and drive-through orders exploded, often within the same enterprise and supported through new digital platforms and artificial intelligence (AI) powered opportunities. This growth has been driven by factors such as increasing fast-food consumption, rising disposable incomes and the expansion of varied-format fast-food restaurants worldwide. One of the traditional strengths of this part of the food business is its excellent ability to combine culinary ingredients from around the world with value-giving components

DOI: 10.1201/9781032724591-13

that come from nearby, like meat from trusted local farmers and spice mixes of exotic origin and sophisticated composition.

Now it seems dining out is coming back to pre-Covid levels (Brehout, 2023). But underlying, some things changed: The options for takeout and ordering from home have increased, many people still work more from home and people have started to eat out for different reasons. In the past, at least breakfast and lunch were moments that were functionally integrated into the working day. People now also want to eat out to socialize, reward themselves and become inspired. Consumer experiences get more attention. Recently, McDonald's Corporation and Google exposed their intentions for a global partnership, where Google's Cloud technology will be implemented to accelerate innovation and deliver better experiences for its customers (Diligence Insights, n.d.).

Because the food service sector is perhaps the most complex and diverse part of the food industry, the food product and service opportunities differ as well, and it makes sense to look at the drivers, trends and opportunities separately. One way to do this is to split the sector into commercial and non-commercial institutional services.

INSTITUTIONAL SERVICES

Institutional food services includes schools, hospitals, prisons, military bases and other organizations that provide food and drinks as part of their services or operations. These institutions have different objectives and requirements than do commercial food service businesses. Another difference is that customers here have sometimes no or only a very limited freedom of menu choice. The "customers in such situations are trapped; they cannot simply go elsewhere and choose an eating spot. Providing food is not the primary purpose of institutional services. And while this sector is not necessarily profit driven, all have to manage their budget. And often that means that the quality of the food offered in this sector immediately depends on the budget provided by the owner. The food offered tends to be more functional: Often the "nutritional requirements" are the primary "design principle" and sensorial or hedonic aspects only follow that.

Therefore, the challenges institutional food services still face, especially when providing meals for large groups in settings such as schools, hospitals, cantinas, mass travel and correctional facilities are:

- *Budget constraints*: Institutional food services often operate under tight financial constraints. The selection of quality ingredients, quick preparation of fresh components, essential training of employees and investment into appropriate kitchen equipment within the planned budget often create an ongoing tension.
- *Menu variety and dietary restrictions*: Balancing the need for diverse menu options while accommodating various dietary restrictions can be complex. Institutions must cater to different preferences, allergies and cultural requirements.
- *Food safety and sanitation*: Special attention is often needed to ensure safe food-handling practices in such institutions. Not all have trained kitchen staff and equipment to manage hygiene, control all preparation temperatures and minimize cross-contamination and waste of expired or spoiled ingredients.
- *Sustainability and environmental impact*: Since these enterprises are often serving large numbers of people, environmental care and appropriate waste management are important to maintain their reputation and effective cost management.
- *Adapting new culinary trends*: Food preferences and dietary trends are changing. International cuisines and plant-based foods are also requested in such locations, and the culinary menu requires flexibility and continuous learning from the kitchen staff.

The last challenge listed can, for example, be seen in the classical company cantina. This is no longer the space where affordable, easy-to-prepare local favourites are offered, so that the employees are full and can continue to do their job. Today, this needs to be "corporate dining" (Albright, 2024).

The main difference is often the quality of the food and the location as such. A company might order boxed lunches for employees, while another one might order food from a local restaurant or deli. Or the old company cantina might be handed over to a professional food management company. Another flexible option is to bring food trucks, if they are able to manage the number of meals needed. Generally, this will bring additional costs. But it helps employees' work satisfaction and loyalty. People feel more rewarded and respected if they are served good-quality food. The employers can show that they are taking care of the health, well-being and needs of their employees. Particularly for Millennials, company culture and work-life balance are important. And providing lighter and healthier diets may not only positively influence employee satisfaction, but also increase work productivity.

In addition, hospital groups have started to turn away from centralized meal preparation within location and instead often heat up freshly cooked meals. That adds costs the customer of the institution must carry them or find ways to charge those costs out.

For example, schools may aim to provide nutritious and affordable meals for students, while hospitals may need to cater to the dietary needs and preferences of patients. Prisons may have strict security and budget constraints, while military bases may have to cope with changing locations and conditions and tough physical challenges for their soldiers.

COMMERCIAL SERVICES

Commercial food services range from high-end gastronomy to family-owned restaurants to food trucks and street food. Table 13.1 provides an incomplete list.

The list in Table 13.1 can be extended further, and characteristic features overlap more and more. They all have in common that they are commercial enterprises, with a primary focus to sell food and beverages. The key attributes are conscious decisions about the eating and serving experience, the main operating hours, the time to service, the mobility, the cuisines, the quality and, of course, the price.

These businesses vary in their size, location, menu, service style and target market. Currently the biggest is McDonald's with more than 40,000 locations, and turnover of more than $25 billion, quickly followed by Subway and Starbucks (Shauna Stuart, 2024). Average street-food vendors are happy with 50 U.S. dollars or less per day (Reddy et al., 2020). Some fast-food outlets may have drive-through or delivery options, while others may have dine-in or takeout services. Some catering companies may provide food for weddings, parties or corporate events, while others may serve meals for schools or hospitals. And food trucks can operate in busy urban areas, at fairs or as company cantinas during lunch hours and as food stands in the afternoon and evening hours at festivals.

Another way to look at it is to split the food service industry into four categories: Full-service restaurants, quick-service restaurants, institutes and others.

Based on a report by Vantage Market Research, the market share of full-service restaurants was 32.6% in 2021, followed by quick service restaurants with 28.9%, institutes with 19.8% and others with 18.7%. However, these shares are expected to change over the forecasted period, as quick service restaurants are projected to grow faster than other types due to the convenience, affordability and variety they offer (Quick-Service-Restaurants-Market, 2024).

The main factors influencing the food service sector are also consumer preferences, economic conditions, health and safety regulations, environmental issues, technological innovations and social trends. These factors affect the demand, supply, quality and profitability of food-service businesses and organizations. For example, consumer preferences may change due to cultural diversity, health awareness,

TABLE 13.1 Typical commercial food service designs

Restaurants	Offer a variety of cuisines and dining experiences, from fast-food joints to fine-dining restaurants. Often focus on main meal dining, quality dishes and scratch cooking.
Cafés	Often casual spots busy throughout the working day where people come for a hot drink, pastries, snacks and sometimes light meals.
Bars	Places where patrons can socialize over drinks. They often serve snacks or appetizers. Drinks are leading, some food to accompany.
Takeaway/delivery services	Focus on providing food for takeout or home delivery. A separate branch in the past, nowadays they are often an add-on business for in-house eating. Serving time is of some importance.
Food trucks	Simple mobile kitchens that serve food at various and changing locations, events or street corners.
Catering services	Companies that provide food and beverage services for events, parties and gatherings. Can range from simple casual to premium food and dining experiences.
Quick-serve restaurants (QSRs)	Well-known, often global, brands offering quick service meals, QSRs typically focus on fast-food items over a small menu because they can be prepared in the least amount of time.
Bakeries	Shops that sell bread, pastries, cakes and other baked goods. Overlap with cafés, but coffee accompanies the food offering.
Buffet restaurants	Places where customers can serve themselves from a variety of dishes, often "all you can eat" at a fixed price.
Fine dining	Upscale restaurants with elegant ambiance and gourmet menus (see also Molecular Gastronomy).
Ethnic cuisine	Specialized in specific regional or international cuisines often from exotic places, like Ethiopian, Caspin, Korean and Caribbean.
Ice cream parlours	In the past, often take-away ice cream shops, they now stretch to sit-in coffee shops and bars.
Food courts	Typically found in shopping malls, offering a large variety of food options at a central casual dining spot.
Ghost kitchen	Delivery-only kitchens, not open to the public, orders coming in from websites and apps. No tables or waitstaff for eat-in diners.
Street food	Sold by a vendor on a street or at other public places from a portable food booth, traditionally for immediate consumption of handheld food.

convenience or ethical concerns. Economic conditions may affect the income and spending of consumers and businesses. Health and safety regulations may impose standards and inspections on food service operations. Environmental issues may raise awareness and responsibility for sustainability and waste management. Technological innovations may improve efficiency and convenience for food service providers and consumers. Social trends may create new opportunities and challenges for food service businesses and organizations.

Originally, restaurants cooked from scratch, making their dishes from pure raw materials. The raw materials were mostly bought fresh, and additional cooking aids were bought form the stores present in the cash and carry markets. Some are still doing that, but many buy different components of their meals from other companies in the food sector. With the rise of the food chain and upcoming system gastronomy, food preparation shifted from cooking to precooked food assembly. Pre-prepared food components are combined in a tailor-made form. Similar to how it was done on old manufacturing lines, the final dish is assembled. The staff working there is much more trained in hygiene and safety than in cooking and food preparations. For the food industry, from ingredient to final branded products, the food service is a very important market. The product and services food manufacturers are providing vary from the products offered in the retail chain. And food products for the system gastronomy are different from those of the traditional gastronomy.

SYSTEM GASTRONOMY

System gastronomy is a term describing the application of scientific and technological principles to the culinary arts. Others call it "chain, quick serve" or "franchise gastronomy". It combines elements of industrial food production with traditional culinary gastronomy. Typical aspects are:

- *Focus on standardization*: Here the processes are consistently defined, allowing better planning of all processes and a more efficient control of all required resources such as goods and personnel.
- *Driven by efficiency*: The purpose of the standardization is to minimize time- and labour-consuming processes.
- *Recognizable identity*: This includes a clearly defined concept that distinguishes system gastronomy from traditional gastronomy, trade and competitors.
- *Centralized control*: Here the key processes are coordinated by a central headquarters.
- *Quick rollout*: The right concept enables a smooth and efficient rollout of a uniform corporate identity in all branches combined with a consistent quality of food and drinks, aiming to create loyal regular customers.
- *Customer satisfaction*: Customers should feel "at home" in every restaurant of the brand wherever they enter it. No surprises. Customers should know exactly what to expect when visiting the restaurant or ordering from the delivery service.

Some of the major players in the global food service market are McDonald's Corporation, Yum! Brands Inc. (KFC, Pizza Hut), Starbucks Corporation, Compass Group PLC, Sodexo SA, Aramark Corporation, Restaurant Brands International Inc., Domino's Pizza Inc., Darden Restaurants Inc. and Chipotle Mexican Grill Inc. These players compete on various factors such as product type, quality, innovation, pricing, service, location and brand recognition.

While most of them are widely known by their brand name, other companies and their branded services are much less visible, despite their global reach. An example is Sodexo, headquartered in France. It provides food services to companies, hospitals, schools, prisons, universities and a range of other organizations in 45 countries, has 430.000 employees and up to 80 million end-customers – those eating (Sodexo in Brief, n.d.). Under their premium brand, Sodexo constantly develops new food products and dishes, guided by the latest trends like plant based, nutritious, healthy and sustainable. There is a uniform corporate identity in all the branches, and they guarantee the expected food quality and a feel "at home" in every location.

A key feature of the food services landscape is the more focussed new food product development. In high-end and Michelin star restaurants, the lead chef may still invest a good amount of time in new food creation. But in most other food service businesses, the food development happens to a large extent outside the location. Students in a quick serve restaurant are not supposed to be creative with the food. Often others are doing that for them, for example the food developers of big food companies, from the food ingredient industry or food development services. A typical food service enterprise would not sell products the consumer can buy in retail stores. They are interested in meal and food preparation solutions helping them to create their own products more efficiently. Quick service restaurants and others may buy their burgers, plant-based patties, carbohydrates and prepared salads from other companies and use branded meal components, sauces, dressings and condiments in their kitchens. Other parts of the meal can be sourced locally. But food service businesses would normally not present the brand name of the products they use in their kitchen to their customers. The brands added value to the business is anonymous, providing help that fits the needs of their professional customers, based on size of the operation, locations, equipment and skill level. This is the "back of house" service they provide. "Front of house", they often proudly present the branded food product they buy. The spices, sauces, dressing, condiments and, of course, the beverages can be recognized by the customer enjoying the food.

MOLECULAR GASTRONOMY

A recent arm of gastronomy aims to create novel and innovative dishes that appeal to the senses, challenge expectations and enhance the enjoyment of food. In a sense they are the most "technology affine" part of gastronomy. This includes the application of tools, techniques and ingredients that are not typically found in traditional cooking. Best-known examples are molecular gastronomy, sous vide, spherification, foams, gels and edible films. Some examples of system gastronomy techniques are making caviar-like spheres from fruit juices, cooking meat in vacuum-sealed bags at precise temperatures, creating airy mousses from cheese or chocolate and forming thin sheets of gelatine from vegetable purees. This perhaps most experimental and creative trend began at the end of the last century with "molecular gastronomy" in high-end restaurants. While "gastronomy" in its original meaning is interpreted around "fine cooking" and the "study of human nourishment" (Brillant & Anthelme, 1848), "molecular" adds natural sciences, based on planned and structured experiments, observation and conclusion.

Here, top class cuisine is no longer art, performed by a great artist, but it is the science of meal preparation. Chefs progressively introduced the use of siphons, liquid nitrogen, rotary evaporators, decanting bulbs, thermocirculators, ultrasonic probes, etc., to prepare exceptional dishes and more. Their kitchens constantly create new, radical innovations. In a sense they have the licence to experiment, and restaurant guests often come with less hunger than curiosity and give permission to radically surprise their expectations and senses. A late differentiation of this trend is the "note by note" cooking and "synthetic" cooking, a meal construction process that even more created the physicochemical reactions in a step-by-step culinary protocol.

Celebrated chefs like Grant Achatz, Heston Blumenthal, Ferran Adrià and Wylie Dufresne constantly develop surprising dishes that massively challenge the conventional expectations of food. Dishes like spherified olive, bursting in the mouth; eggs benedict with cubes of fried hollandaise; cold potatoes with truffle and cheese; see-through ravioli; exploding chocolate; mango caviar or foie gras lollipops use physical and chemical principles to create sensorial surprises. These outbursts of creativity require a lot of experimentation and factual product understanding plus the will of the culinary expert to dive into various types of material science. The results can be experienced in high-end restaurants, in moments where the nutritional aspects are less relevant, but the sensorial curiosity is endless.

A glimpse into these molecular kitchens often reveals an armada of functional food additives and creative small-scale processing equipment. Chefs adopt well-known additives like agar-agar, carrageenans, alginates and lecithins to create surprising structures, astonishing combinations and unique textures. In hindsight, this was one of the few moments where chefs proudly adopted industry principles to make top-end restaurant dishes. Understandably that trend could not spill over to the packaged-food sector. And the technical solution providers for those creative kitchens were not the big, branded food manufacturers, but the food ingredient industry, offering science-based functional ingredients without interfering with the chef's culinary plan or even trying to influence it.

STREET FOOD

While the global food chains clearly express the globalization of the food industry, street food often represents the opposite side of the food service business. It often grows because of globalization, like the QSRs, but they are often also driven by local culture, and culinary heritage:

- *Globalization*: Street food vendors increasingly operate in bustling urban spaces, serving dishes rooted in local flavours, traditions and meal preparation creativity. It's not only daily food of locals, but also a favourite of all the tourists coming from everywhere. The famous Thai street and night markets offer the country's signature dishes in a tourist-adapted version: Less spicy, mainstream spice combinations and only known and meat from well-known animal source.
- *Culinary heritage*: Tourists seek authentic experiences and want to explore local taste and understand local life. The food remains rooted in local contexts.
- *Fusion*: In places that are frequently visited by tourists vendors experiment with components or product formats from other cultures. A well-known example is "bánh mì". The popular finger food aspects of both Vietnamese and French cuisine are nowadays one of the signature dishes of Vietnam. Other examples are Chinese buns stuffed with hot cajun chicken, Mexican taco pizza, Korean BBQ beef quesadillas or red curry risotto.

This part of the food sector is substantial, and it did not start with international travel. Some estimate that today more than 2.5 billion people daily consume street food (Street Food Central, n.d.). In many countries, eating on the go and on the street is a traditional way of food consumption. It's available all over the world, but today's street food capitol is probably Bangkok, where more than 20,000 vendors are offering their quick served food. And it's not only the sheer number of stalls, but also the versatility of the offerings on the streets. The combination of the deep-rooted local habit to pick something up to eat "on the go", snack 4–7 times throughout the day, rather than sitting down for three main meals, might have built the foundation. And nowadays, where Bangkok is one of the top three locations for tourists from all over the world, constantly more customers and cultural influences fuel the scene. The still-increasing attractiveness has its reasons:

- The sheer variety offered in popular spots is overwhelming.
- It's tailor-made. Even for a simple noodle soup, customers can choose any combination of dozens of ingredients seasoned to their personal liking.
- The snack size offers the possibility to try out more than one item. No need to decide for only one item, leaving all the others.
- It's transparent. The customer can watch the preparation, see the quality of the ingredients used, observe the preparation, be reassured that no "nasty" things are added and can interfere anytime with personal wants. Food and its preparation can be explored with all the senses.
- It's cheap. Sometimes the prepared items are even cheaper than the material used when cooking at home.
- It's convenient and often even faster than quick serve fast-food restaurants. It can be ready in less than 5 minutes and without all the negative connotations of "fast-food".
- It's social. The opportunity to interact with street food makers and other customers in public brings people together. For tourist it's a low barrier to dive deeper into exotic cultures and cuisines.
- It creates income. Street food generates essential income for those who do not work in formal economic sectors. People, particularly women with insufficient education, literacy or skills and with a lower income, can find income here.

- It's necessary. In busy metropoles, living space is limited and expensive, so apartments increasingly are built without a kitchen.

This long and impressive list of advantages is accompanying by a couple of disadvantages and watch out points.

- *Hygiene*: Poor hygiene practices of vendors, inappropriate facilities or no potable water can lead to an increased risk of microbial contamination, especially in tropical climates.
- *Health*: Popular street food dishes have high concentrations of fat, sugar and sodium and have unknown quantities of micronutrients.
- *Monitoring*: Vendors are very mobile; authorities have difficulties monitoring compliance with essential food standards.
- *Education and training*: Vendors often have low education, and so they are unaware of good hygiene practices.
- *Location*: Vendors come to crowded places, and people come where the vendors are. This can block traffic and bring food in the vicinity of polluted places.

This branch of the food service industry is aware of these threats. The primary concern is foodborne pathogens caused by microbial infection of the food items through the whole chain from harvest, primary processing, transport, storage, display and preparation to the moment when the food is served. The absence of potable water, inadequate infrastructure and food storage at elevated temperatures causes poor hygiene, and insects, birds and rodents can carry over pathogens (Andrade et al., 2023). The U.S. Food and Drug Administration (FDA) listed the "big 6" pathogens globally: Norovirus, Hepatitis A virus, *Salmonella typhi*, *Shigella* spp., Shiga toxin-producing *Escherichia coli* and nontyphoidal *Salmonella* (FDA, n.d.). The one-person enterprises do not have the option to waste the leftovers at the end of the day.

With the ever increasing prominence of this part of the food business, the local authorities take more care and attention. Governments and international organizations are taking several steps to improve street food safety. Typical steps are:

- *Understanding risks to improve safety*: Studying the types of local foodborne pathogens in street foods and the conditions under which they can survive and grow.
- *Regulatory measures*: Increasing legal control and sanitary surveillance.
- *Comprehensive policies*: Organizations like the Food and Agriculture Organization (FAO) are working on programs to assist authorities in ensuring that street food is safe.
- *National food security policies*: Based on local risks, action plans are developed to strengthen national food securities.
- *Global strategy for food safety*: Forward-looking, evidence-based, people-centred and cost-effective food safety systems combined with education, training and governance.

Particularly in countries that depend on the revenues from tourism, authorities started long ago with food safety campaigns, setting specific regulations in place, offering food preparation and hygiene training and implementing monitoring systems.

Recently, universities have launched ideas for machinery and equipment for street vendors to make the food they are selling safer (Tanadatet, 2019). Right now, that still looks a bit academic, requiring substantial investment, skills and training from a business that is often forced to make cash quickly. More radical, and potentially more effective is a governmental attempt to move flying street food vendors into fixed, centralized food court locations, where food safety and hygiene can be better monitored and controlled, and the business can probably also be more easier taxed. Singapore has already implemented this, and other cities and communities in the region have announced plans to follow that path. No doubt, this will make street food safer, but it also changes the specific culture and atmosphere of flying food stalls.

The locals know this only too well! For them, spoiled food can be much more than a bad night and a visit to the doctor and the pharmacy. It may cost them days without income and extra out-of-pocket costs for doctors, hospitals and medicine.

The street food business is also very innovative; vendors quickly spot when customers are moving elsewhere or when the neighbour offers something new. And particularly in areas where tourists are visiting, the possibility to sell at better prices to them lets hawkers constantly look for new ideas that they can make with their skills and equipment.

The link with the industrial new food product development is more indirect. Street food vendors often get major parts of the food items they need from local retail chains and the nearby cash-and-carry shops. Some of the food items they simply resell; the rest they need for their cooking.

Industrial packaged food items can help them save time and money and expand their menu, attracting more customers and serving more personalized food preparations. These people normally have a good feel for how much they could sell within a day, and buying too much that can perish quickly is a severe economic risk for them. They look for different package sizes that fit their pocket money and with a robust ambient product stability since normally they have no ability to store the food stuff chilled. The sales unit preferably is not too heavy, as they must carry their business day in and day out and generate an amount of packaging waste that they can manage.

A recent study by researchers from the University of Oxford and the London School of Hygiene and Tropical Medicine investigated the use of packaged food in various places in Africa. The study found that street food vendors often buy packaged food items such as bread, rice, noodles, pasta, sauces, spices, drinks and snacks from supermarkets and wholesalers. They use these items to make or complement their offers, like sandwiches, salads, soups, stews, noodles and fried rice (Hill et al., 2019).

In some countries, cash-and-carry chains began to offer product and formats specifically for their vendor customers. Besides product size and formats, they also started to offer product solutions to help customers.

And all the influences that are nowadays shaping the whole food service industry are coming down on them. A very-short-term business, selling on a daily horizon, using the cash received to buy the material for the next day, and willing to move wherever the customers are going, receives more regulation, control and the necessity for a longer-term business perspective.

CASE STUDY: NEW MEATBALLS

Like in most other cuisines of this world, minced meat preparations are part of the national signature dishes of Thailand, known for a long time and very popular, proud representers of the cooking culture. In the Thai kitchen, minced meat from pork, beef, chicken or aquatic animals is used in meatballs for noodle soups (guay tiew), minced meat with rice, herbs and spices (pat ka phrao), meat salad (laab) or simply BBQ meatballs on a stick. Loved by tourists and locals, these dishes are also very popular on the street. And they are cheap. A stick with 4–5 meatballs can sell at U.S. $0.25–$0.35. The competition on the street is fierce, and customers are always looking for the best deal. For the vendors, the margin is very low, and their business is very sensitive to price fluctuations on the meat market. In the tropical climate, the meat perishes quickly, so they normally only buy small quantities of minced meat, from which they make the balls, when needed, without stocking up too much.

They normally buy the meat from a nearby cash-and-carry market, like Lotus, Big C or Makro. These retail chains know what the street food vendors need and they have close contact with the company producing the packaged products they sell. Here then, the R&D department of the branded food maker was asked to develop a more suitable solution for their customers.

First, the partners organized open market visits to observe and then a couple of in-depth interviews to understand the business of the street food vendors, their hopes and concerns so afterwards they could design a new product. The price of the minced meat turned out to be the primary concern, since vendors were not able to pass on the cost straight to their customers. So, the innovation team came up with a set of dry, powdered pre-mixes for those balls, consisting of a very cost-effective mixture of plant-based protein concentrates, binders, natural colouring and spice mixes. Soaked with water and mixed with minced meat, the product could replace up to 50% of the meat in the balls, without negatively affecting taste, texture or aroma. The product development was strictly guided by the raw material costs so that the price in sales would allow street food vendors to retain a better margin for their meatball product.

With all partners wanting to have their margin, this initially turned out to be challenging. Particularly, the herbs and spice mixes turned out to be cost drivers, pushing up the resell price. Several reformulations rounds finally delivered a competitive product nearly matching the original business case that could be presented to the future customers. A selected group of street food vendors were exposed to the idea, the product attributes were presented and a couple of representative calculations were explained, showing the cost saving or margin improvement potential of the new meat mix.

The reaction of the vendors contained a couple of useful learnings. While most of them liked the principle idea, they had doubts with the calculation. In their heads, there were always a good percentage of leftovers that they couldn't find back in the calculation. For them it was difficult to see how they could flexibly manage the usage of the pre-mix with the amount of minced meat needed. And they needed a simpler dosing system to manage the making of the meatballs. And when using and tasting the prototypes, they recognized a very different taste and aroma. And finally, some mentioned that actually their personal seasoning made their meatball dish special and that was the reason why people on the street were coming regularly.

This last point provided the solution. While the vendors were, of course, interested in anything that helped them to minimize food waste and increase cooking convenience, they also believed that "their personal touch" was what made their product different and brought consumers back. For the development team that meant that, besides a very basic seasoning, no herbs and spices were needed. And that helped the business case. After some optimisation work on the package size, so that it better reflected the number of servings the vendors sold within a day or two, the project got market approval and was launched.

CUSTOMER AND CONSUMER

Some important differences exist between retail and food service businesses in terms of the nature of their products, the skills required for their workers and the challenges they face.

Retail products are usually tangible and durable, while food service products are mostly perishable and consumed quickly. This means that retail businesses must deal with inventory management, storage and distribution, while food service businesses have to ensure food safety, quality and freshness. Retail products also tend to have longer life cycles and lower turnover rates than food service products, which can affect the pricing and marketing strategies of the businesses.

Another difference is that retail workers need to have good customer service, sales and merchandising skills, while food service workers need to have good cooking, serving and hygiene skills. Retail workers often interact with customers who are browsing, comparing or purchasing products, while food service workers often interact with customers who are ordering, eating or paying for their meals. Retail workers may also need to handle cash registers, scanners and electronic devices, while food service workers may also need to handle kitchen equipment, utensils and dishes.

A third difference is that retail and food service businesses face different kinds of challenges in their operations. Retail businesses may have to deal with competition from online platforms, changing consumer preferences and seasonal fluctuations in demand. Food service businesses may have to deal with health regulations, food waste and labour shortages. Both sectors may also have to cope with rising costs, low margins and high employee turnover.

QUALITY AND SAFETY

Like basically all parts of our daily life, the food service sector can benefit from the progress in AI and big data. As mentioned, the global food service sector is very versatile. The number may easily pass 50 million and does not include all semi - or non-official street food vendors and ghost kitchens (Statista, 2023). Included are big global food chains, domestic food hubs, food-delivery servicing, hospitals, hotels, prisons, schools and a single person cooking on the street. Depending on ownership, location, the organisation, material sourcing, the economic condition, the training and education of the staff conditions, food quality and hygiene can be very dissimilar. The increase in food-related diseases is a key concern for institutions, not just the World Health Organization (WHO). Much better nutritional and hygienic safety is necessary, and that requires continuous assessment, which at the same time needs to be extremely cost effective.

New monitoring ideas use Internet of Things (IoT) Fog-Cloud technology to improve monitoring and decision-making capabilities. Tracking and real-time review of different food quality–oriented parameters form the core of the idea and allow efficient frameworks for storing, processing. The necessary tools are various types of sensors, including wireless devices, preceptors and radio frequency identification (RFID) technology. Sensing technologies and stochastic data can fill the gaps (Bhatia & Ahanger, 2021).

PRODUCT DEVELOPMENT IN/FOR FOOD SERVICE

Different businesses and different customers also mean different new food products and services. Nobody can afford to stand still. Competition is high in all market segments, and standing still means a step back. Any new idea/excitement a competitor is having can impact the sales of a business. The actors in this sector must always look for new product or service ideas.

Generally, new culinary trends often appear in the food service sector. Ethnic foods, cuisines from other countries and cultures, come through new restaurant openings and finding new customers that have visited their home countries and tried the local delicacies. And the level of culinary experimentation in the restaurant kitchens is higher than in retail. For some of them, it's also easier to experiment and try something new. Leftovers from the day can inspire the creation of a new dish. If the customer likes it, it helps to cut waste and save money. This is not an invention of the food service sector; everybody in the kitchen traditionally tries to minimize food waste. But it requires some kitchen creativity, which would give these chefs a stronger "licence to innovate". It is part of their reputation, to constantly create something new.

New technical solutions also find their way into the wider market through food service. Microwave, deep frying, sous vide, steam cooking or recently three-dimensional (3D) printing are some recent examples. These solutions often require a substantial capital investment into new equipment and professional handling, and they are not suitable for all products and preparations. The flexibility created through speed, versatility and waste reduction initially fits the needs of the food service industry.

But because this space is so diverse, common standardized new product developments are rare. Something that helps the chef in a privately owned restaurant kitchen might have no benefit in a big chain restaurant, like IKEA (sometimes called "the biggest restaurant in the world, with an attached furniture shop").

What differs for basically all of them is that the new product needs to fit the skill and expectation of a person using it. That can be the fully trained, experienced culinary chef, someone who loves cooking and food preparation but without any formal training, or someone who just wants to make a living out of it. But all of them first want a better solution for their busy work schedule, more efficient cooking, and possibilities to safe resources, cut costs and reduce waste.

A chef knows how lengthy it is to prepare bouillon from scratch or to be constantly whipping fresh crèmes for all types of dishes. So, the food industry can offer them solutions that take away effort but always leave chefs with the creative space to give their dishes at least the finishing touch. Basic building blocks, like a set of basic sauces, a common element in many food preparations, are typical new food products for this part of the business. Another example is "sofrito", a flavour base in many Mediterranean and Latin American cooking. The combination of various aromatic ingredients, cut into small pieces and sautéed or braised in cooking oil at low heat, is normally prepared separately and can add so much depth to a typical dish. It's not difficult to make, but it costs time, and therefore a ready-to-use solution is welcomed in professional cooking.

But basically, any solution must fulfil a couple of additional requirements. The product preferably should not contain any food constituents that must be declared on the menu or have a negative image. A declared technical taste enhancer can convey the message that the chef has difficulties making the dish taste great. That would be a hit to the chef's core competency. A much less trained "food operator" may not bother that much. But even here, times are changing. In Southeast Asia, the vast majority of its 10,000 street food vendors have somewhere on their stands a board or sticker saying, "Clean Food – No MSG", despite the fact that most families regularly use this ingredient as a taste enhancer at home.

The need to declare allergens is also nothing the sector looks forward to. But as generally allergen-free cooking is limiting from a culinary perspective, these are often accepted. People having an allergy or a sensitivity know about it, would ask accordingly and choose a dish free from "their" allergen.

New food products for the food service sector are not generally unique. Often the formulation is also available on the retail market, if it comes in a suitable format and size. A restaurant serving 200 dishes per day has a different throughout than a main street 24-hour quick serve. Many food items cannot be stored for long when opened, so the package size has to fit the usage requirements, the possibilities to manage shelf life, minimize waste and provide some cost advantages.

And for those companies producing a selection of formats, this often has a consequence. The production volumes for food solution product formats are generally much lower than those for the retail business. The faster the format changes in the high-speed lines, the more necessary cleaning is to material management, and this can easily lead to a situation where the total manufacturing costs of "bigger pack" food service offerings are higher per weight or volume unit for the retail business. In activity-based costing (ABC), the efficiency reduction in the high-speed lines is predominantly charged to the low-volume product, so a successful food solution format would be able to carry that premium, or the business would be willing to accept lower margins.

New product development for the system gastronomy, like the QSRs, again works differently. Here, the customer provides a detailed product brief of the NFP to the collaborating development team on the partner side. This can include signature features of the chain, an iconic taste or flavour, consistence colour or a particular nutritional profile, etc. The brief sets tough price targets for the new development, so that frequently only very lean and efficient developing companies are able to take that project.

And while it's critical to have first-mover advantages on a new trendy dish or offering, the food service chain normally does not need advice on trends from the food manufacturers. Here the branded food industry is the third-party manufacturer of the food service sector.

Generally, leaders in the Food service do not need advice on trends. From their day-to-day interaction with their customers, they have a very early view on trends and opportunities, so they are not very interested to hear from typical sales force teams passing by about the latest great opportunities.

But as many workers are chefs, proud of what they can create and very open to receiving professional advice, they are happy to listen to new ideas from a credible expert who knows their situation. That's the reason why some of the food manufacturers have invested in a chef network for food solution, which connects them with the people running the kitchen. Former chefs from famous restaurants or other food serving locations often know what the people in food service need. And they bring a reputation that opens the kitchen doors.

The single restaurant chef or a street food vendor somewhere on this planet has other possibilities to find new development opportunities. They can observe what their neighbour is doing, listen to their consumers or scan the internet for new ideas. The bigger chains and global chain account organizations deploy new technologies, especially in social media, to listen to their consumers, join focus groups or simply carry out consumer interviews in their restaurants. The new possibilities in digitized data management offers them tremendous opportunities. All the data from multiple sources, like ticket information, credit card, media, app and geolocation, can be integrated in numerous ways, and correlations and consumer behaviour prediction can be rapidly established through different AI tools.

Another aspect is the "value of a brand" in this business. Food brands, having a marketing and R&D department to be competitive in lean enterprises, have a different cost structure and margin expectations. But is a food service company as a potential customer willing to pay that premium? Here, "front of house" (FOH) and "back of house" (BOH) can make a difference. FOH is the area where the guests are sitting, and here branded products are welcome. The well-known butter or margarine producer or the premium tea brand can give a breakfast table more reputation. Famous condiment brands for sauces or dressing on the table can increase brand visibility and can serve as a quality sign for the food service location. It is a different situation "back of house" (BOH). Here the professional performance of the food item counts. The brand can enter that space, but most chefs would not perceive them as superior to what they prepare, and hence they are not willing to pay the premium. And, apart from a few exceptions, there is no value increase, if they would announce on their menus for the customers FOH that they cooked with a particular packaged food item. An "intel inside" approach does not work here, except for directly sourced farm items and a few very premium brands. A sophisticated dish like boeuf bourguignon may benefit from the announcement that a French red wine has been used in the preparation.

The bigger organisations in the gastronomy sector have their own R&D departments that are responsible for the development of new products and services and the improvement of existing ones. They work together with their suppliers, whom they often brief with detailed ideas for new food products and engage with their consumers to pick up new trends and opportunities early.

The situation differs for solutions requiring broader, more complex or systemic themes like upcoming new regulations or new standards of corporate responsibility. Here the innovation teams from consumer packaged good (CPG) companies take the initiative and alert management about upcoming changes and provide solutions. New concepts, how to reduce food waste, compliance with new labelling guidelines or perhaps shifting menus to plant-based offerings, are then presented proactively.

Food service for hospitals or elderly homes caters to people who are more sensitive to food-borne infections, so the necessary hygiene management in hospital catering follows the normal measures of communal catering. For food product development for hospitals and other locations with more vulnerable consumers, a combination of divergent requirements often shapes the product brief. First of all, these foods must still have the food attributes of variety, quality, aesthetics, taste, texture and aroma. But additionally, these products must match the clinical needs. This "functional profile" can limit the possibilities to match personal preferences, like with typical low-sodium, low-fat, low-calorie or low-carb diets or a smooth texture, particularly if the diet is part of the therapy where it has to meet medical or special nutritional requirements.

Also special and with focus on the "food functionality", but more on the other side of the spectrum, are offerings for members of the military. Regular soldiers have more active lives; they burn more calories and have different nutritional requirements. They are more like athletes. A soldier in heavy combat training may need between 3000–4000 kcal per day, compared to the average man at that age needs about 2500 kcal. And while the general recommendation nowadays tries to limit carbohydrate intake, military guidance suggests that 60% of the diet come from carbs (Military Muscle Tactical Nutrition, 2019). The recommended protein intake is also higher than normal to support muscle formation. And depending on locations and climate, military guidebooks recommend up to 12 g of sodium to compensate for salt loss through sweating and a good amount of iron to support testosterone formation.

In the non-commercial food service sector, the principal priorities often still are:

- *Safety and nutrition*: These are important particularly in prisons, hospitals, long-term-care facilities, military organisations and school meal programs. A food intoxication running through these populations could have devastating effects. Here, presence of allergens in foods is a key concern. Culinary choice is an extra.
- *Food quality*: In these locations, food quality does not need to be superior. People won't come back because of the meal quality. It should be acceptable and not rejected. Transportation meals, like in airplanes, for example, must not be so unacceptable that customers switch carriers.
- *Budgetary restrictions*: Budgets often limit quality and availability of choices. The price of the meal is normally included in the total offer or service people are taking. Food intake is not the main reason, and higher price premium offers are still the exception.

So, while the needs in the different parts of the food service landscape can differ significantly, some demands and challenges are common. The shift to more plant-based dishes and the need to reduce food waste is valid for the total food sector.

EVALUATION IDEAS AND TESTING THE MARKET

The testing and fine-tuning of new food products in food service is much less focussed directly on the final customer or consumer feedback. The owner of the enterprise or the meal preparing operator or chef often has a sharp view on what loyal customers want. These people see limited benefit in general market research data about consumer preferences. They will use their culinary competence to decide about the product to choose, but will closely look at the functional benefits a new product or service offers.

For the food preparation, the decision maker will use their knowledge and skill level as a reference to decide how much value a new product offers. A restaurant considering a refresh of their cuisine through some new exotic dishes may see good value in pre-formulated herb and spice mixes, saving time and money in buying, mixing and storage.

The food service will always think about their customer service. Any new product that ensures a positive dining experience through attentive and friendly service can add value.

More efficient supply logistics helping to save time with material sourcing, sourcing, quality, shelf life, storage and waste management can help them to run their business better.

Finally, as always, any investment into new tools should lead to financial improvement. All possibilities for better cost control and profit optimization help food services to sustain their business (FHA-HoReCa, 2024). Return on investment is a key consideration for them.

DIGITIZATION IN RESTAURANTS

Done well, the digitization in restaurants brings major advantages: Starting with a better customer experience up to a more effective operational management, leading to improved efficiency. Through the collection of data at every stage, business decision making will be facilitated. The implementation of digital menus, self-service kiosks and cashless payment methods can reduce human-to-human contacts, optimizing labour costs, but impacting human connection. Guests used to have a dialogue with a well-informed waiter or even a chat with the chef leaving the kitchen to look after his guests might find such an atmosphere less pleasant (StartUs Insights, 2024). Non-physical food-ordering chatbots and voice bots can take food orders, make recommendations and address restaurant-related inquiries. Using customer preference and behaviour data, such tools can create personalized food recommendations.

But digital tools are not only entering the front office. New solutions optimize restaurant operations. They automate back-office tasks like order creation, reservation management and menu pricing. Some include guest analytics and support inventory management and customer loyalty programs. The aim is efficiency and time and cost savings (see RestaurantOS, 2022).

In principle the food service sector offers good opportunities to implement integrated AI solutions, starting with innovative solutions for time and resource management back of house and the possibility to enhance the guest experience front of house while improving profitability.

Digital ordering or pre-ordering systems using chatbots has many advantages. Big restaurants and those with takeaway can enhance the ordering accuracy, personally engage the guests online and afterwards improve their marketing campaigns. For smaller kitchens, the liaison with the delivery services form is a known possibility. For dine-in, AI will optimize waiting lists or table reservations. Mobile and tablet ordering that send orders directly to a kitchen display system in the kitchen help facilitate communication between front- and back-of-house staff, which can greatly improve order accuracy. How digital the consumer experience in the dining rooms should be is the choice of the owner. From ordering by tablet and table cleaning and menu delivery by robots at the table, much is possible.

AI can improve day-to-day customer service operations in a number of ways. AI-powered solutions for customer service like chatbots can provide automated, personalized responses to patrons anytime, anywhere.

And for routine tasks in the kitchen, robots have started to become an addition and an alternative. Putting patties on a flame grill and turning them does not require an experienced kitchen worker. In other restaurants, robots execute the whole burger-making process already, from mincing up meat to the final burger assembly. Or they bake bread, fry noodles or scramble eggs (Gottsegen, 2023). This saves a great deal of time and energy, and the qualified kitchen staff can focus on non-routine tasks.

The other area where AI provides additional value is in the business forecast. By noting customer behaviour and predicting guest needs and preferences before they arise, businesses can benefit from reduced costs and increased guest satisfaction. And these data do not only need to be from one of the companies. AI engines can collect data from every open data source, like social media interactions and digital restaurant platforms. After cleaning the data and comparing them with historical consumer behaviour, trends can be detected much earlier or even created.

The core of new product development in food service is directly affected by these technologies. They can serve as inspiration for creative product work by finding weak hidden correlations between data and suggesting surprising so far undetected associations, which can spark new insights in consumer needs and product ideas already. But since the gastronomy is so much closer to their customers, careful human observation skills may be more efficient to spot new product or service opportunities. The verbal dialogue, a gesture or facial expressions can quickly make a point, something more difficult for AI to pick up on.

WEIGHT MANAGEMENT IN FOOD SERVICE

Weight gain and subsequent development of non-communicable diseases through excess energy intake is not only a burning issue for the CPG companies selling their products through classical trade and internet. Except for the time during the Covid pandemic, food consumption has gradually shifted in many countries from homemade to food service, eating outside the home. So, more people consume more calories out of home, and that literally. A broader study in the *Journal of the Academy of Nutrition and Dietetics* showed that over the last 30 years the portion sizes in American fast-food restaurants increased by on average 13 g per decade and 24 g per decade for desserts (McCrory et al., 2019).

That is neither new nor surprising. But it brings the nutritional impact of out-of-home eating more into focus for professionals that are concerned about the health of the population. So, NGO and increasingly government administrations start urging the food service to accelerate efforts to provide better solutions for energy intake, starting with portion size control and energy density. Consumers and authorities realize that there are nutritional consequences of eating out more often. Bigger portions have more calories. The responsibility of the food service sector for healthy eating needs to grow quickly. The growing "share of stomach" of the food service sector gradually demands more responsibility for the sector to make healthy eating pleasant. Hence, they have an opportunity to offer their customers healthier tasting solutions without cutting variety. The tremendous ability of the sector to offer more choice to customers (+226% for sides, entrees and desserts in the past 30 years) must be enriched by portion- and energy-controlled consumption.

While the requirement for good nutritional quality is a key design element in new food product development in the "non-commercial" sector, it is nowadays also getting more relevant in the commercial sector! Not only because the customers are asking for it, but as well from the governmental side, since the popularity of eating outside the home is still increasing and hence the responsibility of this sector for healthy and balanced eating is growing.

The second nutritional trend also hitting the food service sector is the shift to plant-based food. Typical vegetarian restaurant dishes like pizza, pasta, rolls and soup do not immediately need new product development. But dishes where the meat protein must be replaced with a plant-based alternative often require sophisticated formulation work and new culinary skills that are not standard in every cooking school. Here pre-formulated solutions from commercial partners offer a great help for the food service outlet.

WASTE MANAGEMENT IN THE FOOD SERVICE INDUSTRY

The second main societal and environmental challenge for the food service industry now loudly knocking on the restaurant doors is food waste and the global environmental care. The UN Environment Programme has calculated in their Food Waste Index Report 2021 that globally about 900 million tonnes of food are discarded every year and more than a quarter of that is produced by food service, compared to "only" 13% by retail, with the majority by households (https://www.unep.org/resources/report/unep-food-waste-index-report-2021). The significant share comes from all types of food services in restaurants, including fast-food chains, cafés, cafeterias, canteens, dining halls and catering (Euromonitor International, 2016) and serves billions of meals every year (Gössling et al., 2011). Consequently, the sector has a critical role in the global food waste challenge. Food waste management in the food service industry is as complicated as in other parts of the business and spans a wide range of factors and activities. There are numerous definitions in the scientific literature for food waste; in most cases it includes "food loss", the post-harvest losses and "surplus food", the excesses produced, but not ingested by any human, even under correct conservation (Lins et al., 2021).

For the single enterprise in the sector, food waste may first be a waste of money and time, and less a social necessity, but for the sector it's a systemic challenge. Finding a solution for a systemic problem first requires data that can be analysed and put in relation. Right now, this is still difficult, since consistent definitions are still missing. Food waste calculations in Switzerland measure calories, while in Sweden, they focus on the weight of waste.

And others believe that a resilient waste management approach needs to capture waste composition, quantity (mass), bulk density (which implies volume), size, moisture content, chemical properties and mechanical properties. Such differentiations sound sensitive but require a huge amount of work.

A different approach, therefore, suggests that food waste be grouped into three categories: avoidable food waste, unavoidable food waste and possibly avoidable food waste. These criteria could be used on a restaurant level in day-to-day operations. Avoidable means food that could have been eaten at some point in time, unavoidable food waste is the fraction of food that is not eaten (for example, banana peels and chicken bones), and possibly avoidable food waste is what could be eaten, e.g., potato skins (Papargyropoulou et al., 2016).

Until recently, food waste has not been part of management practice. Basically, the waste management chain in food services consists of four main steps: collection, sorting, storage and disposal (public or private). The waste is mainly produced in kitchens and back offices, with trimmings and peelings, cooking oils and fats and bones. And front of house waste is often plastics.

Bringing waste management more into business thinking in food service requires a favourable cost analysis, the ability to experiment with different management practices and possibly a disruption of the business model (De Visser-Amundson, 2019). Most straightforward are cost-oriented initiatives, including sequential and gradual alterations to the core business practices based on cost-saving analysis. Material losses occur through overproduction, preparing more food than is sold and preparation inefficiently. Advanced analytics tools connecting sales with production levels have shown to be very effective. Another big factor is spoilage, coming from improper storage and failure to adhere to food safety codes. A focus on proper storage practices, temperature control and usage according to expiration dates can help here. A reasonable level of training for the staff and adherence to safety protocols also helps. Spillage is another common source of food waste in restaurants, often caused by stress in the kitchen or unexperienced kitchen workers. The remedy is training.

The potential gain is massive. Studies estimate that adopting technical innovations to quantify and minimize food waste already could save up to 33.50% (Martin-Rios et al., 2020). And another study estimated that 25.02% or more can be achieved by just creating the awareness and enforcing discipline (De Visser-Amundson & Kleijnen, 2019).

In the eyes of restaurant leaders, food waste is primarily a cost factor, while most restaurants would not be able to tell how much money they wasted. But more and more partnerships for innovation by co-operating with other stakeholders such as suppliers, associations, local authorities and waste management companies are starting. These partnerships state as the main purpose cost minimizing, but they aim to implement better environmental principles.

Such help can come also from big, branded food-making companies. For them the food service landscape is an important part of their business. And they all made public commitments for more sustainable end-to-end business, with measurable progress targets. End-to-end business requires the inclusion of the environmental impact their products and services are having on their customers. Targets like a 50% reduction in waste until mid of the 2020s require a reduction in the food service sector they are catering. Companies like Unilever can leverage the experiences and solutions made in the retail sector and shape it for food service customers (Unilever, 2024). New solutions include plant-based alternatives, low-sugar options, reusable and recyclable packaging, leftover meal creativity, menu planning, online training, marketing and operational support.

In summary, new food development for the food service industry differs remarkably from the food retail business. Truly, new food product development happens only in a minor part of the business, like high-end scratch cooking restaurants and street food vendors on the other end. All others are open to new food innovations helping them run their businesses better and more successfully. The primary customer is a professional businessperson, who needs to run a sustainable business. New innovations from external partners focus on functional advantages in all aspects of the sector from preparation to customer experience. The sensorial decision is often taken by the food service lead. These people are professionals, and they have the competence and confidence to decide on the culinary aspect. Only bigger chains may test new formulations first with internal experts, before test marketing them with their customers. And since in many countries the consumption in food service is further increasing, the expectation from science and regulators to address societal and environmental challenges in this sector, as well, is increasing.

REFERENCES

Albright, D. (2024, June 6). *Everything you need to know about corporate dining | Fooda*. Fooda. https://www.fooda.com/blog/everything-you-need-to-know-about-corporate-dining/

Andrade, A. A., Paiva, A. D., & Machado, A. B. F. (2023). Microbiology of street Food: Understanding Risks to Improve Safety. *Journal of Applied Microbiology*, *134*(8). https://doi.org/10.1093/jambio/lxad167

Bhatia, M., & Ahanger, T. A. (2021). Intelligent Decision-Making in Smart Food Industry: Quality Perspective. *Pervasive and Mobile Computing, 72*, 101304. https://doi.org/10.1016/j.pmcj.2020.101304

Brehout, L. (2023). *Canadians are back to dining out again with restaurant spending reaching pre-COVID levels.* National Post. https://nationalpost.com/news/canada/canadians-are-back-to-dining-out-again-with-restaurant-spending-reaching-pre-covid-levels

Brillant, S., & Anthelme, J. (1848). *Physiologie Du Goût.* University of Leeds Library. https://archive.org/details/b21525882

De Visser-Amundson, A., & Kleijnen, M. (2019). Nudging in Food Waste Management: Where sustainability meets Cost-Effectiveness. In *Food Waste Management – Solving the Wicked Problem.* Palgrave Macmillan, (pp. 57–87). https://doi.org/10.1007/978-3-030-20561-4_3

Diligence Insights. (n.d.). *Food service market, size, share, trends, and forecast 2030.* https://www.diligenceins.com/food-and-beverages/food-service-market

FDA. (n.d.). *Employee health and personal hygiene handbook.* fda.gov. https://www.fda.gov/media/110822/download

FHA-HoReCa. (2024, May 22). *An introduction to the food service industry.* https://fhahoreca.com/blog/food-service-industry

Food Service Industry. (2023, September 20). https://www.futuremarketinsights.com/reports/food-service-industry

Gössling, S., Garrod, B., Aall, C., Hille, J., & Peeters, P. (2011). Food Management in Tourism: Reducing Tourism's Carbon 'foodprint.' *Tourism Management, 32*(3), 534–543. https://doi.org/10.1016/j.tourman.2010.04.006

Gottsegen, G. (2023, April 3). *14 Restaurant robots changing the food industry.* Built In. https://builtin.com/robotics/robots-in-restaurants-food-service

Hill, J., Mchiza, Z., Puoane, T., & Steyn, N. P. (2019). The Development of an Evidence-Based Street Food Vending Model Within a Socioecological Framework: A Guide for African Countries. *PLoS One, 14*(10), e0223535. https://doi.org/10.1371/journal.pone.0223535

Lins, M., Zandonadi, R. P., Raposo, A., & Ginani, V. C. (2021). Food Waste on Foodservice: An Overview Through the Perspective of Sustainable Dimensions. *Foods, 10*(6), 1175. https://doi.org/10.3390/foods10061175

Martin-Rios, C., Hofmann, A., & Mackenzie, N. (2020). Sustainability-Oriented Innovations in Food Waste Management Technology. *Sustainability, 13*(1), 210. https://doi.org/10.3390/su13010210

McCrory, M. A., Harbaugh, A. G., Appeadu, S., & Roberts, S. B. (2019). Fast-Food Offerings in the United States in 1986, 1991, and 2016 Show Large Increases in Food Variety, Portion Size, Dietary Energy, and Selected Micronutrients. *Journal of the Academy of Nutrition and Dietetics, 119*(6), 923–933. https://doi.org/10.1016/j.jand.2018.12.004

Military Muscle Tactical Nutrition. (2019, December 29). *Nutrition for military personnel.* https://militarymuscle.co/blogs/studies/nutrition-for-military-personnel

Papargyropoulou, E., Wright, N., Lozano, R., Steinberger, J., Padfield, R., & Ujang, Z. (2016). Conceptual Framework for the Study of Food Waste Generation and Prevention in the Hospitality Sector. *Waste Management, 49*, 326–336. https://doi.org/10.1016/j.wasman.2016.01.017

Quick-Service-Restaurants-Market. (2024, June 3). www.fortunebusinessinsights.com. https://www.fortunebusinessinsights.com/quick-service-restaurants-market-103236

Reddy, A. A., Ricart, S., & Cadman, T. (2020). Driving Factors of Food Safety Standards in India: Learning from Street-Food Vendors' Behaviour and Attitude. *Food Security, 12*(6), 1201–1217. https://doi.org/10.1007/s12571-020-01048-5

RestaurantOS. (2022). https://www.restaurantos.ca/

Shauna Stuart. (2024, March 21). *McDonald's is the world's largest restaurant chain. And now, there's a new second.* Masslive. https://www.masslive.com/business/2024/03/mcdonalds-is-the-worlds-largest-restaurant-chain-and-now-theres-a-new-second.html

Sodexo in Brief. (n.d.). Sodexo. https://www.sodexo.com/en/about-us/sodexo-in-brief?path=about-us&path=sodexo-in-brief

StartUs Insights. (2024, January 9). *10 Emerging food industry trends in 2024 | StartUs* Insights. https://www.startus-insights.com/innovators-guide/top-10-food-technology-trends-innovations-in-2021/#digital-food-management

Statista. (2023, April 25). *Number of food service units worldwide 2020, by country.* https://www.statista.com/statistics/1240159/number-of-food-service-establishments-worldwide-by-country)

Street Food Central. (n.d.). What is street food? | Street Food Central. https://streetfoodcentral.com/what-is-street-food/?utm_content=cmp-true

Tanadatet. (2019, October 3). *KMIT introduces three innovations to improve Thailand's street food - Thai PBS World.* Thai PBS World. https://www.thaipbsworld.com/kmit-introduces-three-innovations-to-improve-thailands-street-food/

Unilever. (2024, April 18). *Tackling food waste.* https://www.unilever.com/planet-and-society/waste-free-world/tackling-food-waste/

Formulation Development

14

FRESH VERSUS PACKAGED FOOD

Fresh food and processed/packaged food items are sometimes perceived as two different worlds. Here the unmodified natural goodness, and there the designed, composed and constructed edible food item with suspicious additions. But increasingly more people realize that most of today's farm-based food is the result of human alteration, using target-driven mutation and selection. Since the beginning of agriculture and animal husbandry, farmers designed edible plants and animals to produce wanted characteristics and suppress unwanted features. Food design started here, and it started long ago!

Many agricultural goods deteriorate quickly after harvest. The minimization of post-harvest losses and quality maintenance require the active modulation of transport and storage conditions. That is needed since "*all crops are metabolically active organisms, even after harvest, until they are either processed or consumed.*" (Ziv & Fallik, 2021). Variety depends on natural physiological processes, such as respiration, ripening, and senescence, and must be controlled to maintain or sometimes even generate the quality consumers want.

As necessary protection against microorganisms, insects or rodents and the delay of senescence, some of the most popular fruits need post-harvest treatment to reach their desired quality level. Climacteric fruits like apple, mango, papaya, pear, apricot, peach, plum, avocado, plantain, guava, nectarine, passion fruit, blueberry and cantaloupe need a burst of respiration to initiate the irreversible changes in colour, flavour, texture, aroma, volatiles and nutritional content for which each fruit variety is famous. This physiological cascade is intentionally initiated.

Bananas, the most traded fruit in the world with 114 million metric tons, are nicely packed by nature with an excellent protective packaging. But they are inedible at harvest and receive ethylene treatment during transport to the final markets (Hailu et al., 2013). The creation of the consumer-wanted attributes by scientists is the result of a "*genetically programmed highly coordinated process of organ transformation from unripe to ripe stage to yield an attractive edible fruit*" (Perotti et al., 2014). These processes change their carbohydrate profile, the sugar content, and colour, texture and aroma.

The vaporization with ethylene atmosphere is not smart "chemical engineering", but a copy of the controlled initiation of a natural metabolic pathway through a natural present substance. It is a natural processing and completely safe. But it is controlled human intervention. By applying the sophisticated science, specific attributes of the final fruit are designed that way (Maduwanthi & Marapana, 2019).

My point at the beginning of this chapter about "food formulation" is that the use of science and technology in food starts long before sellable food items are "assembled" and that farm-based products, even if they are not packaged and not processed, are the result of human modification. Experts in the food ingredient industry will try to influence the growth and harvest conditions of their ingredients from the viewpoint of the ingredients they are making for their customers, the food service or food retail sector.

DOI: 10.1201/9781032724591-14

NEW FOOD FORMULATIONS

From the ten thousands of food innovations hitting the market every year, many are basically variations of already existing formulations: New packaging formats, environmentally more friendly, healthier products, easier to eat or prepare, with a new design or a new claim. Truly new formulations with very different sensorial experiences or that are radically new are the minority. Putting something into our mouth and ingesting it requires trust. And we trust what we know. But today there are also moments when people are keen to try something new. This is often during holidays when visiting new destinations and exploring new cultures or at home when people book a table in an ethnic restaurant or one with a molecular kitchen. Now the food provider has the "licence to surprise", and the customers are open to new experiences. But for most times in our lives, the food better doesn't surprise to much, if we want to eat it again. Just changing the name of a well-known food offering can already cause strong rejection (Despres, n.d.).

Then there are taste preferences. The more often we taste a food, the more we tend to like it. This process starts already before birth. Negative experiences, for example, in childhood or a sensory overload and an anxiety around an eating situation, keep may people conservative in their food choices. Very often this is seen in children with their preference for sweet taste and rejection of bitter and sour. This food neophobia exists in 50%–75% of all children and probably has evolutionary reasons (Dovey et al., 2008). Often it weakens or disappears later in life but might return as they grow older or it may exist lifelong. The reasons for this neophobia are genetics (like the individual amount and type of taste buds) and environmental factors.

This aspect caught attention in the debate about the potential of insect protein to replace regular meat. While on average around 10%–18% in a population are neophobic, measured through the Food Neophobia Scale (FNS), new data suggest that up to 45% in a population *a priori* reject that source as food (Sogari et al., 2022). For those enterprises working for very good rational reasons on such new food sources, overcoming such rejections is the main hurdle. A single tasting session, even if sensorially fully satisfying, will not change behaviour.

SENSORIAL PERCEPTION (TASTE AND FLAVOUR)

"Taste is king" in foods, and this heavily influences our food choice and acceptance. More precisely it's taste and smell, aroma. The reason for that might be that the smell signal of substances travels extremely fast to the amygdala and hippocampus: These are the parts of the brain where emotions and memories are triggered (Harvard Gazette, 2024). And the taste, or better, the odour released during chewing sends equally fast and strong signals. The smell immediately revives strong and long-buried memories and emotions. That can be the aroma of the favourite food grandmom prepared at home, when we were young. So, the known smell "gets its liking score" before other senses or the rational part of your brain has the opportunity to modulate the impression.

And while constantly new flavour trends and hypes are renewing the markets, "taste remains king!" Functional food, with an added health benefit, needs to taste good. If it doesn't, it might make a career as a supplement, but it will hardly become part of the regular diet. This dominating role of taste in food choice is classically explained by the evolutionary importance of having reliable sensors on the tongue to differentiate the five basic tastes: sweet, sour, salty, bitter, and savoury (or umami). This physiological setup helps to evaluate the foods we need, as well as which ones to avoid. In a simple evolutionary view, sweetness indicates the presence of sugars, signalling calories and instant energy. Sourness flags acids, which can indicate ripeness or spoilage of fruits and vegetables. Salt (sodium), together with other minerals are essential for fluid balance and nerve function in the body, so "salt receptors" on the tongue can detect it in the foodstuff.

Bitterness is more an evolutionary warning system, detecting various molecules, often alkaloids or phenols, of either harmful substances or medicinal plants. Finally, savoury taste is triggered by glutamate, an amino acid and the building blocks of proteins, a valuable and essential macronutrient (McQuaid, 2012).

All these taste sensors are located on the tongue, and when triggered by specific molecules, they signal to the brain. In reality, the simplified "one trigger, one receptor, one signal" concept may be more sophisticated, and some academic discussion is still ongoing about when an impression can be called unique. The famous meatiness or umami impression is linked to unique receptors and stimulated by specific molecules, but whether this creates a truly separate taste impression is discussed controversially (Tepper & Barbarossa, 2020).

And some more news from science: Older textbooks locate the different taste buds in different zones on the tongue (sweet on the tip, bitter in the back, etc.). That seems in its simplicity wrong, but different regions of the tongue may have different thresholds for the different tastes (Munger & The Conversation, 2017). This newer understanding of the distribution of the receptors on the tongue is shown in Figure 14.1. And individuals have very different amounts of taste receptors; depending on the type, this can make them supertasters; and in food development great candidates for descriptive sensory panels (see Chapter 10), or as sensitive consumers food neophobes.

Recently, a few more "taste candidates" have been added to the list, like fat, starch, calcium and metallic. Future science will show if such perceptions are triggered by specific receptors on the tongue.

But besides the individual precepts taste and smell can trigger, they also stimulate the production of saliva and digestive enzymes, which are necessary to break down food and absorb nutrients. And they regulate our appetite and hunger signals to manage calorie intake and body weight.

In discussions these days, we can get the impression that the evolution, the force that brought us to where we are today, now works against us. Our taste buds developed to optimize "food finding", and not "food intake limitation". It will take too many generations before babies are born who like bitter and reject sweet.

The physiological setup is to a large extent common. Taste preference depends on psychology, culture, environment, physiology and genetics. It is not a conscious individual choice, as it is influenced by personal history, identity and values.

And while "taste (and smell) are king", both senses alone do not shape food preferences. All the other senses complement the eating experience. So only when texture, temperature, visual and even acoustic

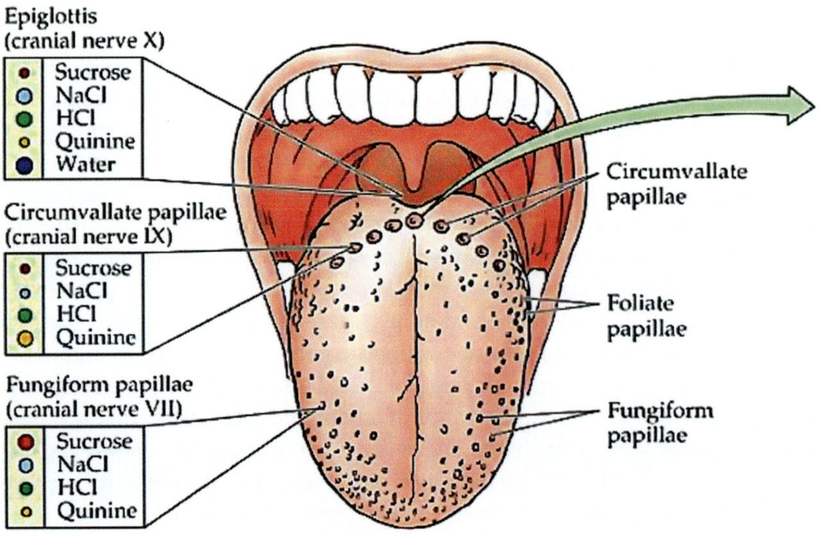

FIGURE 14.1 Distribution of taste receptors and their relative sensitivity. (From Purves et al., 2001)

signals are recognized is the impression complete. Science has started to capture all these influencing multisensory factors under the term "flavour", as the "integration" of all signals arriving through the different sensory channels, into a constructed percept, or mental representation.

Here is an example to illustrate: The percept of an apple is formed by colour, shape, texture, smell, taste and sound, but also by the person's prior knowledge, attitude, mood and experience. And it's not carved in stone; it's open for modulation. And the brain of a person, who has up to now only experienced sweet-tasting apples, will make a mental note after eating a sour one. It's a subjective dynamic mental construction. Or, as it was summarized by Garmt Dijksterhuis: "*In food-related applications of flavour, both bottom-up (sensory) as well as top-down (expectations) processes are at play. Most of the complex interactions that this leads to take place outside the awareness of the perceiving subject.*" (Dijksterhuis, 2016).

And while it may not fully capture science, the "taste is king" verdict is good guidance for food developers. Even with a detailed, thorough investigation about all relevant attributes triggering flavour and then food liking, success is not guaranteed. If the best product development brief is written without an understanding of the consumer context, the project may go wrong. Therefore, sophisticated product-optimization techniques like preference mapping (see Chapter 8) really help to find those sensory characteristics that drive consumer liking. Combined with behavioural observations and attitudinal data, the new food product has a much better chance of success.

CONVENIENCE

The success of a new food product, especially in consumer packaged goods (CPG), does not only depend on "aroma". Convenience is another important factor influencing consumer preferences. It's not competing with aroma but is often a differentiator splitting success from failure. In the past, convenience was narrowly described as the reduction of preparation time for a food item before it can be consumed, like ready-to-eat meals or snacks (Wikipedia Contributors, 2024). And while this truly captures a key aspect of processed packaged food, this does not reflect the full spectrum of relevant convenience for the consumer. Food products not only save time in preparation, but also in acquisition and cleanup and much more.

Making it easier to carry the product away, makes it easier for busy shoppers and hence more convenient. Even in the kitchen, new convenience ideas are not limited to time savings. New clear handling instructions, easy-to-use instructions, pre-portioning, smart storage solutions or the ability to prepare a complicated dish or food items helps people. And in times where environmental aspects and waste management receive more attention, new improved solutions will get more attention. Environmentally friendly packaging and ethically sourced ingredients are becoming more prevalent as part of the convenience trend.

Usually, convenience food is perceived as processed, less-nutritious food. That is changing. There is no reason why convenience food shouldn't offer balanced nutrition or serve specific individual dietary needs. Even wishes for more transparency and traceability from consumers can be served conveniently, through the addition of barcodes, usage of blockchain technology and links to dedicated apps.

Life is changing, and it's changing faster and faster. So many people have so much to do in the daily hectic life and less time to prepare food. There are surely moments when the food provider consciously put time and attention into fresh food preparation. But more often, the busy schedule and the daily routines let people look for convenience.

Only in cultures where cooking at home is the cultural expression of care and affection towards the family do cooks spend more time in the kitchen. In India, a typical food provider spends on average 13.2 hours per week in the kitchen, while an average person in South Korea would only use 3.7 hours per week for cooking (Statista, 2017).

And in western countries, the average food preparation time in the kitchen has decreased from about 60 minutes per day in 1965 to about 37 minutes per day in 2016 (Karen Hamrick & McClelland, 2016).

So, a wider description captures the essentials of food convenience as "... *reduced requirement for resources including time, physical effort, mental effort and skills by the consumer in relation to the planning, acquisition, preparation, storage, transport, consumption or clean-up of food.*" (Bogard et al., 2024).

And cooking has also benefitted a lot from convenience products. A simple bouillon cube or powder is used millions of times daily not only to create a broth, but as seasoning for soups, stews, sauces and gravies, in marinades and dressings or over rice and grains.

People all over the globe use industrial-made bouillon to make wonderful dishes, and most of them would not even know how to make it from scratch. Only a few high-end restaurant chefs sometimes make that effort. Here is a quick reminder of what it takes:

- *Rinse 2 pounds of beef shank and 2 pounds of beef bones and cover with water. Bring to boil, lower the heat and simmer for 2 hours, skim off any foam that rises to the surface.*
- *Add salt, cloves, bay leaves, onions, carrots, celery stalk, and parsley. Simmer for another 6 hours, adding more water if needed to keep the meat and bones submerged.*
- *Strain the broth through a fine-mesh sieve into a large bowl or container. Refrigerate the broth overnight and remove any solidified fat from the surface.*
- *Reheat the broth before using it.*

From Corbin (2007)

It takes two days of hands-on work and lots of chopping, peeling, skimming and sieving to make a stock. And the material is 10–20 times more expensive than industrial cubes, granules or powder that can be used in seconds. What a step up in convenience for those cooking at home!

Reducing the time in the kitchen is only one way to make food more convenient. Yet another angle of convenience is the packaging, sometimes captured under the ease of use, storage, transportation and disposal of food. There is an endless list of ideas bringing more convenience to the consumer though innovative packaging ideas. A few examples are:

- Resealable bags or pouches that allow consumers to open and close the package multiple times without compromising the freshness or quality of the food.
- Single-serve or portion-controlled packages providing the right amount of food for one person or one meal, reducing food waste and overconsumption.
- Microwaveable trays or bowls that enable consumers to heat and eat the food directly from the package, saving time and dishes.
- Easy-open or peel-off lids that eliminate the need for a can opener or a knife to access the food inside the package.
- Squeezable bottles or tubes that allow consumers to dispense the food easily and precisely, such as for ketchup, mustard, mayonnaise or yogurt.
- Reclosable caps that prevent spills and leaks of liquid or semi-liquid foods, such as milk, juice or soup.
- Transparent packages that let consumers see the contents and quality of the food.
- Edible or biodegradable packages that can be consumed along with the food or composted after use, reducing environmental impact and waste generation.
- Smart or interactive packages that provide information or feedback to consumers about the food, such as the expiration date, temperature, freshness, nutrition or cooking instructions.
- Flexible or collapsible (shrink wrap) packages that adapt to the shape and size of the food and take up less space when empty, facilitating storage and transportation.
- Self-heating cans, where a push breaks down a membrane and two components have an exothermic chemical reaction, releasing heat.

Also, variety can be understood as another expression of convenience and offers once again countless options for food product development. Consumer needs and preferences differ; they are not the same all the time. Being able as a company to rapidly bring new variants is a clear competitive edge in the market and one of the key reasons why companies are trying to create their products around common design principles. Variety can also stimulate interest and curiosity among consumers, who may be more willing to try new products if they have a choice. Furthermore, variety can improve the quality and safety of food products, by using different ingredients, methods and technologies that can optimize the nutritional content, flavour, texture, shelf life and environmental impact of the products. Sixty years after the invention in Japan, instant noodles are today a food category that serves globally anyone short of money, time or kitchen. More than 100 million servings of instant noodles were consumed worldwide in 2018 (世界ラーメン協会, 2024). And the list of flavours reaches from chicken, beef, shrimp, pork, seafood over curry, cheese, kimchi, tom yum, and miso to more unique and exotic flavours, such as chocolate, pizza, ice cream and even crocodile. The number of instant noodle flavours is constantly increasing as manufacturers experiment with new ingredients and combinations to satisfy the tastes of consumers. Besides all the different ingredients, different formats are available, such as cups, bowls, packets or blocks. Instant noodles are so on trend, especially in the younger western population, that in summer 2024 the Danish government stopped the sales of spicy Korean variants to prevent possible poisoning by capsaicin (Keane, 2024).

Cereals started as a popular breakfast for people in the United States. Nowadays, it's a regular part of the global morning routine. Cereal comes in various flavours, such as chocolate, fruit, nut and honey; can consist of various grains, like oats, bran, corn, rice, and wheat; and are available in processed formats are flakes, puffs, granola or bars.

A final example is yogurt. Probably nobody knows the number of variants of yogurt existing in the market. Generally healthy and delicious, yoghurt can be enjoyed at any time of the day. Besides an endless palette of flavours, yogurt comes with very different ingredients, like probiotics, fruit pieces, nuts, cereals or granola; different formats, such as cups, tubes, bottles, or pouches; and different consistencies. Yoghurts are a flagship format in all new trends like high protein, vegan, clean label, low sugar and healthy. No wonder this segment has globally already crossed the $100 billion mark (Mordor Intelligence, n.d.).

Offering variety provides an opportunity to bring well-known and well-liked sensorial preferences to new products and formats. And sometimes surprises are allowed! Who would have expected that the salted egg flavour for savoury food products would become a top variant in some of Asia's ice cream markets.

So, while eating at its core is conservative, there are endless opportunities to bring new benefits to the consumer through new food product development (NFPD). Often it is an "add on", with a new or better-served consumer need. Replacing taste with health or convenience might not be a good idea. But adding both to the beloved taste can create a big innovation. For the company selling the new product, it's just important that the new innovation create "incremental sales", either through increased penetration or higher frequency.

FINDING THE RIGHT FORMULATION

Finding a suitable formulation may in the past have been a combination of inspiration and perspiration of a trained individual. It can be an innovative process, where a professional uses experience to design new products. Or early in the morning the inspired chef may walk over to the fresh food market, examine all offerings and then, back in his kitchen, use experience and inspiration to create a gorgeous new dish.

That's a beautiful picture, but more of a niche approach in high-end restaurants. In small enterprises, food labs or small community kitchens something similar may still happen. Many cultures know the expression: "*Many cooks spoil the broth*". So this is not teamwork; instead a single person crafts something new, which is then instantly tasted and tested and step by step further improved. Once a satisfying result is achieved, the formulation and some preparation instructions are captured on a sheet of paper, perhaps in the computer.

This is a trial-and-error improvement approach, often without documentation, that can work in small teams of experienced food developers when they know their target market or customer very well. The organization is lucky to have such people. But two risks exist: If the "knowledge holder" is no longer available, the knowledge may get fully lost. Without truly retrievable documentation about the past development work, the successes, learnings and failures, the development contingency is at risk. Secondly, if the markets move on or a new market should be conquered, the individual "feeling" for the wishes of the consumer/customer may not be good enough.

For these and a couple of other reasons, contemporary product development systems put the focus on complete documentation, compositional accuracy, the generation of useful chassis and predictable commercial information like material costs.

PRODUCT FORMULATION TOOLS

While eating as such is intrinsically conservative, the innovation intensity in the food sector is still increasing (see Chapter 1). More products are coming on the market, and the life cycle is getting shorter. A variety of bigger trends reflecting lifestyle changes provide ample opportunities to innovate.

The process of creating a food mixture by putting together substances in correct proportions is called formulation. Mistakcs in thc formulation can destroy a reputation and damage consumer health through food-borne illnesses or allergic reactions. Various regulations guide the food formulation, such as ISO 22000 and others (see Chapter 10).

One key task to manage food formulation is the recipe complexity. Often 10, 20 or more ingredients in a wide range of quantities need to be put together accurately.

In its simplest form, the formulation work can be made with office-based tools like "Microsoft Excel" or "Google Sheets" to list all ingredients and the quantities in which they are added. But Laboratory Information Management Systems (LIMSs) are in more widespread use, helping to ensure compliance with regulatory standards, maintaining data integrity, improving product quality, giving formulation control and process standardization and helping to integrate the work of different people, processes and equipment in a team approach.

Once all components and their weight/volume are captured, the legally required nutrient profile can be established. Today, the easiest way is to connect to commonly available food composition databases, which provide the nutrient profile of many ingredients. The U.S. Department of Agriculture (USDA) food database system provides the nutrient content for more than 8000 food items in different "industrial" pretreatments, like raw, cooked, peeled, and microwaved (https://fdc.nal.usda.gov/). Choosing the right variety and primary processing step is important since the nutritional content of a food item varies not only by region, variety and climate condition, but of course also as a function of the initial processing. Europe has a similar data system, which is called the "European Food Information Resource" (EuroFIR, n.d.), while Australia relies upon the Australian Food Composition Database, with over 1500 foods commonly consumed in Australia (Food Standards Australia New Zealand, n.d.). Differences in the nutritional composition of the same crop reflect the different growth conditions. And the diversity in climates and different soils often results in unique varieties with different taste and composition.

Generally, a correctly calculated nutrient content based on public sources is sufficient to inform the nutrient table. But for explicit nutrient content claims, an analytical determination provides a more solid base.

In the "manual" formulation approach, food developers will have tried and tested the emerging prototypes, to improve gradually the sensorial aspects like taste, aroma and texture. This step requires time, expertise and experience. This is one reason why some companies have started to develop "prediction" tools, especially for their top seller ranges. The prerequisites are sufficient data about the sensorial outcome of a change in the formulation. So, different variants or prototypes have been developed to generate a sensorial spectrum. For the very important, but complex, senses like taste and smell, a trained panel has captured the different dimensions with the help of a common vocabulary (see Chapter 10). The prototypes are tested with target consumers/customers, where they express their liking, often on a hedonic nine-point scale (Society of Sensory Professionals, n.d.). The results can be captured in a prediction tool and constantly become enriched with more data from tests. Repeating that long enough builds a gradually reliable prediction tool for that product and the sensory aspect investigated.

These models are "more deep than wide" and require consistency in the data gathering. Today, they are used to predict taste perceptions like saltiness, sweetness and umaminess. It doesn't require a lot of imagination to see here another field for more artificial intelligence (AI) applications in food development.

In the next step – when the changes in the formulation are captured – the costs of the formulation, as a sum of the ingredient costs, must be calculated. Manually, that can become a laborious task, using reliable prices, either actual or projected based on sourcing volumes, launch dates or manufacturing location and delivery. The "bill of materials" (BOM) is the list of raw materials and their quantities required to make a product. This BOM is later also used in "enterprise resource systems" (ERP) for the production. Companies selling those widely used systems offer integrated solutions, with product life cycle management (PLM) and ERP, plus the "quality management system" (QMS) (Sengupta, 2023).

PLM

One of those integrated management solutions is "product life cycle management". The software started in the early 21st century as an alternative to paper-based configuration management systems. The first industrial users were automotive and consumer electronics, but today PLM is widely used in bigger enterprises, including fast-moving consumer goods (FMCG) enterprises (ArcherGrey, 2021). "Life cycle" stands for the ambition to manage company products from the very first idea all the way through until they "retire" and become delisted.

The beginning of a product life cycle is the ideation and new product development. The improvement that PLM brings here is first a centralized formulation control and storing. For a new food product, the different functionals like research and development (R&D), quality assurance (QA) and marketing need access to the formulation to ensure the prototype is "in specification", storing and managing recipes. This eliminates version control issues and ensures everyone works with the latest formulation.

PLM fosters enhanced and seamless collaboration between product development, quality control (QC) and marketing teams. Teams can share ideas, go over various formulations and ensure product specifications are met. PLM uses a tight version control, a structured approach to managing changes and supporting the traceability of all changes that have been made since the start (Food Industry Executive, 2024).

The connection with other databases often owned by procurement provides real-time ingredient costs, and additional supplier information supports data-driven decision making and cost-effective formulation optimization (Corley, 2024).

Table 14.1 summarizes the main benefits of an integrated PLM system for a food enterprise versus a more siloed development approach.

While these system functionalities help different functions in the businesses, new elements have been added in the meantime focussing on the specific needs of R&D (BECPG, 2023). The key functionalities are listed in Table 14.2.

More functionalities than listed in Table 14.2 are possible. As such, a single tool does not stimulate creativity, but it can provide more time to be creative and can speed up the final formulation work before

TABLE 14.1 Key functionalities of PLM systems in the NFPD

FUNCTIONALITY	*DESCRIPTION*
Collaboration	Enabling non-co-located, multi-disciplinary teams to collaborate using trusted, up-to-date product information
Document management	Captures, tracks and stores various types of electronic documents such as PDFs, word processing files and digital images of paper-based content
Version control	The reliable tracking of different versions of the same document type
Change management	The change to allowable values for reference tables, physical data stores, data models, data definitions, data structures, data movement, etc.
Bill of materials (BOM) management	The list of ingredients; constituents a material needed to make a product, including rework and losses
ERP and manufacturing execution system (MES) integration	Seamless connection with enterprise resource planning and manufacturing
Data governance	All arrangements, including technical, policy, regulatory or institutional affecting the data
Reporting	Generation of all types of reports from various angles of the product life cycle
Product planning	Organizing the planning phases in a life cycle
Product development	Supports the development phase of a product in its life cycle
Product retirement	Structures the retirement phase in PLM

it becomes locked. Formulation alternatives can be checked quickly on the shift of the nutritional composition, the possibility for an on-pack claim, a cost target or the impact of a material exchange and new material price.

The heart of PLM-based systems consists mostly of diverse calculation tools for the formulation, sometimes in different layers, the packaging and the assembly of both. For the formulation, the ingredient specifications are the basis. These are the full technical documentation of the material bought from the supplier, in which the full composition of the raw material is defined: All components, all physical and

TABLE 14.2 Main system functionalities of PLM-based systems for food and beverage development

FUNCTIONALITY	*CONTEXT AND RESULT*
Nutritional analysis	Calculates the nutritional content of the food product, to meet certain dietary requirements or claims, using databases, a company's own analysis or specific supplier data
Cost calculation	Estimates the cost of the food product based on the ingredients and quantities used, either through projected or actual costs
Regulatory compliance	Checks if the food product complies with different local and international food regulations, and which claims are possible
Shelf-life prediction	Predicts the shelf life of the product format made, the storage conditions based on the ingredients and packaging used
Sensory profile modelling	Models the sensory profile of the food product, predicting how it will be perceived by consumers; often connects with a company's own expert systems, built by consumer feedback
Ingredient compatibility	Determines which ingredients work well together, both in terms of flavour and stability based on a company's own internal data

chemical parameters, the nutritional facts, sensorial attributes, allergens, microbiological limits, quantitative limits for all possible contaminants, all the methods to analyse those parameter and non-technical qualifiers, like bio, organic, and vegan.

The new formulation development normally starts with a scratch recipe or formula, and the software will roll up the cost. If this delivers a product that does not meet the target costs, tools can be used to modify the existing formulations. That is done quantitatively, when "playing" with the amount of different cost-driving ingredients, or qualitatively, by suggesting alternative materials with a similar functionality. The formulation tool will then evaluate the cost factors and adjust the formula as required. Cost changes are not only caused by material price differences but also by modifications in different production line design, leading to lower operating cost (CIMdata News, 2012).

These tools are also helpful in product renovation and value improvement projects, when later in a product life cycle the margin is coming under pressure and material cost reductions are necessary. New formulations can also be framed through different rules, like maximum overall sodium, sugar, fat, minimum vitamin or vegetable content, or replacement rules for dairy fat with plant oils or fats.

At the end of the "experimental phase", a version is "locked". This document is the basis for all the documents needed to make a commercial product.

Such solutions accurately roll up as-designed costs and by linking with enterprise resource planning (ERP) will include the latest, most accurate cost data for the materials. Some modules can suggest design changes to optimize the costs and manage margin erosion coming from commodity price changes.

Together with the packaging specifications, the final product is described. The factory receives it as part of the BOM, the complete list of items, parts, assemblies and other materials necessary to create a product.

A PLM system also performs repetitive tasks and checks, like the recipe approvals and the essential regulatory compliance checks. For companies selling hundreds or thousands of different products or product variants from different locations into different markets, the implementation of PLM is an opportunity to manage all these data-driven processes more robustly and efficiently. Prerequisites are clean and plausible sets of all relevant data. Only if data are unambiguous can the system automatically calculate, composition, predict shelf life, build the nutrient table and ingredient list, draft claims and calculate costs. Food companies that have grown constantly in the past and continuously launched new products may have accumulated many different data from suppliers and their own operations in various formats. Often relevant data have been uploaded in the past as needed but hardly anybody had actively deleted obsolete or wrong data in the systems. The results can be outdated material specifications, different suppliers, different materials under the same specifier/name and outdated costs. Depending on the amount and complexity of data managed in the new system, a massive cleansing exercise of the existing date is necessary before integrated solutions can unfold their full potential. Many companies focus their effort on the accuracy of all data for new and existing products, but not into the time to cleanse and archive their historic data stores properly. Old data do not create profit or business success. But once connected to the new data management systems, they can cause confusion and misleading calculations. Suppliers might have been replaced; products are taken from the market or have been reformulated; manufacturing processes have been modified or the labelling regulation changed. All these outdated data can contradict the actual, valid datasets and lead to wrong calculations.

So, before an enterprise can benefit from the enormous advantages of such systems, they need to ensure in their system feeding data the "single version of the truth".

The Value of Harmonized Data

Some leaders in food see the management of specifications and data as one of the most important tasks for R&D teams, a job description that new food creators and enthusiastic product developers do not easily accept. Managing the specifications has a bit of the smell of boring bookkeeping and nitpicking But all the data captured in ingredient, packaging, processing finished goods, storage and distribution specifications define the corridor in which safe and high-quality products can be made confidently and reliably.

R&D is not the only function managing masses of data. In complex organisations, each function of the business often has its own set of data, which sometimes contradicts others' sets of data. If the R&D formulation tool is a stand-alone solution without an active interphase to the actual material cost tool of procurement, efficient cost optimisation can become a headache. Replacing functional data silos that cause complexity and confusion with full transparency of information across the organisation can facilitate interdepartmental collaboration and cohesion enormously, and that is important for successful product innovation, where time is money and shorter lead times are unchallenged incremental turnover. So, all speed brakes need to be brought to a minimum. Visibility and insight into accurate data are also crucial to maintain consistency across processes while also being aligned with emergent trends.

In fact, more than 85% of consumer packaged goods companies indicate that PLM is critical for digital transformation (Kalypso, 2022). Others estimated that in an FMCG company, PLM can streamline the compliance process and save the company up to 35% of administration costs (Centric Software, 2023).

PLM implementation also helps to compare sourcing units and factories, where similar data often have very different formats, the waste and rework levels are inconsistently captured and local buying has different buying prices than the ones central R&D had calculated earlier.

The price for the efficiency gain is a full data cleansing prior to activation. That requires good data expertise to ensure that the new cleansed data are fully accessible for all programs and tools depending on them.

A costly, but helpful approach to ensure the safe transformation of these massive datasets can be experienced external business consulting partners. They can help to organize the data cleansing, while all the "normal day-to day operations", including new food product development, continue. Implementation can happen in discrete blocks, where one company or a specific R&D team finish the data transformation first. Dedicated consultants like Accenture and others provide such a service. As partners, they need dedicated internal teams, often from R&D, to prepare the migration, including cleansing, testing, simulation and people training.

Thc valuc of thc data and thc optimal level of harmonization and differentiation is determined by the business and its strategy. The case study illustrates this thought.

CASE STUDY: HARMONIZED INGREDIENT SPECIFICATIONS

A multinational food company wanted to harmonize its complex dry ingredient portfolio in Europe. The idea was driven by the wish to cut the data complexity and reduce the losses coming from obsolete stocks, complex storage and warehousing, and to create higher buying volumes with less suppliers. Even a moderate harmonization was calculated to deliver a savings of dozens of millions of dollars.

In the past, the product developers had focussed on variation for all "value-adding" ingredients, like dry vegetables and herbs, pasta, spices and meat pieces. Commodities like onions, carrots, broccoli, garlic, celeriac, potatoes, tomatoes and sweet corn were bought as toasted, fried or air-, freeze-, oven-, spray-, drum-, microwave-, or solar-dried, in various shapes and formats and sizes. Part of the portfolio was bio and organic source food or from local origin means from the region nearby so that the product could carry a provenance claim. Ingredient suppliers had happily presented their new products, and developers had used the different material to create a unique product in the market. Reoccurring VIP, savings initiatives had identified cheaper material alternatives that could be used in parts of the portfolio and that could be sourced cheaper. Now the ingredient portfolio was so complex and versatile that nobody could recognize or understand many of the specified differences. Even commodities like the various flours, salt and milk powders were used in dozens of variants.

To simplify the enormous complexity, the material experts from R&D, quality and procurement in a first step collected samples of all bought materials and their specifications, together with their annual buying volume and spending. Comparing materials side by side gave a first understanding of how unique or common materials were. Non-technical aspects like organic, origin and bio were not included in this step. In that stage, a 75%–85% reduction for the ingredient specifications was seen as feasible without a big risk of losing sales.

Next, they created a full product portfolio overview: All products and product groups using the materials with their total annual sales and profit in total and per market. These data were discussed with the regional business leadership to define relevant business key performance indicators (KPIs), thresholds and cutoffs. To help the decision process, procurement modelled the direct savings potential for different harmonization scenarios. It ranged from 5%–30% of the buying spent.

Once the business target was set, the R&D team invited all country marketing managers for each category and country into the R&D centre to present all their "as sold" and "prepared" products. To help them, similarities were highlighted between the different products and compositions, and suggestions for harmonization were made.

In addition, the leadership team set minimum sales and profit targets for each country depending on size and competitive situation. The country managers then got the chance to "save" some of the designated "delisters" below the cutoff line, by pooling their formulations with products of better commercial performance above the cutoff. To respect the actual market circumstances in the markets, country managers could exchange a low-volume ingredient or formulation with a more common one suggested by R&D.

Managers from the different factories were invited to confirm the technical possibilities of manufacturing with the new harmonized material.

The preliminary result of the harmonized portfolio was presented to all factory managers to confirm their ability to manufacture the new product portfolio with the expected volumes.

For all the materials with extra non-technical specification elements (bio, organic, provenance, etc.), the "incremental claim benefit" was estimated and compared to the calculated oncosts.

All ingredients that were regarded as non-essential differentiation and premiumization were classified as commodities, and R&D was asked to rewrite the specifications to globally traded market standards, so that best prices and spot market opportunities could be taken in the next buying round.

The harmonized differentiating ingredients were reworked as needed to fix differentiating aspects. These were discussed with suppliers to align their capabilities with the now demanded quality. Where needed, new supplier audits were made, assessment methods standardized and certificates for analysis agreed upon.

For lead ingredients in national top sellers with no technical criteria, the decision process was sometimes reverted. In those cases, the marketing teams were asked to bring quantifiable evidence about the added value of the unique material. Where those data were not available, the innovation team was asked to create new campaigns to claim the benefit to the consumer better (for example, "organic", "with sun grown tomatoes", "with bigger pieces/new flavour"), a process called "use it or lose it".

A cross functional steering team under technical leadership was installed to drive the harmonization in the next 3 years, monitor the progress and readjust the approach as needed. From all the savings realized, an agreed part was used to "reinvest" into the brand and key products. This strategy was used to invest into those materials where extra quality would be valued by the consumers, for example, to replace some food additives that had a negative consumer perception.

Since the whole operation basically meant to rewrite all ingredient specifications, the company also agreed to standardize the specification layout and data stores in such a way that the migration into a central standardized data management system was possible. The result was an international, harmonized product formulation tool, with fixed specifications standards and validated datasets, a key step towards "the single version of the truth."

Various processing steps can be defined by their characteristics, and the impact on product ingredient quality and quantity can be predicted. Premixed can be captured and calculated, and changes in physical or chemical properties can be calculated. Nutrient or water loss was used to predict the final product weight and composition.

Depending on the manufacturing process, the formulation can be split; for example, when it's useful to premix some components to ensure proper ingredient blending, this can be done. Or the waste in a processing step can become converted into the agreed amount of rework, which can be used in the next formulation. And it can be recalculated into batches of suitable product lot sizes.

And all these data can feed a labelling function to ensure there are compliant declarations for the different markets; not only for the content information like ingredient lists, quantitative ingredient declarations (QUIDs), allergens and nutrient information be generated, but also the full artwork.

Such software is more than just a powerful analysing and data-capturing tool allowing full documentation of the product development process and provision of necessary legal documents. Food companies making the effort to migrate all their existing data into PLM tools are not only looking for a better data archive and storage system. They want to transform the classical new food product development process in the kitchen, lab and pilot plant into a digitized version. The ambition is "*in silico*" first, meaning that all experiments and trials should be modelled or simulated using a computer before practical work is done.

Compared to other engineering areas, the *in silico* food product development is still in its infant state. The reasons are the lack of captured and well-documented data about all the physical, chemical processes and the natural variability of the food "material" used. Today the data, model and expert systems for food cannot fully replace the experimental trial, but they allow more targeted optimization of relevant interactions before any practical experiment or factory trial. A gluten network formation model can predict the gas retention during baking. Other programs can simulate the effect of cheese composition and ripening conditions on the texture and flavour of the cheese. And again others can model the crystallization of fats in chocolate or their hardening in bouillon cubes, predicting the product texture or the machinability of an intermediate product.

More advanced solutions will likely be generated when AI starts scanning datasets for hidden relations and interactions. Starting with the prediction of interactions between the components at the molecular level and the processing conditions, the scope will become extended to predict *in silico* consumer preferences or the full environmental impact for the whole life cycle. First examples for simple formulations have been reported, but right now they seem far away from the abilities of a "human" formulator (Becker et al., 2023).

A feasible approach of using AI for product formulation has recently been presented by the Singaporean company Hoow Foods (Hoow Foods, n.d.). Leveraging from pharma, this food-technology company takes an existing food formulation and then matches a healthier product profile into the matrix. The tool then uses an iterative learning cycle to optimize this new product matrix. The "intelligent" formulation work here is less ambitious than for the "*de novo*" product creation. A skeleton of tested knowledge, including elements like the hedonistic consumer, are already given, and many physicochemical reactions are known and tested. One-by-one material exchange allows a good prediction of the outcome. The process is schematically presented in Figure 14.2.

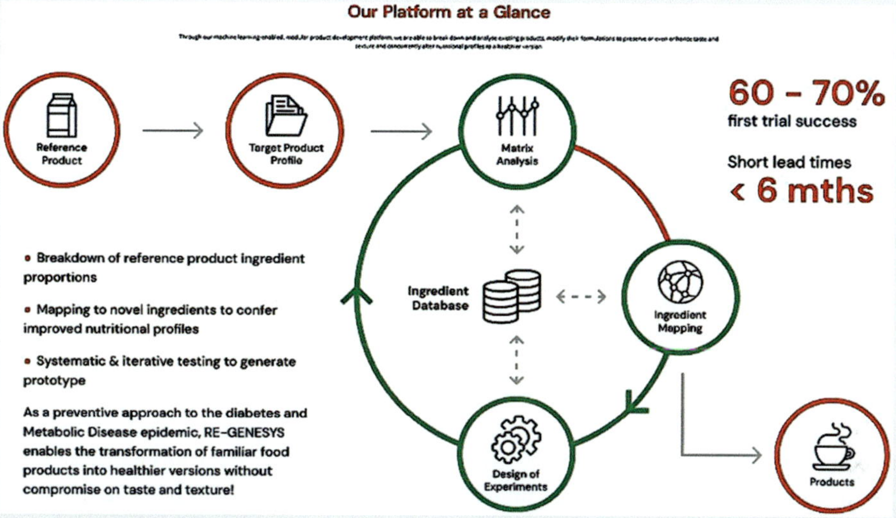

FIGURE 14.2 Food profile improvement through an iterative learning cycle steered by available ingredient datasets. (Taken from Hoow Foods, n.d.)

Such an approach fits straight into the major health and wellness trends for food and vegetables. "Uncompromised pleasure" – health food that can be enjoyed without any sensory compromise – is globally a major growth opportunity. Low-calorie, low-sugar ice cream based on this formulation tool has already been launched.

PROCESSING EFFECTS ON/FROM THE FORMULATION

It is a truism that the commercial production volumes of packaged food require different process technologies than handmade kitchen examples made at home. The food producer has the process experience, the equipment and the trained personnel to transform a prototype into a safe and high-quality commercial product. And the recipe dictates the processing to convert fresh food into edible, convenient, safe and stable food.

Cereals, pasta, meat products and many snack food formats need an extruder, a machine forcing ingredients through a small opening to give them a specific shape. Often, such extruders also apply heat and pressure to the ingredients, changing the chemical and physical properties of the ingredients remarkably. When this is done cold, the original properties are better preserved.

The point I want to make is that industrial food production is not the upscaling of home food preparation processes. The wanted product or result is obtained by different technologies. Sometimes the idea is to make something really new, like a low-fat ice cream, or to create something where the processing steps are preferably minimal or invisible, like for canned soup or juice. But the implications of even the simplest process technology need to be incorporated into the product design thinking.

Dry mixing and blending are simple, physical processes without major modification of the components. The formulation better prevents the behaviour of those ingredients, as they have variable physical properties, different particle sizes, shapes, densities or flowability. If the aim is to have all those ingredients homogeneously distributed in the final product, product composition and processing need to be aligned. Here the right type and design of the mixer or blender is a critical element in the final product quality.

If the formulation just needs gentle mixing of ingredients at low to moderate viscosity and density, a ribbon mixer might be a great choice! For higher viscosity and faster homogenization, a paddle mixer may be better. The good heat transfer may be wanted or not, and the high shear force can change ingredients. A view blender can mix fragile ingredients with good free-flowing properties and simple operation. Or perhaps a plow mixer, a cone mixer, a double cone mixer, a tumble blender, a Nauta mixer, planetary mixer or a fluid bed mixer could be used.

An ordinary pack of dry muesli can easily have 15 to 20 different ingredients like rolled oats; wheat, barley, rye, rice, or corn flakes; various dried fruits and different nuts and seeds. All different forms, weights, sizes and densities can be present in very different amounts in the final mix. The composition, presence of value-giving ingredients and whole nutritional content need to be maintained in all packs. This needs to be considered when designing a formulation. How many mixing/blending steps can be performed during manufacturing setup? Can additional dosing steps for the most valuable ingredients be added through a weighting scale? Is it possible to create different premixes with components of similar properties?

Except maybe for food startup, most product formulations are designed with the strengths and weaknesses of the processing that is available. Very often, the best technology for production of a new product is known, but the existing manufacturing is missing some of those features. It's a hard decision to leave the existing, paid processing equipment unused and instead invest in new technology. While "in theory" manufacturing must be able to assembly the products the consumer wants, the reality demands that the product respects the existing manufacturing. Minor product modifications must sometimes be considered to bring a new idea to market quickly and efficiently. The watch-out point here is to ensure that the required changes are not impacting the buying intention of the consumer/customer.

Many food items cannot be eaten fresh/raw. They need to be boiled, fried, baked, toasted, etc., to be safe. The food industry has a wide array of thermal treatments, including traditional sous vide, low temperature long time, high temperature short time, steaming, baking, broiling, and nonthermal processing methods like pulse electrical field, high pressure, ultrasound, cold plasma, microwave and irradiation. The choice of the right equipment is the search for the best mix between safety, nutrition and food quality.

Clearly, the quality of the final food product is impacted – and sometimes completely transformed – by the processing used. Ice cream is much more than the freezing of fat and sugar and a few aroma-giving substances. Without the incorporation of 30%–50% of air, a block of frozen lollies might appear at the end of the line. And all emulsions require the mixing of two high-energy, basically immiscible phases like water and oil. Whether it's an "oil in water" or "water in oil" emulsion, the physical process creates very different droplet sizes and mixtures. And the stability of the resulting emulsion is massively influenced by type and amount of emulsifier, fat type, antioxidants, antimicrobials, flavours, nutrients, etc.

So, enterprises having such "semi-stable" products in their portfolio try to apply *in silico* model systems to predict the sensorial impression and the shelf-life stability. For a mayonnaise maker, the type and length of the distribution chain can be a major question. This famous emulsion type is also thermodynamically unstable, but sophisticated ingredient treatment and mixing can give commercially available products an ambient shelf life of 15 months or more. Still, these companies have difficulties to predict the stability when the product is distributed via new ways, for example, by ship for export rather than by truck and road.

Food formulation strategies are often the core of new product development. What may finally look like a simple assembly of various components into a final consumable is in fact often the result of a sophisticated development process. Between the different components used, wanted and unwanted chemical or physical reactions can happen.

Depending on the "system" – the food matrix and its composition – an *in silico* prediction has different reliability. But basically, all new food product development (NPD) tools today ensure seamless documentation of all experiments.

Especially all those early formulations that have never made it into a final product offer a great learning pool for future development work. The prerequisite is that the experimental data have been well documented, properly labelled and archived in a way that they can be retrieved, not only for the successor or temporary replacement in the job, but as well-structured input in digital workbenches. Whatever the organization has created as knowledge should be available to facilitate NPD, not only to avoid making mistakes again, but to offer new opportunities. That's why R&D leaders constantly remind their formulation teams to document also the unsuccessful formulation and processing attempts. "If we would only know, what we knew" is the phrase capturing that thought!

But since then, the gain is much higher! With the accelerating options for computer simulation, AI and deep learning, all data can be of enormous importance, not only for an efficient information flow towards launch as a dataset without errors. But it can be used as data pool for new tools, programs or AI to either validate assumptions of the experimenter to detect new, hidden relations or predict product and process interactions better. Will the exchange of one modified starch against a differently modified one result in better water-binding properties, strengthen the gel or cause syneresis? How is the shelf life affected, or how will different mixtures influence the speed and accuracy of the dosing in the filling line? Or will the double or triple fortification of a complex dry ready meal be stable over the full shelf life and preparation time?

Food/formulation-packaging interactions, distribution and shelf life stability are other key questions influenced by the formulation. Several computer-based models are on the market, but the predictability depends a lot on the type of product, the processing, the way it's transported and how it is stored in warehouses, shops and consumer homes.

Aroma-food matrix interactions in complex food products are even more difficult to predict, but it's critical to control the food quality and sensorial match. Only very recently, first academic approaches have been suggested to open the field to machine learning and AI (Anker et al., 2022).

Many more aspects are critical in food formulations, like the price of ingredients, the microbial status, other constituents and contaminants, sensory profiles and shelf life.

Today, the practical experience of the formulator, exact documentation, extra analytical capabilities and various tools and programs define the efficiency of the development process. That's also why innovative tools from other industries have made their way to the NPD of food. These tools are often not specifically designed for foods but have been adopted to fit the purpose.

PRODUCT DESIGN OPTIONS

Food product developers in internationally operating enterprises are aiming at a high degree of standardization and harmonization of product and manufacturing design. These resulting economies of scale and operational efficiency are the basis for attractive and competitive prices. Using the same ingredients, the same or a very similar process technology and an identical line configuration in several factories makes an eventually necessary shift of the manufacturing or the rollout of an existing product into a new market much faster and easier. And faster normally means more sales at a better price in the FMCG market.

The harmonization starts at the ingredient level. And the enabler for harmonized ingredients is well-defined material specifications. Being conscious as a food enterprise of which ingredients can be "commodities" and which ones are better specialties has advantages. A commodity specification allows flexible sourcing and maybe spot market buying, while the risk of obsolete stocks is reduced. And a well-chosen speciality brings product differentiation.

Many functional aspects of a food product are not directly visible for the consumer. These attributes shape the product performance, but they are not the reason why this product is bought. If shelf-life stability, texture, mouthfeel, solubility, preparation time, portability colour or even taste can be created in a product category by a common "chassis" made from commodities, the ingredients and basic formulation can be used everywhere. Image the making of globally relevant dry instant soups. The consistency, mouthfeel, solubility, colour, core ingredient composition and maybe nutrient profile are globally applied.

Just like dosing of some value-giving ingredients makes them different, this does not only make the assembly of the material sourcing more efficient, it allows a faster launch of new variants, roll out into new countries, and central development of new functionalities, like sodium reduction or fortification or new dosing technology.

But the advantages of harmonized product design do not stop here. In the food business, huge amounts of money are invested in market research, where consumers are asked to validate the new product before it is scaled up. In each market, food companies focus on the optimization of their product to achieve "superiority" in at least one key dimension over their direct competitors in the market.

In the past, the decision about the type of research conducted in the different stages of the new product development was done locally. In one market, the leading team decided that the new product line extension just needed a quick in-location test comparing the new offering with an older one or those of the competitor. As a result, some options for further product fine-tuning were recognized, and the product was modified accordingly to maximize the chances in the market.

The product launch was a success, and now the company board of a different market wants to launch as well, but the market research type or target population are not the same, and hence the research work must be repeated. There is no marketing agency conducting exactly that type of work in the new market.

Then the new research also indicated options to fine-tune the product to further enhance consumer liking. Unfortunately, this requires different ingredients and/or different processes. A powdered soup now is better granulated, the size of the pieces should be 3 mm bigger, a chilled product would be better than a frozen one or the starch used for binding doesn't sound very foody in the ingredient list. And in the next market, the story repeats.

As a result, a product planned to be launched in six to eight countries has been fully market tested in each market, locally tweaked and now requires different material and technology in each market. Maybe a bit more salt, a darker colour, slightly modified package size or differently sealed plus a bit different nutrient profile and therefore made with somc locally grown ingredients.

Is such variation now the unavoidable result of a sophisticated local optimization process? Or is it a missed chance for a harmonized product made efficiently and hence at a lower price? It's hard to tell. The consumer research as a basis for the business case will not provide the conclusive feedback where the incremental profit is lower than the costs of the resulting complexity.

Hence, the option to harmonize food products starts already with the market research tools. International companies have global market research departments in their centre. Their task is to look at all market research tools and the experts providing it. From all that is available, they choose those best suited for the foods business, maybe validate it to confirm international suitability and search for service providers/agencies that have offices in the target markets.

The disciplined execution avoids a lot of testing, makes results directly comparable and gives a database that allows the prediction of consumer responses in future markets. AI will most likely further boost the quality of these predictions.

And such tool and partner harmonization makes "things more affordable". A global deal with an internationally active agency gives a more sustainable business perspective and internal synergies so they can offer their services at a better price.

In Chapter 8, the predictive power of a methodology called "preference mapping" in foods was discussed. Unfortunately, this type of research is very time and money consuming. But doing a couple of mappings in a standardized approach in a few countries might allow for a reliable enough consumer preference prediction to launch in new markets without expensive testing. A producer of roasted peanuts with decades of experience in some European and Asian markets plans to launch in the Americas. Sure, they could do preference mapping in all those target markets for $100,000–250,000 each, in the next 1–3 years. Or they could try to predict the sensorial preferences *in silico* and quickly test the predicted profile in one market.

A strategic thinking about the new product design is at least equally beneficial. By defining upfront a "core formulation" design and a "chassis" approach, a lot of time- and money-eating complexity can be circumvented.

Thinking carefully about a product backbone is a bit of a balancing act. To be successful, the local food preferences must be met. But developing a new tailor-made product every time when going to a new market would mean not to use the possibilities an international food business has. Even worse, it may create a level of complexity that could kill the business. Hence, the balancing question is, How global can a product be, and how local must it be? How is that translated into the food formulation and assembly?

Similarly, the manufacturing technology can be strategically predefined. To allow a quick rollout of new variants, the innovation teams have a clear understanding about the physicochemical properties of the material to be used. If the line only has one conveyor belt for dosing, the possibilities to vary the value-giving ingredients are limited by particle size and format density. If additional gravimetric or volumetric dosing options exist, much more local flexibility exists, but at a higher operating cost.

THE FUNDAMENT OF GLOBAL FOOD PRODUCT DESIGN

Food culture globalization refers to the worldwide exchange and adoption of food cultures, cuisines and cooking techniques. The still-increasing global connectivity and the corresponding cultural exchange stimulate that trend.

One result is the exchange of authentic local cuisines. More tourists bring back new food experiences from their travel, and more people move from one country to another. They have to make a living here, perhaps open a restaurant with a local kitchen and find more customers who can have memories from their travels. These kitchens are getting trendy, and after a while the branded food industry will offer packaged versions of the local signature dishes, and after that creative individuals start to recombine and amalgamate elements of different cuisines and create fusions.

But globalization can also lead to homogenization in local food cultures. The spread of the "quick serve restaurants" (QSRs) are an obvious outgrowth. Hamburgers, hotdogs, tacos, pizza, fries and fried chicken are not have very sophisticated recipes. They are not polarizing, but easily please everyone's taste. But it is not only the "mainstream" formulation that made them so successful. They also represent the American way of life, individualism, real work, freedom and busy lifestyles. On the go, snacking, rewarding after a hard day's "work and simply "the real stuff". This culture may globally lose attractiveness, and the health consequences of 3000-kcal sandwiches do not convince other cultures any longer. Instead, Asian cuisine, sushi, hot pot, chicken, pho soup, etc., are becoming popular outside their home markets, but often also as quick serve business via international chains. Maybe that's why the QSRs had so many difficulties to land in countries with deep dining etiquette like France, Italy or Thailand, where food culture demands that people sit down properly around the family table and have a good conversation, instead of hastily eating finger food while driving to the next destination.

All these originally iconic food items now have their local siblings: Japanese-style burgers, burgers with sriracha sauce, hoisin burgers, Ramly burgers and tantamen burgers, etc. Also, here the product chassis is to mostly maintain, allowing industrialized mass production with economies of scale, combined with "local heroes" and "signature dishes".

Although there is a "slow food trend" – the rejuvenation of the traditional conscious joy of eating –the hectic urban lifestyle is still growing and will guarantee "QSRs" from various food cultures a nice place in the future.

Depending on the food format and the target markets, the degree of "the food chassis" varies a lot. For a package of dry soup mix, a global chassis can be set up to create a particular texture, mouthfeel and a few standardized binding systems in a few predefined colours. The manufacturing process, here mixing, blending, filling and sealing, can be standardized by using globally specified ingredients. The different soup variants can be made by using different ingredient blends, inclusions and premixed spice levels.

The degree of chassis in such product can be 80% and more. A clear or creamy chicken soup is now available in most parts of the world. By thinking chassis design, the harmonized building blocks allow a high degree of automatization, support speed and efficiency.

The art in this design thinking is to find the level up to which a product family can have a shared design and where the identity-giving differentiation has to start. The chicken soup in Thailand may get some more basil or chilli and perhaps a different chicken flavour in the last dosing step. While a durian dairy ice cream dessert can be better as a stand-alone solution, where just the base processing technology is similar and even different lines are used to avoid cross contamination.

The benefits of the chassis approach must be measurable. If it saves costs and time, that can be quantified. And production with dry premixes can not only be more efficient but might also protect intellectual property. Now a "black box solution" can be sent to a third-party manufacturer, who is entering a new market without the need to build its own supply chain, while all the little tricks that make the premix superior are not shared.

CULTURAL DIFFERENCES IN CONSUMER PERCEPTION

It is not easy to get reliable sensorial descriptors for food items that can guide a successful new product development. National culture, as an important component of individual identity, influences food acceptability, and all sensorial impressions are influenced by external factors, a mechanism called "priming". Familiarity with certain food products generally leads to higher acceptability, for which then a higher sensory sensitivity exists.

For food globalization, it is necessary to optimize food products across the world (Jeong & Lee, 2021). The probably globally best known condiment is tomato ketchup. When tester describe the product professionally, they regularly use sensorial attributes like bright red colour, smooth and glossy surface, uniform consistency, accompanied by a tomato-like, typically slightly acidic, aroma, possibly with hints of vinegar and spices aroma and a sweet and tangy flavour, with balanced acidity and saltiness and notes of tomato, vinegar, onion, garlic and spices. And the texture should be thick and viscous, but easy to pour out or spread, with a smooth and homogeneous mouthfeel and no grittiness. Even the interaction between product and packaging can create a differentiation attribute: Even the sound of the ketchup squeezed out of the bottle gets attention, and should for example, it might be decided that the sound of the tomato ketchup being squeezed out of the bottle should be more squelching than popping.

Now, when it comes to the Philippines, a different sensorial space becomes relevant. America's culture has influenced the Filipino kitchen since the early 20th century, and dining and, hence, ketchup became very popular. A shortage in tomato supply during World War II inspired a local lady, Maria Orosa, to create a new version of ketchup, made from abundantly available bananas, with vinegar and spices (Rampe, 2022). The war ended, and tomatoes became available again, but in the local cuisine the banana version had found its place, not as a substitute, but as a new experience: The sweeter, less-acidic banana version is now the preferred partner for famous dishes with chicken and spaghetti, while with oily, fatty meat–based dishes like burgers or french fries, the acidic tomato is preferred. Knowing this offers new opportunities for product developers and reminds them how new impressions can become anchored in the local kitchen.

When interviewing consumers in a typical group discussion about new prototypes, it's good to be aware of top-level cultural differences. One of the best known is the one between individualism and collectivism (for an overview see Hamamura et al., 2018). Individualistic cultures, like in North America or Western Europe, show an independent self-construal that casts the self as distinct from others, while collectivistic cultures, for example, in Asia, foster a co-dependent, socially embedded culture. Interdependent panellists, therefore, may more easily feel uncomfortable when they become distracted by others around them. So, they do not want to articulate a different impression, even if they have it. On the other hand, individuals in group discussions in the Western world may not bother too much about the social consensus but like to emphasize minor differences.

So, a common product sometimes needs to be tested differently in different countries and cultures with consumers or in a trained panel. The best-trained panel may not be able to capture all sensory dimensions relevant in one food culture, while it may differentiate far too much in others.

CASE STUDY: SINIGANG

For a launch in the Philippines, a company tested prototypes of a cooking aid with the local target consumer group. They invited them to a central location and prepared a typical well-known soup for them, making sure that everybody got the same plate and started to carefully collect the answers of the testers.

They heard about many likes and some dislikes that they understood and to some extent had expected as well. But besides all those other sensorial impressions, the invited consumers mentioned that lovely balanced "sinigang" note as an essential and preferred aroma direction. Sinigang is a well-known Filipino dish based on a sour broth, and it is made from tamarind, tomatoes and other ingredients. The resulting sensorial impression of sinigang is complex. People in the Philippines love the tangy flavour of the broth as a stimulant for the taste. The trained panel of the company for that product group was placed in another country in Southeast Asia. They had no difficulties to describe various meaty and spicy notes with great precision. But when the culinary R&D team tried to link the translated individual and subjective feedback to the taste and aroma dimension the trained team was giving back, they could not link this local vocabulary unambiguously to the set of words they were normally using to articulate their expert feedback, that the product formulators were needing to fine-tune the product. Only after a few days of retraining were they able to recognize this sour, sharp, acidic impression as sinigang.

BORDERS OF GLOBAL FOOD PRODUCT DEVELOPMENT

The globalization of new food products finds its limitation in a couple of directions. On a less tangible level, food is part of cultural heritage, a way of life that has been passed down and developed from generation to generation. Traditional recipes, cooking techniques or the dining etiquette reflect these values. And while some of those boundaries are fundamental, other are softer, more transient and in flux (see Table 14.3).

TABLE 14.3 Typical limitations in a global product design approach

FOOD DESIGN LIMITATION	*EXAMPLES*
Dietary restrictions	Beef in Hinduism, kosher or halal-shaped dietary preferences, lactose in major parts of Asian populations.
Cultural differences	Different milk, fermented products, fruits and protein sources can cause food neophobia.
Geographical and agricultural patterns	Different regions exhibit distinct food patterns. The Mediterranean diet emphasizes olive oil, fish and fresh produce, while Northern European diets often include dairy, meat and potatoes.
Technology	Countries and populations have different opinions about the use of innovative technology in crop and food production. This drives preference.
Demographic changes	Countries with more aging populations choose different products and convenience levels than those with young populations.
Legal restrictions	Some additives are allowed at different levels in specific food formats or are even forbidden.

For an international "food producer", a good understanding of cross-cultural food liking is important. Through the identification of food features that are common between different countries and cultures, product grade of chassis usage can be found. For that, there are two different approaches that can be applied: Individuals can be invited to specialized sensory-testing facilities or asked to fill out a food liking questionnaire (FLQ).

The latter allows the inclusion of a larger part of the population, while the first gives a deeper, more direct, but personal assessment. Asking people about their food preference with a questionnaire triggers their memory, not the actual sensorial perception. How good the methods correlate depends on a lot of various influencing factors and is still the focus of academic research (Wanich et al., 2018).

Today, it's safe to say that FLQs may serve well in population-based research but are not good enough to help with the internationalization of food product design.

Smell and taste are influenced by contextual variables (Strauss, 2005). And some authors even see taste more as social sense, an activity related to socio-cultural context, and not so much the result of an individual internal reflection (Højlund, 2015).

An example may make this clearer: Japanese subjects express the taste concept "umami" in addition to the classical four basic tastes: sweet, sour, salty and bitter, while people in the Western world do not often use that description. This is not really surprising since that "fifth taste", umami, was recognized first in Japan in the early 20th century as a leading differentiating note in mainstream food items (see Chapter 1). The other way around, people from China or Korea have the tendency to describe coffee taste and aroma in a less detailed way than people from England (Hu & Lee, 2019). These results imply that the sensory perception of food and beverage items may also vary as a function of culture and language.

It's good to know which attributes are necessary to establish the sensory attributes for a new product development. But once all relevant attributes of that sensorial space are known, the fundamental research on what consumers like and how many of them like can start. And this has to include packaging.

REFERENCES

Anker, M., Krupitzer, C., Zhang, Y., & Borsum, C. (2022). Prediction of Aroma Partitioning Using Machine Learning. *Engineering Proceedings*, *37*(1), 48. https://doi.org/10.3390/ECP2023-14707

ArcherGrey. (2021, October 26). *The history and future of PLM*. https://archergrey.com/plm-history-future/).

Becker, D., Schmitt, C., Bovetto, L., Rauh, C., McHardy, C., & Hartmann, C. (2023). Optimization of Complex Food Formulations Using Robotics and Active Learning. *Innovative Food Science and Emerging Technologies/ Innovative Food Science & Emerging Technologies*, *83*, 103232. https://doi.org/10.1016/j.ifset.2022.103232

BECPG. (2023). https://www.becpg.net/industries-en/plm-software-agri-food-industry/

Bogard, J. R., Downs, S., Casey, E., Farrell, P., Gupta, A., Miachon, L., Naughton, S., Staromiejska, W., & Reeve, E. (2024). Convenience as a Dimension of Food Environments: A Systematic Scoping Review of Its Definition and Measurement. *Appetite*, *194*, 107198. https://doi.org/10.1016/j.appet.2023.107198

Centric Software, Centric Software, & Inside Retail & Centric Software. (2023, November 14). PLM is removing compliance bottlenecks for FMCG companies: Here's how. *Inside FMCG*. https://insidefmcg.com.au/2023/11/13/plm-is-removing-compliance-bottlenecks-for-fmcg-companies-heres-how/

ChemLinked. (n.d.). *More global investors look at ice cream market in China | ChemLinked*. https://food.chemlinked.com/market-insights/more-global-investors-look-at-ice-cream-market-in-china

CIMdata News. (2012, April 18). *Optimizing product cost by leveraging InFOR10 PLM Process (Optiva)*. https://www.cimdata.com/newsletter/2012/16/03/16.03.02.htm. https://www.thedailymeal.com/1428962/salted-egg-yolk-ice-cream-exists/?

Corley, T. (2024, January 3). *What is material management in ERP? How to streamline procurement*. Godlan. https://godlan.com/material-management-in-erp/

Despres, R. (n.d.). *Why are some people reluctant to try new foods?* A Moment of Science - Indiana Public Media. https://indianapublicmedia.org/amomentofscience/people-reluctant-foods.php

Dijksterhuis, G. (2016). Multisensory flavor priming. In *Multisensory flavor perception* (pp. 133–153). From Fundamental Neuroscience Through to the Marketplace. https://doi.org/10.1016/b978-0-08-100350-3.00007-9

Dovey, T. M., Staples, P. A., Gibson, E. L., & Halford, J. C. (2008). Food Neophobia and 'Picky/Fussy' Eating in Children: A Review. *Appetite*, *50*(2–3), 181–193. https://doi.org/10.1016/j.appet.2007.09.009

EuroFIR. (n.d.). *Food data*. https://www.eurofir.org/food-information/

Food Industry Executive. (2024, April 21). *The untapped potential of PLM: How food and beverage manufacturers can gain a strategic edge - Food industry executive*. https://foodindustryexecutive.com/2024/04/the-untapped-potential-of-plm-how-food-and-beverage-manufacturers-can-gain-a-strategic-edge/

Food Standards Australia New Zealand. (n.d.). *Australian food composition database | Food Standards Australia New Zealand*. https://www.foodstandards.gov.au/science-data/monitoringnutrients/afcd

Hailu, M. H., Workneh, T. S., & Belew, D. (2013). Review on Postharvest Technology of Banana Fruit. *African Journal of Biotechnology*, *12*(7), 637–664. https://doi.org/10.5897/AJBX12.020

Hamamura, T., Bettache, K., & Xu, Y. (2018). *Individualism and collectivism* (pp. 365–382). SAGE Publications Ltd. https://doi.org/10.4135/9781526451200.n20

Harvard Gazette. (2024, January 10). *What the nose knows*. https://news.harvard.edu/gazette/story/2020/02/how-scent-emotion-and-memory-are-intertwined-and-exploited/

Højlund, S. (2015). Taste as a Social Sense: Rethinking Taste as a Cultural Activity. *Flavour*, *4*(1). https://doi.org/10.1186/2044-7248-4-6

Hoow Foods. (n.d.). *Technology*. https://hoowfoods.com/technology/

Hu, X., & Lee, J. (2019). Emotions Elicited While Drinking Coffee: A Cross-Cultural Comparison between Korean and Chinese Consumers. *Food Quality and Preference*, *76*, 160–168. https://doi.org/10.1016/j.foodqual.2018.08.020

Jeong, S., & Lee, J. (2021). Effects of Cultural Background on Consumer Perception and Acceptability of Foods and Drinks: A Review of Latest Cross-Cultural Studies. *Current Opinion in Food Science*, *42*, 248–256. https://doi.org/10.1016/j.cofs.2021.07.004

Kalypso. (2022, July 8). The future of PLM in CPG: Perspectives from Kalypso, PepsiCo and Tech-Clarity. *Kalypso*. https://kalypso.com/viewpoints/entry/the-future-of-plm-in-cpg-perspectives-from-kalypso-pepsico-and-tech-clarity

Karen Hamrick, K., & McClelland, K. (2016, July 28). *Americans' eating patterns and time spent on food: The 2014 Eating & Health Module Data*. https://www.ers.usda.gov/publications/pub-details?pubid=80503

Keane, D. (2024, June 12). Denmark recalls South Korean instant noodles for being too spicy. *Evening Standard*. https://www.standard.co.uk/news/world/denmark-recalls-samyang-instant-noodles-spice-poisoning-b1164013.html

Maduwanthi, S. D. T., & Marapana, R. A. U. J. (2019). Induced Ripening Agents and Their Effect on Fruit Quality of Banana. *International Journal of Food Science*, *2019*, 1–8. https://doi.org/10.1155/2019/2520179

McGruther, J. (2016). Broth and Stock from the Nourished Kitchen: Wholesome Master Recipes for Bone, Vegetable, and Seafood Broths and Meals to Make with Them. Ten Speed Press. ISBN-13: 978-1607749318

McQuaid, J. (2012, April 15). *Flavor 101: What are the five basic tastes?* Health & wellness flavor 101: What are the five basic tastes? JOHN MCQUAIDAPR 15, 2021. https://parade.com/396983/johnmcquaid/flavor-101-the-five-basic-tastes/

Mordor Intelligence. (n.d.). *Yogurt market insights*. https://www.mordorintelligence.com/industry-reports/yogurt-market

Munger, S. D., & The Conversation. (2017, May 22). The taste map of the tongue you learned in school is all wrong. *Smithsonian Magazine*. https://www.smithsonianmag.com/science-nature/neat-and-tidy-map-tastes-tongue-you-learned-school-all-wrong-180963407/

Perotti, V. E., Moreno, A. S., & Podestá, F. E. (2014). Physiological Aspects of Fruit Ripening: The Mitochondrial Connection. *Mitochondrion*, *17*, 1–6. https://doi.org/10.1016/j.mito.2014.04.010

Purves, D., Augustine, G. J., Fitzpatrickm D., Katz, L. C., LaMantia, A.-S., McNamara, J. O., & Williams, S. M. (Eds.). (2022). *Neuroscience* (2nd ed.). Sinauer Associates. https://www.ncbi.nlm.nih.gov/books/NBK10799/

Rampe, A. (2022, March 16). She invented banana ketchup & saved thousands of lives. Why have we never heard of her? *Food52*. https://food52.com/blog/24700-maria-orosa-profile

Sengupta, A. (2023, August 25). *Integrating PLM, QMS, and ERP to solve manufacturing challenges*. https://research.g2.com/insights/manufacturing-challenges

Society of Sensory Professionals. (n.d.). *Society of Sensory Professionals*. https://www.sensorysociety.org/knowledge/sspwiki/Pages/The%209-point%20Hedonic%20Scale.aspx

Sogari, G., Dagevos, H., Amato, M., & Taufik, D. (2022). *Consumer perceptions and acceptance of insects as feed and food: Current findings and future outlook* (pp. 147–169). Springer. https://doi.org/10.1007/978-3-031-13494-4_8

Statista. (2017, May 8). *Cooking: Hours spent by country 2014 | Statista*. https://www.statista.com/statistics/420719/time-spent-cooking-per-week-among-consumers-by-country/

Strauss, S. (2005). The Linguistic Aestheticization of Food: A Cross-Cultural Look at Food Commercials in Japan, Korea, and the United States. *Journal of Pragmatics*, *37*(9), 1427–1455. https://doi.org/10.1016/j.pragma.2004.12.004

Tepper, B. J., & Barbarossa, I. T. (2020). Taste, Nutrition, and Health. *Nutrients*, *12*(1), 155. https://doi.org/10.3390/nu12010155

Wanich, U., Sayompark, D., Riddell, L., Cicerale, S., Liem, D. G., Mohebbi, M., Macfarlane, S., & Keast, R. (2018). Assessing Food Liking: Comparison of Food Liking Questionnaires and Direct Food Tasting in Two Cultures. *Nutrients*, *10*(12), 1957. https://doi.org/10.3390/nu10121957

Wikipedia Contributors. (2024, April 16). *Convenience food*. Wikipedia. https://en.wikipedia.org/wiki/Convenience_food

Ziv, C., & Fallik, E. (2021). Postharvest Storage Techniques and Quality Evaluation of Fruits and Vegetables for Reducing Food Loss. *Agronomy*, *11*(6), 1133. https://doi.org/10.3390/agronomy11061133

世界ラーメン協会. (2024, March 28). *World Instant Noodles Association*. https://instantnoodles.org/en/noodles/demand/

Packaging Development 15

Most food items that we buy, store and eat are prepackaged. On the aisle, the packaging captures our attention and navigates us. Outside of the shop, we pay much attention to it and may even complain about the amount of packaging waste in our hands now, once we brought the food item safely home. The essential purposes of the packaging material often only scratch our conscious perception.

The packaged food market is globally growing, so the food packaging market is growing as well. This results in an expected compound annual growth rate (CAGR) of nearly 6% in the coming years for a business that is already worth $362 billion (Fortune Business Insights, 2024) (Grand View Research, n.d.a) Convenience-seeking consumers, in their busy, often hectic lifestyles, value good packaging material for all the reasons listed in Figure 15.1. The fundamental purpose of packaging material is to keep the food safe, something most of us were very much reminded of during the last pandemic. In that situation, a lot of formerly unpacked food items became wrapped in the most problematic plastic. And now, as the invisible threat disappears, public, scientist and governments are pushing again for the implementation of more sustainable packaging solutions.

Globally, a hierarchy of solutions are classified, ranging from the most wanted "no packaging waste" to the least wanted "disposal". This puts extra emphasis on new, better packaging in the NFPD. And at its core, this may not be so much radically new consumer benefits, but more the approach to preserve the functional properties of the packaging material, without the environmental impact. Besides the environmental aspects, new packaging functionality, traceability, new contaminant sensing or improved quality preservation are opportunities to create differentiation for food producers.

TABLE 15.1 Core functionalities of food packaging

FUNCTIONALITY	*BENEFIT*
Freshness or extended shelf life	Protects food from external factors like air, light, moisture and bacteria during transportation and sales. And it extends the shelf life of food products.
Shape and texture	Packaging ensures that food retains its original shape and texture. It protects against external shocks and prevents breakage.
Visual attractiveness	Appealing food packaging attracts consumers and influences purchasing decisions. Eye-catching graphical and haptic elements provide important cues; and the written, text gives a lot of factual information
Waste optimization	Optimized packaging protects food, without material overload, minimizing environmental impact and costs.
Information	Displays all product facts from ingredients to producer and the packaging material used.
Brand communication	A specific typography, colour, logo or brand name creates visual impact and helps consumer with shelf navigation.
Convenience	For the consumer to transport, use and store or during the distribution from factory to sales location.

DOI: 10.1201/9781032724591-15

FIGURE 15.1 Hierarchy in the European Union Waste Framework Directive. (From Environment, n.d.)

For the food packaging development, this opens a wider field for gradual improvement or radically new solutions that deserve a discussion in a separate chapter.

DIFFERENT TYPES OF FOOD PACKAGING

The core functionalities of packaging to preserve quality, keep safety, allow good shelf life and deliver convenience have conventionally shaped the packaging development. The results are that today we often find sophisticated and elegantly designed composite structures, where different material properties are combined to optimize the required functional properties. And most of that innovative work is not much valued by the consumer. The shopper takes packaging aspects into their food purchase decisions but are usually not prepared to pay extra for it. After transport and consumption, the single-use packaging material becomes useless waste. This is changing in many markets, since shoppers nowadays know about the consequences of improperly disposed packaging material and about the environmental impact. But often, the consumer also believes that the food industry should take care to make the packaging material reusable, recyclable and part of a circular economy, but without oncosts for the shopper. That is known as "extended producer responsibility" (ERP), as part of the product life cycle (EPR, 2022).

In the past centuries, the foods packaging evolution has seen a shift from "first generation" materials, like paper, glass, metals and cardboard, to more composed solutions, including various types of plastic. The latter were a sort of perfect solution for all wanted functionalities as long as the environment was not "part of the equation". Hence, it is no wonder that more and more food items became packaged, so that in Europe right now each inhabitant generates annually nearly 190 tons of packaging waste and two-thirds of it is comes from food and beverages (Packaging_Waste_Statistics, 2024).

Of these 190 tons, about 41% is paper and cardboard, 19% each are from plastic and glass, 16% from wood and 5% from metal (Deshwal & Panjagari, 2019). Before looking at the opportunities for better, more "future fit" solutions, we first take a short look at the individual strengths and weaknesses of the materials.

MATERIAL PROPERTIES

Before packaging material became a place for consumer information and entertainment, packaging had to control contamination/deterioration by environmental factors in combination with maximum preservation of all sensorial and nutritional attributes. The shielding aspect can be further segmented as protection against elevated temperatures and exposure to moisture, humidity and oxygen. Shielding from physical, chemical or biological contamination from foreign bodies, chemical substances or microbial infections is equally important, particularly in the traditional food distribution systems in the developing and emerging markets.

Against these requirements, the diverse food packaging classes perform differently.

GLASS

The first glass containers may have been made in 1500 years BC. But it took until the early 19th century for glass to become a mass packaging solution. Just like for margarine and bouillon cubes (see Chapter 2), there was an urgent need from the military to supply the marching troops with masses of safe food. It took then another century before Michael J. Owens invented the first automatic glass bottle–making machine. And with the basics established, the engineers in the next decades focussed on efficiency increase of the process, so that today single machines can produce 2500 bottles per hour (FEVE, 2023a).

And glass containers provide many important functionalities. The material is nonporous and impermeable, which means it can safely store liquids for long periods without the risk of contamination. The mixture of silica (silicon dioxide), soda (sodium carbonate) and limestone (calcium carbonate) is chemically inert. It is hence resistant to the transmission of gases and particles and therefore has superior protection properties. It's 100% recyclable and can be recycled many times without loss in quality or purity. The fundamental advantages and disadvantages of glass as packaging material are listed in Tables 15.2 and 15.3.

At the same time, glass containers are heavy, and hence this massively increases transportation costs. And the primary manufacturing is one of the most energy intensive industries, as it requires up to 1600°C to melt, so the production emits significant amounts of carbon dioxide (CO_2), nitrogen oxide, sulphur, dioxide and dust. It's said that for every ton of bottles on average 1.2 tons of equivalent CO_2 is released (Innovations Admin, n.d.). But of course, material recycling and lighter and thinner bottling processes reduce the environmental burden.

TABLE 15.2 The strengths of glass packaging

CHARACTERISTIC	*ADVANTAGE*
Non-porous and impermeable	Material has impermeability, allowing safe storage for a long time.
Non-toxic	No risk that harmful substances would leach into edible products.
Maintains carbonation and scent	Glass retains carbonation and scent well, which is important for carbonated soft drinks and perfumes.
Haptic and aesthetic	More premium touch and feel. Perceived as more valuable.
100% recyclable	Recyclable without losing any integrity or quality after they are repurposed.
Transparency	A functional packaging allowing consumers to see the product.

TABLE 15.3 The weaknesses of glass packaging

CHARACTERISTIC	*DISADVANTAGE*
Transportation costs	Weight and low-impact resistance add cost to the distribution.
High energy consumption	High temperatures for processing and manufacturing glass, resulting in a significant amount of required energy.
Low-impact resistance	Increased breakage risk.
Safety	When glass breaks, it's difficult to detect. Shattered glass can cause injuries.

An easy way to cut material usage is to make the glass container thinner. But that makes it less impact resistant, as the brittle structure breaks easily. And this is a key threat in every factory safety program. Broken glass in manufacturing is not easy to detect. An X-ray can do it, but the installation is expensive and sophisticated.

Glass is often transparent, so the buyer can see the product and perceive the quality, an important benefit. But the light also causes autocatalytic reactions of components, like unsaturated fatty acids, which can dramatically change the product quality. That risk can be reduced through use of coloured glass.

Overall, it is a mixed picture for this widely used packaging material. Right now, the glass packaging market size is estimated at $85 billion and is still growing at about 4% CAGR (FEVE, 2023b). The balance shows freshness and transparency attributes, long quality preservation and recyclability on one side, and safety limitations, costs and initial carbon footprint on the other. In the short term, glass is a valid option for liquid food and many beverages, especially when the distribution lengths of the whole cycle is kept short, so that the distribution costs can be kept low.

PAPER AND CARDBOARD

Paper and cardboard represent about a third of the global packaging market segment and more than 40% of the food packaging. In the old days, these were the materials of choice in the grocery store. Nowadays they are widely used as primary (in direct contact with food) and secondary (transportation and storage of primary packaged food) food packaging.

Primary applications include ice-cream cups, microwave popcorn bags, baking paper, milk cartons, fast-food containers such as for pizza, chocolate wrap, sandwich bags, paste boxes and beverage cups.

Like for most materials, a couple of strong arguments are accompanied by fundamental weaknesses. The advantages are listed in Table 15.4, and the disadvantages are listed in Table 15.5.

The protection and barrier properties are mixed, and heat sealability and strength are low. To improve these attributes, the environmental advantages are compromised. Some papers are impregnated with additive or laminated with aluminium or plastic layers to improve the barrier properties (Deshwal et al., 2019). The chemical substance used here can cause health hazards.

From today's perspective, paper and cardboard are good packaging solutions. The "environmental credentials" are strong. The material can be recycled five to seven times when the appropriate infrastructure is in place to conduct all necessary steps, which include collection, sorting and transportation, shredding and pulping, floatation and de-inking, and then drying.

The fundamental qualifications are very promising, and the recycling rate of paper and cupboard in the municipal waste in the United States has exceeded 63% (EPA, 2024). But the fact that it is separately collected, does not mean recycled. The various coatings, chemical treatments and especially the contamination with food residues limit the recycling potential. So, it is burned or ends up as landfill, and here it usually takes up more space that plastic. And since landfill is mostly dumped under anaerobic conditions, it degrades poorly and the microbes growing under such circumstances often producing methane, which is a very potent greenhouse gas.

TABLE 15.4 The advantages of paper and cardboard as food packaging

CHARACTERISTIC	*ADVANTAGE*
Biodegradability	Decomposes naturally in the environment, reduces waste.
Recyclability	Paper can be recycled a few times.
Bio-based	Paper is made from renewable resources, primarily wood pulp, sustainable when managed responsibly.
Environmental impact	The paper industry contributes a little to global greenhouse gas emissions but can offset CO_2 levels through sustainable forest management.
Customizability/multi-usage	Can be tailored to many specific needs, including size, shape and design, making it a versatile packaging option.
Attractiveness	More appealing for many consumers for food products.
Safety	When handled properly, no major safety risk.
Distribution costs	Lower specific weight versus glass.

TABLE 15.5 The disadvantages of paper and cardboard as food packaging

CHARACTERISTIC	*DISADVANTAGE*
Durability	Paper is not very durable, particularly when exposed to moisture or heavy loads.
Barrier properties	Insufficient protection against moisture, gases and other environmental factors.
Space and weight	Paper and cardboard can be bulkier and heavier and thus cause increased transportation costs.
Health and environmental concerns	The paper production process can use harmful chemicals, like inks, phthalates, surfactants, bleach and hydrocarbons. These can leach into the food chain and the environment.
Limited recycling	Paper can only be recycled a limited number of times, and the diverse nature of the paper industry waste can complicate recycling efforts.
Disposal	Typical disposal methods like incineration, landfilling and pyrolysis impact the environment.
Strengths	Poor resistance, quickly loses strength when wet.

As primary packaging, the materials have limitations, which is one reason why alternatives came up throughout the years.

The major role cardboard plays in today's food distribution comes from its usage as secondary and tertiary packaging, when the primary packaging material forms the initial protection shield and the carton provides extra protection in storage and distribution.

METALS

Cans are the most used metal packaging and are either made from steel or aluminium. Steel-containing cans are made from electrolytic tinplate or electrolytic chromium/chromium oxide–coated steel, while food and beverage packaging made of aluminium is as well used as cans made from 90% aluminium plus other metals like copper, zinc and manganese, or in tubes, trays, lids, foils, or as part of multilayer structures.

Typically, these cans find usage for the packing of fruits and vegetables, soups and broths, different meats, ready meals and, of course, beverages (Foodpackagingforum.org, n.d.). The long history of metals

TABLE 15.6 Advantages of metal packaging in food

CHARACTERISTIC	*ADVANTAGE*
Contamination protection	The robust material structure protects against most environmental challenges.
Shelf life	Possibility of a very long shelf life, keeping food inside unspoiled for a long period.
Recyclability	Metal can in principle be endlessly recycled.
Temperature resistance	Metal withstands high temperatures, so the food can be sterilized within the can.
Traceability	Easy traceability in the production process, possible metal contaminations can be easily detected.

TABLE 15.7 Limitations of metal packaging

CHARACTERISTIC	*DISADVANTAGE*
Dent or rust	Exposure to moisture-damaged metal food containers.
Non-microwavable	Not usable from quick heat in microwave use, as it will cause sparks.
Weight	Heavier than plastic or paper packaging, which increases transportation costs.
Not biodegradable	If not recycled, the material pollutes the environment for a long time.
Changing taste	Particularly acid material can cause ion leakage.
Cost	Metal has a tendency to be more expensive than other alternatives.

used in robust food packaging material stems back to the early 1800s when the first commercial canning started in England (Cosper, n.d.). Today, metal packaging is often coated with an organic polymer to prevent direct contact between the metal and the food, which could otherwise lead to changes in the food properties. The advantages of metal packaging are listed in Table 15.6, and the disadvantages are listed in Table 15.7.

So yes, steel and aluminium both are excellent barriers to oxygen, light and microbes and are recyclable. But metal packaging, especially steel, can rust (oxidize), so its usage in food packaging often requires an extra coating with anti-rust materials, which is suspected to release potentially harmful substances into the food, like, for example, bisphenol A (BPA) (Deshwal & Panjagari, 2019). Generally, the pH tolerance is also more limited. Many liquid or higher-water-content food products are acidic, which can lead to metal ion precipitation. Material costs and weight limit the usage further, while aluminium is expensive, and the production is still extremely energy intensive.

Still the total cans market is estimated to have reached 450 billion cans, with three-quarters of them used for beverages (Business Wire, 2015).

Just to mention it, wood can also be used directly as packaging material but is less prominent in foods. It is more used to transport fresh food items, not as primary packaging, but for pallets. But it is porous and absorbs substances, which is a clear hygienic concern and requires effective sanitation. But the naturalness perception can help premium food items to express their authenticity. French raw milk cheese is traditionally sold in wooden boxes (after being packed in coated paper).

PLASTICS

The most successful food packaging material class of the last century and the most criticized one is plastic. Starting with a production of around 2 million tons in 1950, plastics production reached more than

TABLE 15.8 The advantages of plastic as food packaging material

CHARACTERISTIC	*ADVANTAGE*
Barrier properties	Very effective against most external challenges, including microbes, moisture, oxygen, chemicals and gases.
Shelf life	High barrier properties significantly extend shelf life.
Durable	Good impact resistance, not as robust as metal, but much better than paper and glass.
Lightweight	Low transportation costs.
Transparent	Many plastics can be as transparent as glass; shoppers see the product.
Flexible	Material can be moulded into a variety of shapes and sizes.
Resealable	Many types of plastic packaging can be resealed, keeping food fresh and safe for longer after opening.
Cost effective	Significantly cheaper than glass or metal.
Carbon footprint	Due to plastic being lightweight, the greenhouse gas (GHG) impact is often initially better than for glass or metal.

460 million tons in 2022 (Statista, 2024a). Cumulative until 2019, that's about 9.5 billion tons of plastic or more than 1 ton of plastic for every living person (Ritchie & Roser, 2024). And about 36% of all this is used for food packaging (Huong, 2023). The fundamental chemical reactivity makes plastic packaging not only wasteful, but also harmful for wildlife, biodiversity and even human health, as it accumulates through the food chain and the typically added substances like phthalates, bisphenol A (BPA), styrene, and benzene, can leach into food, water and the environment.

But let's start with the list of advantages which explains the past success (see Table 15.8).

So, the diverse material family of plastics used for food and beverage (F&B) packaging basically "ticks all functional requirements": Low price, safe, hygienic, long storage, directly usable in a microwave and much more. Clearly, in times when environmental aspects were not "part of the total equation", plastic packaging was a breakthrough innovation. And even the carbon footprint leaves room for second thoughts. On a weight base, plastic is bad. But making that comparison per "functional unit" shows a very different picture (McKinsey & Company, 2021). Plastic is lightweight, and so the carbon footprint of some plastics is much better than the one for aluminium or glass.

And the direct functional limitations of plastics are minor, and all disadvantages are related to what will happen after the plastic packaging has done its job. Non-biodegradable, laborious to collect after use, difficult to recycle and polluting the environment. These disadvantages are captured in Table 15.9.

Chemically, plastics are often synthetic or semi-synthetic polymer materials made from fossil fuel–based chemicals like natural gas or petroleum. So, they are a true human invention of the recent past,

TABLE 15.9 Disadvantages of plastic as food packaging material

CHARACTERISTIC	*DISADVANTAGE*
Environmental impact	Plastic is a non-biodegradable material; it does not decompose naturally. Plastic waste can take hundreds of years to degrade.
Pollution	Plastic can return into the food chain (microplastics).
Recyclability	Not all plastic can be recycled, but even when it can, the process is a burden for the environment.
Health risks	Potential health risks associated with certain types of plastics or leaching of used additives.
Smell absorptions	Plastic packaging absorbs some flavours.

which means that nature has not been exposed to the material long enough to integrate it into a natural circular process.

The human "success" story started in the mid-20th century when polyethylene (PE) became a replacement for the packaging for fresh bread. Then followed a more rigid version, polypropylene (PP), which brought numerous new possibilities to protect all types of snacks. Today, many different versions with very different properties exist: high-density polyethylene (HDPE) for rigid containers; low-density polyethylene (LDPE), which is softer and more transparent than HDPE; linear low-density polyethylene (LLDPE); and many ultrahigh, medium- or ultra-low-density variants, all optimized for their specific packaging purposes.

Plastic production consumes a lot of energy and resources, mainly oil, gas, water and land. And the quantity, the billions of plastic packaging units, are the "gorilla in the room". Per single unit, the production emits only a little greenhouse gas, but the totality is estimated to consume more than 10% of the total carbon budget left for the 1.5°C target (Center for International Environmental Law, 2022).

Here is a quick look at the three main types of plastic from a functional perspective:

- Polyethylene (PE) accounts for about 34% of the packaging material used, or 100 million tons. It is a flexible and lightweight material, often used to make bags, wraps, pouches and bottles. Two variants dominate the market: HDPE and LDPE. The HDPE tends to be more rigid and resistant to heat and chemicals, while LDPE is more transparent and softer.
- Polyethylene terephthalate (PET) is a clear and strong material that can be used to make bottles, trays, films and clamshells. PET is also flexible and lightweight, and it has good barrier properties against oxygen, moisture and carbon dioxide. It is the most recycled plastic in the world and can be made into new products such as fibres, carpets and clothing.
- Polypropylene (PP) is today a more than 87 million ton market. The material is heat resistant and microwave safe and is used to make tubs, cups, lids, caps and closures. PP has good stiffness and strength, and it can also be made into flexible films and sheets. It has a high melting point and can withstand sterilization processes such as pasteurization and autoclaving.

The different polymers all have their advantages and limitations. While PE is flexible, durable and tear resistant, PP is harder and has a higher melting point and is more resistant to organic solvents (Tajeddin & Arabkhedri, 2020).

Plastics do not degrade, but they can be recycled. But that is a multistep process that includes sorted disposal, collection, again sorting, cleaning, shredding, melting and finally perhaps moulding. Depending on the type and quality, the plastic can be recycled into different products, such as bottles, containers, bags, fibres or pellets. But up to now only 9% of all plastic waste ever produced has been recycled, while 12% has been incinerated and 79% has accumulated in landfills or somewhere in the environment.

For food packaging, the situation is even more difficult: Often the plastic is contaminated with food residues or other substances, which reduces the material quality and value. And plastics that come from non-food items, like other fast-moving consumer goods (FMCG) products, cannot be recycled for new food packaging due to contamination concerns.

From today's point of view, the problem with plastics got bigger when smart inventors began to combine the technical advantages of the different types of plastic into combined layered structures or films. And now plastic packaging is often a multilayer structure, where material combinations are used to optimize the overall packaging functionality. They are made from thin layers of different materials and laminated or co-extruded plastic types.

A typical laminate structure contains the following elements:

- *Outer layer*: PP or oriented polypropylene (OPP) has excellent optical properties, allowing mixing with various inks, and has excellent barrier properties, particularly against moisture and oxygen.

- *Tie layer*: This layer acts as an adhesive between the outer and barrier layers or between different barrier layers. It helps to improve the bond strength and compatibility of the laminate structure and provides a bridge between different materials. PE is a common tie layer material with good adhesion and flexibility.
- *Barrier layer*: This is the layer created to prevent the permeation of gases, vapours or liquids through the package. It preserves the freshness, flavour and quality of the product by preventing oxygen, moisture or odours from entering or escaping the package. Material solutions are aluminium, polylactic acid, ethylene-vinyl alcohol copolymer (EVOH) and others. Aluminium is very popular, due to its excellent barrier properties against gases, vapour and light.
- *Sealant layer*: This is the layer that forms the innermost surface of the package and provides a seal to prevent leakage or contamination of the product. It also enables easy opening and closing of the package by providing a peelable or resealable feature. Wanted features are a low melting point, but high seal strength and good tear resistance. LDPE is a typical sealant layer material.

Many different plastics are used, and more layers can further improve the barrier properties where needed. But these carefully designed structures enhance the environmental problem.

Single-use plastics, or disposable plastics, are only used once before they are thrown away. And when packaging material ends up in a landfill, the biodegradability becomes very important. Plastics, as said, isn't good at that.

The better alternative is to reuse or recycle the material. Plastic can be recycled, but it must be homogeneous. And that's what laminated structures are not, so they often end up in landfills or incinerators. There are countless quotes expressing the urgent need to drastically cut the plastic pollution in the environment. The Ellen MacArthur Foundation, for example, predicted that "*by 2050, there could be more plastics than fish in the oceans.*" (Ellen MacArthur Foundation, n.d.).

The primary cause for this is single-use plastic packaging. In a global, concerted effort, converters, manufacturers and retailers alike are actively seeking ways to create eco-friendly products and increasingly sustainable supply chains. The initiative had led to the formation of a coalition committed to use "new design rules" for their packaging, which should make recycling easier (The Consumer Goods Forum, 2024). More initiatives from the FMCG industry steer towards a circular economy, an economic system aimed at eliminating waste and encouraging the continual use of resources. The wider design commitment of the FMCG industry aims not only at recyclability, but also reuse or compensation. These are tough targets and firm commitments (Unilever, 2024).

And similar to the demanded changes in the food formulation, like less sugar, calories or salt, the voluntary replacement of non-recyclable plastic packaging material may take too much time so that new regulation or taxation is established to accelerate the switch.

The European Union, the part of the world where each inhabitant on average produces 190 tons of waste per year, promised to be climate neutral by the middle of the decade. The transformation of plastic, especially as packaging, is a key prerequisite for this. So new targets demand that by 2030 all plastic packaging placed on the EU market be reusable or easily recycled. Other countries tackle single-use plastics, or the amount of recycled material.

This strategy not only sets targets but also allocates substantial funds to "*develop smarter and more recyclable plastics materials, enhance recycling processes, and remove hazardous substances from recycled plastics*" (Unilever, 2024).

Today, progress is visible, as the amount of recycled plastic is increasing, but this increase covers the additional demand, leaving the amount of virgin plastic production unaffected, but up to now, this increase is only able to cover the additional demand, not cutting the use of virgin plastic since 2018 (Statista, 2024b). And "recyclable ready" means what it says: The material can be recycled, as there are, in principle, methods to do so. But that requires a recycling stream, a system enabling the collection of recyclable material. Even in the most advanced regions of this world, like Europe, only 30% of the plastic is recycled (EuRIC AISBL, n.d.). In many other markets, it's still virtually zero.

While governments start to introduce plastic taxes, the packaging material producers have made commitments to use "post-consumer used plastics". But the safety concern doesn't make this easy. Now, authorities remain sceptical about plastic reuse due to concerns about potential contaminations in non-food plastic. But a theoretical solution, split collection schemes, does not seem feasible on a larger scale. And even in economies where a recycling system has been established, a substantial part of the collected material is rejected, mostly because of contamination. This plastic then eventually becomes exported to countries with different regulations where it may end up as landfill or is recycled. Until the year 2017, China received most of the post-consumer used plastics. After the legal ban, the streams were diverted to other countries or sent to incineration in the home country (Wen et al., 2021).

"FUTURE FIT" PACKAGING

There is no doubt that experts everywhere in the industry and academia have been working hard for years to develop new solutions with a better balance of core functionality and environmental impact (Bauer et al., 2021).

Different development strategies are possible. Maybe the simplest one is, Use it again! For food packaging that translates to "refillable", returnable" or "transient packaging" (Coelho et al., 2020).

Refillable solutions, which are based on bulk stores for food products, exist on a smaller scale. They encompass the refill of an already used container brought by the customer. The main challenge is the maintenance of the critical hygiene and safety demands in this open filling process. So, the implementation is largely limited to ambient stable food with a long shelf life like dry products such as cereals, nuts and candy.

A compromise is the "refillable outer or secondary packaging". Here the food product is protected by functional primary packaging and the secondary packaging is brought by the consumer or customer to the selling point. Such a concept already has a big potential to reduce packaging material use and distribution costs.

Home care, personal care and cosmetics brands use this concept widely. The outer packaging often has a premium touch and feel and transports the luxurious brand image and therefore represents a good part of the product costs. So, shoppers bringing their "used" premium outer packaging can be rewarded with a little discount.

The "returnable packaging" idea is widely implemented in the food industry, particularly in the business-to-consumer (B2C) market and food service. Beverages like beer, carbonated soft drinks and mineral water are sold that way. National deposit systems strengthen the implementation. The bottle labels and caps often remain single use. Coca-Cola has recently started to sell plastic bottles where the cap remains attached to the bottle (Coca-Cola, n.d.).

A variant of this is the reused transit packaging. This is already the norm in many business-to-business (B2B) interactions, where pallets, crates, dunnage, drums, bulk containers and big bags are used. By driving standardization and system harmonization, impressive cost reduction and cost avoidance has been realized.

Overall, the business projections for reusable packaging are often conservative. For Europe, different scenarios see anything between a 5%–8% penetration until 2030, unless disruptive new solutions are available (Feber et al., 2022).

So, most of the packaging material for foods is not directly used again, predominantly for hygienic reasons. For those materials, a huge bundle of ideas is in place or under development to improve the environmental footprint, while keeping most of the direct functionality. But also here, the "one size fits all" solution does not exist. For a brief look on those ideas under way, a couple of new definitions and concepts have been introduced and standardized to structure the territory. A few important ones are now briefly explained.

Life Cycle Assessment

The life cycle assessment (LCA) evaluates the environmental impacts related to the packaging, describes the energy and materials used and the waste emitted to the environment and evaluates the impact. An LCA covers the entire life cycle of a product, process or activity, including: The extraction and processing of raw materials; production, shipment and distribution; use, reuse and conservation; and recycling and disposal, from "cradle to grave", which goes beyond the older "from farm to folk" for foods. Included is the role of packaging, the end-to-end (E2E) environmental impact, the waste problem, and softer factors such as consumer behaviour. Various enterprises, agencies and academic groups are offering help in the analysis service. The basis is the methodology and guidelines for conducting LCAs of the International Organization for Standardization (ISO) 14040–14044: 2006.

One key direction focussed on the replacement of the classical fossil-based thermoplastic polymers by various types of bioplastics, materials that are part of the natural cycle, having the wanted barrier properties to protect the food. Those can be split into:

- Bio-based and non-biodegradable polymers
- Biodegradable and based on fossil resources
- Bio-based and biodegradable (Salgado et al., 2021)

REUSABLE

Reusable packaging can be refilled or used again for its original purpose. It can be used for storing, transporting or serving food. It therefore needs to be durable, easy to clean and safe for food contact. The classical examples of reusable food packaging are glass jars, metal containers or bamboo utensils for cooking and eating. In an idea of a circular economy, reuse is preferred over recycle.

"RECYCLABLE" AND "RECYCLE READY"

"Recyclable" refers to packaging that has been proven to be recyclable in practice. It's a design principle for food packaging and refers to the standards for recyclability, including successful collection, sorting and recycling along the existing scheme.

"Recycle ready" basically specifies the plastic material that can be used for that purpose. This statement on a food packaging does not promise that the material is recycled, only that it can be, when the necessary infrastructure is in place.

BIODEGRADABLE AND BIOPLASTICS

While the terms "biodegradable" and "bioplastics" sound clean and simple, they can describe very different things. Biodegradable food packaging material breaks down quickly upon disposal, without pollution or other detrimental effects on the environment; it is the type of packaging that can be decomposed by micro-organisms. The best-known biodegradable food packaging is cardboard or paper, or bio-based plastics.

Bioplastics are a mix of biopolymers made from renewable raw materials and biomass, more specifically. Typically, they contain cellulose, soy, starch or lactic acid, so that the material can slowly disintegrate into carbon dioxide and water. Other examples of bio-based plastics are starch-based polymers, cellulose-based polymers, and polylactide (PLA). Polyethylene (PE) can be "bio-based", as it at least partly comes from renewable resources, like sugarcane, and it is recyclable, but not biodegradable (Tajeddin & Arabkhedri, 2020).

COMPOSTABLE

Compostable packaging is organic material that biodegrades in a defined time frame under defined controlled conditions (temperature, humidity, etc.). In contrast to biodegradable packaging, which can take thousands of years to decompose, compostable material decomposes more quickly. Compostable plastics often means that the material can biodegrade under controlled industrial conditions with a subsequent composting step. It does not mean it can be home composted. Typical examples of compostable packaging materials include corn starch, bamboo, mushroom, cardboard, paper and corrugated wraps.

The environmental impact and material properties of compostable packaging varies by the source. Paperboard mostly comes from cellulose fibre. If compostable coatings, additives and inks are used in the process, the entire paperboard-based package can be composted. However, there is something to watch out for here: "*Packaging made from food does not necessarily mean it is compostable nor a sustainable use of resources*" (AMCOR, 2023). Bioplastics can also be compostable. The most prominent example is polylactic acid (PLA).

BIOPOLYMERS

Whatever the material and the reusable conditions are, biopolymers seem a major route to replace plastic. In a simple sense, biopolymers are materials found naturally, which means nature has had the time to find ways to reuse the material. All the different materials have specific strengths and weaknesses. The "must haves" for all these materials are "non-toxic, biodegradable and renewable". The "nice to haves" are low cost; flexible; transparent; excellent water, ultraviolet (UV) light, oxygen and aroma barrier; mechanical and antimicrobial properties; good elasticity; film-forming ability and good processability.

None of the currently available options can fulfil all those criteria and have a full match of the plastic functionality. Depending on the food material to be packed, compromises need to be made without impacting the consumer experience too much. A different classification of biopolymers is based on their origin.

Natural Biopolymers

These are extractions from polysaccharides, lipids and proteins. Polysaccharide polymers consist of cellulose, alginate, starch, chitosan, pectin, gum, carrageenan, pullulan or derivatives of these substances. These materials have limitations: Poor water vapor barrier, because of their hydrophilic nature, making them susceptible to moisture; low mechanical strength; and long-term stability. It's difficult to form films out of biopolymers because the material is often brittle and, as such, has susceptibility to breakage.

Several attempts are underway to improve the weaknesses without losing the environmental benefits (Panou & Karabagias, 2023).

Protein-based films tend to have much better mechanical characteristics and are good barriers to oxygen gas. They often are of plant origin, such as wheat gluten, soy and corn zein, or animal origin, such as collagen, whey, casein and gelatine. But their hydrophilic character and their water vapor barrier quality also limits their wider applications.

Synthetic Biopolymers

These materials are the result of chemical processes of biological monomers. Typical materials are aromatic copolymers, polyesters, polylactides, aliphatic copolymers and polylactides. They are made from renewable bio-based monomers, for example, polylactic acid (PLA). Advantages are that PLA is promising due its wide availability. Like all biopolymers, it's recyclable and compostable but has poor mechanical and barrier qualities.

Microbial Biopolymers

These biopolymers are made by micro-organisms. Polyhydroxyalkanoates (PHA) are polyesters synthesized through the bacterial fermentation of sugars and lipids. They are biodegradable, exhibit good UV light resistance and are water insoluble, but often have a low pH resistance (Perera et al., 2023).

To improve the technical weakness of biopolymers, so-called reinforcement agents are used, materials that are added to packaging materials to improve the packaging functionality. Nanoparticles have shown their functionality here.

Cellulose nanocrystals (CNCs) are a type of reinforcement agent that can be used to improve the mechanical properties of packaging materials. They come from natural sources like wood, cotton and other plant-based materials. They meet the key criteria of being biodegradable, renewable and having very low toxicity, so they are an attractive alternative to synthetic reinforcement agents. But consumers often remain sceptical; they fear migration of those particles into the food, leading to food contamination.

RECYCLING STREAMS

So packaging material that is not reused or degraded must be recycled; that means it needs to enter a suitable recycling stream for plastics. But there are so many different types of plastic, so in 1988 the U.S. Plastics Industry Association introduced a "voluntary plastic container coding system". This numbering allows the normal citizen to identify the plastic type and be able to segregate the different plastics, the prerequisite for the recycling step.

RESIN IDENTIFICATION CODING SYSTEM

The Resin Identification Code (RIC) is a set of symbols appearing on plastic products that identify the plastic resin out of which the product is made to enable its reuse after being recycled. In its original form,

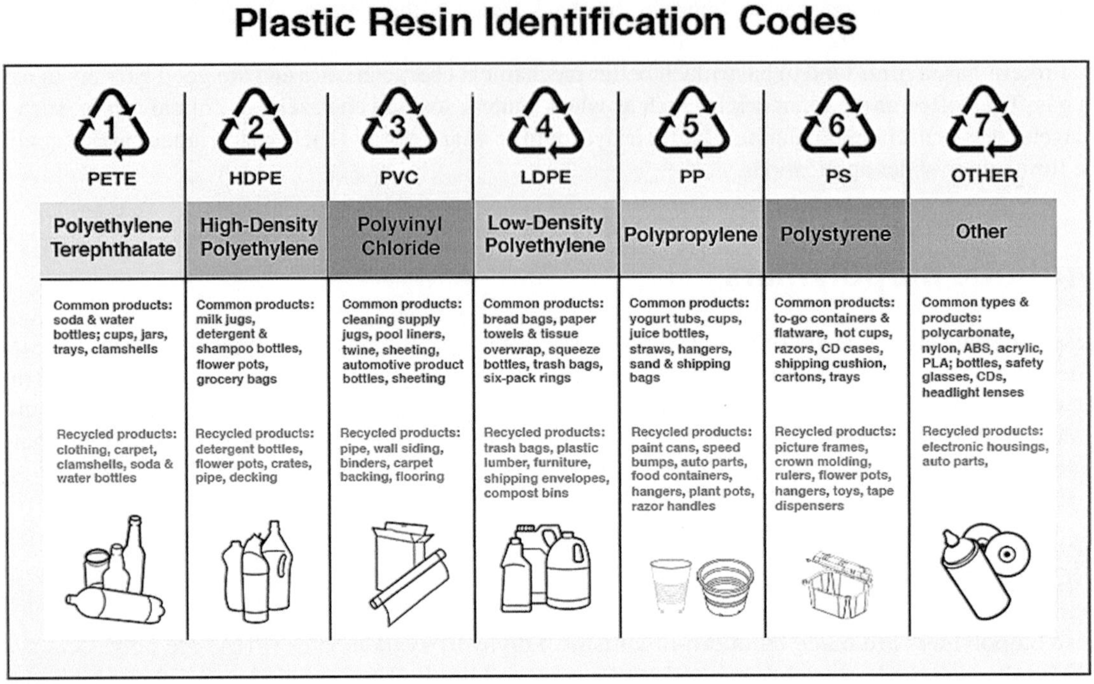

FIGURE 15.2 Resin Identification Code. (From https://www.thongguan.com/plastic-food-packaging-symbols-and-what-they-mean)

the symbols were used as part of the RIC system and consisted of arrows that cycle clockwise to form a triangle. The following numbers exist (see also Figure 15.2):

1. Polyethylene terephthalate (PETE)
2. High-density polyethylene (HDPE)
3. Polyvinyl chloride (PVC)
4. Low-density polyethylene (LDPE)
5. Polypropylene (PP)
6. Polystyrene (PS)
7. (OTHER)

The numbers do not only indicate the different resin types. They also specify the general ease and cost-effectiveness of the material recycling: 1 classifies the easiest, and 7 is the most difficult. Today, laminates used in food often still fall under 7.

The recycled resins are used for new packaging material, so that the production of new, virgin plastic can be reduced. But the amount of recycled plastic used in new packaging is still low. In 2020, only 5%–6% of the 40 million tons of plastic waste generated in the United States was recycled, and the global recycling rate for plastic bottles was 27.2%.

The corresponding regulation is unfortunately not harmonized. Local jurisdiction has its own standards, requirements and criteria for the quality, safety and performance of recycled plastic in new packaging. In the European Union, recycled plastic in new packaging must comply with the Regulation (EC) No 282/2008 on recycled materials intended to come in contact with foods. It specifies that recycled plastics can only be used for food contact if they come from recycling processes that have been evaluated and authorized by the European Food Safety Authority (EFSA). The regulation also lays down requirements for traceability, labelling and compliance documentation of recycled plastic materials and articles that come in contact with foods. Other countries, such as Japan, China, Canada and the United States, have established similar, but not identical, guidelines for recycled plastic materials that come in contact with foods.

The plastic bought by consumers and customers are very often composites, plastic laminates. These materials pose an extra challenge when it comes to recycling. Laminated plastic is the most difficult material to recycle. Since it is made up of multiple layers of different materials, separation is needed before the different materials can be individually recycled. In the meantime, several industrial processes are in upscaling to achieve the separation. (https://www.feica.eu/information-center/good-practices/food-packaging-2). And the packaging material producers have developed structures that can be recycled.

IMPLEMENTING SOLUTIONS

Bio-based packaging is a feasible choice for a variety of food products: fruits and vegetables, grains, nuts, cereals, bread and pastries and some dairy products. Other products need packaging that can "just" be recycled, for which the plastic chosen must be "recycle ready." That material turns into recycled, when a sustainable recycling stream has been implemented, which means a profitable business that effectively protects the environment.

The transformation would be easier if consumers would accept some compromises in the packaging. First, would they be willing to pay a little extra for such packaging? And indeed, consumer research in the past has shown consumer acceptance to pay a little extra for a "more responsible" solution, maybe up to 10% more for food products in such new packaging (Brunn, 2020). But market research knows that this could also be "lip service", where they express agreement in words but do not follow through with action. And the economic distortion in the past years and the inflation in food prices may force many consumers to be even more price sensitive.

Another option would be to simply accept a shorter shelf life. But again, before a food company launches such a product, it would want to be sure that the consumer doesn't mind the cut in the shelf life in the home stocks and doesn't turn to the competition next time. And this option would require detailed discussions in the whole distribution with all involved partners. The trade normally does not like to order products that have a reduced shelf life or require storage at a lower temperature. So today, it often means that packaged food needs to be made with the same efficiency as the current, not recyclable, ones.

For preformed packaging material – that is, packaging that is shaped into a specific form before being filled – the effective integration into the process without efficiency loss seems easier. A pouch or a plastic bottle made with a different material does not interact as much with the product-filling step in the line. The final sealing or capping step may need some adjustments, and the movement of the product in the line may need fine-tuning steps, but not a complete reset of all key parameters. This sounds not too complicated, but since the old materials have gone through a long evolution of improved "machinability", the slightest material modifications can have a massive impact on the operation efficiency. And a few percent loss in the operational efficiency of the manufacturing can make the difference between commercial success and failure. Often, machine operators and technicians have fine-tuned the interplay between material and machine step by step to maximize the operation efficiency, to achieve a processing efficiency that allows a competitive price on the market.

The packing foil that is used in form and fill (FFS) or fill and seal (FS) packaging machines has various configurations. Those lines can be horizontal or vertical, making flat bags, three-side sealed pillow pouches, four-sided sealed pillow pouches, double-flat bags, twin pouches, bottom fold/gusset pouches, stand-up pouches/with a spout, contour pouches and other formats. They pack solid, powder and liquid products at any size, volume or weight. The capacity of an FFS packaging machine can range from 10 to 250 pouches per minute, per single lane. And some machines have eight or more lanes.

And each of those lines has been optimized for "their" specific format and "their" composite material: PE, PP, PE ALU, PET, OPP, metallized OPP with 5-, 6-, 7-, 8-, 9-, 10-μm Alu layers, glossy or matt. Each combination has its own settings. Some foil structures are more rigid than others, some are easier to seal and some are easier to cut. Machines have been ordered and installed over the years. Some are very old and others are new, but each is specifically adjusted for their purpose or reconfigured.

The message is that if a major food company commits to switch to new, more environmentally friendly packaging material, this may require an immense trial program. Often these factories have 30, 40, 100 or more lines, where the new foils behave differently, often unpredictably, which directly impacts the machineability. Every technical step of forming, filling, sealing and cutting must be tested on every line short term and in the long run, before a factory can predict the operational efficiency and hence the price it must charge for making the product. The amount of waste generated, the change over time and the quality of the pack must be confirmed. Product performance compromises are as well not an option. The fundamental property of the packaging needs to be maintained. If the material is thinner, it still needs to be as robust as the old one and remain undamaged during storage, distributions and home use. Reduced shelf life through thinner layer and reduced barrier properties will lead to more pressure in the distribution system and negotiations with the trade. And if the packaging looks a bit less flashy, the brand logo perhaps a bit less vibrant and the picture of the food less appetizing, marketing may have concerns. And consumers often accept a shorter remaining shelf life only at a lower price, something the trade cannot easily swallow and will try to charge back to the food manufacturer.

And with all of this, something that on first sight looks quite straightforward becomes a massive operation and immense trial program that may cost years and millions of euros.

TRIAL COSTS

Packaging material and (primary) packaging machines went through a coevolution of optimization. Today these processes run fast and efficiently. Horizontal form fill machines can fill 200 bags or more per minute. Packaging machines are highly sophisticated and expensive. For the manufacturing, the efficient run of such machinery is critical for their operations. Changing the primary packaging material, for example, to introduce better recyclable packs, can be a major exercise. Even long relationships with the material suppliers, long experiences with the machinery, harmonized and fully validated specifications and digital "*in silico*" pretesting cannot avoid a substantial testing program in the factory. After *in silico pretesting*, lab and perhaps pilot plant trials, short, medium and long run trials with the new material are needed to ensure that the operational efficiency of the packaging line can be set. Such trials can last for days or more than a week to confirm the feasibility. They test the speed at which the packaging can be filled with food product; the quality of the new packaged product; the time and ease to load the next packaging material batch; the amount of waste created; the ability to pack into secondary packaging, for example, a carton; and the performance during distribution, in the trade and with the end user. And the extrapolation of the full results of one packaging line to the next is also difficult. Often each machine has been adjusted and individually modified in the past to optimize the production of a single stock-keeping unit (SKU) or format. Once the new material has been fully tested on one line, there is no guarantee it will run down the next line with the same efficiency. Leave alone that these machines are often used for many years, while the supplier has constantly improved the machine performance and design through little changes. That can lead to situations where machines of the same type, but different "vintage", run in a factory beside the ones from different suppliers. Bigger food factories sometimes have more than 100 filling lines. Replacing packaging material on them and for all formats is a massive undertaking with trial costs easily exceeding $1 million and an implementation period of 1–2 years. And more testing also means less sellable product.

Computer and data modelling today can help to limit the amount of testing needed, but especially for the high-speed lines or those filling difficult food products or filling under difficult circumstances, the *in silico* predictability has clear limitations still.

CASE STUDY: RECYCLE READY

A leading producer of condiments and seasoning mixes in Indonesia decided to move their current sachet and pouch packaging portfolio from regular laminates to new, "recycle ready" material. Existing suppliers were asked for their material offerings, and after some initial lab tests and detailed discussion of the technical specifications, two suppliers were selected as partners of choice. To serve all relevant sectors and shopping occasions, a wide portfolio of lots of different SKUs had been brought to market in the past, ranging from 9-, 20-, 22-, 100-, 200-, 220-, 230- and 600-g single packs in various flavour notes.

The renewal task was substantial! The powder sachets were made by horizontal and vertical form-fill-seal (HFFS and HFFS, respectively) machines, where the pouch is produced in the line, filled and finally sealed at high speed. An initial investigation revealed that more than 100 different lines would need to be piloted, tested, readjusted, upscaled and commercialized, an impossible task for the technical team in a foreseeable time span. The packaging R&D team, technical department, experts manufacturing and quality control department started to investigate the technical specification of each machine, reviewed the operational performance and started to discuss the technical parameters with their machine and packaging material supplier. For the first learnings, one of the "simpler" lines was chosen for a first short run trial. Here "simpler" meant relatively new, recently commissioned, slow running, fresh seal and cutting systems. For both trials with the day shift of very experienced machine operators, both the machine supplier and foil suppliers sent their technical experts.

These trials were not a complete success, but over time, several adjustments were made during the trials to the machine, and the materials indicated an operating window for production and new opportunities for further improvement.

Over the next 2 weeks, data were collected, compared and analysed. While speed and efficiency were not the focus at this stage, both new foils started to show differences, for example, in the smoothness with which the laminate was flowing through the machine, and in the quality of sealing and cutting. The draft material and processing specifications were adjusted accordingly.

Together with the procurement department, the team made a decision on the material, asking the laminate supplier for some further technical improvements, and then moved swiftly into the long run trial, where the material was used for several shifts or days, mainly under the guidance of the machine operators, and the R&D and technical departments only interfering when absolutely needed.

All line parameters were collected, and QA took all their representative samples to assess the new material against the agreed-upon quality standards. Once all data were analysed, the leadership of the factory accepted material and processing specifications for that single line at a slightly lower operational efficiency, a slightly higher material waste level and of "out of specification" finished products, so they could still be "shipped to trade".

In the meantime, the production samples from the first run had been through 6 months of accelerated shelf-life testing, and the R&D team could confirm that the new packaging material protected the formulation as good as the old one, so basically there was no aroma loss, colour change, or loss in nutrients. And the "going to market" operation had tested the robustness of the new products over the whole distribution lengths, with all its different stores and shop types, the bumpy roads in the countryside and the sometimes-drastic tropical climate.

The remaining, still more than 100, lines were clustered into groups, depending on supplier, basic processing principles, age and past performance and level of expected complexity during material change.

It was a very long and winding road. Many unexpected problems were found during the change to the recycle ready material. Sometimes the sealing would not work, as the seal was not completely closed on one or two sides or was burned through. The team detected that the "sealing window" of the new material, which describes the temperature span during which the pouch could be properly sealed, was much narrower than the old one. So, particularly in the older machines, the sealing jars could not keep the sealing temperature constant enough or equally over the full lengths. They needed to be exchanged, and a more sophisticated temperature regime was established. Or the missing aluminium layer in the recyclable foil changed the stiffness of the material so that it could no longer be cut in line efficiently, and so new knives were needed. Or the foil flow from the reel was not smooth, and the line speed needed to be adjusted.

Over the years, thousands of man-hours were spent; millions in material, analysis and machine parts invested; and an enormous complexity and somewhat lower efficiency in the production process was brought to the factory, all to get a product to the market that the consumer would hardly recognize as "better for the planet", while the country is still busy working on the setup of an efficient material collection scheme.

SMART PACKAGING AND NEW BENEFITS

While the conversion into environment-friendly materials is the key challenge for food packaging, upcoming new science and technology offers new benefits for packaged food. A few of them are listed in Table 15.10.

These functionalities are often captured under the term "active packaging", as they are made by the intentional addition of a wanted function. For some of them, the idea is to locate the protective action in the packaging rather than the food product itself (see also under "antioxidants" in Chapter 12). With this, the food inside the package remains unchanged. Other scavengers react with oxygen to prevent spoilage.

TABLE 15.10 "Smart" food packaging functionalities

PACKAGE FUNCTIONALITY	*MODE OF ACTION*
Antimicrobials	Organic salts, metal oxides, bactericides, enzymes and other macromolecules hindering microbial growth.
Antioxidants	Preventing fat oxidation by addition of antioxidant compounds or elimination of substances promoting oxidations.
Moisture absorbers	Absorbing moisture and controlling humidity in the headspace of the package or removing liquids leaking from food.
Modified atmosphere packaging	Creating a preserving gas atmosphere in the pack, often used for fresh food.
O_2 scavenger	Substances quickly oxidizing and thereby absorbing oxygen. Newer solutions are no longer based on iron.
CO_2 scavengers and generators	Growth inhibition for aerobic microbes and gas phase completion after O_2 scavenging.
Ethylene scavenger	The phytohormone ethylene promotes the ripening, softening and senescence of fresh fruits.
Flavour scavengers	Substances capturing off flavours from lipid or protein degradation in the pack to avoid consumer rejection at opening.
Microwavable, self-heating and cooling	Increased convenience through endo- or exothermic reactions or magnetic field influence.
Indicators	Signalling changes in temperature, time, pH or microbial conditions through a sensorics change, often colour change.
Sensors	Sensors designed to detect pathogens, chemical polluters or unwanted biological reactions.
Data loggers	RFID systems to trace the products along the whole food chain,

One of the most popular oxygen scavengers in the food industry is Ageless from Mitsubishi Gas Chemical Company. The sachet contains iron powder and salt, which react with oxygen and moisture to form iron oxide and water (Daniel & Lambert, 1993).

An antimicrobial agent inhibits the growth of micro-organisms. They are incorporated into plastic or paper packaging materials to reduce bacterial and fungal contamination. These agents often disrupt vital microbial functions, damaging essential proteins, the cell membrane, the oxidative activity or the DNA. These substances are added during the manufacturing process and provide a lasting protection for the product surface. Older applications are based on silver ion, zinc or copper; newer antimicrobial technologies use organic compounds (Simeonova, 2024).

A different class of substances act as indicators, where, for example, a label shows how long a product has been exposed to specific conditions known to impact the quality and safety of a product. A temperature-sensitive liquid is broken up by squeezing, and the liquid flows through the opening and into a transparent window that shows the temperature range. In case the temperature exceeds a particular limit, the colour changes. A variant uses dye adjacent to a microporous membrane. After activation, the liquid dye comes out of the blister and into contact with the membrane. As long as the temperature of the product is below the threshold temperature, the dye remains solid and cannot move through the membrane. Above the threshold temperature, the dye becomes liquid and moves through the membrane (Timestrip, n.d.).

Edible films or coatings are "smart" in a different way. They can be consumed along with the food, so no packaging waste is produced at all. A thin layer of edible material creates a protective coating or layer on foods. They are made from food-grade biopolymers, such as polysaccharides, proteins, lipids or their combinations. The main purposes are to prevent moisture loss and extend shelf life. The environmentally friendly shell can also be used to deliver antimicrobials, antioxidants, colouring agents, flavours and nutraceuticals. Organic aromatic oils like clove, lemon, cinnamon, tea tree, lavender, oregano and peppermint have a natural antimicrobial activity against a wide range of microorganisms, and they have been approved by the U.S. Food and Drug Administration (US-FDA) as safe additives for food applications. To broaden the spectrum of those antimicrobial actives, without impacting taste and aroma, applications use nano emulsification of those oils (Kong et al., 2022).

Radio frequency identification (RFID) is not new and acts more as a data logger, transmitting data about the "journey" the food product has made. It enables location tracking, temperature and freshness monitoring, or provides consumers with information about the origin, ingredients and nutritional value of food products. Newer developments are microwave-safe and do not need to be removed from the food package prior to microwave heating, and they can be recycled in paper recycling. Today RFIDs also play a role in traceability and transparency of the food industry and can prevent food fraud, contamination and recalls. In foods, tags are not widely adopted, due to their still high costs, a lack of standardization and interoperability and data security concerns (Fuertes et al., 2016).

An alternative is the blockchain technology, an electronic record book that captures all transactions securely, imminently and transparently. Blockchain technology provides a transparent view of the entire food supply chain, enabling E2E tracking of food products from the farm to fork (consumer) and beyond. The technology improves food safety through detection and prevention of contamination, fraud and waste. The packaged food industry has started to use it, either the origin of an ingredient, like the farm a meat product is coming from, or the finished product lifespan. First pilots were technically successful but did not raise much consumer interest (Santana, 2022). In an increasingly global food supply chain, product tracking is much easier. And it helps to reduce the risk of counterfeits, a severe problem for premium branded goods. And claims that are relevant for consumers, like sustainably or organically sourced, kosher or halal, can be verified by the consumer or customer.

In combination with AI and big data, there are more possibilities to optimize the E2E food supply chains (Ellahi et al., 2023). It's likely that the food packaging innovation agenda will benefit from new technology applications. The core proposition is clear. The currently still used very efficient and effective plastic formats need a substantial functional refresh, where all the great product functionality of food plastic packaging is maintained, while in an E2E view, from cradle to grave, the negative impact on the environment is abolished. A variety of new technological developments, especially in material sciences, provide chances for new solutions as fully functional food packaging under circumstances fitting all consumers' demands.

REFERENCES

AMCOR. (2023, February 28). *More-sustainable-packaging-five-common-questions. Amcor.com/insights/blogs.* https://www.amcor.com/insights/blogs/more-sustainable-packaging-five-common-questions

Bauer, A., Tacker, M., Uysal-Unalan, I., Cruz, R. M. S., Varzakas, T., & Krauter, V. (2021). Recyclability and Redesign Challenges in Multilayer Flexible Food Packaging—A Review. *Foods, 10*(11), 2702. https://doi.org/10.3390/foods10112702

Brunn, M. (2020, April 23). *Sustainable packaging: Consumers willing to pay more - RECYCLING magazine.* RECYCLING Magazine. https://www.recycling-magazine.com/2020/04/23/31232/

Business Wire. (2015, July 27). Research-markets-global-metal-cans-market-2015. https://www.businesswire.com/news/home/20150727005538/en/Research-Markets-Global-Metal-Cans-Market-2015

Center for International Environmental Law. (2022, February 1). *Plastic and climate: The hidden costs of a plastic planet - Center for International Environmental Law.* https://www.ciel.org/plasticandclimate/

Coca-Cola. (n.d.). Coca-Cola begins introduction of attached caps across entire portfolio to boost collection and recycling, and help prevent litter. https://www.coca-cola.com/gb/en/media-center/coca-cola-attached-caps-great-britain

Coelho, P. M., Corona, B., Klooster, R. T., & Worrell, E. (2020). Sustainability of Reusable Packaging–Current Situation and Trends. *Resources, Conservation & Recycling. X, 6*, 100037. https://doi.org/10.1016/j.rcrx.2020.100037

Cosper, A. (n.d.). *Evolution of metal cans for food packaging.* https://www.desjardin.fr/en/blog/evolution-of-metal-cans-for-food-packaging

Daniel, V., & Lambert, F. L. (1993). *Ageless(tm) oxygen scavenger: Practical applications.* Permission-on-file. https://cool.culturalheritage.org/waac/wn/wn15/wn15-2/wn15-206.html

Deshwal, G. K., & Panjagari, N. R. (2019). Review on Metal Packaging: Materials, Forms, Food Applications, Safety and Recyclability. *Journal of Food Science and Technology/Journal of Food Science and Technology, 57*(7), 2377–2392. https://doi.org/10.1007/s13197-019-04172-z

Deshwal, G. K., Panjagari, N. R., & Alam, T. (2019). An Overview of Paper and Paper Based Food Packaging Materials: Health Safety and Environmental Concerns. *Journal of Food Science and Technology/Journal of Food Science and Technology, 56*(10), 4391–4403. https://doi.org/10.1007/s13197-019-03950-z

Ellahi, R. M., Wood, L. C., & Bekhit, A. E. A. (2023). Blockchain-Based Frameworks for Food Traceability: A Systematic Review. *Foods, 12*(16), 3026. https://doi.org/10.3390/foods12163026

Ellen MacArthur Foundation. (n.d.). *Open letter.* https://www.ellenmacarthurfoundation.org/plastics-open-letter

Environment. (n.d.). Waste framework directive. https://environment.ec.europa.eu/topics/waste-and-recycling/waste-framework-directive_en

EPA. (2024, May 9). *Facts and figures about materials, waste and recycling.* https://www.epa.gov/facts-and-figures-about-materials-waste-and-recycling

EPR. (2022, July 15). *The food & drink federation.* https://www.fdf.org.uk/fdf/what-we-do/environmental-sustainability/plastics-packaging/extended-producer-responsibility/

EuRIC AISBL. (n.d.). EURIC AISBL – Recycling: Bridging circular economy & climate policy. In *Plastic Recycling Factsheet.* https://circulareconomy.europa.eu/platform/sites/default/files/euric_-_plastic_recycling_fact_sheet.pdf

FEVE. (2023a, January 5). *What is the history of glass packaging? - FEVE.* https://feve.org/about-glass/history/

FEVE. (2023b, July 14). *Facts & product details - FEVE.* https://feve.org/about-glass/facts-product-details/

Feber, D., Grünewald, F., Haag, J., Nordigården, D., Pley, M., & Spengler, J. (2022, October 28). *Reusable packaging: Key enablers for scaling.* McKinsey & Company. https://www.mckinsey.com/industries/packaging-and-paper/our-insights/reusable-packaging-key-enablers-for-scaling

Foodpackagingforum.org. (n.d.). *Factsheet_metal_V1.* Retrieved June 22, 2024, from https://www.foodpackagingforum.org/fpf-2016/wp-content/uploads/2021/06/FPF_Factsheet_Metal_v1.pdf

Fortune Business Insights. (2024, June 3). *Industry-reports/food-packaging-market.* Fortunebusinessinsights.com. https://www.fortunebusinessinsights.com/industry-reports/food-packaging-market-1019410

Fuertes, G., Soto, I., Carrasco, R., Vargas, M., Sabattin, J., & Lagos, C. (2016). Intelligent Packaging Systems: Sensors and Nanosensors to Monitor Food Quality and Safety. *Journal of Sensors, 2016*, 1–8. https://doi.org/10.1155/2016/4046061

Grand View Research. (n.d.a). *Food Packaging Market Size, Share & Trends Analysis Report by type (Rigid, Semi-rigid, flexible), by material (Paper, plastic), by Application, by region, and segment Forecasts, 2023–2030.* https://www.grandviewresearch.com/industry-analysis/food-packaging-market#:~:text=The%20global%20food%20packaging%20market%20size%20was%20valued,rate%20%28CAGR%29%20of%205.7%25%20from%202023%20to%202030

Huong, N. (2023, December 4). *How much plastic does food packaging use? KimEcopak.* https://www.kimecopak.ca/blogs/news/how-much-plastic-does-food-packaging-use#:~:text=Here%27s%20a%20breakdown%20of%20the%20numbers%3A%201%20Europe%3A,plastics%20produced%20are%20used%20in%20packaging%20Weitere%20Elemente

Innovations Admin. (n.d.). *Rising energy costs and glass shortages make returnable glass bottle life a priority – Innovations Food.* https://innovationsfood.com/rising-energy-costs-and-glass-shortages-make-returnable-glass-bottle-life-a-priority

Kong, I., Degraeve, P., & Pui, L. P. (2022). Polysaccharide-Based Edible Films Incorporated With Essential Oil Nanoemulsions: Physico-Chemical, Mechanical Properties and Its Application in Food Preservation—A Review. *Foods*, *11*(4), 555. https://doi.org/10.3390/foods11040555

Kravchenko, S. (2020, November 13). *Código de identificação de resina hdpe isolado em branco.* Dreamstime. https://pt.dreamstime.com/c%C3%B3digo-de-identifica%C3%A7%C3%A3o-resina-hdpe-isolado-em-branco-fundo-s%C3%ADmbolo-reciclagem-image201874130

McKinsey & Company. (2021, July 28). *True packaging sustainability: Understanding the performance trade-offs.* https://www.mckinsey.com/industries/packaging-and-paper/our-insights/true-packaging-sustainability-understanding-the-performance-trade-offs

https://doi.org/10.1016/j.foodqual.2023.104996

Packaging_waste_Statistics. (2024, March 22). https://ec.europa.eu/eurostat/statistics

Panou, A., & Karabagias, I. K. (2023). Biodegradable Packaging Materials for Foods Preservation: Sources, Advantages, Limitations, and Future Perspectives. *Coatings*, *13*(7), 1176. https://doi.org/10.3390/coatings13071176

Perera, K. Y., Jaiswal, A. K., & Jaiswal, S. (2023). Biopolymer-Based Sustainable Food Packaging Materials: Challenges, Solutions, and Applications. *Foods*, *12*(12), 2422. https://doi.org/10.3390/foods12122422

Ritchie, H., & Roser, M. (2024, February 2). *FAQs on plastics.* Our World in Data. https://ourworldindata.org/faq-on-plastics

Salgado, P. R., Di Giorgio, L., Musso, Y. S., & Mauri, A. N. (2021). Recent Developments in Smart Food Packaging Focused on Biobased and Biodegradable Polymers. *Frontiers in Sustainable Food Systems*, *5*. https://doi.org/10.3389/fsufs.2021.630393

Santana, N. (2022, February 18). *Supply chain traceability for Unilever Knorr: A Scantrust case study.* Scantrust Anti-counterfeiting, Supply Chain Traceability, and Customer Engagement with a Product Digital Identity. https://blog.scantrust.com/farm-to-fork-traceability-supply-chain-awareness-and-consumer-engagement-help-knorr-regain-market-share-despite-aggressive-competitor-pricing/

Simeonova, M. (2024, February 23). *How antimicrobial technology works | BioCote Ltd.* BioCote. https://www.biocote.com/how-antimicrobial-technology-works/

Statista. (2024a, May 30). *Global plastic production | Statista.* https://www.statista.com/statistics/282732/global-production-of-plastics-since-1950/

Statista. (2024b, April 26). *Virgin and recycled plastic in packaging used by major FMCG companies 2018–2022.* https://www.statista.com/statistics/1343227/virgin-recycled-plastic-packaging-used-companies-globally/

Tajeddin, B., & Arabkhedri, M. (2020). *Polymers and food packaging* (pp. 525–543). Elsevier. https://doi.org/10.1016/b978-0-12-816808-0.00016-0

The Consumer Goods Forum. (2024, June 20). *Packaging design.* https://www.theconsumergoodsforum.com/environmental-sustainability/plastic-waste/key-projects/packaging-design/

Timestrip. (n.d.). *Timestrip PLUS - Temperature indicators by Timestrip.* Irreversible Temperature Indicators - Timestrip. https://timestrip.com/indicator-labels/temperature-indicators/timestrip-plus

Unilever. (2024, May 8). *Plastics.* https://www.unilever.com/sustainability/plastics/#:~:text=Reduce%20our%20virgin%20plastic%20footprint%20%E2%80%93%20by%2030%25,25%25%20recycled%20plastic%20in%20our%20packaging%20by%202025

Wen, Z., Xie, Y., Chen, M., & Dinga, C. D. (2020). China's plastic import ban increases prospects of environmental impact mitigation of plastic waste trade flow worldwide. *Nature Communications*, *12*, 425. https://doi.org/10.1038/s41467-020-20741-9

Major Trends in New Food Development 16

For this chapter, I will keep the split between the pull from the market and the push coming from new technical opportunities. It will start with the perspective of consumers or customers, their emerging or increasing wants and needs. Good high-level data about trends and market insights can be found on the websites of the global market research and industry analysis companies. Euromonitor, Nielsen, Mintel, Kantar, Statista, Mordor intelligence, and many others provide a lot of information for free on their sites or in their newsletters. Their source of information is their own fieldwork and/or the analysis of publicly available data. For an initial view on markets and opportunities, these data are often sufficient and freely available, while companies often buy the detailed reports to shape their new product development plans.

In the second part of this chapter, the perspective will turn around, to discuss how new enabling sciences and technologies can satisfy those needs! Better technology meets a need better. And a better solution is the archenemy of the existing good one. An effective, cheap and comfortable foody protection against the common cold has long been a wish of many. But solutions like chicken soup, antioxidant cocktails or mega doses of vitamin C fulfil this need only to an extent (see Chapter 2: Food Myths and Misconceptions). Thus, the question is, What are the technologies that have the potential to meet those needs truly better?

Increasingly, opinion formers are becoming the transmission belt between consumers' needs and science and technology as a "solution provider" in foods. These leading scientists, subject matter experts and opinion formers are articulating their opinions. They connect in both directions. They are receivers of the emerging science, transform it and send it to the interested consumer or customer. New evidence is seldom final, easy to understand, always true and without any controversy. Do sweeteners help to lose or maintain weight? Are all additives bad? Does everybody benefit from reduced sodium intake? The opinion leaders act as interpreters and navigators. They offer their (substantiated) opinion and help consumers to make choices.

And the other way around, they demand solutions from science, even if the markets have not yet articulated them and the food industry hasn't yet spotted the opportunity. These experts are often foresight driven and use their ability to anticipate future developments through a combination of analysis and intuition. The discussion about the environment and the need to protect it is such a topic. Long before the public became conscious, a small group of experts, called the "club of Rome", demanded already in 1972 a radically different utilization of the resources on this planet. They predicted the end of the economic growth model that societies followed in the centuries before. At the end of this chapter, I also want to look at them and the role they play in new food product development (NFPD).

LEADING CONSUMER AND MARKET TRENDS

Consumer needs are identified by observation and the extraction of relevant insights. All the famous market research agencies are a leading example source for that. They have local, regional and global data about those trends. It's no surprise they often find the same ones, but they differ in their context giving and explanation. Towards the end of 2023, the agencies Mintel, Ipsos, Kantar, Nielsen, HRA, etc., reported

DOI: 10.1201/9781032724591-16

consistently global trends: Plant-based food, sustainability, reuse food, zero waste, circular food, small treats, indulgence, regional and seasonal products, clean and minimally processed, health and wellness, personalized food, and nutrition.

Their individual opinion around the facts gives a differentiated, often very valuable view on new food product (NFP) opportunities. Mintel for example offers the top three overarching trends: "a*) trust into the process, b) a new definition of aging and c) the optimisation of our nutrition!* " (Zegler, 2024).

This sounds somewhat different to the overall trend consensus! But a closer look shows that these needs are basically a more pronounced articulation of the general trends. Their "trust into the process" for foods incorporates "minimal processed", "authentic origin", "transparent", "natural", "better for me" and "minimal waste". Their second trend, "new aging", is based on the health and well-being trend. What it adds is the idea that especially the "golden ager" is an important group. This group is a growing part of the population in many counties and is often affluent enough to spend money for a longer active and healthy life. Finally, the third one, "optimization of food nutrition", captures the wish for better nutrition for all, convenience and the "personalization of food" that is "good for me", but also "good for planet."

These trends portray emotional consumer wants in the food and beverage space. They do not prescribe a particular new food product. The trend insights leave it to the inventors to find the new product, its format, size, prize, composition and nutritional profile. This happens in the ideation session, where the trend insights are the input, often generating a couple of concrete product ideas (see Chapter 6). Often, the enterprise would choose those products, which they could make most competitively! That's the ability – or the right – to win.

It also works the other way around! By thoroughly observing all food and beverage product launches, one can look for commonalities, common attributes and clustering them sensibly.

With that approach, Mintel can extract, for example, that "*The link between the gut microbiome and overall health is being increasingly recognised by science and consumers. This has been reflected in the market. Mintel's GNPD data shows that 22% of global food/drink/healthcare introductions with a digestive health claim in the past five years also had an immune health claim headlines.*" (Mintel, n.d.b).

Such trends stay for longer! They become shaped and modified over time, but they exist long enough to allow the time to plan and launch a new product. A bit more volatile are sensorial trends, like flavour. Here a community of flavourists, chefs, trend-spotters and futurologists predict the trends for the next year or next season. A bit like the fashion industry! They anticipate flavour trends, where moods and atmospheres condense into a new sensorial impression. So, a health trend, for example, mental health or relaxation, could materialize as a low-sugar product with a subtle sweetness, lighter and more floral. All notes associated with well-being and relaxation, or good sleep, are suitable sensorial expressions, carrying wellness credentials. And if the health and wellness is so important for those caring about aging – why not bring back (less) sweet old memories and launch hibiscus or lavender as floral notes. And for those seeking adventure, excitement and novelty, most of the exotic kitchens of this world are characterized by their flavour or new flavour pairings.

Even "instagrammability," as a flavour trend offers new trend opportunities, especially when they are part of an appealing visual. The flavour can awaken the memory of gorgeous food visuals, shared or promoted via social media. To what extent this all is the discovery of new trends or its creation is sometimes more difficult to tell.

The nice thing is that most flavour trends are easily actionable elements in NFPD. Often, they can be added to an existing range, without major changes in the product design or the manufacturing. In principle, they are a low-risk range extension. But since it normally takes at least a few months to launch a new flavour variant, it is good to ensure that the new flavour is indeed a longer-lasting trend and not a quickly fading hype.

The broader trends are a better starting position for NFPD. Based on firm wants, they stay for a couple of years, so that companies can even decide to first invest into the building of new technical skills, which then enable a product fulfilling the need. Many of the breakthrough innovations coming under the "health and wellness" trend umbrella were only possible after serious strategic upfront investments into the firm's own R&D. Unilever´s cholesterol-lowering research or Danone's effort around probiotics are examples.

Functional Food – Health and Wellness

In the middle of the 1990s, consumer trend scouts began to predict increasing consumer interest in food with an extra health benefit. For the food industry, this was an excellent opportunity to find new growth opportunities in an otherwise quite stagnant market. But to establish an attractive claim, with a competitive reason to believe, new skills were needed in foods. And medical and pharmaceutical industries had those skills. So, it was no surprise they saw new opportunities beyond supplements. The companies had experience with "over the counter" ("OTC") medicine and now saw a good possibility to "stretch into foods". What was missing was the market understanding, the market access, food brands, the tools to find subtle consumer insights, and the practical knowledge of how to make not only healthy, but tasty new food products.

But also, the "clinical program" for functional food showed peculiarities. These trials are not simply "light clinical ones", in the sense of being less time and money consuming, and therefore less risky. Pharmaceutical trials test the efficacy and safety of new drugs, first with healthy participants to confirm safety, and then – as final proof – with patients. For those trials, the participants were normally thoroughly diagnosed for the particular health condition.

Functional foods are well tested with healthy people, not only for safety, but also for efficacy. But here the participants were less well screened and diagnosed.

The best trial design is in most cases double-blind, placebo-controlled trials to ensure the validity of the results. But they differ in length of the trials, the number of participants needed, the type of data collected and the effect size that can be observed. Finding reliable biomarkers is sometimes a problem for functional food. The food matrix around the active and modulating factors out of the individual physiological conditions like subclinical inflammation, gut microbiome and individual genetic background further complicate the outcome. (For Overview: Granado-Lorencio & Hernandez-Alvarez, 2016).

Finally, in functional food, the "active" is often much lower concentrated than in prescribed medicine. Food products are freely available. Everybody can buy and consume them anytime, anywhere – from curious young children, pregnant women, to health-compromised elderly persons. Hence, the safe amount of bio-actives in food is lower than in medicine, which in turn makes it more difficult to prove the promised effect.

Despite these difficulties, the functional food market has quickly grown to an impressive size. Leading market research and advisory firms estimated the market at more than $280 billion in 2021, and it is expected to grow on average above 8%, adding each year more than $20 billion (Grand View Research, n.d.). The estimated size, value and growth differs between sources, surely also because a sharp separation between the functional food sector and adjacent sectors is difficult.

There are different ways to cut the "cake" of the big health and wellness (H&W) market: Traditionally by food category, like dairy, bakery, soy, meat and fat, or by the "beneficial active" like minerals, proteins and amino acids, prebiotics and fibres, and bacteria and plant sterols (Allied Market Research, n.d.). It is also possible to split it by "health benefit": As described by leading brands, it could be "blood pressure", "cardiovascular health", "weight management" "sports nutrition", "clinical nutrition" "sleep and relaxation" and others. And sometimes they name the target tissue as "reason to believe" like bone, joint, heart, gut, brain, liver, and skin health. Or they simply speak about mental health, cognitive performance, immunity, strengthened natural defences, etc.

There is yet another way to differentiate the H&W market. As we have seen in Mintels 2024 trends, segmentation by age group is possible. Science would agree that the body needs different things in different amounts in different life stages. So, it makes sense to offer solutions for infant nutrition, child health, women of childbearing age, school children, young urban professionals and golden agers, etc. Mintel just called out "*Age Reframed: Healthy ageing will be redefined by debunking 'old' stigmas and prioritising extending consumers' healthy years*" as one of the top three food and beverage trends in 2024 (Mintel, n.d.a). The Covid pandemic did not slow that trend down. Rather the opposite! Anything that could help the body and mind against this infectious agent gained extra interest when targeting, for example, "immunity", "natural defences" or "relaxing sleep".

Whichever way the segmentation is done, the sustained growth is powered by the combination of "market pull" and "technology push". People all over the world, not only GenY and GenZ, understand that right food choices are key to a healthy lifestyle and preventive health. Medical treatment is an emergency and can often be prevented by right food choices. This is not a new insight. It is rather the modernization of "an apple a day, keeps the doctor away" – even though this one is clinically difficult to prove. And besides the all-time favourites, science and technology can produce nutritionally superior food products, with well substantiated and trusted functional evidence.

The functional food sector is not a "tree with low hanging fruits", but often more "a bumpy road" towards a competitive claim. The Covid-19 pandemic left clear marks on the global consumer trends in foods. Eating healthy was always a good advice, captured already in the 2000-year-old statement "let the food be your medicine". But the pandemic caused a sharp increase in the interest in healthy food. Either proactively with better defence mechanisms, like "improved immunity" or "strengthening the natural defences", or by helping with the cure, like "mood and relaxation", better sleep" or "reduced stress".

At the time of writing this (early spring 2024), there is still debate about which foods should be defined functionally and which ones not. The gravity centre of the dispute has shifted to the question of whether naturally beneficial bioactive-containing food should be included in the category or not (see also: Alongi & Anese, 2021).

In summary, foods and beverages with a health benefit remain a top innovation area. According to Euromonitor International, "*56% of global consumers actively look for healthy ingredients in their food, and 30% are willing to pay a premium for health and wellness*" (Euromonitor International's Voice of the Consumer: Lifestyles Survey 2022, 2022). Or in other words, in the big battle for a "fair share of the consumer stomach", tasty but yet also healthier food created the opportunity to take share from the regular food sector and gave the chance to charge a bit more for it. Hence, a clear growth opportunity and a perspective for better profitability! And that is what investors like!

But not to forget: "Taste is king", and it still needs to be real food! A recent systematic literature review took a deep look at consumer preferences and the purchasing behaviour of food products with nutritional claims and health claims (Ballco & Gracia, 2022). The authors took the effort and analysed 125 articles in detail! And the results just underlined what a plethora of consumer research over the years had detected: The importance of "foodiness"! Consumers are simply not willing to sacrifice sensorial pleasure, in particular, taste, for any health promises. Their purchase decision is positively affected by motivation, nutritional knowledge and familiarity. And things like brand trust, packaging design, and of course price, play an important role in motivating the purchase. But taste rules! If it doesn't taste good, it will not be bought again!

The added "active" and food matrix serving as the carrier need to fit together – first, on the true functional level. The food product preferably works synergistically with the active, making the absorption better. Antioxidants like vitamin C improve iron absorption, and omega fatty acids are great in combination with antioxidants from vegetables. The opposite, antagonistic effects must be avoided. A combination of vitamins E, K and A in the same food can influence the absorption of an essential mineral. And fortifying tea or flour with extra iron is a questionable idea, since different components in the matrix complex the iron. Adding then vitamin C can change the picture again. Now these common and affordable staples are good carriers (Hurrell, 2021). Furthermore, consumer perception plays a role. Whether there is sufficient scientific proof or not, some combinations are perceived as good: Dairy and calcium, antioxidants and fruits and vegetables, a vitality shout and instant energy, tea or chocolate, and mood and relaxation.

Cognitive and mental health are another hot spot in this area, More and more people wish for more mental power to do their jobs, peak perform in leisure time or more resilience against daily stressors. The list of functional candidates, like B6, B12, zinc, caffeine, taurine, L-theanine or ginseng, has been stable for centuries, but recently, actives named "adaptogens" or "nootropics" made waves in this area.

Briefly, adaptogens are botanicals believed to help the body resist stressors of all kinds. These plants have been used for healing traditions in different cultures for centuries. Popular representatives are ashwagandha, panax ginseng, holy basil or schisandra. Chemically, these actives mostly belong to flavonoids, terpenoids and phenylpropanoid glycosides groups (Esmaealzadeh et al., 2022). These substances got their name, due to the assumed ability to "adapt" their function to the physiological needs. Today, only preliminary evidence suggests a promotion of mental and physical well-being.

Similarly, "nootropics", such as L-theanine m γ-aminobutyric acid, ritalin or caffeine, are more correlated with executive functions like enhanced focus, brain power and alertness. They are assumed to influence dopamine neurotransmitter levels (Malík & Tlustoš, 2022). Recent market research indicates that particularly GenZ feels frequent performance pressures. So, they are open to trying novel ingredients like those nootropics and adaptogens, as they are looking to prioritize brain health, improve attention span and productivity, and regulate mood.

The market for those products is expanding, either as supplements, over-the-counter (OTC) or prescribed medicine, or in food and often beverages. And several brands have launched products, often with hints of the natural, but exotic origin, and some older narratives about their effects.

Convincing scientific evidence for the efficacy of these products is still limited. Results are often coming from cell and animal studies. For those aiming to obtain a specific claim here, there is most likely still a very long way to go to identify credible biomarkers and go through well-designed human interventions.

CASE STUDY: FOOD FORTIFICATION I: PRODUCT DESIGN

A fast-moving consumer goods (FMCG) company in Thailand saw the opportunity to enrich some of its popular products with a critical micronutrient: Iron! Their bouillon cube seemed a very suitable carrier. The product was cheap, had a great penetration, especially in lower income groups, and was consumed regularly, as it was used in many popular meals like a sort of curries and soupy snacks.

At the start of the project, the team consulted the local Food and Drug Administration (FDA) and the Department of Health (DoH) about the nutritional situation. Under the voluntary fortification scheme for packaged foods, iron was a priority, since iron deficiency (ID) is a nutritional deficiency with serious complications such as weakness, fatigue, tiredness, heart problems, pregnancy complications, and developmental delays in children (Petry et al., 2016).

The authorities have set the nutrient reference value at an intake level of 15 mg per day, and from those 15 mg per day, 15% was sufficient to allow an on-pack iron content claim. For the bouillon cubes, this had to be correlated to individual intake levels, since such cubes would be used in the dish preparation. A typical 10 g block is made from salt, sugars, starch, vegetable fats, hydrolysed vegetable proteins, herbs and some spices. It would normally be used with 500 mL of liquid and would then serve on average two people. It was further assumed that households may consume bouillon containing dishes on average one to two times a day. Bringing this together leads to a calculation that 2.5 mg per cube was sufficient to make an on-pack claim.

Besides the concentration of the fortificant, the iron source required attention. Some sources like ferrous sulphate showed a good availability for the body, but also a remarkable impact on the sensorial attributes of bouillon and the shelf life at the intended concentration (Dold et al., 2020). An alternative complex, ferric pyrophosphate (FPP) performed better in sensorial tests but showed lower bioavailability. Fortunately, the team had evidence from cell line studies that the addition of extra sodium pyrophosphate (Na-PP) could improve the situation (Eilander et al., 2019).

This iron source was expected to match the narrow corridor between good bioavailability, minimal to no sensorial impact on the cube and the product made from it and minimum oncosts, since a price increase was expected to be not feasible. Now shelf life and consumer tests had to confirm the product design idea.

CASE STUDY: FOOD FORTIFICATION II: REGULATIONS

In the same country, the national FDA has established mandatory and voluntary fortification schemes. The mandatory scheme was implemented to combat the high prevalence of nutritional deficiencies in the population. Details differ per country depending on dietary habits, usage of local staples and nutritional status (Gayer & Smith, 2015). Thailand mandates fortification of salt with iodine, margarine and condensed milk with vitamin A, and rice with folate. Manufacturers of those goods are often supported by the government.

The voluntary scheme offers private companies the option to fortify their products to promote their sales. From a public health perspective, this helps to close additional, more specific, nutritional shortages in the population. For the manufacturers, it's an opportunity to differentiate with a new product and a new advertising story from competitors.

Hence, companies producing top-selling seasoning powders to help with cooking saw an opportunity to fortify their product range with iron, an important nutrient, not included in the mandatory scheme. So, they started to examine the necessary regulations.

For overall market regulation, the FDA had created four risk categories: "Specifically controlled food", "food with quality or standard requirements", "food with labelling requirements" and "general food".

The first question was how to identify the appropriate food category for the fortified cube. Discussions with the authorities revealed that a "cooking aid" – that's what a seasoning powder is – that contains more than 10% monosodium glutamate (MSG) or taste enhancer would fall into the "controlled food category", which means that premarket approval was needed. Reduction of the MSG content was regarded as unfeasible, due to the sensorial impact. An exchange with another taste-giving ingredient, like hydrolysed vegetable protein (HVP) or yeast extract, was deemed not helpful, as the product classification remained questionable and there was a significant cost impact.

After discussion, both parties agreed to take 100 g as the serving size, so the technical team in the company added 20% of the Nutrient Reference Value (NRV) for iron (2.71 mg) to 100 g of the powder to be able to claim 'contains iron' on pack.

That launch would be useful for the company if it was able to communicate positively about the extra nutrient. Earlier research has known that shoppers, often moms with kids, were generally aware about the importance of iron for health. Thus, the simply "content claim" on iron could trigger enough interest from the buyer.

When towards the end of the development phase company representatives presented the case to the FDA, new concerns came up. The on-pack usage suggestion for the seasoning powder for the popular "stir fry" dish application was about 1 teaspoon – or about 4 g – per 300 g of meat/vegetables.

Taking that as a reference point would lead to an intake of only 0.11 mg or 0.7% NRV per serving, very low compared to the as-sold references of 15%. In addition, typical seasoning products contain salt (sodium) and MSG, ingredients the authorities want to see consumed only in moderation.

So, while as such neither the low-to-moderate iron content nor the overall amount of sodium and MSG due to their dilution in the final dish was a problem, the combination brought the FDA to the conclusion that more support would be needed for a nutrient content claim for iron in a bouillon cube. They did not find sufficient scientific evidence that the suggested fortified formulation would be beneficial and qualify for a nutrient content claim on pack. So, they suggested an extra intervention study to create such evidence. But the design, planning and execution of an appropriate intervention study takes time, a year or two. And for the company such a delay of a long planned "product rejuvenation" was not acceptable and the additional resources needed for the study not budgeted. They felt that they had to "rejuvenate" their bouillon business much faster and therefore decided to put a new claim about better "taste" on pack, for which they had the evidence from ongoing consumer research work already.

A few general learnings can be taken from this: First, while the technical enrichment with packaged food products with nutrients offer differentiation opportunities for companies and brands, market regulations about what can, must or should not be done differs significantly per country. Secondly, simply adding nutrients to a food product following nutrient labelling guidelines is often not sufficient. The whole food "carrier", its composition and the consumption context need to be brought into the equation. Modern nutrition regulation will ensure that only nutritionally "clean" food items can make nutrient claims. And finally, not every case or product fortification is regulated by the book. In this case, a powder or granule pack for a popular seasoning mix is differently regulated than basically the same product as a bouillon cube, even though both products are used the same way for the same dishes.

CASE STUDY: FOOD FORTIFICATION III: CONSUMER

A couple of years back, a food company decided to invest into the development of health food products in Vietnam! Their technical team researched the scientific literature, read statistics and talked to relevant opinion leaders in and outside the government. They heard about a governmental decree for the mandatory fortification of a few staples, including salt, wheat flour and vegetable oil (Thi Thanh Huong, 2018). Even though the decree was still pending, extra initiatives from the private food sector were welcome. In their product portfolio, the company had liquid seasonings, popular cooking and seasoning sauces like soy, fish or oyster sauce. These condiments were widely used to bring taste and aroma to otherwise quite bland food items like rice or green vegetables. So, instead of fortifying the staple rice with iron – a technically difficult, not too effective and costly process – the in Vietnam popular and affordable fish sauce was an attractive alternative.

And for the company this was an opportunity to launch a truly differentiating alternative in this competitive market segment. Even key opinions from in and outside the government liked the idea.

An experienced project team started making various prototypes of fortified fish sauce and tested various iron sources, known for good bioavailability, but not impacting colour, taste, viscosity and mouthfeel, a well-known side effect of iron fortification. Samples were placed in accelerated shelf-life tests to confirm the product stability in distribution and storage in the tropical climate. All prototypes were tasted by an internal trained panel before presenting the best ones to several consumers, representatives of the target group, women of lower-income groups with children who cooked at home (see also Chapter 8).

To bring the idea alive, the team created various storyboards, explaining the importance of sufficient iron in the diet. A local market research agency recruited regular and loyal users of the fish sauce brand in rural and urban locations to conduct consumer research. Most women reacted positive to the idea of fish sauce with extra iron, so they were also asked to try the new product over their rice and green vegetables.

It was a surprise for all to hear that about half the women stated they could taste the iron and therefore would not want to use it in their cooking, while the trained panels hadn't noted any sensorial difference before when testing the dish.

So, the team went back to work to find the root cause of the perceived off taste, which they obviously missed before. New employees, sensitive to bitter notes ("supertasters"), were invited into the panel, and various vegetable varieties were tested with differently produced iron-fortified fish sauce. At the end, most panel members tasters couldn't sense any deviation in the fortified product, while two of them said that they may taste a light bitter taste.

Now there was only one way forward! Extra resources were made available to design a more sophisticated consumer product test. Four combinations were offered: (1) fortified product with concept mentioning iron, (2) fortified, but not mentioning iron, (3) standard fish sauce and the storyboard explaining the benefits of iron concept and (4) standard fish sauce without fortification but with the concept talking about benefits of iron.

Again, about half of the participants mentioned a slight bitter, metallic taste. Surprisingly, even participants that tasted the unfortified fish sauce together with the storyboard explaining the benefits of iron reported an iron off taste! How was there an iron off taste without any iron in the product? The team had no idea how to progress from here!

But launching a silently fortified product, not talking about the "iron benefits", and the risk that loyal brand users would switch to a competitor's product due to the off taste, was not an option! So, the project was stopped, and the fish sauce relaunch communicated instead just simple "superior taste"!

Months later, when the R&D team visited a few rural health centres in the Mekong Delta for a different project, one team member saw that local woman of childbearing age, were receiving iron and folic acid supplements (IFA). When talking to them, the women said that they were happy to receive the IFA, since it protects their own and their children's health, even though it's a bitter tasting medicine. But surely it was not a taste that they would want in the food they prepare for their families.

Even though not further investigated, this at the end explained the project failure. The prevailing bitter taste of the IFA might have created such a strong memory of "bad iron taste" that even the naming of the word "iron" triggered the bitter metallic perception. And perception is reality!

Naturalness and Authenticity

"Natural food" is another long-term trend in packaged foods and shows a few notable aspects. It's a consumer want, but it's mainly satisfied by perception. Science can hardly be used to split "natural" food from "non-natural" (see Chapter 11). Consumers have very different perceptions, when food is natural. But food manufacturers use different signals to trigger those perceptions. Some consumers deem natural food as organic, minimally processed and ideally locally sourced, while others base their perception on taste, freshness or the overall appearance of the product. Consequently, food manufacturers use natural ingredients; try to avoid any artificial, non-natural ingredients; and highlight the traditional roots of the recipes. Others inform about their usage of innovative new technologies, a better nutritional profile, or a more environmentally friendly sourcing or processing. This diffuseness leaves the door open for misleading concepts, unclear interpretations and increasing mistrust.

Recently, key stakeholders made attempts to frame the concept better. The International Organization for Standardization (ISO) published in 2017 the "technical specification (TS) 19,657" as "*Definitions and technical criteria for food ingredients to be considered as natural*" (ISO, n.d.). This document is seen as a first step towards clearer rules. The document suggested a few principles:

- The ingredients should be derived from a natural source, such as a plant, an animal, a microorganism or a mineral.
- The ingredient and the product should not undergo any chemical modification or processing that alters its essential nature or properties (physical modification is possible).
- The ingredient should not contain any artificial additives, such as preservatives, colours, flavours or sweeteners.

This consensus document is not binding, but it immediately started to form the base for the business-to-business (B2B) discussion between food ingredient industry and consumer goods manufacturers about the "natural requirement" of the used ingredients. The perception of naturalness happens on at least three levels. First the level of how the product and the ingredients are grown, then how the products are made and finally what the product is and what it contains.

"Subject matter experts", scientists, technologists, regulators and producers share their professional viewpoints, while others try to get a deeper understanding of the consumers' perceptions (Battacchi et al., 2019).

Not surprisingly, farming techniques, ranging from organic farming to usage of genetically modified new plant varieties, impact naturalness. And similarly, processing steps can affect the perceived naturalness. But not always. The usage of chemical processes normally destroys the naturalness perception, while a physical or biological processes does not impact the perception.

Research has indicated that perception of naturalness is also prompted by the packaging type (D'Astous & Labrecque, 2021). This is not necessarily logical, since a good packaging and the food packed should not interact with each other. But as food packaging is the visible surface, the outer impression is taken as a representation of the food item. A posh, shiny, top-quality outer packaging will not affect the formulation quality inside. And an extra lacquer layer should not reduce any barrier property the packaging should have.

The touch and feel (haptic) of a matte packaging finishing cues naturalness more than the posh and glossy packaging surfaces, as long as no stronger, conflicting cues are present (Marckhgott & Kamleitner, 2019).

And the positivity induced by natural packaging also influences the expected tastiness, which in turn increases purchase intent. This is a good reason to use a matte and not shiny packaging outside, even if the product shoot may come over a bit less flashy.

Transporting naturalness via a suitable packaging is a balancing act. Plant-based proteins are one of the fastest growing segments in the food industry, and it's environmentally responsible, healthy and more natural. Yet, the primary protein extraction process often includes wet processing, in which strong chemicals are applied. Even old all-time favourites like soybean oil or sunflower oil are made through hexane extraction, while the oil is generally accepted as natural. There is no indication that any of these or similar processes are negatively impacting the protein composition. But what will happen to the consumer trust, once those technical details become known to the wider public?

More transparency about packaged, industrial food is necessary. Many processes that are safe best practices in the food production require a "demystification". A lot of industrial processes are very different from homemade preparations, cooking, baking, etc.

Today, the discussion around these themes is often "expert conversation" between industry representatives, academia and authorities. They agree which ingredients can be used for a product, which processes can be applied and how that can be explained to the public. For the layperson, that often appears intransparent. At the same time, it's not easy for a layperson to fairly judge the strength of underlying science and technology. Here, trained and well-respected "key opinion formers" and "influencers" can mediate the necessary discussion. Today, it is a wide-open field to classify all physical, chemical and biological processes as either natural or not.

Only a very small amount of food is simply natural. For some food items, we can be glad when they are processed (see case study), and for many others, the industrial influence started long before the harvest.

If the food industry wants the trust of the consumer, that the food they make is strictly controlled, safe and nutritious, the details must be discussed between all partners to avoid the impression of an expert "horse trading" behind closed doors.

CASE STUDY: NATURAL CHEESE: CAMEMBERT

Camembert, an iconic soft and creamy French cheese with its characteristic bloomy rind made from unpasteurized milk, has delighted gourmets all over the world for many centuries and is a flagship of French Agri-*Culture*.

Coming from the northern part of Normandy, its first appearance probably dates to the 11th century, but more traceable sources see it made first in 1791 in the village of Camembert by Marie Harel, a farmer from Normandy, after a passing priest from Brie shared cheese-making tips with her. And the legend continued that "a piece of Camembert" was the last wish of King Ludwig XVI, before going to the guillotine.

Today, the original Camembert de Normandie is protected by designation of origin and must be made from raw milk in Normandy.

The first step is to inoculate warm cow milk with mesophilic bacteria and then add rennet, an enzyme mix produced in the stomachs of ruminant mammals. The mixture coagulates, and the curd is cut and salted and put into the typical cylindrical camembert moulds. The moulds will be turned by hand a few times in the next 2 days at 16°C, before the surface is sprayed with a suspension of the mould *Penicillium camemberti*, and then left at 13°C to ripen for a couple of weeks (Harel, Marie (Fl. 1790) | Encyclopedia.com, n.d.).

The cow milk must come from Normande cows (or today at least 50%), who are less productive but give the typical high-fat milk.

Once ready, a regular 250-g piece sells in Europe for between 7–12 euro. But this pride price is not the reason why this cheese today is threatened by extinction. Original camembert must be made from raw milk, a process forbidden in the United States and some other countries, due to safety and hygiene concerns. The camembert there comes from pasteurized milk and is not identity protected.

Another treat for this past crown of a gorgeous meal is the nutrition profile. The bit of calcium does not outweigh the substantial number of calories, salt and saturated fatty acids. The consumer guiding "Nutri-Score" (see Chapter 11) may make it a rare meal companion.

And even from the environmental side, eyebrows go up: Cheese is made from cow's milk, and they digest grass by "enteric fermentation", as science would say! By-products are a lot of methane with their burps and manure, nowadays know as an extremely potent greenhouse gas, while in Normandy the usually wet summers are getting drier, with less juicy grass.

The alternative comes from safe milk, produced industrially at a much lower price for the global market and designed and targeted to local taste preferences. And even in France, startups around Paris develop cheese without milk that is lactose free and vegan. They use modern biotechnology with traditional cheese-making to create an animal-free, planet-friendly, vegan cheese. Here microbes, fed with plant-based sugars in large fermenters, produce casein-like protein, which is harvested and purified as a starting block for the traditional cheese-making process, curdling, draining, melding, salting, and ripening (Standing Ovation, n.d.).

So, how will the future of original camembert look like? Will it proudly survive in a premium niche market to please true – and sufficiently wealthy – gourmets, while most soft cheese lovers are served with good, "safe by design" equivalents from the global food industry or vegan fermenters?

How will Normandy cows be fed juicy Normandy grass if it rains less in the coming years? Or, a bit less poetically: Will the new EU regulation aiming at much better packaging recyclability put an end to the iconic wooden box protecting the soft cheese during transport and storage, just because it cannot be conventionally recycled (Euronews, 2023)? And the latest findings of high levels of bird flu H1N5 influenza virus only underline the risk of unprocessed milk (National Institute of Allergy and Infectious Diseases, 2024).

Millennials and Generation Z

Consumers are getting older. Every year there will be fewer baby boomers and more Millennials and Gen Z, with increased purchasing power globally. Hence, it is important for the food business to respond adequately to the information wants of their younger consumers. It seems that more and more people follow scientific advice more often. One of the main sources of advice on good food is the dietary guidelines issued by governments and health organizations. Recent data from a pan-European study by the "European Institute for Innovation and Technology" showed that Generation Z (born between 1997–2010) is *"keenly interested, knowledgeable and engaged when it comes to their eating habits, and especially how this links to their wellbeing and mental health*" ("OUR FOOD, OUR FOOD SYSTEM," n.d.). And the generation before, the Millennials (born between 1981 and 1996) have prepared that way earlier. A 2018 study by the International Food Information Council found that millennials are more likely to prioritize health and wellness when making food choices than previous generations. The study found that 80% of millennials consider health benefits when selecting foods, compared to 64% of baby boomers, born between 1946 and 1964 (Texas Health Resources, n.d.).

Now, Gen Z is said to be the healthiest and most environmentally friendly generation the world has ever seen. And as mentioned before, they distrust big brands and corporations. What they demand is authentic and transparent interactions with businesses: They do their own research and rely on peer reviews and recommendations by friends and influencers. And they say they value social and environmental responsibility over profit.

One critical task for the big global brands is to review their operations, create transparency and modify their communication accordingly.

NEW ENABLING SCIENCE AND TECHNOLOGY

New technical solutions are the enabler of consumer wants. In the FMCG sector, the search for new products or services often starts with the question of how many consumers and which ones (target group) articulate a not-well-met need. Often, they are not fully happy with what is currently available. Being further asked, they may demand "health food" or "guilt-free indulgence" or "real food". Depending on what they know already, consumers are then able to articulate their needs more specifically and request "heart-healthy food", "foods supporting mental performance", food that is "good for people and planet", etc.

On that level, science and technology kicks in, and R&D may start to look at which technical developments could fulfil the need better than what is on the market. Are there new ingredients boosting immunity? Which part of the immune system do they target and how? Are those ingredients food grade, sufficiently regulated and available? How good is the scientific evidence to make a new claim?

And the answer about the "strengths of the enabling technology" will differ for different areas! "Better for people and planet"! There are many new technical solutions to offer immediately "delicious plant-based food, sustainably sourced". Good food promoting "weight loss" has been on the market for many years, but the true effectiveness is often moderate to low, while more people truly want and need that benefit. A breakthrough food technology seems not at hand, but new insulin- and diabetes-influencing drugs, like GLA-1 analogues, offer interesting opportunities.

I use the next pages to discuss a few promising science and technology areas, where major progress has been made with the potential to fulfil consumer needs better.

Replacing Traditional Meat

Today, most consumers on this planet still want animal-based proteins. In 2020, they ate 574 million metric tons of meat, seafood, dairy, and eggs, equivalent to nearly 75 kg per person. There is a huge differences between the countries: Hong Kong and the United States consume 128 kg or more, while India and Bangladesh consume 5 kg or less (World Population Review, n.d.).

At the same time, the concerns about the environmental costs of farming all these animal masses, how these farmed animals are treated, and concerns about the health consequences of eating so much conventionally farmed protein, are increasing.

Meat consumption must change, and fortunately various alternatives are already on the market, and they are gaining "more share of the global stomach". The alternatives are coming from different directions: Some are well-established and improving step by step, some are "kissed awake" again and others are radically new. Since science and technology are very active in this area, I want to briefly introduce a few very different ones here.

Plant-based, vegan, vegetarian and non-dairy

There are many good reasons why the plant-based proteins in the meantime have moved out of their niche into the mainstream. In 2020, 3 million metric tons of alternative proteins were already consumed a lot but were still just 2% of the animal protein market (bcg.com/publications, 2021). It is forecasted that alternative proteins will have a rapid growth to 97 million metric tons by the middle of the next decade, but this will still only be about 10% of the total protein market.

Out of this, more than 70% is coming from plant-based protein, nearly 24% from microbial sources and just some 7% from animal cell–based protein. The Boston Consulting Group expects average revenues of $3 per kilogram, making this a $53 billion market in 2023 and a $290 billion market by the middle of the next decade (BCG, 2022).

Only about one-third of these substitutes are truly vegan, while the rest are simply plant-based. That means they still can contain animal products in low amounts, while vegetarians strictly exclude any animal-based constituents. For them, this is often a non-discussable part of their wider ethical and environmental lifestyle, while plant-based consumers focus more on their own health benefits.

And in those markets where plant-based products mimic animal meat, meat eaters are not "converted" to become vegan, but establish a "flexitarian" lifestyle, where they continue to eat meat from time to time, but much more consciously and much less. A Euromonitor study from 2021 reported already more than 40% of all consumers claim to be flexitarian versus only 3%–5% being vegan in major Western markets and China (Euromonitor International, n.d.).

Technically, various technologies can be used to generate a variety of plant-based protein concentrates and isolates. Plant proteins can come from different raw materials including protein-rich seeds and protein-rich by-products of the food industry. Some of the best sources of plant-based protein include oilseeds like soy; legumes like lentils and chickpeas; cereals like wheat, rice, millet, maize, barley, and sorghum; pseudocereals like buckwheat, quinoa, and amaranth; nuts like almonds; and seeds like flaxseed, chia, pumpkin, sesame and sunflower.

The first challenge is the modification of plant-based protein flours, protein concentrates or protein isolates to match sensory properties of animal proteins. To get to a meat-like fibrous protein structure, high moisture extrusion is often used. And the addition of structure-forming agents, emulsifiers or stabilizers helps further. Basic taste, shelf life and to an extend the texture are influenced by the choice of plant proteins. After wet extraction and separation, milling and drying follows to get to a dry powder.

Another aspect frequently discussed is the protein quality, measured by the amino acid composition. Animal proteins are considered complete proteins, meaning they have all nine essential amino acids that the human body cannot synthesize on its own. Plant proteins from different sources lack one or more of these essential amino acids, which makes them incomplete. But this can be easily repaired through a combination of different plant protein sources.

Technically, a lot of progress has been made in the past few years in the space of meat alternatives. Right now, soy, wheat or pea isolates or concentrates are the leading materials used to produce plant-based meat, due to their gelation properties and ability to build fibrous structures, combined with good nutritional value and a good amino acid composition. And they are affordable!

The basic formulation for plant-based meat is similar: 20%–50% plant-based protein for the good nutritional profile, overall structure, colour, and texture; 5%–25% polysaccharides from starches, flour or fibre for consistency and water binding; and finally, 1%–5% plant-based fats and oils for flavour, texture and mouthfeel.

The differentiating work starts with the flavouring, the addition of taste- and aroma-giving ingredients, the extra colourants, emulsifiers, thickening agents, acidulants, texturants and any fortification to match the nutrition profile of meat in micronutrients (Boukid, 2020).

Once finally formulated, such a plant-based patty shows a long – and not necessarily appetizing – ingredient list, with many ingredients consumers would not expect in a burger patty. In other words, plant based is not "clean label", an attribute increasingly demanded in processed food. Such a long ingredient list shifts the perception of the product from good, healthy food towards constructed, designed, processed or even chemical foodstuff.

No wonder the simplifying of the ingredient list, avoidance of allergens and fully matching meat texture are the priority areas for the next generation of plant-based alternatives. The expected growth rates, the momentum even in countries where meat traditionally is "centre of the plate" and all the irrefutable scientific evidence of the meat industry have urged the global players in the food industry to team up with entrepreneurs and startups, forming alliances and partnerships to satisfy the upcoming demand.

The other big exchange wave is the exchange of dairy. The market for non-dairy alternatives exceeded $25 billion in 2022 and is expected to grow on average annually more than 13%, which makes this segment even faster growing than the market for meat alternatives (Fortune Business Insights, n.d.)

The drivers behind this trend are like those for meat replacers. Increasing awareness about the ecological footprint, here additionally the ruminant green house gas (GHG) emissions, has caused people to have concerns about animal welfare and their own health.

Today, a substantial number of people believe that they are lactose sensitive or intolerant, and any unpleasant feelings and digestive problems are accredited to lactose or, more precisely, the inability to produce lactase, the enzyme necessary to break down milk sugar. And reports about allergens and hormones raise additional concerns.

The list of dairy alternatives includes cereals, legumes, nuts, and seeds; all of them can be converted into plant-based milk, cheese, cream, yogurt, butter and ice cream. The basic production processes of these alternatives are wet milling, filtration, the addition of supplementary ingredients, sterilization, homogenization, potentially supplementary ingredient addition, aseptic packaging. Facultative extra steps are cold storage dehulling, roasting, dry grinding, steeping in dilute acid and the addition of some enzymes (Plamada et al., 2023).

The product formulation challenge is sometimes bigger, due to the specific organoleptic and nutritional profile of dairy. Dairy products have unique sensory and textural properties. Milk fat consists of 98% triacylglycerols, monoacylglycerols and diacylglycerols, plus free fatty acids, phospholipids, sterols, and carotenoids. These bring the unique flavour and texture. A plant alternative needs to match that. Flavour addition helps, particularly in those markets not familiar with soy. Here, the "beany taste" needs coverage.

Bringing non-dairy alternatives to the market has tight legal boundaries. In the European Union, the market for agricultural products is defined by Regulation (EU) 1308/2013 and in there "milk products" are defined as products that are exclusively derived from milk. And "milk" is specified as the "*mammary secretion obtained from one or more milkings*" (CodEx Texts | CODEXALIMENTARIUS FAO-WHO, n.d.). And the regulation stretches into cheese, butter, yoghurt, etc. Long time back "standards of identity" were introduced to protect the nature, composition and essential characteristics (e.g., manufacturing, appearance) of a high-value food item. The intention was consumer protection against fake products misusing the name. Those were times when quality food was characterized by fat and calories. The U.S. FDA introduced Standards of Identity in 1939, the year World War II started. Nowadays the nutritional situation has changed for most of the population, and milk fat composition and fat content are no longer straightaway good food. Calories are surely no longer scarce. Those that want to follow a healthy diet envisage problems to identify contemporary, better alternatives, as these are not allowed to carry the name of the original they rejuvenate.

The necessity to care much better for the environment is another driver to manage the resources on this planet. Science and technology make a very substantial contribution here. Alternatives to the "traditional way of making products" have been developed, upscaled and brought to market. Some still show weaknesses, but such alternatives will find adoption. Plant-based meat is much more sustainable than conventional meat: 47%–99% less land use, 30%–90% less greenhouse gas, 72%–99% less water and 51%–91% less aquatic nutrient pollution than conventional meat (The Good Food Institute, n.d.).

Generally, plant-based alternatives require more active product design. Today, they still need many functional additives, pumping up the ingredient list. But science and technology focus now on plant proteins that naturally have the functionality. All these different beans, soy, lentils, chickpeas, edamame, wheat, quinoa, potatoes, nuts or combinations have their specific strengths and weaknesses. New mixtures, new variants, different growth conditions and changed primary processing have the potential to cut the list of additives in animal protein alternatives.

Mushroom-based meat analogues

Edible mushrooms often have the perfectly meat-matching essential amino acid profile, so their proteins meet dietary requirements, and they have an economic advantage over animal and plant farming, as they can be cultivated quickly, even on very poor soil. But the mushroom "kingdom" still is a wide and often unexplored territory. Mushrooms are a rich source of bioactive compounds, ranging from those with suspected antioxidant, antitumor and antimicrobial properties to potentially detrimental, life-threatening substances, like phenolics, phytates, tannins and powerful enzyme inhibitors (Ayimbila & Keawsompong, 2023). So, the perception ranges from being "Nutritional Heaven" to "Pandora's bio box".

Filamentous fungi have the advantage that they excrete the mycoprotein into the surrounding medium, an attractive option in large-scale, continuous-flow fermentation. Since the proteins are secreted, they contain all the posttranslational modifications needed for full protein functionality.

The history of mycoprotein started already in 1985, when Marlow Food, a joint venture between Rank Hovis McDougall (RHM) and Imperial Chemical Industries (ICI), "entered the food market under the brand name "Quorn". Both companies used a strong patent position to block competition until its expiry in Europe in 2005 (Wikipedia Contributors, 2024). While the technological position was strong, the consumer appetite was underdeveloped. Global meat protein shortage was more an academic discussion topic, and environmental concerns were smaller and focussed on topics like acid rain.

The "technical performance" of this meat protein replacer was astonishing: Perfect amino acid profile, similar textural performance and great scalability, but the consumer perception and regulatory environment lowered the innovation speed. In those years, leading authorities, for example in the United States, had doubts about the safety of mycoproteins and delayed the market approval. A few reports about allergic reactions and asthma attacks then undermined the consumer trust; even so, a potential allergenicity was much lower than that of other well-known food items such as shellfish or soy.

Today, many new food innovating companies are active in this field. In some key markets, like the European Union with its Novel Foods, regulation is still difficult, as the fungi and the process must either prove earlier safe consumption in Europe or create the evidence through demanding research programs. That is still an effective entry barrier for new protein sources (foodnavigator.com, 2023). Overall, it is astonishing that the amount of mushroom protein as a powerful healthy and environmentally friendly alternative to animal meat on the markets is still relatively low.

Lab-grown meat

Much newer is the idea to replace farmed animal meat with lab-grown alternatives. The fundamental advantages of this technology versus livestock farming are striking: First, its "real" meat. It's made from animal cells but without the burden of the environment impact and without the livestock suffering. The fact that this meat is "made" under laboratory cleanliness could be taken as a safeguard to prevent contamination with infectious pathogens, a well-known threat in animal meat production. And in the meantime, the technical progress in tissue engineering–based cellular agriculture has reached a level where the upscaling into commercial production has started and where bioreactors are beginning to harvest muscle cells.

So, the approach is technically undoubtedly smart, but what about the commercial, environmental, cultural, and health-related aspects? While the upscaling is still in its early stages and elsewhere new creative formulations start to combine plant and lab meat, the first results from consumer research indicate reluctance (Van Loo et al., 2020).

In a study, 1800 consumers in the United States were asked about their preferences when having the choice between plant-based, lab-grown and farm-raised beef patties. The authors converted the quantitative answers into virtual market share. And the outcome showed 72% chose farm-raised beef, 16% plant-based (here, pea protein) meat alternative, 7% plant-based (animal-like protein) meat alternative, and only 5% lab-grown meat. Any additional technical information in favour of the technology increased the preference for non-farmed meat only slightly. And the majority did not even recognize non-farm-raised alternative as "beef". And so, the authors concluded; "*Not only is lab-grown beef the furthest from being technological and commercially feasible, but it is also the least desirable of the products studied*". Food and eating are emotional; just providing rational arguments for a behaviour change will not pull over consumers.

Some hope for lab-grown meat is coming from the feedback of the younger generation in that study, as it tended to be more positive. So, an enterprise aiming to tale advantage of such emerging trend in the younger generation, would need good patience to gain good share in a market with the 3rd highest meat consumption in the world.

And on the regulatory front, the signals are mixed. End 2020 sales permission has been given in Singapore and in 2023 in the Netherlands for the first lab-grown meat, while the Italian government is considering a full ban. Their agriculture minister stated that "*Laboratory products do not guarantee quality, well-being and the protection of the Italian food and wine culture and tradition, to which part of our tradition is linked.*" (Kirby, 2023). You don't have to be a prankster to think about the influence of the Italian agriculture lobby. In a country with such a long, proud history in genuine food quality, globally praised for the Mediterranean diet, the idea cell culture and modern biotechnology has no easy landing.

Time will show the success of lab meat. Several startups are working on innovations in that space, as the technology can be applied for different species or cell types, including marine species. Recently, lab-grown meat has been cleared by the FDA for sale in the United States. Two companies, Upside Foods and GOOD Meat, got the permission to sell their chicken products. Some reports state that the costs of producing a kilogram of such lab-grown meat would today be at about $100, much lower than the $800–$1000 quoted 1 or 2 years back. But we are still, of course, miles away from price parity with farm-grown meat.

So, for now, that leaves a niche role as "curious" alternative dish in a few high-end experimental restaurants for price-insensitive, experimental gourmets. A bit like the "molecular kitchen" gastronomy (see Chapter 13).

As an added note, even the environmental benefit of lab-grown versus farm-grown meat is not undisputed. According to a recent, not yet peer reviewed, study by researchers from the University of California, Davis (UCD), and the University of California, Holtville, the currently applied production methods of lab-grown meat could end up being way worse for the environment than beef farming – and not the promising sustainable alternative. Their end-to-end life-cycle assessment of current meat-growing processes found cultured meat production could emit between 4–25 times more carbon dioxide per kilogram than regular beef, particularly through the hidden energy requirements from the necessary strict purification procedures (Risner et al., 2023).

If that becomes confirmed, the lab-meat makers will have to put in extra effort to reduce the environmental footprint, without creating extra costs, as that would further open the price gap between both alternatives.

Possible hybrid solutions between plant and lab-grown may technically be exciting (texture) and help to manage the price premium. The consumer acceptance of such "designer food" is also better not taken for granted.

Single events may shift the needle into a different direction: A fatal disease outbreak in farm-based cattle may turn the fermenter into a safe alternative, while a similar infection in the fermenter would probably make it "Frankenstein Food".

Insect protein

Analysts show rosy growth projections for the insect protein market with impressive double-digit growth forecasts for the coming years. Cereal-based foods and flours are seen as primary application for human eating (Industry-Reports/Insect-Based-Protein-Market, n.d.). In reality, about 80% of insect protein is used as animal feed, and a stretch into mainstream food categories is not easy.

The rational side of the equation mentions the common practise of eating insects in many cultures, their variety, a great composition of high-quality protein, essential amino acids, unsaturated fats, vitamins, and minerals and the environmental potential of such food production (Cunha et al., 2023).

Against all these good arguments is that the various types of insects can be powerful vectors for foodborne disease transmission. Some allergen risks have also been noted in the meantime. Good food regulation and policies will be needed improving the safety of insects in food. The health risks will probably be per species and the substrate used for their farming, harvesting and processing (Acosta-Estrada et al., 2021).

But the biggest hurdle to become a true alternative to farm-based protein is the perception of the consumer. Particularly, western societies show a strong unwillingness to eat insect-based food. Their reaction is characterized by food neophobia and disgust. And the fact that maybe up to 2 billion people, especially those in developing countries with a tropical climate, from time to time eat specific insects, should better not be seen as a signal for mass market potential. For now, even the production of insect protein does not seem to be very cost competitive.

Bacterial protein

Further out there are investigations under way to understand how bacteria can be used to make protein. The idea is the fermentation of so-called hydrogen oxidizing bacteria (HOB) (Lin et al., 2022). These organisms, also called "knallgas" bacteria, use hydrogen and oxygen as the electron donor and acceptor. These special bacteria use hydrogen and oxygen to build biomass through carbon dioxide fixation, in a biochemical pathway called the "Calvin-Benson-Bassham" (CBB) cycle. These bacteria can make biomass including single-cell protein theoretically with a conversion efficiency about 20 times better than in the photosynthesis of plants. So, the potential is enormous, a theoretical yield of up to 70% of single-cell protein from the bacteria, the possibility to make also environmentally friendly bioplastics out of those proteins, or use them to make biofuel or clean waste material biologically.

2 Finnish companies have joined forces and started to work on commercialization, and recently they got approval from Singapore's Food Agency to test their Solein® protein in the confectionery category (foodnavigator.com, 2024).

The underlying science is exciting, but the industrial upscaling of a fermentation with hydrogen and oxygen (knallgas) requires extremely robust safety procedures. And once this is done, the regulatory approval, in Europe under "Novel Food", provides the next challenge.

This rather short list of "alternative protein ideas" shows the enormous spectrum of new technologies used and under exploration to replace meat from traditional feedstock farming by more sustainable protein sources. Cleaner plant-based alternatives seem to have the lead at least for a while, but other options could become a surprise or conquer an attractive market niche. The future success will be shaped by all the perceptions people have about food, ranging from rational aspects like nutrient content, healthiness, environmental footprint, and processing to sensorial and pure emotional reactions.

Personalized Nutrition

Healthy and nutritious food products are the biggest global food trend, reaching a market value of $24.75 billion in 2024 (Statista, 2024). And nutrition is the academic discipline stating what is good and nutritious and what is not or less good. But the nutritional guidance is still general, while health and wellness is personal. People prefer to know what is "good for me" instead of what is generally good for people. Too much salt is unhealthy. It can raise blood pressure and cause cardiovascular diseases. But a minority does not seem to be affected in that way (Choi et al., 2015). Is the saying "a calorie is always a calorie" true? Yes, perhaps thermodynamically, but not in the context of human energy uptake (SciTechDaily, 2024). Is it true that "sugar raises insulin levels"? Yes, but differently in different individuals and under different situations and conditions (Martin et al., 2020).

In the past, nutrition provided population advice: Folate for pregnant women, more iron for women of childbearing age in lower-income groups, more calcium for golden agers and vitamin D for those not getting enough sun. Today, science and technology are entering a stage in which nutrition advice is becoming personal and can provide tailored advice for individuals in a particular moment.

Nearly 10 years back, Zevi et al. presented research showing a high interpersonal variability in blood glucose levels after an identical meal (Zeevi et al., 2015). And when the team started to feed a machine-learning algorithm with individual data and blood parameters, dietary habits, anthropometrics, physical activity and information about gut microbiota, they were able to predict an individual's postprandial glycaemic response significantly better.

Powered by new enabling technology like high-throughput (HTP) screening, culturing, metabolomics, sequencing, etc., science is now able to take a much closer look at the "microbiota" of a person – that's the totality of all microorganisms in the body. In the meantime, it's a well-accepted idea that a balanced microbiota (called eubiosis) is the fundament of being healthy, while disruption of that balance (called dysbiosis) contributes to the pathogenesis of many non-communicable diseases like obesity, diabetes, and gastrointestinal and cardiovascular diseases. And the ability to analyse individual microbiomes allows much better understanding of the microbiome, diet and human tissue, presenting an entry to microbiota-based personalized dietary recommendations (Yeşilyurt et al., 2022).

And not only in a clinical setting. Startups in the "nutrigenomics" area, like Gini, Pinto, Gutyx, Foodmarble and LSEE, begin to offer DNA-based nutrition, exercise and mealtime goals; microbiome analysis with dietary and lifestyle choices; or the analysis of specific blood biomarkers (StartUs Insights, 2023).

And even larger FMCG companies start their pilots in personalized nutrition. In 2018, Nestlé started a "Wellness Ambassador" project in Japan, the first country in the world with a functional food regulation. The project used artificial intelligence, DNA testing and meal analysis to create personalized food products (Siegner, 2018). Subscribers were asked to provide meal photos using a specially designed AI app. Home kits for blood and DNA generated the data for personalized capsules for teas, smoothies and snacks.

Our understanding about the immense diversity of effects food can have on the individual microbiome, health and well-being has only started. Regularly publications appear describing effects on gut and cardiovascular health, immune modulation, various types of cancer or depression, anxiety, schizophrenia, Alzheimer's disease and autism. There are more and more opportunities for functional food for a personal benefit.

Relations between personal diet and individual health are often complex, but big data and AI offer much bigger chances to detect them. Many of those billions of bacteria, yeasts, viruses, fungi, protists, eukaryotes and phages in the intestine are influenced by factors in the diet, the lifestyle, taken medication and the actual environment. And the individual effect prediction may not emerge as a simple algorithm or something like a "core human microbiome", something that might even not exist (Sharon et al., 2022).

But the capturing more physiological parameters, such as those collected by smart digital devices, will enable more personalized solutions.

And it's not only personal health! A collaboration between a brewery and a DNA test enterprise promises a beer serving personal preferences based on saliva analysis (François, 2024), a solution for those people who do not want to taste too many beer varieties or who are not sure they have found their favourite one yet.

Bringing it back to new food product development, all the impressive scientific progress in all those "omics", like genomics, proteomics, metabolomics, expressomics, interactomics, epigenomics, transcriptomics, microbiomics and phenomics, try to find reliable biomarkers and principles. AI can then find predictable personal cause-effect relations. Once that is done, the new product requires personalization, which also means that classical strongholds of the food industry like upscaling and mass production economies of scale need new answers.

Any tailormade solution needs to hit the unmissable "profit and margin sweet spot". Highly customized products will be more expensive than less customized ones. Late customization is a strategic direction. But still, the price a company must charge to cover all costs and make a profit has to meet the willingness of the buyer to pay for it (BCG, 2020).

Biotics

Since more than one generation, the idea of "good for you" microorganisms in the human body is gaining more momentum. What started once with probiotics, the living "good" bacteria, is now accompanied by prebiotics, synbiotics and parabiotics. All of these are available as food products or supplements on a background of constantly increasing scientific evidence about their effects for good health and well-being. Conclusive predictions will remain a challenge, as the about 40 trillion microbial cells in the human microbiome make us and our metabolites actually a minority in our own body (Sender et al., 2016).

While the "microbiome" is a relatively new word, the habit of ingesting microbes for good health and well-being is very old. Traditional fermented products, like yoghurt, kimchi, sauerkraut, miso, nem chua, ogiri, kombucha and chungkookjang, are part of the healthy local diet of every culture. The approach has shifted from the product characteristics to the microbe strain or species responsible for that transformation. Currently, the celebrities are bacteria belonging to the families of *Lactobacillus* and *Bifidobacterium* or members of the yeast family making all-time favourites of bread, beer and wine.

The World Health Organization (WHO) defines probiotics as "*live microorganisms which when administered in adequate amounts confer a health benefit on the host*" (Contributors to Wikimedia Projects, 2024). This is a suggestion that is still discussed! The European Food Safety Authority has not adopted this definition, as for them the statement implies a health claim that is not proven and not measurable. Most reported health benefits are prevention and treatment of diarrhoea, support for the immune function, improved digestive health, lowerer blood pressure and balanced cholesterol levels.

But potential health benefits are seen as strain specific; they are not an intrinsic property of a particular bacteria family. Some of the most common and well-studied probiotic strains are:

- *Lactobacillus acidophilus*: Can help prevent and treat diarrhoea, vaginal infections, eczema and allergies and support immune system function and lactose digestion.
- *Lactobacillus rhamnoses*: May help prevent and treat diarrhoea, especially in children and travellers. A strain can also reduce the risk of urinary tract infections, respiratory infections and anxiety.
- *Lactobacillus helveticus*: May lower blood pressure, improve bone health, enhance cognitive function and mood, and modulate inflammation.
- *Bifidobacterium animalis*: Can contribute to digestive health, nutrient absorption, immune system function and bowel regularity. It can also fight off harmful bacteria and viruses.
- *Bifidobacterium lactis*: May support healthy cell growth and immune system and respiratory function, protect the gut lining from damage and reduce inflammation and oxidative stress.

It's likely that science will identify many more "good bacteria" in the near future. And the big rest of the microbiome is not just "quiet inhabitants". Some of them can impact health negatively: They can build up "bad" cholesterol, stimulating chronic kidney disease, metabolic syndrome, causing obesity, colitis, Crohn's disease, celiac disease, colon cancer and irritable bowel syndrome (IBS), heart disease and type 2 diabetes and trigger inflammation, autoimmune diseases, rheumatoid arthritis and osteoarthritis and may even impact the central nervous system.

Microbial misbalance is the situation to avoid, and restoration is the solution. Hence, the first promising approach is to eat living probiotics so they can (re)colonization in the intestine system and outgrow the detrimental ones.

The initial difficulty is the "delivery route". By ingesting microorganisms with food, the probiotics quickly land in an extremely harsh environment, the stomach, where they face low pH, bile salts and digestive enzymes. It's estimated that up to 99% of the ingested probiotics are killed here. Not very effective, but science can help again. Some strains have naturally a much better acid stability, for example, because their thick peptidoglycan layer protects them. And for others, encapsulation helps. Even simply swallowing them outside the meal might be a way to increase survival rates. These and a few other tricks, like co-cultivation with other microorganisms, the immediate addition of prebiotics or other protective agents, and strain optimisation through genetic engineering, can allow for the design of functional probiotic applications where more "beneficial" bacteria in the food arrive alive at their destination. Clearly all these strategies need careful safety evaluation, proof of effectiveness, and cost-effectiveness before they can be applied to commercial products (Castro-López et al., 2022).

An elegant way to circumvent the problems with the acid baths in the stomach are prebiotics, basically non-digestible components of food, often fibres, promoting the growth of probiotics already living in the gut. They provide the growth requirements for the good microbes. The scientific definition of prebiotics is: "*A prebiotic is a substrate that is selectively utilised by host microorganisms conferring a health benefit*" (Danielle, 2022). This definition assumes that prebiotics have a sort of strain specificity, and do not serve other, less beneficial ones in the microbiome.

Again, this is not a new "invention" of scientists or food technologists. Many are naturally present in many different sorts of plant-based material, including onions, artichokes, leeks and apples.

Common types of prebiotics include inulin, oligofructose, chicory root, dandelion greens, Jerusalem artichoke, cocoa flavanols, but also polyphenols and others. The simplest action mode assumes prebiotics provide the feed that makes the "good" bacteria grow faster, and so they outgrow the "bad" ones. Fibres are carbohydrates that we, the host, cannot absorb or get energy from. When beneficial microorganisms in the digestive tract can utilize them, they are prebiotics. Otherwise, fibres are just "bulking agents", also beneficial, but without microbial activity.

The next health concepts are "synbiotics", combinations of prebiotics and probiotics. Typical examples of synbiotics are bifidobacteria with fructooligosaccharides (FOS), or lactobacilli with FOS, inulin or galactooligosaccharides (GOS), or also yogurt with whole-wheat cereal or fruit. The idea is charming: Combine the "good bacteria" with their preferred feed first protecting them during their dangerous journey and then, once they are settled in the dietary tract, provide them nutrients. First introduced in the mid-1990s, the assumed health benefits are largely the same for pro- and prebiotics, since they are a combination of both (Gibson & Roberfroid, 2008).

Many *in vitro* and *in vivo* studies show the ability to modulate the gut microbiota by growth of beneficial bacteria, regulation of gut microbial metabolites and improvement of the intestinal barrier. The symbiotic concept seems more holistic, as it combines the advantages of pro- and prebiotics. While the idea is convincing, the final evidence for improved efficacy for clinical conditions compared to pre-and prebiotics is still missing (Li et al., 2021).

While the intake of living probiotics is generally safe, a few concerns have been raised, such as strain-specific behaviours, including niche-specific action, risk of developing antibiotic resistance or possible transfer of virulence genes.

These and other arguments brought a new class of "biotics" into the spotlight: The "postbiotics", non-living substances or metabolic products secreted by probiotics, for example, enzymes, short chain fatty acids, vitamins, peptides, or other metabolites, possibly employing beneficial effects for human health and wellness. Their reported benefits are antimicrobial, anti-inflammatory and immunomodulating. The International Scientific Association for Probiotics and Prebiotics (ISAPP) tried to capture the essence of a postbiotic as "*a preparation of inanimate microorganisms and/or their components that confers a health benefit on the host*" (Rafique et al., 2023).

This definition is somewhat suboptimal, as it includes all excreted substances of living beneficial microorganisms, but also everything that is released after cell lysis. And this creates an overlap with the concept of "parabiotics", the collection of all inactivated or dead probiotic bacteria or their cell components. Both para- and postbiotics may also be suitable for immunocompromised people who cannot take live probiotics. The proposed health benefits and molecular actions again overlap with those of other biotics (Nataraj et al., 2020; Viswanathan & Muthusamy, 2022). The emerging evidence points towards anti-inflammatory, antiproliferative immunomodulatory, antioxidant, anti-obesogenic and anti-hypertensive effects. Underlying physiological mechanisms are not well understood, and the whole area will need more research to disentangle the various effect of strains and molecules on the host pathway. The idea in the background is to link more specific effects to substances to target those better. The progress in genome analysis and engineering, imaging techniques, microfluidics, computational modelling and experimental design and all the other omics have allowed these techniques to become powerful tools to accelerate our understanding here.

The huge opportunity, but also the massive challenge, is that the composition of the microbiome varies greatly between individuals and throughout life; it's deeply personal. That indicates the great potential for a truly personalized nutrition. There is huge scope for new findings in the future. The identification of individual fingerprints set a base for "personalized solutions", ideally through a reliable marker.

And even though not everything is proven, this group of functional foods is taking a leading position in personalized nutrition. The digestive system is the prime and natural route to provide the body with essential molecules. If pre-, post-, syn- and parabiotics are as active as good bacteria, they can apply to a wide range of food products. They are more robust, survive the stomach passage better and survive all the necessary food processing steps needed to make the products safe.

And while global market sizes for the latest ideas, like syn-, para- and postbiotics, are not easy to find, the probiotics market is estimated to be worth more than $75 billion and the prebiotic one was worth $6.2 billion in 2022. Both are projected to have healthy double-digit growth in the coming years (Industry-Analysis/Prebiotics-Market, 2024).

Finally, the "good" also provide new tastes, exciting aromas and flavours combined with great digestibility by their action in traditional food fermentation.

Precision Fermentation

Precision fermentation refers to the combination of fermentation and precision biology, two processes allowing the production of complex organic molecules by simple microorganisms. It differs from traditional fermentation in that it can produce molecules not naturally made by microbes, but also modified ones. And it can control the quality and quantity of the molecules more efficiently than with traditional fermentation.

Technically, fermentation is the cultivation of microorganisms such as bacteria, yeasts and fungi to break down more complex substrates into simpler ones. It's one of the oldest food processing technologies to preserve or convert food and drinks through acids or alcohol generation (Teng et al., 2021).

Precision fermentation is already industrially applied to produce taste-giving ingredients like MSG, chymosin and citric acid (Hilgendorf et al., 2024). More recently, the technology has raised more interest in the fermentation of plant-based products as dairy and meat alternatives, including cheese, yogurt and ice cream. The fermentation often does not only provide a non-animal-based alternative but helps to improve the organoleptic properties and nutritional profile of the respective alternatives (Boukid et al., 2023). The production of animal-free proteins to mimic meat, seafood, eggs and dairy is currently a key application area.

Starting in the late 1970s, genetically modified microorganisms were used to produce a variety of food ingredients via fermentation that were scarce and conventionally sourced from animals and plants. The classical example is rennet, the chymosin-containing enzymes mix used to split milk into solid curds and liquid whey, the basic ingredients of cheese. Before, rennet was extracted from the stomach of ruminant mammals, like calves and lambs. Another example is the human insulin production by recombinant *Escherichia coli* bacteria. Newer applications do not require such heterologous gene transfer, where gene(s) from one species are transferred to another one. In the biomass fermentation, the ability of microorganisms to rapidly multiply under optimum conditions and produce a very high protein content, sometimes more than 50% dry weight, is used. Examples of such products include Marmite, the famous bread spread from the United Kingdom, made from yeast extract; fermented bean paste; or the mycoprotein from a filamentous fungus to create meat analogues or the production of riboflavin.

The innovation potential for foods is enormous and shows the characteristics of "future fit" products, since many applications utilize food or plant waste to generate valuable molecules. Today, there are more than 140 companies focussing on innovations of alternative proteins.

The main challenge for this technology is the well-known genetically modified microorganisms' aspect. That includes general consumer acceptance and the regulatory processes implemented. Consumers often demand more innovation and sustainability in food production, but "unfamiliar" technologies are often treated with scepticism, especially if they are perceived as radically different.

A recent study in a few European countries focussed on the question of how the consumer acceptance for precision fermentation technology could be improved (Banovic & Grunert, 2023). The authors concluded that providing reliable information about the benefits of the technology, like "naturalness" or "sustainability", can positively influence the acceptance. But with technology-sceptical consumers, who are afraid or even "technology neophobic," this approach may not work, as the "*... consumers do not reject precision fermentation because of specific characteristics or risks, but rather because they categorize it as just another instance of new food technology.*"

Important is the trust in the source of the information. That indicates a role for the company and brand. If they are perceived as credible, the message is trustworthy. Trust also has a rational element, so full transparency about the technologies and their potential risks and benefits can help to build it. Food companies will have to judge for themselves if their brands have that trust. New, innovative startups may have a better perspective in sceptical societies than the long-established players.

But it remains a fine line to send the messages in such a way that they are believed and not perceived as "greenwashing". "Trust comes on foot and leaves on a horse".

Microalgae

Microalgae, a wide class of aquatic cellular organisms, also carry the broad promise as a solution for many challenges the food chain is facing. It can be used as food ingredients to enrich the nutritional value and sensory quality of different products, for example, in bread, pasta, beverages and supplements. Some of them carry the promise of better human health and well-being, by improving the immune system, lowering cholesterol, preventing oxidative stress, reducing inflammation and modulating gut microbiota. Other, non-health-related properties include thickening, emulsifying, foaming and colouring, improvement of texture and appearance of food products.

The list of positive attributes is continuous with the environmental benefits: Low environmental impact, reduction of greenhouse gas emissions, less water and land use and wide growth flexibility. No wonder this group of organisms has caught so much attention in science and industry (De Oliveira & Bragotto, 2022). In 2021, the market reached $21 billion and is expected to increase about 10% every year in the coming decade (Transparency Market Research, n.d.).

However, challenges remain! There are safety concerns and regulatory issues that need to be addressed. Some of them have a history of safe use for consumption, which will make their inclusion into food stuff from the regulatory perspective easier. For others, the government regulations are more difficult. And the commercial scale-up of microalgae production is complex and requires advanced technology. Right now, production is still largely small-scale. And once again, do not forget the overall low consumer acceptance for unfamiliar food items.

For product developers, the sensory acceptance deserves attention. Algae often have distinct flavours and colours and contain compounds that strongly affect the taste, smell and shelf-life of products.

Hence, labelling microalgae as the next "big thing" in food science and technology may be a bit too optimistic. This extremely diverse group of aquatic organisms ranges from microscopic single-celled organisms to large multicellular seaweeds, made up of different prokaryotes and eukaryotes. Today already more than 50,000 different types of algae are identified, many only poorly described.

Strains from Spirulina, Chlorella, Dunaliella and some others are used in foods. Potential identified hazards for these range from toxins and allergens to environmental contaminants used to harvest. Due to their history of safe consumption, Spirulina and Chlorella are approved as ingredients in many countries and frequently used in formats like beverages and bakery items, but mostly in rather low levels (2%–10%) to manage the sensorial impact. With a protein content of up to 70% dry, they are also candidates as an alternative protein source.

The biotech company Algenuity patented a *Chlorella vulgaris* strain with reduced chlorophyll content and, hence, a more neutral, taste, smell and colour. This creates an option to apply this healthy and sustainable protein source soon in a wide range of food products in the future (Lukauskaite, 2021).

Three-dimensional (3D) printing

The scientific literature is filling up with articles hailing the benefits this new technology can bring to the food makers. It can be described as a rather typical "technology push" situation, leveraging a radical new technology in foods, where so far, as no research has ever picked up a consumer wish to "please print my food". But what has been voiced clearly in the market was the demand for more food personalization and unique inspirational artistic culinary food design.

For the total food industry (retail, food service and ingredient industry), 3D printing is a potentially powerful technical solution. 3D, or additive, printing promises production flexibility, allowing for quick changes and a flexible response to individual customer requests. In a sense, a prototype of real-time handling of information combined with personalized "late manufacturing", on the "last mile" to the consumer (Bravi & Murmura, 2021). Further advantages mentioned are the sustainability advantages, for example, in food waste reduction, and the reduction in food mileage and transport (Alami et al., 2024).

No wonder market analysts promised already 5 years back a sunny future for this technology due to the potential to make complex structures and interesting textures or the possibility to personalize the nutritional composition, for example, for hospitals or elderly homes (Global-3d-Food-Printing-Market-2023.html, 2018). The global market is currently manageable. In 2023, it was estimated at $0.52 billion, but with a 30% average annual growth decade (Food 3D Printing Market Insights, n.d.). Compared to the estimated $12,500 billion food market, it is still a niche segment.

With all this great promise for people and planet, what limits the fast rollout of this process technology? Technically, 3D food printing is not the same as 3D printing, for example, for plastics. Food ingredients have complex properties, such as rheology, texture, flavour and chemical interactions, that can be affected in the printing process. Moreover, printable food material needs to be cooked while or after printing, which brings an extra technical challenge and some product limitations.

Besides this, the safety and quality of the 3D printed food needs to comply with the food safety and quality standards like every other food.

The hardware, the 3D printers and the ingredients used are still expensive. Specialized materials are required that may not be widely available or affordable. And the 3D food printing process is slow and labour-intensive, which makes it difficult to fit the economies of scale the food industry wants.

Initial consumer research revealed mixed acceptance (Feng et al., 2022). Participants with "food technology neophobia" were not too enthusiastic; they did not like the idea of eating printed food.

Others even started to prefer 3D printed food items over conventional technologies in categories like chocolate and candy. Those were the formats first printed about 10 years ago. Now startups explore categories like pizzas, various sauces, pasta, or meat and sushi extrusions.

BeeHex, a NASA-funded printing company, sequentially layered pizza dough with cheese for further garnish (3d-Printed-Food/Layered, 2024). Their lead claim is that it takes only 5 minutes! That's not faster than what a trained pizza baker or chef can achieve without investing into expensive equipment that requires maintenance, and there are texture limitations.

Production efficiency and scalability are also the key hurdles for the Mooji Meats startup company (Mooji Meats, n.d.). They print Wagyu beef cut from plant protein and/or cultured meat cells into a layered structure resembling muscle tissue.

A single-head 3D food printer starts at around $500; more complex ones with several heads and the ability to use different pastes quickly reach $8000. But food materials often have different melting points. So, the whole heating and cooling of the nozzles is more complex and costs time. And after all that, cooking or proper heating is still necessary, since the machine is only melts the product.

So, until scalable technical solutions are available, 3D food printing may remain more a niche application, using the inherent strengths to make unique and creative designs, like individual cakes. This is great for high-end restaurants, bakeries, confectioneries, patisseries and ordered food. So, it's possible that consumers will accept the new technology in the nearer future in the form of a gorgeous, personalized birthday cake, a moment when price points are less important.

OTHERS

Many process technologies are further improved to make safe food products (elimination of microbials) without massive alteration of the nutritional and other sensorial attributes of the treated food product. High hydrostatic pressure processing, ultra-high-pressure processing, pascalization or cold pasteurisation all aim at effectively killing microorganisms.

The fundamental advantages of non-thermal preservation techniques are striking: Longer shelf life, no need for chemical preservatives maintaining most of the valuable, but vulnerable flavour volatile and aroma compounds. The techniques can be applied to liquids and solid foods with relative high moisture content.

Generally, minimal processing technology still has a couple of practical disadvantages: It may not inactivate spores effectively and cannot be used for dry products. And the technology can still change the texture or colour of food products or denature protein. The necessary equipment is expensive.

The practical complications in heat transfer and the resulting non-uniformity of processing still limit the wider application of this technology (Naveena & Nagaraju, 2020).

The promotors of these processing technologies should seek an open dialogue with consumers early enough so that a negative about a difficult-to-understand food process technology is not building up (see also the section Ultra Processed Food in Chapter 17).

KOFS, KOLS AND INFLUENCERS

All these great new technologies will benefit from an educated dialogue between the food producer and the consumer. "Credible voices" in that discussion are important to build trust in products made with new technology: Those individuals or organisations have sometimes different names, and their roles and impacts can overlap:

- *Key opinion leaders (KOLs)*: Individuals who are considered experts in a specific field or topic. Their opinions carry weight, and they are respected by the public due to their expertise and reputation.

- *Key opinion followers or key opinion formers (KOFs)*: The abbreviation is used inconsistently for either "key opinion follower" or "key opinion former". A key opinion former is interchangeable with the KOL, while a key opinion follower is more the next level multiplier in the communication cascade from expert to layperson.
- *Influencers*: Individuals who are often not trained experts and primarily characterized not by the knowledge they have but by the group of followers they have. This group is a phenomenon of social media. If they use expertise, they may overlap with key opinion followers, but often add a "personal view".

And even so, food is first a direct and very personal experience; the view from another person or authority is important to build one's own opinion.

Conventionally in foods, experts from the food chain act as endorsers for a product, concept or process. The most obvious and perceptible group is celebrity chefs:

- Gordon Ramsay is a famous chef and TV personality. He has received 16 Michelin stars and numerous awards, and has appeared on popular TV shows like *Hell's Kitchen*, *MasterChef US*, and *Kitchen Nightmares.*
- Guy Fieri is an Emmy-award-winning American restaurateur who has appeared on TV shows, has written several cookbooks and owns more than 70 restaurants.
- Jamie Oliver is a famous British chef and advocate of healthier eating, with an unconventional and approachable style.
- Masaharu Morimoto, sometimes called the "master of Japanese cuisine", is famous for fusion of traditional techniques with a modern flair.
- Vikas Khanna is a Michelin-starred chef and global champion of Indian flavours.
- Hooni Kim is an ambassador of modern Korean cuisine, known for authenticity and creativity.

This list could be extended endlessly; every country, ever kitchen, every culinary direction has their culinary leads. People listen to what they say and do, watch their shows, buy their books and maybe visit their restaurants. This forms a view about great food, exclusivity and quality. But these chefs also influence shopping behaviour, most directly when they give their name and face to a commercial product. Some famous chefs lend their names to more packaged food items.

Marco Pierre White, a chef with high culinary reputation, has given endorsements for the mass cooking brand Knorr, making bouillons. Other chefs endorse Som Thai drinking vinegar, all-purpose rubs, tomato sauces or baking mixes. For the fit between product and brand, the congruence between the celebrity's endorsement and the product being recommended seems most important (Calvo-Porral et al., 2021).

Chefs normally lead opinions about the "art of cooking" rather than "the science of cooking." They articulate and express the sensorial aspects.

Next to them are the key opinion followers and influencers. They are countless. Their impact is not necessarily based on culinary expertise, but often by passion and the size of their audience. They create content, mainly on social media, and can also serve the food industry to reach their consumers. A service like "Kolsquare" helps to find the most suitable influencer (Suche Und Finde Influencer, Die Zu Deiner Brand Passen - Kolsquare, n.d.). Their propriety credibility estimates the quality and authenticity of an influencer's audience by using elements like engagement (likes, comments and shares), audience authenticity (bots, inactive followers, purchased likes and purchased followers), popularity and creativity. Factors like the influencer's popularity (prominence, creativity, uniqueness of content) can also influence the score and compliance with ethical standards. George Clooney's years-long endorsement for Nestle's Nespresso brand built a robust barista image for their coffee capsules (Influencity, n.d.).

And when food quality dimensions are becoming more complex and difficult to judge for the normal consumer, like typically health aspects or environmental care, the credible endorsement of key opinion leaders shape public opinion even more and can set trends.

Nutrition and the question what health is and which food is more or less healthy or nutritious, is a main topic, vividly discussed in the public domain and, and consequently influencing shoppers. Susan Powter, Marion Nestle, David Katz, Michael Pollan, etc., are trained scientists, giving advice to the interested public. They may not be the spearhead of a particular scientific discipline, but they summarize the latest scientific findings about food, diet and the impact on health. And they are less the "judges" on the latest science, but the "advocates" of important aspects of it: Plant-based food, weight management, gut health, food industry critics and much more. These more "mainstream" scientists invest into discussion with the interested public, but they also stay abreast of the latest scientific developments in their expertise fields. Their expressed views can significantly influence consumer choices, food trends, or even policy decisions related to food safety and health.

The final group are the science leads in their expertise field: Examples are people like Walter Willet, renowned epidemiologist and nutrition expert; Frank Hu, on nutrition, obesity and diabetes; or Dariush Mozaffarian, a cardiologist researching healthy fats and evidence-based approaches to prevent cardiovascular diseases. Another example, is Linus Pauling in 1970 with vitamin C, before the science started to prove him wrong (see Chapter 2).

Other leading experts are not focussing on the health science, but on the societal and environmental impacts, like Vandana Shiva, sometime called the "Gandhi of grain", or Raj Patel, an economist focussing on the causes and consequences of global hunger, inequality and injustice in the food system.

These people may be less known to the public, but they influence the ongoing scientific work, the leading experts in their field and the policy makers and regulators.

These experts are science driven, not naturally supporters of the food industry, often sensitized by the influence and profit orientation of the food industry and their lobby organizations. They may find it more important to counteract the industry, maybe "demasking" some practises in foods. Or they see advantage to liaise with those they regard as credible to promote specific products or practices.

For the "food companies", key opinion formers are an important target group. They can shape the product and technology development, give advice and guidance or even collaborate. Then, they influence the product design or at least the enabling science and technology, including how to maintain crop yield more sustainably, how to use regenerative agriculture, how to diversify the global chains, how to make healthier products, and how to find responsible food labelling systems.

These scientists want to be heard and do not want to leave an impression that they promote the economic success of a product. A conversation with them is rational and science based. On many long-standing topics, the final verdict is not made. Butter or "margarine", which one is the healthier fat? Which constituent raises which cholesterol levels? What does that mean for cardiovascular health? Are carbohydrates helpful in weight management? If so, which ones? Is insulin a good marker, or can we identify markers for the effects of probiotics?

In most science fields groups or schools of opinion exist, discussing with each other the available evidence and the validity of the data.

A food company that wants to develop a functional food product may therefore first want to identify those leading experts that might support their claim intention, a process sometimes called opinion former mapping. These experts can help to design the best supporting experiments for the proposition. They may execute them and are willing to explain and defend the results obtained.

Some see the interaction between a company and an external KOL as "sneaky tactics", where scientific evidence is only a function of commercial success. In the recent broad discussion about the role of processed food in the rise of non-communicable diseases (NCDs), leading industrial representatives are under suspicion to directly influence the scientific consensus through a couple of "unfair" tactics (Sacks et al., 2017).

The ultimate goal of functional food is to create a real consumer benefit that shoppers value and reward. To be really competitive here, the "health claim" has to be really novel and differentiating versus competition. And this can mean to act at the forefront of science, an area where expert consensus does not exist and different opinions exist.

And here comes another group listening to KOLs into the picture; the non-, inter- and governmental authorities. To encourage new regulation, like a new health claim, or to prevent a potentially limiting regulation, like a warning label for salt or a new tax on sugar, the arguments brought to the debate by external scientific leaders have more impact than the speech of a company scientist. This person would be regarded as biased and may not have the chance to share arguments.

To an extent, the question then is whether experts and politicians see the food industry as part of the problem or the solution on an issue. The food sector and the industry have made a major contribution to feed more than 8 billion people. Or are they responsible for the rise of non-communicable diseases through too many unhealthy offerings in most diets? Does this sector therefore need tight control and monitoring? Or does it need stimulus and help to manage the problems of the "Anthropocene" on this planet. New alliances, like the 2030 Agenda for Sustainable Development, with its 17 Sustainable Development Goals (SDGs) and 169 individual targets, are framing a holistic and universal approach connecting scientists with companies, governments and society (United Nations, n.d.).

Maybe the question is even whether the challenges can be successfully solved without the food industry. Most of the nutritional and environmental scientific findings have not led to the change needed. Much of what the UN bodies, the Food and Agriculture Organization (FAO), the International Panel on Climate Change (IPCC) or national experts communicate as consensus is not fully reaching the consumer. By using core skills of the industry, deep consumer understanding and powerful communication, the chances increase that better products will hit the market and consumers will change their behaviour where needed. Different types of diets, such as veganism or gluten-free, can be promoted, without becoming a source of misleading information. It's not the magic stick for healthy diets or environmental responsible behaviour. Ultimately, people's food choices and behaviours are influenced by a complex interplay of individual and contextual factors and vary over time and across situations. But trying it without active participation of the food industry will limit the chances of success.

New partnerships are the way forward. A model in which the industry is seen as stubborn, profit oriented, not interested in societal and environmental conditions, and where honest scientists are battling for the scientific truth, consumers are only marginally interested and regulators bravely battle to establish common rules, seem outdated.

In the food industry, the necessity to earn that "license to operate in the future" is clearly understood. It is a substantial extra effort that will not be rewarded with quick incremental turnover or profit. But it builds the fundament for future business.

REFERENCES

3d-Printed-Food/Layered. (2024, February 20). https://www.3dsourced.com.

Acosta-Estrada, B. A., Reyes, A., Rosell, C. M., Rodrigo, D., & Ibarra-Herrera, C. C. (2021). Benefits and Challenges in the Incorporation of Insects in Food Products. *Frontiers in Nutrition*, *8*. https://doi.org/10.3389/fnut.2021.687712

Alami, A. H., Olabi, A. G., Khuri, S., Aljaghoub, H., Alasad, S., Ramadan, M., & Abdelkareem, M. A. (2024). 3D Printing in the Food Industry: Recent Progress and Role in Achieving Sustainable Development Goals. *Ain Shams Engineering Journal/Ain Shams Engineering Journal*, *15*(2), 102386. https://doi.org/10.1016/j.asej.2023.102386

Allied Market Research. (n.d.). *Functional food market size, share, competitive landscape and trend analysis report by ingredient, product, application: Global opportunity analysis and industry forecast 2021–2027.* Allied Market Research. https://www.alliedmarketresearch.com/functional-food-market

Alongi, M., & Anese, M. (2021). Re-Thinking Functional Food Development Through a Holistic Approach. *Journal of Functional Foods*, *81*, 104466. https://doi.org/10.1016/j.jff.2021.104466

Ayimbila, F., & Keawsompong, S. (2023). Nutritional Quality and Biological Application of Mushroom Protein as a Novel Protein Alternative. *Current Nutrition Reports*, *12*(2), 290–307. https://doi.org/10.1007/s13668-023-00468-x

Ballco, P., & Gracia, A. (2022). Tackling Nutritional and Health Claims to Disentangle Their Effects on Consumer Food Choices and Behaviour: A Systematic Review. *Food Quality and Preference*, *101*, 104634. https://doi.org/10.1016/j.foodqual.2022.104634

Banovic, M., & Grunert, K. G. (2023). Consumer Acceptance of Precision Fermentation Technology: A Cross-Cultural Study. *Innovative Food Science and Emerging Technologies*, *88*, 103435. https://doi.org/10.1016/j.ifset.2023.103435

Battacchi, D., Verkerk, R., Pellegrini, N., Fogliano, V., & Steenbekkers, B. (2019). The State of the Art of Food Ingredients' Naturalness Evaluation: A Review of Proposed Approaches and Their Relation With Consumer Trends. *Trends in Food Science and Technology*, *106*, 434–444.

BCG. (2020, March 12). *Winning-formula-in-personalized-nutrition*. https://bcg.com/publications/2020/Winning-formula-in-personalized-nutrition

BCG. (2022, July 8). *Climate-change-sustainability/alternative-proteins*. https://www.bcg.com/capabilities/climate-change-sustainability/alternative-proteins

bcg.com/publications. (2021, March 24). *The-benefits-of-plant-based-meats*. https://www.bcg.com/publications/2021/the-benefits-of-plant-based-meats

Boukid, F. (2020). Plant-Based Meat Analogues: from Niche to Mainstream. *European Food Research & Technology*, *247*(2), 297–308. https://doi.org/10.1007/s00217-020-03630-9

Boukid, F., Hassoun, A., Zouari, A., Tülbek, M. Ç., Mefleh, M., Aït-Kaddour, A., & Castellari, M. (2023). Fermentation for Designing Innovative Plant-Based Meat and Dairy Alternatives. *Foods*, *12*(5), 1005. https://doi.org/10.3390/foods12051005

Bravi, L., & Murmura, F. (2021). Additive Manufacturing in the Food Sector: A Literature Review. *Macromolecular Symposia*, *395*(1). https://doi.org/10.1002/masy.202000199

Calvo-Porral, C., Rivaroli, S., & Orosa-González, J. (2021). The Influence of Celebrity Endorsement on Food Consumption Behavior. *Foods*, *10*(9), 2224. https://doi.org/10.3390/foods10092224

Castro-López, C., Romero-Luna, H. E., García, H. S., Vallejo-Cordoba, B., González-Córdova, A. F., & Hernández-Mendoza, A. (2022). Key Stress Response Mechanisms of Probiotics During Their Journey Through the Digestive System: A Review. *Probiotics and Antimicrobial Proteins*, *15*(5), 1250–1270. https://doi.org/10.1007/s12602-022-09981-x

Choi, H. Y., Park, H. C., & Ha, S. K. (2015). Salt Sensitivity and Hypertension: A Paradigm Shift from Kidney Malfunction to Vascular Endothelial Dysfunction. *Electrolytes & Blood Pressure/Electrolyte & Blood Pressure*, *13*(1), 7. https://doi.org/10.5049/ebp.2015.13.1.7

CodEx Texts | CODEXALIMENTARIUS FAO-WHO. (n.d.). https://www.fao.org/fao-who-codexalimentarius/codex-texts/en/

Contributors to Wikimedia Projects. (2024, June 26). *Probiotic*. Wikipedia. https://wikipedia.org/wiki/Probiotic

Cunha, N., Andrade, V., Ruivo, P., & Pinto, P. (2023). Effects of Insect Consumption on Human Health: A Systematic Review of Human Studies. *Nutrients*, *15*(14), 3076. https://doi.org/10.3390/nu15143076

Danielle. (2022, July 6). *Prebiotics - International Scientific Association for Probiotics and Prebiotics (ISAPP)*. https://isappscience.org/for-scientists/resources/prebiotics/

D'Astous, A., & Labrecque, J. (2021). The Impact of Responsible Food Packaging Perceptions on Naturalness and Healthiness Inferences, and Consumer Buying Intentions. *Foods*, *10*(10), 2366. https://doi.org/10.3390/foods10102366

De Oliveira, A. P. F., & Bragotto, A. P. A. (2022). Microalgae-Based Products: Food and Public Health. *Future Foods*, *6*, 100157. https://doi.org/10.1016/j.fufo.2022.100157

Dold, S., Zimmermann, M. B., Jeroense, F., Zeder, C., Habeych, E., Galaffu, N., Grathwohl, D., Foman, J. T., Merinat, S., Rey, B., Sabatier, M., & Moretti, D. (2020). Iron Bioavailability from Bouillon Fortified With a Novel Ferric Phytate Compound: A Stable Iron Isotope Study in Healthy Women (part II). *Scientific Reports*, *10*(1). https://doi.org/10.1038/s41598-020-62307-1

Eilander, A., Funke, O. M., Moretti, D., Zimmermann, M. B., Owojuyigbe, T. O., Blonk, C., Murray, P., & Duchateau, G. S. (2019). High Bioavailability from Ferric Pyrophosphate-Fortified Bouillon Cubes in Meals is Not Increased by Sodium Pyrophosphate: A Stable Iron Isotope Study in Young Nigerian Women. *The Journal of Nutrition*, *149*(5), 723–729. https://doi.org/10.1093/jn/nxz003

Encyclopedia.com. (n.d.). *Harel, Marie (fl. 1790)*. https://www.encyclopedia.com/women/encyclopedias-almanacs-transcripts-and-maps/harel-marie-fl-1790

Esmaealzadeh, N., Iranpanah, A., Sarris, J., & Rahimi, R. (2022). A Literature Review of the Studies Concerning Selected Plant-Derived Adaptogens and Their General Function in Body With a Focus on Animal Studies. *Phytomedicine*, *105*, 154354. https://doi.org/10.1016/j.phymed.2022.154354

Euromonitor International's Voice of the Consumer: Lifestyles Survey 2022. (2022, May). *Euromonitor International's Voice of the Consumer: Lifestyles Survey 2022*. https://www.euromonitor.com/

Euromonitor International. (n.d.). *Going plant-based: The rise of vegan and vegetarian food | Euromonitor*. https://go.euromonitor.com/sb-packaged-food-210330-rise-vegan-vegetarian-food.html

Euronews. (2023, November 23). *'We have saved our Camemberts': Wooden boxes saved from recycling rules in last-minute amendment.* https://www.euronews.com/green/2023/11/16/could-new-eu-recycling-laws-mean-the-end-of-camemberts-traditional-wooden-boxes

Feng, X., Khemacheevakul, K., De León Siller, S., Wolodko, J., & Wismer, W. (2022). Effect of Labelling and Information on Consumer Perception of Foods Presented as 3D Printed. *Foods*, *11*(6), 809. https://doi.org/10.3390/foods11060809

Food 3D Printing Market Insights. (n.d.). https://www.mordorintelligence.com/industry-reports/global-food-3d-printing-market

foodnavigator.com. (2023, July 4). *EU novel foods regulation 'hindering innovation' in meat alternatives sector.* https://www.foodnavigator.com/Article/2023/07/04/EU-novel-foods-regulation-hindering-innovation-in-meat-alternatives-sector?utm_source=newsletter_daily&utm_medium=email&utm_campaign=04-Jul-2023&cid=DM1084306&bid=149248368

foodnavigator.com. (2024, January 23). *Protein 'made from air' enters snacks category.* https://www.foodnavigator.com/Article/2024/01/23/Solar-Foods-Solein-protein-in-Fazer-chocolate-snack-bar?utm_source=newsletter_product&utm_medium=email&utm_campaign=06-Feb-2024&cid=DM1118108&bid=317019366

Fortune Business Insights. (n.d.). *Dairy-alternatives-market.* Retrieved July 10, 2024, from https://www.fortunebusinessinsights.com/industry-reports/dairy-alternatives-market-100221

François, W. (2024, March 1). *Five food and beverage personalization trends to watch.* https://hospitalityinsights.ehl.edu. https://hospitalityinsights.ehl.edu/food-and-beverage-industry-personalization-trends

Gayer, J., & Smith, G. (2015). Micronutrient Fortification of Food in Southeast Asia: Recommendations from an Expert Workshop. *Nutrients*, *7*(1), 646–658. https://doi.org/10.3390/nu7010646

Global-3d-Food-Printing-Market-2023.html. (2018). https://bisresearch.com/industry-report/global-3d-food-printing-market-2023.html

Granado-Lorencio, F., & Hernandez-Alvarez, E. (2016). Functional Foods and Health Effects: A Nutritional Biochemistry Perspective. *Nutritional Medicine*, *23*(26), 2929–2957. https://doi.org/10.2174/0929867323666160615105746

Gibson, G. R., & Roberfroid, M. (2008). *Handbook of prebiotics.* CRC Press. https://doi.org/10.1201/9780849381829

Grand View Research. (n.d.). *Functional Foods Market Size, Share & Trends Analysis Report by ingredient (Carotenoids, prebiotics & probiotics, fatty acids, dietary fibers), by product, by application, by region, and segment Forecasts, 2022–2030.* https://www.grandviewresearch.com/industry-analysis/functional-food-market

Hilgendorf, K., Wang, Y., Miller, M. J., & Jin, Y. (2024). Precision Fermentation for Improving the Quality, Flavor, Safety, and Sustainability of Foods. *Current Opinion in Biotechnology*, *86*, 103084. https://doi.org/10.1016/j.copbio.2024.103084

Hurrell, R. F. (2021). Iron Fortification Practices and Implications for Iron Addition to Salt. *The Journal of Nutrition/The Journal of Nutrition*, *151*, 3S–14S. https://doi.org/10.1093/jn/nxaa175

Industry-Analysis/Prebiotics-Market. (2024, July). https://www.gminsights.com/industry-analysis/prebiotics-market

Industry-Reports/Insect-Based-Protein-Market. (n.d.). Retrieved July 10, 2024, from https://www.fortunebusinessinsights.com/industry-reports/insect-based-protein-market-100780

ISO. (n.d.). ISO/TS 19657:2017. https://www.iso.org/standard/65717.html

Kirby, B. P. (2023, March 29). *Italy moves to ban lab-grown meat to protect food heritage.* https://www.bbc.com/news/world-europe-65110744

Influencity. (n.d.). *KOLs vs influencers: What they are & what differentiates them.* https://influencity.com/resources/guide/what-are-kols-and-what-sets-them-apart-from-influencers

Li, H., Zhou, D., Gan, R., Huang, S., Zhao, C., Shang, A., Xu, X., & Li, H. (2021). Effects and Mechanisms of Probiotics, Prebiotics, Synbiotics, and Postbiotics on Metabolic Diseases Targeting Gut Microbiota: A Narrative Review. *Nutrients*, *13*(9), 3211. https://doi.org/10.3390/nu13093211

Lin, L., Huang, H., Zhang, X., Dong, L., & Chen, Y. (2022). Hydrogen-Oxidizing Bacteria and Their Applications in Resource Recovery and Pollutant Removal. *Science of the Total Environment*, *835*, 155559. https://doi.org/10.1016/j.scitotenv.2022.155559

Lukauskaite, A. (2021, April 9). *Patent filings reveal companies embracing microalgae for sustainable foods.* HGF. https://www.hgf.com/de/neuigkeiten/patent-filings-reveal-companies-embracing-microalgae-for-sustainable-foods/

Malík, M., & Tlustoš, P. (2022). Nootropics as Cognitive Enhancers: Types, Dosage and Side Effects of Smart Drugs. *Nutrients*, *14*(16), 3367. https://doi.org/10.3390/nu14163367

Marckhgott, E., & Kamleitner, B. (2019). Matte Matters: When Matte Packaging Increases Perceptions of Food Naturalness. *Marketing Letters*, *30*(2), 167–178. https://doi.org/10.1007/s11002-019-09488-6

Martin, S., Kempf, K., & Rommelfanger, J. (2020). *Wie Insulin uns alle dick oder schlank macht: Wissenschaft, neuster Stand: wie Sie Ihre Fettverbrennung selbst zu aktivieren.*

Mintel. (n.d.a). *2024 Global Food and Drink Trends | Mintel.* https://www.mintel.com/food-and-drink-market-news/global-food-and-drink-trends/

Mintel. (n.d.b). *Trends impacting the healthy eating market | Mintel.* https://www.mintel.com/insights/food-and-drink/global-trends-driving-evolution-of-healthy-eating/

Mooji Meats. (n.d.). https://www.moojimeats.com/

Nataraj, B. H., Ali, S. A., Behare, P. V., & Yadav, H. (2020). Postbiotics-Parabiotics: The New Horizons in Microbial Biotherapy and Functional Foods. *Microbial Cell Factories, 19*(1). https://doi.org/10.1186/s12934-020-01426-w

National Institute of Allergy and Infectious Diseases. (2024, May 24). *High H5N1 influenza levels found in mice given raw milk from infected dairy cows.* https://www.niaid.nih.gov/news-events/high-h5n1-influenza-levels-found-mice-given-raw-milk-infected-dairy-cows

Naveena, B., & Nagaraju, M. (2020). Review on Principles, Effects, Advantages and Disadvantages of High Pressure Processing of Food. *International Journal of Chemical Studies, 8*(2), 2964–2967. https://doi.org/10.22271/chemi.2020.v8.i2at.9202

Our Food, Our Food System. (n.d.). Retrieved July 10, 2024, from https://www.eitfood.eu/media/news-pdf/Our_Food%2C_Our_Food_System_-_EIT_Food_report_.pdf

Petry, N., Olofin, I., Hurrell, R., Boy, E., Wirth, J., Moursi, M., Angel, M. D., & Rohner, F. (2016). The Proportion of Anemia Associated With Iron Deficiency in Low, Medium, and High Human Development Index Countries: A Systematic Analysis of National Surveys. *Nutrients, 8*(11), 693. https://doi.org/10.3390/nu8110693

Plamada, D., Teleky, B., Nemes, S. A., Mitrea, L., Szabo, K., Călinoiu, L., Pascuta, M. S., Varvara, R., Ciont, C., Martău, G. A., Simon, E., Barta, G., Dulf, F. V., Vodnar, D. C., & Nitescu, M. (2023). Plant-Based Dairy Alternatives—A Future Direction to the Milky Way. *Foods, 12*(9), 1883. https://doi.org/10.3390/foods12091883

Rafique, N., Jan, S. Y., Dar, A. H., Dash, K. K., Sarkar, A., Shams, R., Pandey, V. K., Khan, S. A., Amin, Q. A., & Hussain, S. Z. (2023). Promising Bioactivities of Postbiotics: A Comprehensive Review. *Journal of Agriculture and Food Research, 14*, 100708. https://doi.org/10.1016/j.jafr.2023.100708

Risner, D., Kim, Y., Nguyen, C., Siegel, J. B., & Spang, E. S. (2023). Environmental impacts of cultured meat: A cradle-to-gate life cycle assessment. bioRxiv (Cold Spring Harbor Laboratory). https://doi.org/10.1101/2023.04.21.537778

Sacks, G., Swinburn, B. A., Cameron, A. J., & Ruskin, G. (2017). How Food Companies Influence Evidence and Opinion – Straight from the Horse's Mouth. *Critical Public Health, 28*(2), 253–256. https://doi.org/10.1080/09581596.2017.1371844

SciTechDaily. (2024, January 8). *Is a calorie a calorie? Not always, when it comes to almonds.* SciTechDaily. https://scitechdaily.com/is-a-calorie-a-calorie-not-always-when-it-comes-to-almonds/).

Sender, R., Fuchs, S., & Milo, R. (2016). Revised Estimates for the Number of Human and Bacteria Cells in the Body. *PLoS Biology, 14*(8), e1002533. https://doi.org/10.1371/journal.pbio.1002533

Sharon, I., Quijada, N. M., Pasolli, E., Fabbrini, M., Vitali, F., Agamennone, V., Dötsch, A., Selberherr, E., Grau, J. H., Meixner, M., Liere, K., Ercolini, D., De Filippo, C., Caderni, G., Brigidi, P., & Turroni, S. (2022). The core Human Microbiome: Does It Exist and How can We Find It? A Critical Review of the Concept. *Nutrients, 14*(14), 2872. https://doi.org/10.3390/nu14142872

Siegner, C. (2018, September 5). *Nestlé's personalized nutrition pilot taps AI, uses consumer DNA.* Food Dive. https://www.fooddive.com/news/nestles-personalized-nutrition-pilot-taps-ai-uses-consumer-dna/531560/

Standing Ovation. (n.d.). https://www.standing-ovation.co/

StartUs Insights. (2023, October 12). *5 top personalized nutrition companies | StartUs Insights.* https://www.startus-insights.com/innovators-guide/5-top-personalized-nutrition-startups-impacting-the-healthcare-industry

Statista. (2024, May 22). *Functional and natural health food: market value forecast worldwide 2023–2033 | Statista.* https://www.statista.com/statistics/502267/global-health-and-wellness-food-market-value

Texas Health Resources. (n.d.). *Study shows younger generations are more health-conscious than previous generations.* https://www.texashealth.org/areyouawellbeing/Health-and-Well-Being/Study-Shows-Younger-Generations-Are-More-Health-Conscious-Than-Previous-Generations

Suche und finde Influencer, die zu deiner Brand passen - Kolsquare. (n.d.). https://www.kolsquare.com/de/produkte/suche-und-finde-influencer

Teng, T. S., Chin, Y. L., Chai, K. F., & Chen, W. N. (2021). Fermentation for Future Food Systems. *EMBO Reports, 22*(5). https://doi.org/10.15252/embr.202152680

The Good Food Institute. (n.d.). *Environmental impacts of alternative proteins | GFI.* https://gfi.org/resource/environmental-impact-of-meat-vs-plant-based-meat/

Thi Thanh Huong, N. (2018, June 7). *Joint-Press-Statement-who-and-UNICEF-call-Vietnamese-authorities-enforce-food.* https://www.unicef.org/vietnam/press-releases/

Transparency Market Research. (n.d.). *Algae market.* https://www.transparencymarketresearch.com/algae-market.html

United Nations. (n.d.). *Transforming our world: the 2030 Agenda for Sustainable Development | Department of Economic and Social Affairs.* https://sdgs.un.org/2030agenda

Van Loo, E. J., Caputo, V., & Lusk, J. L. (2020). Consumer Preferences for Farm-Raised Meat, Lab-Grown Meat, and Plant-Based Meat Alternatives: Does Information or Brand Matter? *Food Policy, 95*, 101931. https://doi.org/10.1016/j.foodpol.2020.101931

Viswanathan, K., & Muthusamy, S. (2022). Review on the Current Trends and Future Perspectives of Postbiotics for Developing Healtheir Foods. *eFood, 3*(6). https://doi.org/10.1002/efd2.47

Wikipedia Contributors. (2024, May 22). *Quorn.* Wikipedia. https://en.wikipedia.org/wiki/Quorn

World Population Review. (n.d.). *Meat consumption by country 2024.* https://worldpopulationreview.com/country-rankings/meat-consumption-by-country

Yeşilyurt, N., Yılmaz, B., Ağagündüz, D., & Capasso, R. (2022). Microbiome-Based Personalized Nutrition as a Result of the 4.0 Technological Revolution: A Mini Literature Review. *Process Biochemistry, 121*, 257–262. https://doi.org/10.1016/j.procbio.2022.07.012

Zeevi, D., Korem, T., Zmora, N., Israeli, D., Rothschild, D., Weinberger, A., Ben-Yacov, O., Lador, D., Avnit-Sagi, T., Lotan-Pompan, M., Suez, J., Mahdi, J. A., Matot, E., Malka, G., Kosower, N., Rein, M., Zilberman-Schapira, G., Dohnalová, L., Pevsner-Fischer, M., … Segal, E. (2015). Personalized Nutrition by Prediction of Glycemic Responses. *Cell, 163*(5), 1079–1094. https://doi.org/10.1016/j.cell.2015.11.001

Zegler, J. (2024). Globale Lebensmittel- und Getränketrends [Report]. https://insights.mintel.com/rs/193-JGD-439/images/Mintel_Globale_Lebensmittel_und_Getraenketrends_2024.pdf?version=0&mkt_tok=MTkzLUpHRC00MzkAAAGTQ7CgEg1jJPB_kDiM6xyFgNVUJkT076wc7rroBcGxc2g2bfmILaS73hTSECom-ExmoWeZfsEefsQFvJjrnDl36BIGShZ2ZF57P5xINEQ5bFJb

The "New Product Development" Reality 17

Similar to "big pharma" or "big oil", "big food" is nowadays also no longer perceived as "beautiful" by many. And "big" was never "quick" – an important ability to succeed in fast-moving consumer goods (FMCG) markets! Big food brands got big because they applied new technologies efficiently and effectively to make food products safe, convenient, available and affordable (see Chapter 1).

Today, where people feel the world is changing faster every day, faster decision making in the FMCG sector is even more important. Charlie Chaplin expressed the same feeling in his film *Modern Times* in 1936.

With speed being the lead currency, new supply chain models are required; systems must be able to produce faster, adapt more quickly, but still be capable to upscale efficiently. And since the commercial outlook is more volatile and uncertain, the ability of enterprises to predict "their return on investment" fades. Massive investment into new lines, equipment or factories, before the product has even hit the market, is riskier.

The development of smaller, more flexible manufacturing units and/or a wider network of co-packing/third-party manufacturing can form a solution. The three-dimensional (3D) food printer is a nice piece of technology enabling late finishing and personalized food (see Chapter 16). But they make only simple structures and textures at rather low volumes and with an "ultra-processed food" image. So, great for birthday cakes, but not a broad solution.

The innovation speed, the ability to react to new needs and opportunities faster than the competition, will gain further importance. Innovation cycles are getting shorter and shorter: A look at the recent history of social media platforms makes one dizzy: The World Wide Web needed 7 years to reach 100 million users, Facebook needed 4.5 years; Instagram, 2 years and 6 months; WeChat, 1 year and 2 months; TikTok, 9 months; ChatGPT, 2 months, and recently Threads, 5 days (Rao, 2023).

Innovation speed is not that explosive in foods (yet) but shows similar acceleration patterns in growth areas. Here is a quick look at the animal protein alternatives market of the last 15 years.

- *1985*: Marlow Foods and Imperial Chemical Industries (ICI) founded Quorn, the mushroom protein alternative in the United Kingdom.
- *1997*: Quorn entered the American market.
- *1998*: Yves Potvin founded Gardein ("Garden Protein").
- *2009*: Beyond Meat began working on plant-based meat alternatives.
- *2010*: Jaap Korteweg started The Vegetarian Butcher in the Netherlands.
- *2013*: Sophie's Kitchen began, specializing in plant-based seafood alternatives, and JUST Egg started to make plant-based egg substitute.
- *2014*: Pinnacle Foods acquired Gardein
- *2016*: Impossible Foods launched the Impossible Burger; Tyson Foods invested in Beyond Meat; Hampton Creek introduced Just Mayo, a plant-based mayonnaise; Good Catch debuted with plant-based seafood alternatives; Nestlé acquired Sweet Earth; and Beyond Meat launched "Beyond Sausage".
- *2018*: JUST expanded into plant-based dressings and mayonnaise; Lightlife introduced a new plant-based burger; Oatly, a Swedish company, became popular with its oat milk products;

DOI: 10.1201/9781032724591-17

Nestlé introduced the Awesome Burger, competing directly with Beyond Meat and Impossible Foods; and Unilever bought The Vegetarian Butcher.

- *2020*: Miyoko's Creamery created a new type of plant-based cheeses, and Good Catch launched plant-based seafood alternatives. A few months later, NUGGS came with plant-based, protein-packed nuggets; and Heura Foods, a Spanish startup, brought new chicken and beef alternatives.
- *2022*: JUST introduced JUST MILK, while Atlas Food Co. entered the market with plant-based bacon made from mycelium. Vevan Foods created new lines of plant-based cheeses.

A short and incomplete list! But it shows how innovation speed in an initially niche market segment, existing for more than 20 years without much momentum, erupted suddenly, creating more and more solutions for a simple need. Now, at the beginning of 2024, this market has passed the $50.4 billion turnover mark, and the number of new product solutions is constantly growing (TBRC Business Research PVT LTD, 2023).

This development reveals a few characteristics. First, sometimes it takes a long time before a new "idea" becomes a real trend. Being first can then also mean, being too early. The Imperial Chemical Company" (ICI) had the market for nearly 20 years (the time span for patent protection by the way). Consumer interest in the main market really started around 2010–2012 (try Google trends on this one). But once a base technology is there that meets existing needs, the markets burst. The technology becomes widely applied, modified, broadened, and creates more and more solutions. These new formats are often coming from startups; once their products are mature, big players are engaging (share taking or take over) and use their market strengths.

This is still very much in line with innovation cycles theory by Joseph Schumpeter from 1942, that a new wave is faster and steeper than the one before (Ziemnowicz, 2020). The reasons are technological advancements, globalization, competition and market pressure, network effects and shorter product life cycles.

Faster launches require new approaches for prototyping and idea testing and innovation portfolio management. A valuable innovation pipeline has a "now, new, next" approach, including the fast, often incremental innovation (now), structured innovation efforts to push the breakthrough innovation (new) to market, and the serious and consistent focus on the true next game changers (next).

The new product development reality puts more pressure on time. Where the big brands in the past had 0.5–1.5 years to launch a low-risk "quick win", now new ideas to keep brands fresh and relevant need to come faster. And 2–5 years for a new platform idea, or 5 years and more for the next game changer is also no longer fast enough.

And while each week earlier in the market with a new product promises time of unchallenged sales, speed cannot be gained through a little less accuracy. The already low hit rate in new food product development (NFPD) cannot be further lowered, since the costs associated with late failure are simply too high. Failure needs to be recognized early in the innovation process.

Innovation requires a continuous cross functional management. Product tests need to be done, manufacturing options must be scaled, research collaborations started, new skills acquired, and necessary funds justified and secured. Full involvement and gatekeeping by company leads is necessary, as these are the "big bets". The older, heavy complex and sometimes slow. Innovation project management (IPM) gatekeeping process can be managing the core. If a company has a truly big innovation product or process idea, a structured, slow and pretty controlled project development can be the right choice to minimize risks and manage the idea to launch.

The traditional top down, command-and-control structure cascading from leadership to middle management to the front line does not fit. Instead, a network of cross functional teams, able to advance more autonomously to meet their project goals promises greater success. Here, leaders become coaches.

All other, smaller and perhaps more speculative ideas, still being an opportunity too good to let go, require more agile operating running at lower risk. Moving fast and deciding quickly increases the risk of failure. But a higher risk-taking level is necessary, and so the companies are looking for early indicators/milestones to "fail fast and learn quickly".

The top-line estimates about food innovations that failed are estimated between 66%–86%. And the list of reasons for this remarkable high number is substantial: lack of a clear innovation strategy, poor understanding of customer needs, inappropriate resource allocation, a non-supportive organizational culture, poor execution, ignoring competition, resistance to change, misaligned incentives, lack of preparation, misunderstanding the target market, prioritizing of technology over solution provided, bureaucracy, lack of problem identification, lack of ideation or inability to generate good ideas, stuff without profit and making profit without scaling (interObservers, n.d.).

This impressive list is sharp on the "what not to do": Avoid an unclear innovation strategy! Do not misunderstand your consumers! Provide the right and enough resources!

But to move from failure to success, the following questions require an answer. How do we know, we have a competitive strategy? How do we produce what consumers want? How do we calculate the right resources, without wasting them?

These questions can be difficult to answer., but an unobstructed view, good observation skills and intrinsically flexible processes are a fundament of professional experience and will be helpful in answering them. A broad spectrum of resources and the ability to convert observation into NFPD are also important. The huge majority of NFPD is not a breakthrough but is based on earlier launches the company made. That already is a very valuable source of market understanding, the unmet needs of the consumer and the price they are willing to pay for it. And since the enterprise has done it before, there is enough confidence in how to realize the idea.

New product development builds on experiences and learnings from the past and sometimes also from failure in the past. Management advice like "learn from your mistakes", "fail fast" or "encourage sharing mistakes" are popular and not brand new. Albert Einstein once said, "Anyone who has never made a mistake has never tried anything new." But, promoting innovation projects while decreasing the likelihood of their failure remains a daunting task for both researchers and managers (Astebro and Michela, 2005; Henard and Szymanski, 2001). Learning from past innovation mistakes and setbacks is among the most promising ways to deal with this issue.

Management literature investigated the "learning from failure topic" widely, and in 2021 Rhaiem et al. performed a systematic review of the published studies and generalized the findings (Rhaiem & Amara, 2019). For those hoping for a quick "checklist", the outcome may have been less favourable. The authors draw the picture of an "*integrative framework that brings together various conceptualizations of this phenomenon (as dependent, independent, and mediating variables), as well as the determinants driving or hampering it*".

Making mistakes and realizing them as such is no prerequisite for future success! And they do not directly lead to new, better product ideas. They lead more to better insights that can then lead to a better idea. Heinz's unsuccessful green ketchup has probably sharpened the company's understanding of consumer preferences and the role of product appearance a lot. And Colgate´s frozen meals will have told them a lot about their brand identity and the importance of expertise areas.

Sometimes failure is just the result of unforeseen circumstances, bad timing, a lack of value-added offerings, poor quality, a change in strategy, scope creep or too little consumer focus. But, as we should not forget, a suboptimal project management will cause project failure.

The easiest way to find those insights is a "postmortem" analysis. In contrast to the post launch review, this is done on projects that have been stopped. There is no longer the intention to "repair" the performance of a product in the market, but to frankly analyse why the idea failed. The observations and conclusions should be documented, analysed and archived.

Today's world is sometimes described using the acronym "VUCA":

- *V* = *Volatility*: Different types and dynamics of change, fast and unexpected
- *U* = *Uncertainty*: Poor predictability and the expectation of surprise
- *C* = *Complexity*: Multiple forces and many confounding issues and unclear cause/effect chain relations
- *A* = *Ambiguity*: A fuzzy reality, potential misreads and mixed meanings (Wikipedia Contributors, 2024)

The acronym appeared first in leadership studies in the late 1980s and was picked up by the U.S. Army to describe the situation in the world where the archenemy, the Soviet Union, was collapsing. For management leadership theories, a pure "phenotypical description" was not sufficient. They needed a root cause for systemic and behavioural failure and then suggested mechanisms to manage it. This needs an understanding of how people look at a situation when they make decisions, accompanied by suggestions to make it better:

- Anticipate the problems shaping the situation
- Understand the consequences of problems and actions
- Realize the interdependence of variables
- Prepare for different realities and new challenges
- Interpret and address key opportunities

More details can be found in number of books on the topic, for example in Mack et al., 2016. The food business is VUCA, safety regulations change quickly, consumer preferences shift immediately and dietary preferences and lifestyle choices transform and split. The whole food supply chain is often a complex network involving numerous stakeholders, from farm to fork. And will big data and artificial intelligence (AI) be able to make confident decisions, despite the ambiguity in the food industry? Will new technologies like genetically modified organisms (GMOs) or lab-grown meat be accepted by consumers and regulated constructively by authorities?

So, no magic stick! But many learnings about what to watch for, what to do and what to avoid.

While innovations are the life blood of the business, the required innovation in food depends on a number of factors: The "product life cycle" describes the time from the moment the product is introduced into the market until it's removed from the shelves. In other words, it's the moment when the decision is made to stop advertising the existing product, reducing the price, putting it into new channels or refreshing the packaging, and instead launch a successor.

CLASSICAL VALUE CREATION MODELS UNDER THREAT

For many years the Food industry has successfully created global brands, which delivered on average nearly 15% return to the shareholders.

The underlying value creation model that has been used since World War II in the 1940s builds on a couple of simple principles. The classical "brand builder" started in the Western markets, and they used continuous NFPD for sustainable mass market growth. Their product margins were better than the ones of non-branded market players, and a firm relation between brand owner and retail business ensured the efficient detection of new consumer wants. The industrial key skill of "continuously improving organisational efficiency" delivered cost reduction, which created the funds for further growth. Years later, when the Western markets became more saturated, the companies stretched their business into the developing economies, often through mergers and acquisitions to build their local market presence, followed by another round of reorganization to improve efficiency and economies of scale.

This technical implementation strategy ensured attractive prices and sufficient profit. But when the consumer started to demand more personalized solutions, "mass customization" was needed. This approach had to combine economies of scale and the ability to provide personalized offers. The solutions came through a good interaction between the branded food makers and their ingredient suppliers. Variety had to come in a way that the efficiency of the existing manufacturing processes was not too much affected. New production lines are a significant investment and need a long time before they reach an efficiency level, allowing a competitive consumer price at a good margin. More extensive "manufacturing" was only possible where labour was cheap or the product could carry a price premium.

The smart answer of the food ingredient industry was to provide variety, creating ingredients that could be added without major line changes. New robust diversification in colour, taste, aroma, texture and nutrient profile at a small oncost became feasible. Personalization could be "plugged in" easily.

These new solutions often used the potential of synthetic chemistry, since subtle natural extraction processes would be too costly and are often not sustainable. These man-made solutions were rigorously tested for safety in consumption and certified by the relevant authorities. The final discussion with the regulators led to a settlement where names on the label had to be scientifically correct, but not so chemical that consumers would stop buying them, something that could damage their trusted brands.

Today, the picture is changing! Younger consumers look differently at big mass brands, companies and channels. They favour new smaller and authentic brands and trust less the promises made by big food brands.

Generations Y and Z, the Millennials and digital natives span the birth years from 1982–2010 and nowadays represent already 50% of global consumers. They are more selective in their buying behaviour. When they are convinced about the offerings, they are willing to pay extra, not all the time, but for items important to them. The classical marketing campaigns do not reach them. They get their stories and perceptions from new channels and their personal environment. As digital natives, they are used to getting information through the net and all the social media there. New fresh, smaller brands seem more innovative and communicative. They represent a "plant-forward" diet, whole, organic or natural foods, that gives them solutions to cope with the challenges of today's busy and unpredictable world, fighting fatigue, providing energy, improved sleep and emotional, mental or gut health benefits (Food Insight, 2023).

They still like comfort and quality with a focus on affordability. They look at food as part of their lifestyle, expecting effects on their health and even their identities. Millennials love to snack, as some of them graze instead of eating big meals a few times a day. So, they want more portable snack foods that are resealable and easy to open.

They are not willing to compromise care for people and the planet for the convenience they need. Local food is an attractive solution. It seems familiar, has a visible origin and is not coming from a big global player through untransparent processes.

This authenticity can conflict with the image of global brands. Big is no longer beautiful! They want to know what really stands behind a brand and are committed to authenticity and naturalness.

So, the big brands are losing share. Their reputation is burdened by perceptions that they add sugar, salt, cheap fat and artificial ingredients to keep their products prices competitive and seduce consumers to buy more. New younger enterprises came in with perceived healthier products, often combined with a personal statement from the entrepreneurs about their convictions. These new brands build a sense of intimacy, making then more relatable, so shoppers feel they know them, their story and the people behind them. For their interaction with the consumers, they no longer focus on the linear media. They are professionals in social media and are able to create a "tribe" of loyal followers, sharing their convictions.

In the beginning those new brands may have sold mainly through the internet; they were more a stitch for the growth ambition of the big brands. But today, the modern trade has recognized their potential and is very willing to give them more than their fair share on the shelf, a more critical challenge for the big brands.

Once again, a strict focus on cost reduction doesn't seem to be the answer. In past years, models like zero-based budgeting were introduced to ensure better expenditure transparency and spending discipline than annual budgeting. Marketing became challenged to prove much more the value of their campaigns and were often judged inefficient. And still, these initiatives were able to boost only a short-term increase in profit, but no longer-term good organic growth.

An alternative for big brands is to acquire smaller brands. Financially, this is often no major problem, but it doesn't simply solve the problem with the brand image. Buyers of small challenger brands notice the change of ownership and can react sceptically: For them, it's less about the possibility to scale and nurture the challenger brand, but an undermining of the original values or just another trick to make money.

And this doesn't help to maintain the value of the big old food brands. Worth billions of dollars, these companies have spent millions of dollars over the years to build a strong brand image. Now they need to rebuild the core, recharging purpose, authenticity, consistency and transparency.

The "old" value creation model was frequently also built on the globalization of suppliers. Global food producers have made a lot of effort to qualify new ingredient suppliers from new corners of the world. Especially China was very successful in growing crops according to international trade standards at a scale that made the country extremely attractive for ingredient buyers. Today the country is not only by far the biggest producer of vegetables globally but is as well the biggest exporter (Vegetable Exports by Country US$000 2021 | WITS Data, n.d.). The value of this globalization has suffered. In light of the trust in local food systems, the environmental benefits of short supply chains and, last but not least, the sudden breakdown of the global distribution chains during the Covid pandemic have convinced many food makers to shift more to local and regional sourcing. And policymakers also urge them to build more resilience into the food system.

Consumers have a good understanding of where authentic farmed foods are coming from. Strawberries in January from Mexico, Egypt or the southern United States may look terrific but seldom taste and smell great. Authentic crops also have a season. If they are freshly available all year round, this freshness is not authentic.

These are all reasons, why it's now so difficult for established global food brands to achieve the top line growth their investors are expecting, and so they shift their investments more towards the new brands and players.

As one result, food companies have started to shift their value creation model. They are looking for better strategies to manage their product and brand portfolio and constantly search for new operating models.

Nestlé, the biggest foods company in the world, owns more than 2000 brands, out of which 31 had sales of over 1 billion euros in 2022 (Brands, n.d.). Similarly, Unilever, with about 50% of their turnover in foods and refreshment, holds 400 brands, of which 14 have sales of over 1 billion euros (Unilever, 2023). It is similar for other big players. Focus is needed; not all brands can be kept. The most valuable brand assets are used for mass brand execution. They use their consumer insights experience, and their ability to enter new markets. They are the core of the global business, and in a way "too big to fail". And they must be constantly kept healthy. But the smaller brands or those without a good growth potential may then face "disinvest" or a "milking strategy", so the net extraction of the money earned. The profit made their will be reinvested into more promising initiatives.

Secondly, brands must regain their reputation with the new consumer generation. Promising "care for the planet" is not only a non-negotiable, but a possibility to recharge the big brands with new, contemporary value. This must happen credibly; just claiming it too loudly can backfire. The public is sensitive to "greenwashing", where the degree of environmental caretaking is exaggerated.

And the communication needs to change. The classical approach is to identify the conditions needed to make a "new claim", doing the work, even if it requires years, and then claiming it on the package. You can't do that for "environment friendly" or "100% sustainable", and not because it's so time consuming and costly! It's because the target can never be finally met. What is perhaps superior sustainable practise today is standard tomorrow and inferior the day after. It's a commitment for continuous improvement. Accordingly, there is an opportunity to talk about the progress being made and the steps to follow. This isn't a catchy claim, but it is an opportunity to continuously talk about the ongoing effort. And anyway, who would believe that, for example, the ecological transformation of multibillion dollar businesses would just happen like the annual balance statement? Isn't it more credible to report about the ongoing progress against the ambitious targets? Telling the story and giving the effort a face is more tangible than the reporting of a particular milestone now met.

GLOCAL

The signifier "Glocal food" intertwines local and global aspects. This is not a one-dimensional approach but an attempt to use the best of both worlds, origin of food, its production, processing, distribution, and consumption. There is value in the local food sources and their traditions, the wish to strengthen the regional economy and to reduce the environmental impact. But there are also global opportunities in bringing dietary diversity, innovation and opportunities for new foods and taste experiences. This is not simply a response to the strong disruption of the global food chains in the last years but reflects the desire for a more meaningful, less economic relationship between locally produced and globally imported foods.

Best known are adaptations of a traditional local food item or dish, to the preferences of new consumers. Most of the tourists visiting Thailand every year can be happy that they do not get the famous signature dishes served with the typical Thai chili pungency. And flagship food products like sushi, pizza and curry, originated, of course, in Japan, Italy and India, respectively, but have been modified for the local taste in many countries. Something like Hawaiian pizza is not a favourite in Italy.

And while perhaps the Mongols were the first to "export" minced meat consumption on their way to Europe, the hamburger is an American invention and so successful that it's difficult to count the number of burgers eaten per day on the planet. But it is by far not the same burger everywhere. The dish has a common chassis: Mostly a flat round cake of minced meat, fried or grilled and served in a split bun or roll with various condiments and toppings. What makes the burger glocal are the adaptations made to local consumer preferences. The Indian style burger uses, unlike American-style beef, lamb, and chicken, a vegetable patty together with traditional Indian spices, like ginger and garlic pastes, coriander, cumin, and garam masala, and is served with naan bread and a cooling raita sauce (The Spruce Eats, 2022). In Indonesia, the majority of the population is Muslim; therefore, international quick chain restaurants replace pork with fish, which need to be halal certified, perhaps with the exception of in Bali. In Israel, burgers must be kosher. Japan has the teriyaki burger, and Hungary has the Roeztiburger. Sometimes quick service restaurants (QSRs) also modify the chassis and replace the bun with flatbread (Turkey), pita bread (Morocco, Portugal), toast (Sweden), rice patties (Philippines) or just rice (Indonesia). But all of those are adaptations of a very successful global food product design.

The burger is no longer an export of the American way of life, but the modern quick serve finger food version of local top dishes. The chassis is here not only a common design and manufacturing principle, but a way to incorporate local cuisine into a unique format.

Even Coca-Cola, perhaps the most iconic representative of the American way of life, and globally made from a secret syrup formulation, is by far not the same everywhere. Bottlers all over the world add different types and amounts of sugars or sweeteners and carbonate differently. Estimates that the company may have in total more than 1000 different formulations may be a little high but show the level of customization needed for success globally (Pendergrast, 2013).

The local variants are the result of thorough market research and an understanding of the needs, preferences and expectations of the customers. These products highlight one aspect of the glocal principle: A common chassis type of product design, made by standardized manufacturing combined with an in-depth understanding of local preferences, often recognized through contemporary research methods like preference mapping.

Other approaches are less visible. Food ingredients, described by harmonized global specifications, and made by similar, preferably identical manufacturing processes can simplify the production of various food products enormously. The advantages are numerous. For an international operation, the minimization of the number of ingredients needed simplifies stock management and creates buying leverage. Cutting complexity enables massive cost savings through less waste, harmonized storage, better utilization, less-complex quality controls, less documentation, and overall, less management time needed, etc.

Ingredient harmonization leads to higher buying volumes, which often brings better prices and better service. The dispersing ingredients supplier market reacts to an increasing demand for globally harmonized ingredients with consolidation, mergers and acquisitions, leading to new economies of scale and lower costs.

And an international active food producer can shift materials from one country or factory to the next, in case the manufacturing is stopped somewhere, or the product is delisted. There is no longer a need to destroy preordered material or resell at a low price before the shelf life ends.

The rollouts of successful products into new markets are easier and faster. There is less need to localize the product specification, finding new suitable suppliers. This is helpful to manage third-party manufacturing. Launching products in a new market can be done by using toll manufacturers. This avoids heavy and risky upfront investment into their own new lines and factories. By sending their own harmonized materials, the product quality standards are easier to control for a branded goods maker. And it is a good way to protect proprietary knowledge, in situations where a patent protection is not possible. By sending premix, a company can avoid giving the toll supplier the little tricks and details that make this mix competitive. Coca-Cola's "secret concentrate" and the thousands of independent bottlers and franchise partner are a clear example.

Obviously, there is a limit to such harmonization. Particularly farm-based ingredients are influenced by the local environment they have grown in: Different soil, different climate, many different other factors. An onion from Egypt is very different from an onion from Germany or China. And French basil and Thai basil are worlds apart. Particularly characterizing ingredients like herbs and spices often show a very local sensory profile. It is technically much easier with synthetically made additives. These products are the result of carefully controlled closed processes. The results are the same everywhere, in contrast to the farmed goods grown under an open atmosphere.

But synthetic or non-natural ingredients are under pressure. They need to be replaced by natural alternatives. So here, a new balance needs to be found between the standardization of "cost drivers" and more natural, authentic food products.

Finding the sweet spot between local, unique and global standardization is the key balancing act! Basically, each product format and manufacturing process needs to find the optimum between local character and global principle. An iconic Magnum ice cream deserves a higher level of standardization than a chilled ready meal. People like authentic local food but also international formats. The one thing that may have less potential in the future are local food products made from internationally sourced, globally harmonized ingredients. Value-giving ingredients can come from far away, but they need to have a reason, a story. Quinoa, the famous future grain crop from the Andes region, was produced only in a few countries until 1990. Nowadays, it is grown in at least 123 countries, including places like, Australia, Russia, Mongolia and Iceland (Alandia et al., 2020). But it has kept an authentic image.

SPEED

It does not require a crystal ball to state that new product development (NPD) and reformulation will remain the lifeblood for food and beverage manufacturers. We can expect even more and faster innovation cycles. To maintain a competitive edge, staying on the forefront of upcoming consumer trends, new market opportunities and new technical solutions, new products are needed faster. The key to success is to minimize the time from the moment where data and observation bring a new consumer insight to the moment of launch. This is not easy; the still increasing regulatory complexity often requires extra development effort, costing energy and time. Doing it like in the old days does not work well any longer. New tools are necessary to ensure speed in insights detection, prototyping, product formulation, regulatory approval production and distribution. More speed is needed but with full legal compliance, with superior quality, and delivered efficiently, be it over bumpy roads to remote places somewhere in in the countryside, via local trade or e-commerce.

Luckily, these tools exist. Big data and AI can do that. Earlier, we spoke about the options to use AI to shorten or avoid time consuming and costly factory trials (see Chapter 8). And since longer computer programs use the food composition and the origin of the material to form ingredient lists, check compliance with local laws and suggest claim content, product design and development workflows exist and they unlock faster innovation cycles.

Automated product formulation tools are software applications that help to create new products or improve existing ones. These tools use algorithms and data analysis to identify the optimal combination of ingredients, additives and other components that meet the desired product specifications. The formulation process can be done automatically, which reduces the time and cost associated with manual experimentation and trial-and-error methods. These tools use machine learning algorithms to predict material properties based on their chemical composition and structure. They also provide real-time feedback on the impact of changes to the formulation on product performance.

Today, these solutions work already for the simpler structured products like liquids and powders (CPI, 2024), enabling manufacturers to create new formulations, optimize existing ones and analyse the performance of materials. The tool uses machine learning algorithms to predict the properties of materials based on their chemical composition and structure. It also provides real-time feedback on the impact of changes to the formulation on product performance.

For prototyping, more and more professional facilities are opening, where research and development (R&D) people can quickly create a limited number of product prototypes for initial consumer testing (for example, see Service – Foodinnopolis, n.d.). Companies no longer need to invest in pilot plant equipment and trained staff and accept long lead times before being able to put meaningful prototypes in front of the consumer.

Big data sciences and AI allows, to a large extent, "*in silico* first" production trials, replacing laborious and costly factory trials. The implementation of all type of sensors and monitoring advice in the production line not only help to optimize the actual productions but deliver masses of data as the basis for various types of material and process predictions, which can feed into the modelling tools and improve them further.

But back to the core: The product formulation. Since a couple of years, automated and semi-automated product formulation tools have been on the market. Already presented in greater detail in this book is "SAP PLM", a broad life cycle management software product for development, quality, compliance and traceability processes. Other software products are "Food Solver", "Lascom2", "Batch Master", "ANFIS" or "ESHA Research".

These tools are not replacing human creativity, experience and intelligence, but support them. They can help in creating low-fat yogurt with a creamy texture and a popular flavour, a gluten-free bread with good volume and crumb structure or a vegan burger with balanced nutritional composition.

Besides the compositional and sensorial aspects, these tools identify cost drivers and optimize them or help to comply proactively and in the initial design phase with regulatory and labelling requirements. The R&D team can now decide in the beginning if they use the most cost-effective additive, a flavouring, emulsifier or colouring, allowed in one market or solutions that can be launched in more countries. They can easily simulate scenarios to test product compliance in multiple markets for the future, maximizing both and optimizing resources and market opportunities.

COMPLIANCE

The development of new products will not get simpler, and innovation cycles will accelerate! With more food and beverage (F&B) brands expanding globally, more compliance hurdles need to be successfully managed. Since the costs of even an "average" food product recall is estimated to be between $10–$100 million, spanning from notification, product retrieval, storage, replacement, and product destruction, it's critical to avoid such an event (foodnavigator-usa.com, 2023).

The chance of smoother compliance processes can be improved. A product developer may make a decision to add a slightly cheaper flavouring or colouring agent into a formulation that would still allow for a clean label or natural claim in the first launch market, but not in the ones to follow. This is good if the R&D team can easily see how a change in the composition influences the claim in various markets. Later when the product enters the next markets, the wider compliance check prevents costly late-stage redesigns due to uncovered compliance issues that were discovered too late. Today, these systems are more and more data driven and AI powered, building more resilient innovation cycles influencing design practices.

Modern *in silico* and data systems assist companies in adhering to legal requirements and industry standards, down to the certifications recognized by the Global Food Safety Initiative. They validate the safety and quality of food products, including attractive consumer claims like organic or non-GMO. Some of them are also able to validate non-unambiguous claim options, like "natural" or "clean label". Automated compliance checks liberate valuable resources from "non-value-added work", such as searching for regulations, gathering data, finding exemptions and discussing compliance.

The fast product development is supported by alerts about ingredient risks, upcoming regulation and an effective formulation management. Enabling the formulators to judge product compliance reduces the time needed to "repair" formulations for regulatory or supply reasons in a back-and-forth process between regulatory and product R&D, a clear step closer to the ambition of "design once, deploy everywhere".

Ingredient supply chains today are often global, which not only offers cost advantages or supply security but also the opposite. A way to flag potential ingredient risk early can save a lot of time in late product design phases to "firefight" late-emerging issues related to supplier conformity, geopolitical events or climate conditions

The implementation of more efficient development systems in a food company to keep pace with the markets is often a cultural change: Instead of the traditional, rather intuitive- and creative kitchen product development with prototyping and step by step upscaling, now fast cycles of product testing and redesign are applied, where the usage of big data sets, complex computer simulations and AI predictions become the center of the development process. This requires substantial effort into data quality (cleansing) and people training. Doing that later for new "claim and regulatory compliance systems" means doing it again. Therefore, companies sometimes decide to implement digital end-to-end (E2E) solutions combining product life cycle management (PLM), enterprise resource planning (ERP) and compliance tools in one go. Even so, that may bring a transition period of a year or more, where both processes, the old way of doing NPD and the future one, are co-existing. The need for retraining and change management is smaller.

The new consumer trends, health and well-being, clean label, naturalness and authenticity, instant availability, transparency, global-local (glocal) and plant based, give space for many new innovations in the food and beverage universe.

New technologies, like alternative protein, high-pressure processing, isochoric freezing, nutraceutical and personalized nutrition, provide the chance for superior solutions for existing needs. A better sodium, fat or sugar replacer or a new natural additive may need fast reformulation in many products to remain competitive, or a new tax or regulation may make reformulation necessary.

So, to win, future food companies have no choice. Their new product development processes need to become faster, more successful, less resource consuming and also more sustainable.

"FUTURE FIT" FOOD

The global consumer wants for future food are relevant in most F&B markets, and they are the guidance for the work of food developers. The remarkable number of opportunities science and technology offer today, open several opportunities for truly differentiating products. And the disturbing news about the environmental conditions impact consumers, science and the regulations.

What started in the 1970s with a lonely caller in the desert is today a concern for most people on this planet (Meadows et al., 1972). But now science can do more than describe the problem and prognosticate the trends. It can offer solutions. And food regulation is getting stricter on presumed health or environmental aspects and can use new taxation to cut the usage of critical food constituents like sugar, salt, fat, trans fats and additives. More and more an "avoiding the baddies" is an essential first step for food manufacturers, opening the door to "include the goodies". And the concept of food for better personal health and well-being is no longer mostly reactive and preventive but progressively more proactive or even curing. The knowledge about the microbiome has increased massively. And despite the enormous complexity of the situation, modern technologies have started to offer more individual solutions. The gut microbiome is no longer a population of potentially 3600 different species, but a transient selection of a few hundred in a single human gut (Leviatan et al., 2022). Modern physical, chemical and biological methods allow fast and accurate identification of those species. And many interactions between those species and the processes in the human body have been detected. In its totality, it is far from complete. But all the new "omics" potentiate our possibilities to understand more and different interactions. Big data systems and AI will help to uncover new and important correlations.

So, it is a good time to make new and differentiating food products. The scientific findings and consumer research will not straightaway deliver the next new food product specification. But the trends are clear enough to allow predictions of what and how this will influence the next generation of food products.

Plant Based and Flexitarian

Plant-based proteins will continue to replace animal-based ingredients! Today's alternatives are often already sensorially convincing, and it's likely that they will get even better. Animal protein like meat, egg and dairy will not disappear, but more people in meat-centred cultures will return to a more conscious animal product consumption. That can be seen as a return to typical meals of the older generation, when meat on the table was an event in the weekly diet, not the unconscious daily norm. The ethical concerns and the environmental facts will make meat what it was before the mass industrialization: A premium piece of food, consciously prepared and eaten with attention in the more special moments of the week. Something special to enjoy and understood as a luxurious way to create health calories and good nutrition. Quality beats quantity. The origin of the animal-based food will be value giving. *"More Wagyu - less Schnitzel!"* The future might be more flexitarian than vegetarian! Reported numbers differ, but some 40% and more of the respondents globally say they are flexitarian, while only 5%–7% claim to be vegetarian. The culinary shift in our meat-based food cultures pushes meat out of the "centre of the plate". Less and less will the next meal choice be framed by the meat: "*I have chicken breast, meatballs, sausage or steak. What should I take with it?*" And where the meat is literally centre of plate, like in burgers, it will be replaced with plant-based alternatives. Right now, QSRs are starting to offer plant alternatives cheaper than the meat variants (Dodhiya, 2021). More people will be willing to try alternatives and, when sensorially satisfied, will switch.

Today, people avoid meat also for ethical reasons. They are concerned about animals' welfare and the environmental impact of large feedstock farming. A more conscious meat consumption would give consumers the possibility to pay a little more for their meat; that's money the sector can invest in the animal husbandry while reducing the size of their operations. This reverts the industry trend to maintain profitability through a continuous striving for lower costs, something deeply ingrained in the food sector (see the biggest trick of all in Chapter 2). It allows for competition for the most effective ways to improve farming conditions.

Minced meat replacements, sausages and chicken tenders often already match the meat original, and other plant-based compositions can replace animal-based ones in day-to-day spreads, premium ice creams, plant-based yoghurts or non-dairy cheeses. Plant-based alternatives are offered even for scrambled eggs.

It doesn't seem that the technologies behind plant-based solutions have reached their full potential yet. In two areas, better solutions can be expected:

1. The first area is the "making" of all these plant proteins, from farming up to isolation and purification of the food grade material. New varieties, better breeds, softer processing, more resourceful and locally grown plants will lower the environmental burden. The shift from

traditional low-moisture meat analogues to high-moisture alternatives and combinations of different plant proteins is one option (Lee et al., 2023).
2. The second area with massive space for improvement are all the "additives" needed in meat- or dairy-based analogues. Today's formulations often have a long list of non-natural functional actives. Science is focussing on a deeper understanding of ways to "convert" the mostly globular plant proteins into the often-elongated muscle structure. That will reduce the need for additives imitating the effect. And the strong demand for natural colours, flavours, buffers, preservatives and linking agents (stabilizers, emulsifiers, etc.) will also help to provide "cleaner" solutions.

The innovation potential for plant-based products exists in many corners of the food universe, some more evolutionary, others more radical. A sensorially matching steak alternative, a burger patty made from ecofriendly protein extraction, a good nutrient profile, less additives and allergen-free will move the needle further. And shifting the communication towards a more mass-market consumer-centric approach creates additional momentum. Labelling alternatives "vegan" or "vegetarian" in bold front of pack can limit sales, as not everyone feels their wants are addressed by that. And the shift to plant based should not mainly be an opportunity for the big multinationals to achieve their committed environmental agenda. The benefits are not mainly business to business (B2B) but business to consumer (B2C) (Taking-Alternative-Protein-Trends-Mainstream, 2023).

Vegan, Plant Based (or Healthy)?

Technically, the difference sounds evident: "Vegan" means the exclusion of all animal products, including all steps in the manufacturing and development of such products. And "plant based" focuses on just the ingredients present in the final product. Some debate still exists in how much animal-derived products such as cheese and honey, or yeasts and mushrooms (scientifically not part of the plant kingdom), can be part of a vegan lifestyle. The vegan movement started after the World War II as an ethical movement, aimed to avoid any harm to animals (Piccoli et al., 2014). The broadening towards a plant-based diet came centuries later, at the begin of the 1980s. That brought new arguments, and people were joining the movement not only for the sake of chicken and their souls, but because they wanted to ensure their own health.

Today, both the ethical arguments and a wide array of scientific evidence see an animal-centric diet as a danger for the human future on this planet. The nutritional advice supports the shift to plant-based food.

Today, the dietary advice for strict vegetarians is to watch their diets for sufficient intake of vitamin B12, zinc, calcium, iron, vitamin D and omega fatty acids. The question is less, "Is a vegan diet healthy?", and more, "How does one eat a healthy vegan diet?" That's where enriched and fortified new food products can provide help and where technology can make the food better (NHS Website, 2023).

For a sustainable living, such dogmatism is not needed. Consciously cutting the consumption of animal products is an alternative. In the 2010s, the "flexitarian movement" became popular, aiming to educate and influence people about the benefits of eating plant-based foods and alternative meat products, while eating high welfare meat only in moderation (Grant, 2020). Today between one-third to one-half of consumers in Western societies describe themselves as flexitarians (Malek & Umberger, 2021).

Three product development trends can be therefore proposed.

1. *Superior enriched vegan products.* Fortification is a very efficient technology to compensate for shortfalls in nutrients. But it's an industrial intervention and therefore regularly perceived as not natural. Products with superior bioavailability or bioaccessibility of nutrients create an argument to overcome such reluctance.
2. *Biofortification*: This is an effective alternative to regular fortification. Here an organism, mostly plants, are enriched through selective breeding. This can happen effectively through

targeted genetic modification, triggered mutagenesis or classical cross breeding. All of this is an intended technical intervention, but the public acceptance differs significantly.

3. *Plant-based food.* If one of the missions for the food industry is to help people shifting to a predominantly plant-based diet, the question of how to communicate and market such products is important. The team of Patrycja Sleboda reported an interesting experiment at the end of 2023. They offered gourmet baskets to more than 7000 consumers in the United States that were labelled “vegan”, “plant-based,” “healthy,” “sustainable” or “healthy and sustainable” (Sleboda et al., 2024). What they found, was that “vegan” was the least attractive claim, followed by “plant based”, and both were beaten by “healthy” and “sustainable”.

So, it may be worth discussing in innovation teams if the pure vegan proposition has the best market potential. There is a defined market segment for those consumers driven by absolute principles. Finding alternatives for whey, casein, lactose, gelatine, natural flavourings, etc., can drive formation costs and lower product performance, while being acceptable for the big group of people wanting to move away from animal-based food, without being dogmatic.

Known Heroes

It is an all-time truth that we eat to supply our body with nutrients. And macronutrients like fat, carbohydrate and protein are the main building blocks and energy providers our body needs. When food, particularly industrial food, became more and more abundant for more people, eating became much more than a fundamental necessity.

Shortly after the World War II, the American physiologist Ancel Keys began to systematically study the role of nutrition for health. In a sense, he was one of the first scientists recognizing that something like atherosclerosis was not an inevitable consequence of aging and that a high-fat diet could be a major risk factor for coronary heart disease. Based on data from a prospective, observational heart study in the town of Framingham, he published in 1957 data showing that high blood pressure, obesity and hypercholesterolemia were associated with coronary heart disease (CHD). And 6 years later, he established a correlation between cholesterol levels and high systolic blood pressure as a risk factor for CHD (Montani, 2021).

Now it became nutritional consent that illness and premature death are not the unavoidable path coming with age but that this is something that is influenced by what we eat. Fat became the first enemy, especially the saturated fats. The others, the unsaturated fats, were identified as healthy. And the appropriate diet with lots of plant foods, essential nutrients, fibre and unsaturated fats had already been “invented” long ago: The Mediterranean kitchen!

In the food industry, the health concerns around saturated fats were the opportunity for the margarine producers to take revenge for the century-long discrimination of their products in the dairy states of the nations. Through the 1970s and 1980s, the “fat battle” between butter with saturated fats and margarine containing unsaturated ones dominated the nutrition and health discussions. The debate is not settled yet!

But slowly the expert interest shifted to a new nutrient group: The carbohydrates and their health impact! Science needed a few years to establish first –statistically robust – correlations between carbohydrate intake and longer-term health. Quickly, carbs took centre stage in the nutritional debate about good and bad food and consequently found reflection in new food product development. So, they became the hero ingredient in NPD, and their usage peaked a couple of years back.

Then, the scientific view on this large nutrient group became more differentiated. Clear evidence came up that too much of simple carbs, like regular sugar, are correlated with the risk of diabetes, obesity, heart disease, tooth decay and cognitive functions and recently also various types of cancers (Dewi & Diana, 2021).

The times where fat was the hero for taste and energy delivery are long gone. But the healthy, natural, mono- and polyunsaturated fats of the Mediterranean diet retain the health status they got since

Ancel Keys' times. And even the old pariahs, the saturated fats, have regained a reputation as part of a heathy diet.

Meanwhile, the whole family of carbohydrates lost credibility. The short and simple members bring great taste, can do wonderful things as additives in replacing fat in formulation and cutting calories. But the receipt for such short-term pleasure is the risk of a longer-term health impact. Food processing was made responsible for the simple carbohydrate content. It converted the not so delicious longer-chain carbs into appetizing short ones. Products with too many carbs and especially too many processed, simply carbs were no longer healthy.

A good level of dietary fibres in new food products became the next trend in nutrition, and it has not reached full potential yet. Not digestible by the human body, soluble or insoluble fibres offer endless opportunities to create new tasty food products. More importantly, there is mounting evidence that representatives of this chemical class influence energy metabolism, cardiovascular health, immunity, gut and brain health.

And here the processing can also help. With smart and sophisticated food processing, different types of fibres can be extracted from a huge variety of fruit and vegetable waste (Teshome et al., 2023). These resistant starches not only provide proven health benefits, but also allows a wide range of technological functions for the food industry. One of the latest is the applications of resistant starches in gluten-free products (Nova Science Publishers, 2021).

More Protein?

Today, the last of the three macronutrient groups, the proteins are enjoying huge popularity, less as the natural "goody" component of meat, fish and dairy, but now preferably as a plant-based building block. The new superfood is "high protein", giving "more power", "inner energy", etc. The extra boost is available in basically all product types, desserts, yoghurts, pizza, bars, drinks, beer, etc. On a purely nutritional basis, this extra is not needed. From a nutritional standpoint. people in developed and developing economies are not in danger of protein or amino acid shortage. It sometimes seems forgotten that the body benefits from a diverse diet and not striving for the maximization of protein intake. And while nutrition science is still discussing if 10% or 20% of energy intake, or 50 or 60 or more grams of protein, or 0.83 g per kilogram of bodyweight are a good value (Harvard Health, 2020), many people in more prosperous countries are already getting much more protein than needed. The average balanced diet in most cuisines provides more than the normally recommended 0.8–1.8 g protein/kg bodyweight.

And a bit of a differentiation discussion has also happened. The differentiation is not straight into "good" and "bad" proteins but into those with essential amino acids and those with more non-essential ones. Also, here the old wisdom applies: "Too much of it is not good!" Weight gain, digestive problems, heat problems, maybe even kidney damage or cancer are possible.

For the consumer, the extra addition of protein in many normally low protein food types can provide a good feeling, rather than a rational nutritional argument. The so-called healthy cheat moment gives permission for an extra portion of indulgent food items, like pudding, cakes and bars, since now they have less fats or sugar and more protein, and all of that without affecting the pleasure. This has been thanks to the skills of food developers who were able to maintain the full taste, texture, look and aroma as the "unhealthy stuff". The addition of proteins in many popular food formats might swing and become perceived as constructed, designed, artificial, so industrial. High-protein balls, snack bars, protein drinks and chips are not very natural!

Today the oncosts for some extra protein, especially from plant-based foods, are insignificant but provide the opportunity to sell products at a premium. Protein-rich snacks can fit the indulgence moments in the office, the reward after a job is done or after a long, hard day. Or the emotional "reconnect" moment when the always hungry kids are back from school and can't wait for the next decent family meal!

These protein-enriched snacks give a "better feeling", and they provide the permission to snack. But they are designed and engineered, so in a sense they are ultra-processed food (UPF).

In that sense, protein enrichments do not smoothly fit into a cleaner, simpler, more authentic and transparent NPD.

Sugar and Carbs

As said before, sugar is an "old friend" of humans that turned into an "enemy"! Long before the first humans, evolution created receptors to sense sweet and bitter taste, essential for survival. Bitter warned about poison; sweet signalled instant energy! Simple sugars entered the bloodstream quickly, triggering insulin release to activate sugar uptake into cells needing that energy, like muscle cells. This was a fantastic response mechanism in times when sufficient energy was the survival insurance. Enough energy meant vitality. Still today, many cultures and religions use sugar in celebrations, rituals and as comfort foods, providing instant energy, enhancing flavour and preserving food. Religion protects "Eid al-Fitr", the sugar feast, from critical nutritional review. But no doubt, today too many people have easy access to too much energy, and this sugar abundance is raising serious health concerns including weight gain, tooth decay, diabetes and heart disease. The problem with the receptor is that the "signal" from the taste buds has built a "mental percept" in our brain, an expectation linked with positive emotion and that is endorsing repetition.

Not long ago, sugar was the "hero ingredient" in many products and proud name giver of famous snack and energy-providing brands. Today, brands like "Dextro" just serve in "niche moments", when the brain and body are getting short of instant energy in moments of extreme effort and in conjunction with sports nutrition.

For the rest, sugar intake, too much and too often, derails the "instant energy system". The sugar in the bloodstream is no longer adequately taken up by the cells, more insulin is released, and after a while sugar levels in the blood are increasing, damaging tissue and organs. Hippocrates statement that "the dose makes the poison", applies for all nutrients. And high dietary sugar consumption is harmful (Huang et al., 2023).

Chemically, sugars are the "hydrates of carbon". The "simplest" ones are mono- and disaccharide. They generate those sweet taste signals that could direct the brain of our ancestors to the instant-energy sources. The longer-chain molecules are the oligo- or polysaccharides. They contain proportionally similar energy levels, but the body needs more time to break them down into the single units that enter the bloodstream. This leads to a slower and more gradual rise of blood glucose levels and a less rapid insulin response.

"Complex carbohydrates" are healthier, but the term is used inconsistently in science. Chemists call all molecules with sugar units "carbohydrates"; nutritionists tend to use the term only for the longer-chain molecules, as they focus more on digestibility.

An appropriate way to differentiate the digestibility of carbohydrates is the glycaemic index (GI) of food. Technically, that is the incremental area under the 2-hour glucose response curve after consumption of 50 g of carbohydrate. Foods with a GI between 100 and 55 are seen as high GI, and those with a GI below 55 are seen as low GI. An even better value is the "glycaemic load (GL)", the multiplication of GI with the grams of available carbohydrate in the food, divided by 100. This was first introduced in the 1980s (Jenkins et al., 1981).

For many years, these measures shaped the diabetes research and led to a series of popular diet programs, like Montignac, Glyx and LOGI.

With more studies coming out, the pure focus on blood sugar and GI needed refinement. Diabetologists knew that the insulin level is a key variable in energy metabolism. And while blood sugar and insulin levels are correlated, other food constituents are as well. Low-insulin-stimulating foods and low glycaemic foods are not the same. Some dairy products indeed have a low GI but a high insulin response, while some fruits have a high GI but a low insulin response. The conclusion was that measurements are needed that reflect insulin levels, not just blood sugar. At the same time, insulin prevents the release of stored fatty acids from adipose tissue, thereby blocking fat breakdown.

And other non-nutritive factors, like physical activity, medication, infection, stress, sleep deprivation, microbiome constellation or other hormonal changes, influence insulin levels. These "lifestyle factors" and their influence on insulin levels became predictors of increased diabetes risk and obesity (Kolb et al., 2018).

With today more than 2 billion people on this planet being overweight or obese and the numbers still rising, new non-pharmacological and dietary interventions are urgently needed. Simply telling people what to eat and what not to eat often doesn't work. Particularly older people find it hard to change their diet and simply give up food they love. More promising are approaches that fit into the daily diet, maintaining pleasure with all the emotional-and social aspects connected to it. But for that, the new food must be perceived as real day-to-day food and not a temporary substitute in a dietary intervention.

Guided by this idea, researchers have started to validate solutions (Kempf et al., 2023). One case is from Germany, where bread – like in other Western European countries – is a daily commodity that provides about one-third of the daily carbohydrates. So, the researchers invited 80 older, overweight people to pick their fresh bread from the well-known local bakery. There, half of them received a bread with a medium-carbohydrate and low-insuli-stimulating composition, which was made out of oat flakes, sunflower seeds, flax seeds, chia seeds, psyllium husks, chopped almonds, baker's honey and Rhinish field beans. The control group received a whole-grain rye bread. All were asked to eat as much of the bread as they usually did, but no other bakery products.

Three months later the intervention group had on average lost nearly 2 kg compared to the control group. And while these 2 kg initially do not look too impressive, the encouraging aspect is that this happened under unconstrained conditions and without any perceivable study design. The individuals were tested before and after but had no contact at all during the trial phase.

I mention this here as an example of how the next generation of food product development can use healthy design principles to help people eat healthier but without the need to change their food preferences. Furthermore, this shows that this new food does not need a new "magic" ingredient, an exotic herbal extract or chemically synthesized molecule, to be healthier. It just needs a smarter combination of well-known commodity ingredients. That means no complicated safety testing or long regulatory approval. New products and combinations can be made and quickly tested in an uncomplicated way with non-invasive sensors, measuring blood glucose (as an insulin marker) or ketone bodies in the breath (as fat degradation markers).

A good "functional" food product must fulfil the promise it is making to encourage people to include the product in their dietary routines and keep it there. This extra effort and the money spent need a reward – not immediately, but at least after a while.

This case and the "low keto" idea show similarities with the hypothesis of "ultra-processed food" (UPF) – discussed in more detail later in this chapter. Both blame the presence of simple sugars, too much processing and too simple digestibility. And they praise dietary fibre. But both concepts differ in their esteem for fat and fruits, and more importantly where UPF constraints the packaged food choice massively, the "low insulin" concept does judge food processing as such or specific modern food formats. It judges the nutritiousness on the basic of an analytical assessment.

The attractiveness of "Keto diets may further increase with newer and simpler non-invasive mobile tools to control the insulin response quickly, so consumers can monitor effects easily. An example of more personalized nutrition.

Sweeteners

This is another controversial field. Non-nutritive sweeteners (NNS) have been known for about 150 years. Their key feature is they trigger sweet taste perception but provide low or no calories. So, they should be the heroes in the fight against sugar. The story of saccharin started in World War I, where it first helped to smooth the global sugar shortage but then became suspected to cause cancer. Later, after serious safety testing, its reputation was restored, before it has now been questioned again (American Chemical Society, n.d.).

Today, many more alternatives with low or no calories are available; and they could be the powerful tools in this global overweight and obesity wave.

Browsing through the past three decades of research on the health effects of NNS remains inconclusive regarding its safety. This meta-analysis, the crown of evidence-based health care, should help to make final conclusions. But here, one analysis tended to see that sweeteners can help lower energy intake and body weight when they replace sugar in the diet (Rogers & Appleton, 2020). While another concluded that sweeteners have no benefit in weight management but may instead increase the risk of diabetes and cardiovascular diseases (Toews et al., 2019).

So, we are in a situation where even the best statistical analysis cannot dig out causality behind the correlation. These contractionary conclusions reveal the difficulty in coming to a general conclusion. Maybe this is because the effect is different for different people. But without any idea of what this difference is, it is hard to prove.

But while it's not possible to say who would benefit and who not, new physiological effects additionally complicate the question. There are observations that the combination of artificial sweeteners and easily digestible carbohydrates can trigger insulin levels more than the same amount of carbohydrates alone (Mathur et al., 2020). That's a blow for the popular habit of balancing the consumption of a portion of french fries with a Coke Zero! Different research sees evidence that artificial sweeteners influence the individual microbiome, leading to decreased satiety, and thus promoting more caloric consumption and weight gain (Pearlman et al., 2017).

Again, this is not an indication of weak science. It is just an indication that in the total picture important confounding factors are missing. And these differ between different individuals. So, it is difficult to provide guidance for the personal diet.

And in that situation, the World Health Organization (WHO) released advice not to use any type of non-sugar sweeteners for weight control and garnishes the note with an additional warning that aspartame usage may cause cancer (UN News, 2023). Such a statement, of course, made it quickly into the headlines of the mass media. And not everyone would read through the whole text to find as well that "*... An adult weighing around 70 kilograms (150lbs) would need to consume more than 9–14 cans of soft drinks to go beyond the recommended intake...*".

Wait a moment! 9–14 cans per day? That amount turns the alert into a warning for a very small part of the population. But the news worried many. And what may cause cancer at a high level of intake could also do it at much lower levels. The fact that another leading food authority in the United States, the FDA, immediately rejected the WHO statement, may not help to calm down the discussion about food safety.

Some may still remember that in the early 1990s the same International Agency for Research on Cancer (IARC) of the WHO declared coffee a potential carcinogen. Twenty-five years later, the expert group was finally able to correct their warning. Now they saw sufficient evidence for various health benefits associated with moderate coffee consumption (Szejda, 2016). The latest is that since then top athletes regularly drink coffee to improve their performance.

So, what is now the advice for future food development of sweet products? It is unlikely that the human desire for sweet-tasting food will disappear soon. It's too deeply engrained in our taste preferences. But it's also not realistic to hope that the global overweight and obesity epidemic will come to an end. And sugar is likely contributing more than fat or protein. Therefore, public health authorities and governments should intervene to cut the sugar intake in their populations. Today, already 108 countries apply sugar taxes, often on sugar-sweetened beverages (WHO, 2023).

The non- or low-caloric sweeteners are the alternative to sugar. They have good taste without calories, and hence many F&B producers offer those alternatives, while people are getting more suspicious about them (Southey, 2024).

As a more balanced approach, "low", instead of "no" may gain traction. Some sugar will be used where it's important for the taste, but only there. And the powerful artificial sweeteners like aspartame, acesulfame-K, saccharin or sucralose may face the same headwind as other additives and get-tough competition from natural low/no caloric sweeteners like stevia or monk fruit extract. And newer data point towards the potential environmental impact of artificial sweeteners, since research shows how stable they are and how they accumulate in the environment, like in drinking water. That brings them closer to plastics (Naik et al., 2021). So, as soon as the price gap between them and natural non-caloric sweeteners starts closing, their future may come to an end. And product formulators will have to rejuvenate all their formulations again.

New Heroes

So, each of the three types of macronutrients had and have their focus in the nutritional debate. First the fats, then the carbohydrates and now the proteins. Black-and-white views about their goodness then became more nuanced, and the new food products changed their composition accordingly. Today the proteins, especially those from plant origin, receive a lot of goodwill. It's not likely that the nutritional discussion will end here. Also, proteins have calories, and a too high of an intake is not beneficial for the body. It is possible that like with all the other macronutrients in the past, the protein trend will lose attractiveness over time.

The question is as follows: What will happen to the other substances that the body needs in higher amounts and that are essential: Fibres, water and air/oxygen?

We have already discussed fibres earlier in this chapter and in the "biotics" part in Chapter 16. They will keep their positive reputation in the coming years, and their contribution in food goes far beyond the nutritional aspects.

Water, however, is the most important and versatile molecule in food and beverage. In association with foods, its either free, absorbed (affected water) or bound. When water is located on the surface of macromolecules like starches, pectin, cellulose and proteins it is called absorbed water and creates colloidal systems. Bound water is important for the conformational stability of protein and activity of enzymes.

And of course, the role of water in cooking, sterilization, concentration, dehydration, drying, or freezing in the food chains can hardly be overestimated, but it's not a food constituent, whose role as an active will further improve or change dramatically in the future. The food packaging industry needs to understand the transport of water through packaging materials, since the interactions of water and packaging materials affect food quality and the organoleptic properties. It can be used as liquid, frozen or gas, is essential for health and nutrition and is extremely cheap, but it is not a molecule that conveys new competitive benefits. In food matrixes, water is normally an element that needs to be managed and controlled, but not a functional active.

The maybe most underestimated and underexplored food constituent is air (foodnavigator.com, 2021). Looking at it from a functional perspective shows an ingredient that is available in endless amounts, its clearly material saving and its healthy and extremely cheap. It is also recognizable and product modifying, not just a neutral, sensorially undetectable ingredient. It can be used to increase volume at the same weight. Well-known examples are most types of bread where air is incorporated into the dough through kneading and proofing, whipped cream, cakes, cheese, popcorn, aerated chocolate, soufflés, carbonated soft drinks and ice cream.

Less than 100 years ago, aeration was used to force air into a frozen mixture of sugar, cream and a few aromatic molecules to create ice cream. Today, ice cream is aerated by forcing pressurized air into the mixture as it freezes, creating tiny bubbles that give it its characteristic smooth texture. No calories, minimal costs, more volume at the same weight and new textures are promising attributes for NFPD.

The search for new beneficial actives beyond those macronutrients will continue to shape future product formulations. Stephen L. Defelice, the founder and chairman of the Foundation for Innovation in Medicine (FIM), first used the term "nutraceuticals" to describe a class of edible bioactive compounds said to promote health and well-being, supposedly helping in the prevention and treatment of disease. Different fibres, herb and spice extracts, secondary metabolites like flavonoids and carotenoids, vitamins, minerals, enzymes and all types of "biotica" fall under this fusion of nutrient and pharmaceutical. It has been used heavily since then in the press and scientific publications, but it is "just" a marketing term. There is no general regulation or legal definition available for them, and new candidates can be easily added to the list. But the term represents the solid confidence that there are many natural components beyond the classical nutrients that can bring important health benefits to humans. It just needs good scientific work to establish when, under which circumstances, in which amount to whom! And that is the clear ambition of personalized nutrition.

The main task of new food product development is not primarily to create new sensorial experiences for consumers but to bring the well-known, beloved taste, texture, smell, aroma and appearance in different, more convenient, cheaper and heathier formats. This is not new and began in the packaged food industry long ago often with the addition of a beneficial ingredient/nutrient. In the past, sometimes food product entered the market claiming their nutritional goodness, while the overall profile of the product was rather the opposite. A well-known candy brand started loudly with the promise of "an extra portion of vitamin C", releasing moms and grandmoms from the tension of deciding if they should give something that is healthy or enjoyable. But nutrition and health experts felt the need to react and said that this sweet is basically sugar and calories, which are bad for energy intake and teeth health. So they started to demand that the impression that these treats are good for health has to be stopped.

Ultra-Processed Food

The term "ultra-processed food" (UPF) was introduced in 2009 by the Brazilian nutrition researcher Carlos Augusto Monteiro. In 2017, the idea was grown into a food classification scheme, the "NOVA system". The core thought here is that the "nutritional goodness" of food products is massively influenced by the extent and purpose of industrial processing and not primarily the pure nutritional composition (Monteiro et al., 2019).

In the beginning, food additives, salt, sugars and fats as key constituents of ultra-processed foods were critically discussed: Then the assumed effects of food processing on the accessibility, palatability and convenience were added and the non-availability of essential ingredients in an ultra-processed food matrix and the addition of "chemical" "non-food" substances were also built in.

That led to a new food system with four classes:

1. *Unprocessed or minimally processed foods*: These are natural foods that have not undergone any significant alteration. This includes technologies that remove inedible or unwanted parts, including drying, crushing, grinding, fractioning, filtering, roasting, boiling, non-alcoholic fermentation, pasteurization, refrigeration, chilling and freezing. Typical items are fresh fruits, vegetables, grains, nuts, eggs and milk.
2. *Processed culinary ingredients*: These are the result of minimally processed food items and can include some additives. Examples are oils, butter, sugar and salt. They are often items used to season and cook food from scratch.
3. *Processed foods.* These are foods that have been preserved or modified by adding salt, sugar, oil or other substances. Typical for this class are a combination of items of class I and II to enhance the sensorial properties. Examples are vegetables, cheese, bread and cured meats.
4. *Ultra-processed foods.* These are formulations made through chemical or physical industrial processes, either as ingredients like hydrogenated oils, high-fructose corn syrup and artificial flavours, or as finished products like soft drinks, snacks, ready meals and fast food. The actual food packaging can also push items into this class. But the use of biotechnology or GMO does not inherently make food UPF.

Implicitly, this scheme contains a correlation with the nutritional composition, the excess of high energy, fat and sugar, combined with too low levels of other nutrients, fibres or antioxidants. A recent description tries to capture the core of ultra-processed food as: "*Industrial formulations typically with 5 or more and usually many ingredients. Besides salt, sugar, oils, and fats, ingredients of ultra-processed foods include food substances not commonly used in culinary preparations, such as hydrolysed protein, modified starches, and hydrogenated or interesterified oils, and additives whose purpose is to imitate sensorial qualities of unprocessed or minimally processed foods and their culinary preparations or to disguise undesirable qualities of the final product, such as colorants, flavorings, no sugar sweeteners, emulsifiers, humectants, sequestrants, and firming, bulking, de-foaming, anticaking, and glazing agents* (Steele et al., 2016).

The research here tries to build the correlation between UPF intake and the onset of non-communicable diseases (NCDs), starting with overweight, diabetes, cardiovascular diseases, various types of cancer and mental diseases. That NCDs are globally a major problem is undisputed. That the product and process development in the food industry are a major cause is a hypothesis. The research fundament for UPFs is predominantly observational studies (retrospective case-control studies or prospective cohort), which are generally good to identify patterns, associations or trends but are not suitable to provide causality.

Several ways are suggested for how UPF are affecting health and well-being. First, the excessive energy present, "empty calories", satisfies the hunger but does not provide the essential vitamins and minerals. The high processing makes UPFs food "hyperpalatable", which manipulates the consumer to eat even more.

The idea to define food quality through more aspects than just the nutritional composition is interesting and deserves respect, particularly because in some Western countries more than 50% of the dietary energy intake comes from foods classified as "ultra-processed" in the above scheme (Pagliai et al., 2020).

The first human intervention studies in this area have published their findings. Hall et al. used a carefully controlled clinical setup: Participants could eat for 2 weeks "*ad libitum*" (as much as they wanted) UPF, made in such a way that the macronutrients content was very close to the freshly prepared food the control group got (Hall et al., 2019). After 2 weeks, the participants swaped seats and received the diet of the other group.

The result: Participants on the ultra-processed diet consumed on average about 500 kcal/day more and gained 0.9 kg more weight than the group on the freshly cooked one. This indicates that the unlimited, exclusive consumption of UPFs leads short term to some weight gain. From there to the identification of disease trigger in regular diets is still a long way off. Doubts are allowed!

This is particularly true because new big cohort studies found only weak correlation between ultra-processed foods intake and all-cause mortality (Fang, 2024). Interestingly, when the authors started to split the UPF into different subgroups, they found stronger associations with mortality.

It seems to me that the initially viable thought that the nutrient profile isn't the only thing that shapes the "goodness of food" went too far and is now dogmatically pressing the evaluation of food items into a too rigid, unclear and narrow concept. Starting from valid concerns that some food processing can negatively impact the food quality, more and more additions and exceptions have been built in to keep the original thought alive. That has blurred the clarity of the concept and weakened it.

The nutritional composition remains the basis, and processing can damage – or sometimes improve – that nutrient profile. Heat destroys vulnerable vitamins, hydrolysis and biomass conversion break down complex carbohydrates and hydrogenation transforms fatty acids. But freezing can increase folate content and bioavailability, vitamin C (E300) addition reduced oxidation, processing produces semi-skimmed milk or lactose-free milk (Deshmukh-Taskar et al., 2005). And even a yellow banana has a different GI and insulin response than a greener one. It doesn't need processing, just ripening.

But it's important to identify those aspects that truly influence the healthiness of food items. The idea that generally all modern processing and product formulation deteriorates the original food quality seems wrong. Such dogma hinders approaches to develop processes that improve the goodness of the food (Ditchfield et al., 2023; Marino et al., 2021).

Instead, it's more valuable to critically review the effect of individual processing steps and formulations. Focussing the search on descriptive processing indicators compensates the structural weakness of the NOVA system, that it: "*... does not really reflect the intensity of the processes used but is a mix of technological considerations based more on socio-cultural aspects than on physical-chemical ones occurring during food processing...*" (Braesco et al., 2022).

Another difficulty with the NOVA system is its inaccuracy. The class descriptors are not precise enough to guide a clear classification. The authors of the study just mentioned "tested" the functionality of the NOVA system by asking 400 knowledgeable evaluators to classify a large series of food items in the NOVA system. The result was a low consistency amongst the evaluators, even when additional info was given. Often, items were placed in different classes, and the overall confidence of evaluators was not high.

That makes NOVA not only less suitable as a consumer guidance tool, but it will also complicate the finding of important correlations between food attributes and health. If epidemiologists apply this poorly validated NOVA system, they will have much more difficulty finding robust correlations.

A different angle to evaluate contemporary food quality could be the glycaemic load (GL) or the insulin response. UPFs tend to be higher in their GL, triggering fast spikes in blood sugar levels, followed by rapid insulin activation, which at least in overweight people blocks fat degradation.

To an extent, the UPF debate shows signs of an anti-food industry campaign, condensing concern into a "simple take home message" that demonizes all processed food, All the thoroughly executed safety experiments done in the last decades do not find recognition here. Instead, "hyper-palatability" is introduced as a sort of perfidious strategy of the food industry to sell more products on the back of consumer health. It is a polarizing campaign, dividing experts and not useful to split facts from fiction.

Which substances and environmental and physiological effects could cause overeating? If being "more palatable" is a reason, is that because the products are so convenient, so tasty, smell well and look better, or all of these together? Or are there aspects in these products that reduce or circumvent appetite control? Finding these answers could help to identify new design principles for 21st century food.

In the search for better food design principles, the "Yale Food Addiction Scale" (YFAS) is sometimes mentioned (Gearhardt et al., 2023). While a phenomenon like "food addiction" is still controversially discussed, findings that especially the combination of high levels of refined carbohydrates and added fats can encourage overeating are worth a thought. Also, the idea that some additives may act as reinforcer should be considered (Saffari et al., 2022).

Today, in a world too rich in calorics, a physiology optimized over thousands of generations to crave for energy, struggles to provide good intake guidance.

Generally, appetite and energy intake are modulated by a diverse range of factors, "central" and "peripheral" ones, direct and indirect feedback loops. A complex neuronal regulation may have developed through evolution to guide us to ingest a good balance of nutrients from different foods. A complex mechanism called "sensory specific satiation and satiety" (SSS) has guided us well in the past, but today it becomes derailed by the presence of highly palatable food (Hetherington & Havermans, 2013).

But the idea that generally additives and in the NOVA system even packaging triggers overeating and other NCDs is difficult to digest.

A final thought: Plant-based proteins as meat and dairy replacers in food are the "new heroes". It is helpful to know that these proteins are truly heavily processed into ingredients like protein isolates (PI) and protein concentrates (PC). That earns them a classification as ultra-processed foods in the NOVA system.

But then, these "future proof" food items not only use lots of highly processed alternative proteins, but also a series of different additives, and they are considered "much better for people and the planet" than their animal-based counterparts (Tan et al., 2023).

People that have tried to create meat alternatives know that it requires a profound processing and formulation expertise to make sensorially competitive alternatives. Products with good consumer preference have a complex and rather long ingredient list, with lots of additives. There is no plant-based beef patty, chicken nuggets or minced meat without those ingredients.

Interestingly, one of the archenemies in a healthy diet, the red meat, does not fit well into the NOVA scheme. It's not much progressed and doesn't have chemicals or additives, but is proven to be detrimental to longer health, when consumed too much.

Thinking about it, the merit of the ultra-processed debate is that it has opened up the discussion about healthy food attributes to beyond the pure nutritional composition. But this cannot be replaced by vague characterizations of product and process development.

The NOVA classification might actually have more cultural appeal! It combats the still increasing demand for more convenience food and is a plea to people to return to cooking more at home and from "scratch", with fresh local ingredients. That's a great intent, and people will happily accept that advice to the extent their modern hectic lifestyles and their cooking skills allow. That this can be a replacement for packaged foods, I doubt.

CASE STUDY: TFA

Margarine is technically more processed than butter. It's made by fractionation, interesterification or hydrogenation. It was first made already in 1869 by a French scientist named Hippolyte Mège-Mouriès, in very different times and in response to a challenge by his Emperor Napoleon III, who needed a butter substitute from beef tallow as an affordable energy supplier for the armed forces and lower classes.

Nowadays, butter and margarine in their full fat versions have at least 80% to a maximum of 90% fat. But both have a different fat profile. Real butter made from cream or milk contains mostly saturated fatty acids, while margarine is made from plant-based oils and has more unsaturated fats.

After World War II, the need for pure fat lost importance, while cardiovascular diseases increased, and the consumption of saturated fats was seen as one reason for this. That brought up a call to control the disease not only be medical intervention, but in addition by a diet-based approach. Parts of the medical society started asking for a food-based intervention. So, the leading margarine producer, Unilever, launched a new margarine in 1962 and called it "Becel", this brand name actually presented the scientific program: The initials BCL stand for "Blood Cholesterol-Lowering". And margarine has 80% less "bad" saturated fats than butter and instead contains "good" omega-3 fatty acids.

In the beginning, Becel was therefore only sold in pharmacies, Later the product made its way into the chilled cabinets in grocery stores.

To make the plant oil spreadable on bread, more shelf life stable and solid at room temperature, the oils needed be hardened, a process also known as hydrogenation. The full hydrogenation would saturate all healthy unsaturated fatty acids. So, partial hydrogenation was seen as the solution. But then, evidence appeared that the partial fat hardening would create trans fatty acids (TFAs). These were shown to be directly correlated with increased low-density lipoprotein (LDL) concentration, which is strongly associated with an increased risk of atherosclerosis and atheroma formation (National Academies Press, 2005). Hence, to preserve the original health benefit of plant oils in margarine, a new process was needed. It had to keep the consistency soft, but solid enough for spreading on bread and usage in many baking applications, while maintaining "good" unsaturated fatty acids, but no "bad" unsaturated fats, like TFAs. New ways to structure vegetable oil in such a way were successfully developed and quickly introduced (Thaís et al., 2021). All that lead to different plant-based oils, using alternative methods of modifying, including enzymatic processes and the addition of antioxidants, such as tocopherols, ascorbyl palmitate, or citric acid to prevent or delay the formation of TFAs during storage or heating.

The addition of fat-soluble vitamins supported the health proposition. And in 2000, the company launched "Becel Proactive", of the same composition but enriched with phytosterols, a plant-based active known to actively reduce heart disease. After dozens of human intervention studies in different populations, the product was the first to become approved as novel food with a claim to actively lower HDL cholesterol by about 10%, when consumed daily for 2–3 weeks (European Food Safety Authority, 2012).

This a bit older, but prominent, example shows that a simple equation "highly processed food = bad food" is not useful. A lot depends on the technology applied, the details of the process and the food matrix used. Moreover, this example shows that the food industry can help to improve public health through product reformulation. The previous case study showed that this can occur not just by "preventing disease ("an apple a day, keeps the doctor away"), but even by actively lowering a clear risk factor for chronic diseases.

Any broad disapproval of modern food processing hinders companies and food developers to develop "state of the art" new food products based on the latest science and technology. People don't want to revert their living conditions into those of the last centuries. The daily diet needs to be healthy, versatile, tasty, convenient and affordable. And consumers will not choose only one of those criteria; they want all of them!

CASE STUDY: JUST CUT SALT

This final case is more a parkour ride through the long journey of salt (sodium) reduction. The starting point is that salt has essential physiological functions in living systems, so like for all the other tastes, evolution developed a sensor on the tongue to detect the scarce salt sources in the environment. Nature also established pathways to excrete salt, in case too much is taken up. But the capacity of that mechanism is insufficient to manage in today's world the salt balance for a major part of the population.

Health professionals are crystal clear: "We must consume less salt". Too high of an intake has for most people a negative effect on blood pressure, which is the leading cause of cardiovascular disease and stroke. Statistics count more than 1.8 million deaths per year and a loss of more than 44 million "disability adjusted life years" (DALYs), caused by too high of a salt consumption (more details: Campbell et al., 2022).

Understandably the WHO demands a minimum 30% salt intake reduction target as one of the key steps to reduce such premature death and economic loss. Recommendations say less than 5 g of salt (or better: 2.300 mg of sodium) per day, versus a current average intake of up to 12 g/day in some Asian countries caused by cultural habits, food availability, the local cuisines and other factors.

People making food know that sodium not only brings saltiness but boosts flavour and brings a variety of more technological functionality, such as structuring meat or bread, preservation, spoilage prevention and enzyme activity. And beyond table salt, other "functional" ingredients, like sodium bicarbonate, sodium benzoate and monosodium glutamate (MSG) add sodium to the equation (for overview see Busch et al., 2013). In addition, salt (better: sodium) is interacting with the surrounding food matrix, modulating other taste senses and influencing the food texture, leading to a different mouthfeel (Fonseca-Bustos et al., 2023).

Every household has salt and other sodium-based taste enhancers in the cupboard. But the food industry is targeted, since data show that up to 75% of the daily salt intake is estimated to come from packaged/processed food. On Western food menus, a moderate 140-g slice of pizza can contain 700–1000 mg of sodium; a cup of canned soup, 700 mg; a serving of roasted ham, average of 1100 mg; and a tablespoon of soy sauce >1000 mg. In Asia, soy sauce is the cooking and table sauce and provides much more than simple saltiness.

Food companies started more than 20 years ago to reduce the sodium content in their product portfolios. Newer preservation techniques are available, so salt as a preservative to keep shelf life and storage conditions is replaceable.

The bigger problem is consumer preferences. Companies developed prototypes with 20%–30% sodium reduction and gave it to test consumers. Sometimes they added information about the health benefits of sodium reduction, depending on food type and format. Some variants then scored on par with the regular, high-salt variants. But often, beyond a 15%–20% cut, the reduced salt content was noticed and disliked. Particularly in savoury products, those with meaty, umami notes, like soups, bouillons, sauces and ready meals, testers also mentioned an overall reduction in "hearty aroma" and "rounded mouth feel".

Food developers tried alternative ingredients, like lower sodium salts, potassium chloride or various taste enhancers in numerous trials. But one way or the other, all of them influenced the original taste profile: In the case of potassium chloride, a peaky, disturbing bitter note appeared.

Step-by-step product-specific formulation changes with sophisticated flavour and aroma cocktails enabled a 15%–25% range before the liking scores started to go down.

To reduce sodium further, other solutions than salt replacers were needed. In some cases, the product format offered new options. Products like lasagne could be made where the layers had an uneven distribution of salt content and delivered the "preferred" saltiness pat in lower levels (Lawrence et al., 2009).

A similar design-based approach for uneven distribution helped with typical condiments like mayonnaise, technically an oil-in-water emulsion.

This worked for a limited range of food products. Hence, process engineers started also to focus on the salt crystals themselves to see which shapes would support saltiness at lower sodium levels. Many dry-salted food producers knew that a good number of crystals on the product surface were lost in the packaging, so they had to compensate with extra salt to match sensorial expectations. Costwise this was a minor problem, but the salt levels declared on the packaging raised criticisms.

A more physiological approach then started to focus on the optimal diffusion of free Na^+ ions to the salt receptors of the oral cavity during the food consumption. After many tests and optimization of the physical properties of the salt crystal, again some improvements were made, often through smaller particle size, lower density, hydrophobicity or better flow properties (Hurst et al., 2021). Besides various shapes and sizes, even hollow salt crystals, encapsulated salt or smart spatial distribution were tried. But none of those physical-chemical approaches is the game changer (Vinitha et al., 2022).

Since purely technical approaches were often insufficient, the potential of behavioural pushes and non-taste-related sensory strategies was tested, including health education, public awareness campaigns and multi-component strategies combining education and awareness (Spence, 2022). A recent meta-analysis calculated on average a 10%–13% reduction potential from behavioural change interventions (Khalesi et al., 2021). That would be on average 1 g/day less.

Yet another non-receptor-based arm of research uses the "signal detection reminder method" to boost saltiness perception at lower levels (Esfanjnai & Mohebbi, 2023). Here, sensory science is used to evaluate how different factors, such as odors or tastes, affect the perception of a particular sensory attribute, like saltiness. It's anchored in signal detection theory, to better distinguish between the actual sensory signal and the noise or other interfering factors.

Various test designs acting on taste-aroma interaction showed the substances with "salt congruent odours" like cheese, soy sauce, sardines, meat, soy sauce or MSG would be effective.

The inclination of individuals to associate signals from a specific modality, like vision, with taste is known as "cross modal correspondence" and has been widely researched. It has also shown potential to be used in product and food packaging visuals (Escobar et al., 2022).

Last, but not least, researchers proposed a relationship between the saltiness perception and the microflora composition in the oral cavity (Cattaneo et al., 2019). And very recently, science has started to use quantitative analysis between saltiness perception and electronic tongues to establish more general correlations (Xia et al., 2024). Such experimental designs may allow the systematic testing of various combinations of known and unknown factors and the AI-facilitated solution search.

So, despite the depth and versatility of ways to provide a fully satisfying method to cut sodium intake at the preferred taste, there is no solid evidence that any smart combination of tools would result in a sodium reduction of more than 20%–25%, and hence remain below the target minimum 30% reduction the WHO is demanding.

Cutting sodium in food products has another practical limitation. Regular salt is one of the cheapest ingredients in food formulations. And every replacement with an alternative molecule, especially treated crystals or a more complex food product design, adds costs, while the consumer is not willing to spend a little more on such products (Di Vita et al., 2016). For the food producers, that means they can either find a way to save elsewhere for the oncosts or accept a margin dilution, a strategy most stakeholders do not like.

So the easy-to-apply quick-fix approach does not exist. Responsible companies and brands that cut sodium levels in their products beyond the perceived liking of their loyal (= heavy) users, risk turnover and support the commercial success of competitors who use higher salt levels.

And here yet again, most attempts to reduce the sodium content at the preferred saltiness level use sophisticated food processing tools, making the products "UPF".

Finally, since science and technology have not yet fully solved the problem, regulators have started to act with claim regulations, warning labels or a salt tax for processed food and restaurants. And responsible governments could put this extra money into research for longer-term programs to reduce salt intake at the population level, a long-term profitable investment into the health care system.

Better Storytelling

The recent communications around "good" or "bad" or "healthy" or "sustainable" food teach a lesson about the right way of talking. Content matters, but so does "how something is told". Not only since the last global pandemic, the safety and healthiness of food products is top of mind for many people. And all reports unanimously say that the trend will continue, and many new food product developments will include this dimension in their product design.

The consumer cannot simply sense how healthy a food or beverage is. Health has no specific taste, colour, form, smell or texture. The "official information boards" like ingredient lists and especially nutrition tables provide information, but frequently individuals lack the ability and/or motivation to decode the information. And beyond the nutrition and nutrient content claims, stronger factual health promises are difficult to make.

Usually, food makers trigger lay beliefs about healthy food, sending cues that signal this message (Chan & Zhang, 2022). Well-known "health cues" are blue and green colours, lighter and muted colours, a matt packaging surface, lightweight and not too much emphasis on great taste, as healthy food is intuitively linked to being less tasty. Warm food is connected with rich food/more calories, and cold food the opposite. How much subconscious effect those signals are having differs hugely between individuals, depending on personal experiences, knowledge and other factors.

But cues can possibly be used to make a health story stronger. Consumers are looking for healthier products, and the potential of foods is enormous, but the tough regulation in the health territory makes the building of a competitive claim risky and time and resource consuming. So how can one product stand out in the masses of competing products?

I guess the times are over where research and enterprises could hope to prove relatively simple new health benefits that are simply equally true for all potential consumers or on population level. A lot of evidence created in the biosciences indicates more individual differences, underlined opportunities for personalized health and nutrition, less generally true statements, while the logic in the according regulation expects resilient general proof.

So, the communication about which food product is healthy becomes more personal. Different "opinion camps" exist, likewise in science and technology. "Is fat healthy?" "Are carbs better than fats"? Do additives cause overweight? "Do people need extra vitamins A, C, and E, calcium, iron, folate"? "Can prebiotics prevent some cancers"? "Do antioxidants improve immunity and resistance to disease"? "Is butter healthier than margarine"?

It may take many more years before a scientific consensus is reached (if ever) in these human health-related areas. But the scientific discussions show the different opinions and the open questions.

The leading experts in a research field run complex studies and experiments, carefully balance their preliminary conclusion and do not forget to mention that "more research is needed", since there are still so many questions unanswered. That also means that it's often too early for a regulation to allow a new health-related claim for the food product. The newly created evidence may be promising but not conclusive and stimulate debate between experts. This is an exciting discourse for someone "trained in the art" and with sufficient time to listen. But, what is the take-home message for all the others looking for healthy food?

Here, food brands with a nutrition and health intention can refresh their personality. Instead of waiting for the final scientific consensus, they can take a position in the debate. "Yes, we believe in the benefits of antioxidants!" "Simple sugars block the degradation of fat stores!" "Food processing is not bad"! "MSG is not unhealthy."

Here the brand represents an educated opinion. It takes a defendable position in a still controversial dispute. And the food brand openly admits that there are other views or unclarified questions. Such an approach shows transparency, honesty to listen and the will to learn. It's more dialogue with the consumer than lecturing. And it is often no disadvantage for the brand or company to be more opinionated. Some food companies have already started to take a more confrontational position in such debates (Globescan, 2022).

It's a chance to revitalize a brand purpose. It brings back the spirit of the "old" food brands with their promise about safety and reliable quality standards (see Chapter 1). Many consumers, especially the younger ones, know that what makes food "good" can hardly be captured in a short, legally brushed statement on a package. But when the brand invests in the dialogue, provides the arguments from both sides and defends its position, it will build credibility and trust, the two important attributes many of the old food brands have lost. So, food brands should invite more people to join the conversation.

And one can wonder how the official health authorities can help here best: The short story earlier in this chapter about aspartame in summer 2023 is first and foremost a telling example on how not to do it. The lack of scientific alignment that became obvious can ignite an already sceptical public view about the science. The safety experts from WHO and the FDA involved in this assessment are all respected researchers and are right to flag a potential risk at a particular consumption level after adding a safety margin. But how should the layperson draw their own conclusions?

So, direct messaging from science experts to "normal" consumers can lead to misunderstandings. And this is the moment where opinion formers and influencers get responsibility and relevance. If they are adequately "trained in the art", those experts or trusted personalities can summarize the "state of affairs" around a more complex topic and bring it into an understandable story or take-home message. That may even include the willingness to work through the relevant scientific publications to draw their own conclusions in an understandable language. They do not drive scientific knowledge further but give their personal view on what gives or takes energy, makes people more resilient, helps with weight loss or improves mental performance.

In the past, the open dialogue about the opportunities and risks of new food technologies did not really happen. Science, regulators and industry were sitting together and had a conversation between themselves. The topics discussed were regarded as too complex for the public.

And the food makers could be afraid that a rather detailed technical debate about aspects of the applied technology could further damage the naturalness and healthiness perception of their products. It seemed safter to "negotiate" the details with the experts and then later cover this piece of new science in a vague wording that would trigger the perception of "foodiness", but surely not raising concerns. Then more sceptical consumers began to whisper about the mechanics behind the curtains. And as a result, the mistrust of the public grew, so that now it is a major task to close the gap between the public understanding about how food is made nowadays and the expert knowledge about the possibilities and limits of a food technology.

For new food product development, communication is important. Humans are conservative by nature when it comes to food, which is not surprising for something that we "put into our mouth and swallow".

Unknown, strange food can trigger reactions like food neophobia, the fear to eat specific foods, a phenomenon that can be found everywhere, for different reasons. It can start in small children rejecting sometimes even healthy items like fruits and vegetables, a behaviour that in the human evolution may have helped to avoid toxication.

It is also seen in the avoidance of food from different cultures, unusual and unfamiliar food, novel food, healthy food as an alternative version of already known food, or food of unknown origin and processes. Anybody who wants to open the culinary richness to other cultures may receive those reactions. It exists in basically all cultures and age groups (Siddiqui et al., 2022).

But today, where new food sources and technologies promise opportunities to "feed the world more sustainably" like insect protein, micro-algae, jellyfish or cultured meat insects, such impulsive food fear is no longer a minority or luxury problem but a topic requiring attention and good conversation with the consumer. Scientists and startups focussing on sustainability are very interested to understand the underlying psychological mechanisms and how to overcome them (Wendt & Weinrich, 2023).

Also here, initial evidence suggests that an understandable story about "the nature" of food innovation, a trust in the company or the endorsing brand are possibilities to overcome the hurdles.

Food Demonstrations

Packaged food demonstrations are typically events where food manufacturers showcase their new food products. The food gets centre stage, and the marketing story is served as a side dish. The primary purpose is to increase consumer awareness and trigger purchase intention for the food products. Conventionally, these demonstrations are a typical "point of sale" marketing tool. Companies hire agencies, which hire temps who after a quick training invite passing shoppers to taste the product. It is a good opportunity for consumers to try new stuff and for the company to have a chance to recruit new users. The "persuasion principle of reciprocity" is used here; the feeling of people that they must return a favour given to them (Isabella, 2022). When someone hands you a free sample of chocolate, it makes you feel obligated to return the favour, and maybe buy more of it, even if you had no plan to buy chocolate.

There is a lot of sophisticated psychology in these interactions. The lay theory, for example, uses the beliefs and assumptions that consumers have about a product and how they influence their decision making and behaviour. This one starts with the understanding of how the food item is perceived. If it's associated with pleasure or luxury, consumers are not very open to hearing something about science and technology. But if it triggers the rational decision process, an exciting scientific or technical story can enhance product appeal (see also Dual Processing Theory in Chapter 11). Science and technology can help the sales of organic food, to explain why those are healthier than conventional food or why expensive fruit juices are of higher quality than cheap ones.

As said, normally these product demos are a marketing activity. R&D and the product designer are less involved. But this is a twofold missed opportunity. First, it's an opportunity to get direct shopper feedback and get their opinion and understand the consumer's point of view. Food developers know their products so much that they sometimes do not realize the obvious. Observing and listening in such situations to testers can provide important insights. And there is also an active role. R&D can help to present the product and tell the story. Particularly, when the product promises a "rational" benefit, a health promise, a new convenience trick, a more sustainable packaging or a smarter aroma release. Technical people, that were involved in the product development, can talk more deeply and convincingly about this.

Joining a point-of-sale promotion event allows R&D people also to get a better flavour of what leads the decision making on the consumer side and what affects their choices, preferences, satisfaction and loyalty. And sometimes the findings are surprising. Recent research in Sweden investigated how in-store product demonstrations influence subsequent purchasing (Otterbring, 2016). When asked, people believe that restriction will negatively impact the amount of money they spend and the number of products they purchase. This fits the theories on behavioural consistency postulating that people are likely to continue with the behaviour they are engaged in at present, particularly in public. Interestingly, the study showed the opposite; people spent more money, bought many food items and more expensive ones. Explanations could come from observations that consumers whose decisions are blocked by a restriction become more motivated to buy. And the theory of compensatory consumption argues that customers whose freedom has been restricted often buy and consume those products, signalling a win over the rejection and regaining their freedom.

There are many theories and little tricks experienced marketers, market researchers and salespeople are aware of to convince customers to try the "right" product. Directly experiencing this is the opportunity for R&D to build suitable triggers into the product and packaging design.

And food product demos can happen in many more channels. Companies are more and more including them into the launch plans as part of the rollout package. Typically, R&D is then asked in later phases of the project management to think of ways to bring the "unique selling proposition" (USP) alive in the various communication channels.

A typical virtual product demo is very short and crisp and focusses on one aspect. That is relatively easily done for convenience-related benefits: A running clock to underline a short preparation time, an easy opening or closing of the packaging, etc. It is a bit more challenging to capture the key aspects of taste, aroma and texture in a visual. Here creativity is helpful! Sometimes good ideas can be borrowed from the analytical research. Which method is used to measure texture, solubility, the amount of value-giving ingredients in a product, the lightness or airiness of the item or the rich taste? That's why technically trained people, like those from R&D or quality assurance, are so valuable here. They know the experiment and technical equipment that measure such hidden attributes.

And visualizing this in a short demonstration is powerful support for the product launch campaign. It makes it tangible and adds credibility as it shows "science in action". To make it even more powerful, the experiment can be shown in a "competitive setup", where "the competitor" is also tested but comes in only as second best.

The impact can be increased if the "experiment" is somewhat surprising and with a little bit of a "light touch", not too serious. Older examples show how "Milkybar" keeps on swimming on a bowl of milk, while competitors sink to the bottom, or how many nice "fat eyes" are created on the surface of a bowl of soup that is cooked with one bouillon cube and not the other.

Augmented Reality

A new variant of storytelling for foods comes with the rise of augmented reality (AR), and the potential of artificial intelligence (AI). Today, consumers far beyond GenZ and Millennials are using smartphone apps to receive tailormade information quickly and easily. E-mail planning apps and electronic shopping lists are by far not the end of the path for the food business. Already today there are numerous applications helping the consumer to receive nutritional information, menu options and details about the origin or source of food. Retailers and restaurants start to use AR to present their products, with additional information in more interactive ways through the combination of virtual modifications that are real-world things in real time. By using the smartphone, customers get a realistic 3D view of the place they are in plus all the information the shop, restaurant or brand wants to provide (AugRay, 2023). The new reality is not only playing out in outside locations, but in the home kitchen as well, helping to create better dishes or using the leftovers in the fridge much better. Hellman's "Fridge Night" is a free, simple, practical app that scans the fridge at home and suggests meal preparations based on the items identified, a smart way to help people be more resourceful with their food, have less waste, save money spent on food and be environmentally conscious (Hellmann's, 2022). The app will also not forget to include numerous suggestions to use the brand's famous condiments. For the product developers, this trend means that they have the chance to provide much more information to the interested consumer about the product, the ingredients, the manufacturing, the origin, etc., than could be put on packaging or which needs to be filtered out of website content. And as AR basically shows the food in a particular context, all the data connecting the product with the reality can be fed in. Scanning food items in a supermarket can not only give all information about the product but suggest accompanying items, special offers and new meal completion suggestions.

In a way, this is technology supported "*optimized eating*", a further extension of the "*unguilty pleasures*" trend that has built up in the recent past, where people are no longer willing to compromise pure pleasure with nutritional or environmental deficits.

REFERENCES

Alandia, G., Rodriguez, J., Jacobsen, S., Bazile, D., & Condori, B. (2020). Global Expansion of Quinoa and Challenges for the Andean Region. *Global Food Security*, *26*, 100429. https://doi.org/10.1016/j.gfs.2020.100429

Astebro, T., & Michela, J. L. (2005). Predictors of the Survival of innovations*. *The Journal of Product Innovation Management*, *22*(4), 322–335. https://doi.org/10.1111/j.0737-6782.2005.00129.x

AugRay. (2023, April 13). *Augmented reality and its benefits in food and beverage sector.* Augray Blog. https://www.augray.com/blog/augmented-reality-in-food-and-beverage/

Braesco, V., Souchon, I., Sauvant, P., Haurogné, T., Maillot, M., Féart, C., & Darmon, N. (2022). Ultra-Processed Foods: How Functional Is the NOVA System? *European Journal of Clinical Nutrition*, *76*(9), 1245–1253. https://doi.org/10.1038/s41430-022-01099-1

Busch, J. L. H. C., Yong, F. Y. S., & Goh, S. M. (2013). Sodium Reduction: Optimizing Product Composition and Structure Towards Increasing Saltiness Perception. *Trends in Food Science & Technology*, *29*(1), 21–34.

Campbell, N. R. C., Whelton, P. K., Orias, M., Wainford, R. D., Cappuccio, F. P., Ide, N., Neal, B., Cohn, J., Cobb, L. K., Webster, J., Trieu, K., He, F. J., McLean, R. M., Blanco-Metzler, A., Woodward, M., Khan, N., Kokubo, Y., Nederveen, L., Arcand, J., … Zhang, X. (2022). 2022 World Hypertension League, Resolve To Save Lives and International Society of Hypertension Dietary Sodium (salt) Global Call to Action. *Journal of Human Hypertension*, *37*(6), 428–437. https://doi.org/10.1038/s41371-022-00690-0

Cattaneo, C., Gargari, G., Koirala, R., Laureati, M., Riso, P., Guglielmetti, S., & Pagliarini, E. (2019). New Insights into the Relationship between Taste Perception and Oral Microbiota Composition. *Scientific Reports*, *9*(1). https://doi.org/10.1038/s41598-019-40374-3

Chan, E., & Zhang, L. S. (2022). Is This Food Healthy? The Impact of Lay Beliefs and Contextual Cues on Food Healthiness Perception and Consumption. *Current Opinion in Psychology*, *46*, 101348. https://doi.org/10.1016/j.copsyc.2022.101348

CPI. (2024, March 1). *Formulation, materials and manufacturing.* https://www.uk-cpi.com/technologies/formulation-materials-and-manufacturing

Deshmukh-Taskar, P., Nicklas, T. A., Morales, M., Yang, S., Zakeri, I., & Berenson, G. S. (2005). Tracking of Overweight Status from Childhood to Young Adulthood: The Bogalusa Heart Study. *European Journal of Clinical Nutrition*, *60*(1), 48–57. https://doi.org/10.1038/sj.ejcn.1602266

Dewi, N. U., & Diana, R. (2021). Sugar Intake and Cancer: A Literature Review. *Amerta Nutrition*, *5*(4), 387. https://doi.org/10.20473/amnt.v5i4.2021.387-394

Ditchfield, C., Kushida, M. M., Mazalli, M. R., & Sobral, P. J. A. (2023). Can Chocolate Be Classified as an Ultra-Processed Food? A Short Review on Processing and Health Aspects to Help Answer This Question. *Foods*, *12*(16), 3070. https://doi.org/10.3390/foods12163070

Di Vita, G., D'Amico, M., & Lombardi, A. (2016). *Evaluating low sodium content in food: The willingness to pay for salt-reduced bread.* Agricultural Economics Review.

Dodhiya, M. (2021, April 3). *42% of global consumers are flexitarians and driving mainstream plant-based shift, new report says.* Totally Vegan Buzz. https://www.totallyveganbuzz.com/news/42-of-global-consumers-are-flexitarians/

Escobar, F. B., Wang, Q. J., Corredor, A., & Velasco, C. (2022). The Taste of Visual Textures. *Food Quality and Preference*, *100*, 104602. https://doi.org/10.1016/j.foodqual.2022

Esfanjnai, A. F., & Mohebbi, M. (2023). A Signal Detection Reminder Methodology to Evaluate Effects of Odor–Taste-Taste Interactions to Enhance the Perceived Saltiness. *Food Research International*, *165*, 112464. https://doi.org/10.1016/j.foodres.2023.112464

European Food Safety Authority. (2012, May 16). *Plant sterols and stanols and blood cholesterol.* https://www.efsa.europa.eu/en/efsajournal/pub/2693

Fang, Z., (2024). Association of Ultra-Processed Food Consumption With All Cause and Cause Specific Mortality: Population Based Cohort Study. *BMJ*, e078476. https://doi.org/10.1136/bmj-2023-078476

Food Insight. (2023, February 27). *2022 food and health survey spotlight: Generation Z.* https://foodinsight.org/spotlight-generation-z/

Foodnavigator.com. (2021, December 8). *The biggest ingredient of future food: Air?* https://www.foodnavigator.com/Article/2021/12/08/The-biggest-ingredient-of-future-food-air

foodnavigator-usa.com. (2023, October 2). *Accelerate-your-new-product-development-NPD.*

Fonseca-Bustos, V., Madera-Santana, T. J., Martínez-Núñez, Y. Y., Robles-Ozuna, L. E., & Del Carmen Montoya-Ballesteros, L. (2023). Techniques of Incorporation of Salty Compounds, Food Matrix, and Sodium Behaviour and Its Effect Over Saltiness Perception: An Overview. *Journal of Food Science and Technology.* https://doi.org/10.1007/s13197-023-05861-6

Gearhardt, A. N., Bueno, N. B., DiFeliceantonio, A. G., Roberto, C. A., Jiménez-Murcia, S., & Fernandez-Aranda, F. (2023). Social, Clinical, and Policy Implications of Ultra-Processed Food Addiction. *BMJ*, e075354. https://doi.org/10.1136/bmj-2023-075354

Globescan. (2022, June 29). *Five Trends in communicating food sustainability to consumers.*

Grant, J. (n.d.). *2020: The year of the Flexitarian.* sustainalytics.com. https://www.sustainalytics.com/esg-research/resource/investors-esg-blog/2020-the-year-of-the-flexitarian

Hall, K. D., Ayuketah, A., Brychta, R., Cai, H., Cassimatis, T., Chen, K. Y., Chung, S. T., Costa, E., Courville, A., Darcey, V., Fletcher, L. A., Forde, C. G., Gharib, A. M., Guo, J., Howard, R., Joseph, P. V., McGehee, S., Ouwerkerk, R., Raisinger, K., … Zhou, M. (2019). Ultra-Processed Diets Cause Excess Calorie Intake and Weight Gain: An Inpatient Randomized Controlled Trial of Ad Libitum Food Intake. *Cell Metabolism*, *30*(1), 67–77. https://doi.org/10.1016/j.cmet.2019.05.008

Harvard Health. (2020, March 30). *When-it-comes-to-protein-how-much-is-too-much.* https://www.efsa.europa.eu/en/press/newwww.health.harvard.edu/nutrition/when-it-comes-to-protein-how-much-is-too-much

Hellmann's. (2022, December 21). *Fridge night.* https://www.hellmanns.com/uk/fridge-night.html

Henard, D. H., & Szymanski, D. M. (2001). Customer Satisfaction: A Meta-Analysis of the Empirical Evidence. *Journal of the Academy of Marketing Science*, *29*(1), 16–35. https://doi.org/10.1177/0092070301291002

Hetherington, M., & Havermans, R. (2013). Sensory-specific satiation and satiety (pp. 253–269). In: *Satiation, Satiety and the Control of Food Intake.* Cambridge, UK: Woodhead Publishing.. https://doi.org/10.1533/9780857098719.4.253

H. Meadows, D., L. Meadows, D., Randers, J., & W, Behrens III, W. (1972). THE LIMITS TO GROWTH: A REPORT FOR THE CLUB OF ROME'S PROJECT ON THE PREDICAMENT OF MANKIND. https://archive.org/details/TheLimitsToGrowth?

Huang, C., Liang, Z., Ma, J., Hu, D., Yao, F., & Qin, P. (2023). Total Sugar, Added Sugar, Fructose, and Sucrose Intake and All-Cause, Cardiovascular, and Cancer Mortality: A Systematic Review and Dose-Response Meta-Analysis of Prospective Cohort Studies. *Nutrition*, *111*, 112032. https://doi.org/10.1016/j.nut.2023.112032

Hurst, K. E., Ayed, C., Derbenev, I. N., Hewson, L., & Fisk, I. D. (2021). Physicochemical Design Rules for the Formulation of Novel Salt Particles With Optimised Saltiness. *Food Chemistry*, *360*, 129990. https://doi.org/10.1016/j.foodchem.2021.129990

interObservers. (n.d.). *Why innovation fails: 7 key obstacles explained.* interObservers. https://interobservers.com/why-innovation-fails-causes-solutions/?utm_content=cmp-true

Isabella, G. (2022). Persuasion and the six Cialdini Principles applied in social marketing (pp. 1–8). In: *The Palgrave Encyclopedia of Social Marketing.* Palgrave Macmillan, Cham. https://doi.org/10.1007/978-3-030-14449-4_16-1.

Jenkins, D. J., Wolever, T. M., Taylor, R. H., Barker, H., Fielden, H., Baldwin, J. M., Bowling, A. C., Newman, H. C., Jenkins, A. L., & Goff, D. V. (1981). Glycemic index of Foods: A Physiological Basis for Carbohydrate Exchange. *The American Journal of Clinical Nutrition*, *34*(3), 362–366. https://doi.org/10.1093/ajcn/34.3.362

Kempf, K., Röhling, M., Kolb, H., & Martin, S. (2023). Impact of a Low-Insulin-Stimulating Bread on Weight Development—A Real Life Randomised Controlled Trial. *Nutrients*, *15*(5), 1301. https://doi.org/10.3390/nu15051301

Khalesi, S., Williams, E., Irwin, C., Johnson, D. W., Webster, J., McCartney, D., Jamshidi, A., & Vandelanotte, C. (2021). Reducing Salt Intake: A Systematic Review and Meta-Analysis of Behavior Change Interventions in Adults. *Nutrition Reviews*, *80*(4), 723–740. https://doi.org/10.1093/nutrit/nuab110

Kolb, H., Stumvoll, M., Kramer, W., Kempf, K., & Martin, S. (2018). Insulin Translates Unfavourable Lifestyle into Obesity. *BMC Medicine*, *16*(1). https://doi.org/10.1186/s12916-018-1225-1

Lawrence, G., Salles, C., Septier, C., Busch, J., & Thomas-Danguin, T. (2009). Odour–taste Interactions: A Way to Enhance Saltiness in Low-Salt Content Solutions. *Food Quality and Preference*, *20*(3), 241–248. https://doi.org/10.1016/j.foodqual.2008.10.004

Lee, J., Kim, S., Jeong, Y. J., Choi, I., & Han, J. (2023). Impact of Interactions between Soy and Pea Proteins on Quality Characteristics of High-Moisture Meat Analogues Prepared Via Extrusion Cooking Process. *Food Hydrocolloids, 139*, 108567. https://doi.org/10.1016/j.foodhyd.2023.108567

Leviatan, S., Shoer, S., Rothschild, D., Gorodetski, M., & Segal, E. (2022). An Expanded Reference Map of the Human Gut Microbiome Reveals Hundreds of Previously Unknown Species. *Nature Communications, 13*(1). https://doi.org/10.1038/s41467-022-31502-1

Mack, O., & Khare, A. (2016). Perspectives on a VUCA World. In: Mack, O., Khare, A., Krämer, A., Burgartz, T. (eds) *Managing in a VUCA World*. Springer, Cham. https://doi.org/10.1007/978-3-319-16889-0_1

Malek, L., & Umberger, W. J. (2021). How Flexible Are Flexitarians? Examining Diversity in Dietary Patterns, Motivations and Future Intentions. *Cleaner and Responsible Consumption, 3*, 100038. https://doi.org/10.1016/j.clrc.2021.100038

Marino, M., Puppo, F., Del Bo, C., Vinelli, V., Riso, P., Porrini, M., & Martini, D. (2021). A Systematic Review of Worldwide Consumption of Ultra-Processed Foods: Findings and Criticisms. *Nutrients, 13*(8), 2778. https://doi.org/10.3390/nu13082778

Mathur, K., Agrawal, R. K., Nagpure, S., & Deshpande, D. (2020). Effect of Artificial Sweeteners on Insulin Resistance Among Type-2 Diabetes mellitus Patients. *Journal of Family Medicine and Primary Care, 9*(1), 69. https://doi.org/10.4103/jfmpc.jfmpc_329_19

Montani, J. (2021). Ancel Keys: The Legacy of a Giant in Physiology, Nutrition, and Public Health. *Obesity Reviews, 22*(S2). https://doi.org/10.1111/obr.13196

Monteiro, C. A., Cannon, G., Levy, R. B., Moubarac, J., Louzada, M. L., Rauber, F., Khandpur, N., Cediel, G., Neri, D., Martinez-Steele, E., Baraldi, L. G., & Jaime, P. C. (2019). Ultra-Processed Foods: What They Are and How to Identify Them. *Public Health Nutrition, 22*(5), 936–941. https://doi.org/10.1017/s1368980018003762

Naik, A. Q., Zafar, T., & Shrivastava, V. K. (2021). Environmental Impact of the Presence, Distribution, and Use of Artificial Sweeteners as Emerging Sources of Pollution. *Journal of Environmental and Public Health, 2021*, 1–11. https://doi.org/10.1155/2021/6624569

National Academies Press. (2005). *Dietary reference intakes for energy, carbohydrate, fiber, fat, fatty acids, cholesterol, protein, and amino acids*. https://doi.org/10.17226/10490

NHS Website. (2023, February 10). *The vegan diet*. nhs.uk. https://www.nhs.uk/live-well/eat-well/how-to-eat-a-balanced-diet/the-vegan-diet/

Nova Science Publishers. (2021). *The food industry: Perceptions, practices and future prospects*. https://doi.org/10.52305/qfsm8429

Otterbring, T. (2016). Touch Forbidden, Consumption Allowed: Counter-Intuitive Effects of Touch Restrictions on customers' Purchase Behavior. *Food Quality and Preference, 50*, 1–6. https://doi.org/10.1016/j.foodqual.2015.12.011

Pagliai, G., Dinu, M., Madarena, M. P., Bonaccio, M., Iacoviello, L., & Sofi, F. (2020). Consumption of Ultra-Processed Foods and Health Status: A Systematic Review and Meta-Analysis. *British Journal of Nutrition, 125*(3), 308–318. https://doi.org/10.1017/s0007114520002688

Pearlman, M., Obert, J., & Casey, L. (2017). The Association between Artificial Sweeteners and Obesity. *Current Gastroenterology Reports, 19*(12). https://doi.org/10.1007/s11894-017-0602-9

Pendergrast, M. (2013). *For God, Country, and Coca-Cola: The Definitive History of the Great American Soft Drink and the Company That Makes It*. Basic Books.

Piccoli, G. B., Vigotti, F. N., Leone, F., Capizzi, I., Daidola, G., Cabiddu, G., & Avagnina, P. (2014). Low-Protein Diets in CKD: How can We Achieve Them? A Narrative, Pragmatic Review. *Clinical Kidney Journal, 8*(1), 61–70. https://doi.org/10.1093/ckj/sfu125

Rhaiem, K., & Amara, N. (2019). Learning from Innovation Failures: A Systematic Review of the Literature and Research Agenda. *Review of Managerial Science, 15*(2), 189–234. https://doi.org/10.1007/s11846-019-00339-2

Rao, P. (2023, July 23). *Threads-100-million-users*. www.visualcapitalist.com. https://www.visualcapitalist.com/threads-100-million-users/

Rogers, P. J., & Appleton, K. M. (2020). The Effects of Low-Calorie Sweeteners on Energy Intake and Body Weight: A Systematic Review and Meta-Analyses of Sustained Intervention Studies. *International Journal of Obesity, 45*(3), 464–478. https://doi.org/10.1038/s41366-020-00704-2

Service – Foodinnopolis. (n.d.). https://foodinnopolis.or.th/en/service/

Saffari, M., Fan, C., Chang, Y., Huang, P., Tung, S. E. H., Poon, W. C., Lin, C., Yang, W., Lin, C., & Potenza, M. N. (2022). Yale Food Addiction Scale 2.0 (YFAS 2.0) and Modified YFAS 2.0 (mYFAS 2.0): Rasch Analysis and Differential Item Functioning. *Journal of Eating Disorders, 10*(1). https://doi.org/10.1186/s40337-022-00708-5

Siddiqui, S. A., Zannou, O., Karim, I., Kasmiati, N., Awad, N. M. H., Gołaszewski, J., Heinz, V., & Smetana, S. (2022). Avoiding Food Neophobia and Increasing Consumer Acceptance of New Food Trends—A Decade of Research. *Sustainability, 14*(16), 10391. https://doi.org/10.3390/su141610391

Sleboda, P., De Bruin, W. B., Gutsche, T., & Arvai, J. (2024). Don't Say "Vegan" or "Plant-Based": Food Without Meat and Dairy Is More Likely to Be Chosen When Labeled as "Healthy" and "Sustainable." *Journal of Environmental Psychology, 93*, 102217. https://doi.org/10.1016/j.jenvp.2023.102217

Southey, F. (2024, January 2). *Do-consumers-prefer-sugar-or-non-nutritive-sweeteners*. www.foodnavigator.com.

Spence, C. (2022). Behavioural Nudges, Physico-Chemical Solutions, and Sensory Strategies to Reduce people's Salt Consumption. *Foods, 11*(19), 3092. https://doi.org/10.3390/foods11193092

Steele, E. M., Baraldi, L. G., Da Costa Louzada, M. L., Moubarac, J., Mozaffarian, D., & Monteiro, C. A. (2016). Ultra-Processed Foods and Added Sugars in the US Diet: Evidence from a Nationally Representative Cross-Sectional Study. *BMJ Open, 6*(3), e009892. https://doi.org/10.1136/bmjopen-2015-009892

Szejda, K. (2016, February 16). *A-cup-of-coffee-who-no-longer-says-it-can-cause-Cancer*. https://theconversation.com. https://theconversation.com/raise-a-cup-of-coffee-who-no-longer-says-it-can-cause-cancer-60096

Taking-Alternative-Protein-Trends-Mainstream. (2023, February 22). https://www.bcg.com/publications/2023/taking-alternative-protein-trends-mainstream

Tan, W. K., Zhu, J., Li, J., & Ong, C. N. (2023). Soy, Soil and Beyond. *Soil Ecology Letters, 5*(4). https://doi.org/10.1007/s42832-023-0190-6

TBRC Business Research PVT LTD. (2023, April 25). *Global plant-based food market 2023–2027: Market size, share, trends and segments as per the Business Research Company's Plant-Based Food Global Market Report 2023*. GlobeNewswire News Room.

Teshome, E., Teka, T. A., Nandasiri, R., Rout, J. R., Harouna, D. V., Astatkie, T., & Urugo, M. M. (2023). Fruit By-Products and Their Industrial Applications for Nutritional Benefits and Health Promotion: A Comprehensive Review. *Sustainability, 15*(10), 7840. https://doi.org/10.3390/su15107840

Thaís, J. S., Badan Ribeiro, A. P., & Barrera-Arellano, D. (2021). Margarines: Historical Approach, Technological Aspects, Nutritional Profile, and Global Trends. *Food Research International, 147*, 110486, ISSN 0963-9969.

The Spruce Eats. (2022, August 22). *Desi Indian-style burger for something different*. https://www.thespruceeats.com/indian-style-burger-1957599

Toews, I., Lohner, S., De Gaudry, D. K., Sommer, H., & Meerpohl, J. J. (2019). Association between Intake of non-Sugar Sweeteners and Health Outcomes: Systematic Review and Meta-Analyses of Randomised and non-Randomised Controlled Trials and Observational Studies. *BMJ. British Medical Journal, k4718*. https://doi.org/10.1136/bmj.k4718

UN News. (2023, July 17). *Aspartame sweetener 'possibly' cancer-causing, WHO agency reports*. https://news.un.org/en/story/2023/07/1138732

Unilever. (2023, November 17). *Our brands*. https://www.unilever.com/brands/all-brands

Vegetable Exports by Country US$000 2021 | WITS Data. (n.d.). https://wits.worldbank.org/CountryProfile/en/Country.

Vinitha, K., Sethupathy, P., Moses, J., & Anandharamakrishnan, C. (2022). Conventional and Emerging Approaches for Reducing Dietary Intake of Salt. *Food Research International, 152*, 110933. https://doi.org/10.1016/j.foodres.2021.110933

Wendt, M., & Weinrich, R. (2023). A Systematic Review of Consumer Studies Applying the Food Technology Neophobia Scale: Lessons and Applications. *Food Quality and Preference, 106*, 104811. https://doi.org/10.1016/j.foodqual.2023.104811

WHO. (2023, December 5). *Global report on the use of sugar-sweetened beverage taxes, 2023*. https://www.who.int/publications/i/item/9789240084995

Wikipedia Contributors. (2024, July 10). *VUCA*. Wikipedia. https://en.wikipedia.org/wiki/VUCA

Xia, X., Zhang, Y., Yu, J., Song, S., Zhang, F., Hayat, K., Zhang, X., & Ho, C. (2024). Quantitative Analysis of Perceived Saltiness Through Combination of Sensory Evaluation and Electronic Tongue and Its Method Validation. *Food Analytical Methods*. https://doi.org/10.1007/s12161-024-02607-8

Ziemnowicz, C. (2020). Joseph A. Schumpeter and innovation (pp. 1517–1522). In: Carayannis, E.G. (eds) *Encyclopedia of Creativity, Invention, Innovation and Entrepreneurship*. Springer, Cham. https://doi.org/10.1007/978-3-319-15347-6

Licence to Operate (in the Future) 18

The phrase "licence to operate in the future" here does not label the formal legal requirements to act in the food business but refers to the ability of a company to conduct its business activities in a way that is accepted by its key stakeholders, such as customers, regulators and employees. It's a soft, unwritten and intangible approval from the stakeholder with typical elements like "*Food Safety, Environmental Impact, Animal Welfare, Labour and Human Rights, Impact on Health and Business Ethics*" (Sterling & Charlebois, 2017). The priority of these elements is a choice of each company and depends on the character of the business. The producer of gluten-free cereals will have a different focus than an egg producer or the maker of vegan emulsifiers. But overall, evidence-based information will give the traceability to create the necessary end-to-end (E2E) transparency.

This "licence to operate in the future" lets a company decide on its strategic perspective regarding elements that are not legally required, voluntary activities a company is doing now. The consent of the stakeholder to the communicated plan builds the fundament for the business prosperity in the future.

More than 50 years ago, the "Club of Rome" first articulated the "limits to growth" on this planet and put an end to a business understanding that further economic growth could be endless (Club of Rome, Meadows et al., 1972). Instead, the conscious management of resources came into focus. Initially it was more a cautioning statement, less a concrete action plan. "Sustainability" was still not discussed much outside the experts' circles, when the last edition of this book was published.

Today, GenY, Millennials and GenZ are not the only ones demanding companies take responsibility for life on this planet. "Sustainability" has become a non-negotiable. It's not sufficient to comply with all working and manufacturing regulations. Companies must take a broader social responsibility towards society. For this, two terms are used:

- Environmental, social and governance (ESG)
- Corporate social responsibility (CSR)

Both concepts pronounce the socially conscious and sustainable business operations. The main difference is that CSR is more internally focussed and a self-regulating business model. The companies are aware of the impact their business is having on society, which includes the environment, economy and people. CSR focusses on corporate volunteering and engagement with charities. Companies are making themselves more accountable to themselves, their stakeholders and the public. The CSR program also highlights the values and goals of the business.

ESG also investigates the environmental impact and wider relationships of the company, including employees, suppliers, customers and the wider community. It includes the setting of measurable goals plus the regular auditing of the progress made and often the company's leadership, the executive pay, shareholder rights, audits and the way internal controls are handled (governance). And the company performance against these criteria serves as guidance for investors. ESG requires more concrete measures of sustainability, like concrete carbon emission.

CSR allows a company to articulate its responsibility; ESG evaluates the business in addition. Here, a good ESG performance can also be used to attract new capital.

For the total food industry, sustainability is a key issue. The most meaningful counts to measure the environmental impact are greenhouse gas emissions (GHGs), land use, freshwater usage, eutrophication

DOI: 10.1201/9781032724591-18

and animal and plant diversity (Ritchie et al., 1970). The E2E footprint from farm to fork and waste assigns globally 26% of GHGs to foods and 70% of freshwater withdrawal.

The major companies have fully accepted their responsibility and have set out ambitious goals to make their businesses "future fit", reviewing their operations from various angles, including sourcing, packaging, waste management and carbon footprint. Here are a few examples:

- *Unilever*: This global fast-moving consumer goods (FMCG) giant aims to halve its environmental impact by 2030 and achieve net-zero emissions by 2040. Agricultural goods will be sustainably sourced by 2030, eliminating deforestation from its supply chain by 2023. Additionally, it strives to improve the health and well-being of more than a billion people by 2030 and enhance the livelihoods of millions of smallholder farmers and workers.
- *Nestlé*: One of the world's largest food companies, Nestlé has committed to reaching net-zero emissions by 2050 and cutting greenhouse gas emissions by 50% by 2030. It has also pledged to make 100% of its packaging recyclable or reusable by 2025 and to support regenerative agriculture practices among its farmers. Furthermore, it aims to provide nutritious and affordable food for all, and respect human rights and labour standards throughout its operations.
- *Danone*: This French dairy company has set a goal of becoming carbon neutral across its entire value chain by 2050 and reducing its water consumption by 60% by 2030. It also aims to increase the share of organic products in its portfolio to 15% by 2025 and promote biodiversity and animal welfare. Moreover, it seeks to foster social inclusion and diversity among its employees and stakeholders and empower local communities through its social business initiatives.

Fulfilling these commitments is a massive extra effort, and any "payback" can only come later. Before that, resources need to be invested, and companies should think about ways to talk about intentions and efforts in an engaging, trust-building way.

"SELLING" SUSTAINABILITY

There is consensus that every organization that wants to operate in the future needs to transform those parts of their activities that need a change. Whether this transformation is called "circular economy", "eco-friendly", "carbon neutral", "zero waste", "green economy", "carbon offset", "carbonomics", "upcycle" "recyclable", "reusable", renewable ", "regenerative" or "sustainable" is central for those experts and suggests the detailed actions and targets in line with the underlying principles.

It is often mentioned that for the food sector in general the top priorities are the shift from animal- to plant-based food and the move into packaging materials that are reusable, recyclable or compostable. Less broadly discussed, but equally important, is the urgent need for a massive diversification of current plant-based staples (WBCSD, n.d.). While over 50,000 plants are edible, today only three of them, namely, corn, rice, and wheat, account for up to 51% of the world's caloric intake (Pariona, 2019). The green revolution was extremely successful in boosting yields of key crops. But as a result of that crop focus, the soil is now often depleted, and climate change can further impact future yields.

It was not the consumer calling it out; it was an early warning from science. Serious concerns about the detrimental impact the human population has had on this planet came out in the late 1960s of the last century. And in 1972, the expert team working for the "Club of Rome" used simulation to calculate various economic scenarios and what they concluded was a shock: "*If the present growth trends in world population, industrialization, pollution, food production, and resource depletion continue unchanged, the limits to growth on this planet will be reached sometime within the next one hundred years. The most probable result will be a rather sudden and uncontrollable decline in both population and industrial capacity*" (Meadows et al., 1972). After that statement, the "limits of growth" became vigorously discussed in politics and expert circles.

The word "sustainability" first appeared around 40 years ago, when in 1987 the former Norwegian prime minister Gro Harlem Brundtland, while leading a UN commission, first used "sustainable development" to underscore the urgent need for a much deeper understanding of the impact current human activity was having on the environment. The report was called "*Meeting the needs of the present without compromising the ability of future generations to meet their own needs*" (Sustainability for All, n.d.). Today, sustainable development has become a concept in the UN with 17 main goals, including "No poverty", "Zero hunger", "Good health and well-being", "Affordable and clean energy", "Climate action", "Partnership" and "Responsible consumption and production". Each of the 17 goals has clear, time-bound quantitative targets, indicators and regular progress reports (The 17 Goals, n.d.).

In contrast to claims like "naturalness" and "clean label", "sustainability" has a scientific definition and tangible goals, time frames and targets.

In 1992 during the UN "Earth Summit" in Rio, the industry adopted the idea and started to build the relevant capacities. Sustainable development became the fundament on which the business had to build its future. Two years later, the social entrepreneur John Elkington released his "3P message": "*People, Planet, Profit*", *has to become the triple bottom line of sustainability*" (Jeurissen, 2000). For long-term success, profit-driven enterprises had to equally focus on their financial performance (profit), but also on their social (people) and environmental (planet) impact.

Nowadays, most food companies have detailed sustainability action plans. Unilever's purpose is "*to make sustainable living commonplace*", which it articulated and detailed in "*The Unilever Compass for Sustainable Growth*" (Unilever, 2024). Nestle states that "*Creating Shared Value (CSV) is at the heart of our purpose: to unlock the power of food to enhance quality of life for everyone, today and for generations to come. We have sought to build our business by taking decisions that are not just good for us and our shareholders, but for society and the planet as well.*" (Nestlé Global, n.d.b). Action plans from other big and well-respected companies sound very similar, and all transparently articulate their time-bound target and auditing progress. And the targets in the areas of climate action, nature regeneration, waste-free world, health and well-being, positive nutrition, etc., are challenging.

For foods research and development (R&D), these commitments normally bring extra effort, including the responsibility to redesign products with new – and different – ingredients and choosing processes that are in line with plans to reduce the environmental footprint.

Sometimes surprisingly, changes that look simple are a major challenge. The shift in the packaging material from current plastic pouches to recycle-ready pouches surely helps the environment. But how massive the effort can be has been described in the case study in Chapter 15.

A similar complex challenge exists for many food companies in the people and health area, and with the salt (sodium) levels in food products. A 5% cut in salt often is possible without major reformulation work and perceivable changes in saltiness. Once the new artwork with the new ingredient list and nutrition table is printed, these better variants can be introduced as a step towards sustainability. But getting to the often-demanded 30% more reduction is complicated. A company making savoury products would have to decide if it dared to make a 25% or more reduction step or instead make smaller steps over a longer period of time. In the first case, the company can make a "reduced salt" claim but will launch a product that loyal consumers perceive as different. In the second case, the company commits to a continuous rejuvenation program for their product portfolio, a commitment that will bind reformulation resources. In some product categories, the salt reduction leads to a reduction in declared weight, something shoppers can perceive as "shrinkflation". As another consequence, the Global Trade Item Number (GTIN), the barcode identifier for packaged products in the trade, will need to be changed. For a single food product format, that's not too costly, but if many stock-keeping units (SKUs) are affected, it adds up. And the code change triggers changes in the trade, so the retail partner will want to see compensation for the effort that has been caused.

Alternatively, the food producer could decide to compensate the missing salt with an ingredient that does not change the sensorial profile, a filler. That's a bit more formulation work, plus maybe some extra consumer research to ensure the changes are not perceived by the consumer.

Maltodextrin can help here, which will be easy if it's already part of the formulation. Its position in the ingredient list may change a bit [due to a change in its quantitative ingredient declaration (QUID) amount]. It still needs an artwork change, but this modification can be simply applied with the next print run.

If the ingredient is not in the formulation yet, it must be declared on the package to inform consumers. It starts to become more difficult if the original product carries a "naturalness claim". While maltodextrin is of natural origin, once it's processed, its addition does not qualify for a "naturalness claim" in many regulations. Alternative native starches exist, but need extra formulation work and oncosts, diluting the product margin.

This is just an example of how an obviously small change plan can lead to unexpected hurdles, each of them perhaps small, but in sum can turn into a situation that makes improvement sometimes difficult.

The move to more sustainability is an E2E approach, from cradle to grave, including consumers, their households and their behaviour. But consumers often do not accept their part of the responsibility. A survey carried out in the 27 member states of the European Union in late summer 2020 among more than 25,000 citizens showed that many consumers see it as the food producer's responsibility (European Union, n.d.).

And more sustainability sometimes creates extra costs. Are their options for food innovators to manage those? Trend research says that even after the pandemic customers still demand healthy, sustainable food (PWC, 2023). But these reports also signal that customers are exhibiting cost-cutting behaviours, due to concerns about inflation and the "*cloudy macroeconomic climate*". A PwC study counted that half of the 9180 consumers interviewed in 25 different countries were "*either very- or extremely concerned about their own personal financial situation. And 96% of surveyed consumers intend to adopt some type of cost-saving behaviour over the next six months.*"

In moments where consumers feel the inflation in food prices, they are less willing to pay extra for sustainable products. Instead, they shift the accountability for that to the producers and manufacturers. And more costs with no possibility to charge shoppers leads to margin dilution, a factor impacting the third "P" in the sustainability approach, built on "people, planet, profit"! What are the options then?

The Boston Consulting Group (BCG) asked more than 3000 participants in eight different countries about their attitudes and behaviours around sustainable food. On the top level, only 10% said that sustainability would be in their top three buying reasons (BCG, 2023). So, directly claiming sustainability appeals only to a clear minority. But when the researchers correlated the data with results of an earlier study, they found that the majority of the remaining 90% of consumers link sustainability with "healthy", "high quality" and "filling" food! So, it could be that approaches to make sustainability "more personal", like claiming that food products are more nutritious and are tastier, can help consumers in their buying decision, even with little oncost.

"Sustainability" is a holistic scientific framework that for consumers is difficult to understand. And more and more on-pack claims, logos and labels praising sustainability might tire the consumers or be perceived as "greenwashing".

And while the consumer is more and more bombarded with "hollow" sustainability claims, leading experts start to refocus the necessity for the food sector towards "regenerative agriculture". The goal is similar: Production of enough safe and nutritious food for the whole population, without exhausting the capacity of the planet, as formulated in the leading EAT-Lancet report (Willett et al., 2019). Here, the emphasis is on soil and soil quality and "*the farming that uses soil conservation as the entry point to regenerate and contribute to multiple ecosystem services*" (Schreefel et al., 2020).

The food sector adopted that new focus and started various new farm-based approaches. As for sustainability, the "better solutions" depend on the crop, the climate, the soil and other environmental conditions. It's often farm practice.

New practices can include usage of "cover crops", "minimal soil disturbance (no-till), diversified crop rotations, field buffers and crop and livestock integration. The opportunity is to create tangible "stories" about tomato farmers in Spain, rice growers in Thailand or the ecosystem restoration in Africa.

SUSTAINABLE/REGENERATIVE FOOD PRODUCT DESIGN

How important it is to bring that need for change in global agriculture to full life can be seen by the fact that today 75% of the world's food comes from just 12 plants and 5 animal species. And in those just three (wheat, rice and maize) provide nearly 60 percent of calories in the entire human diet (Food and Agriculture Organization, n.d.). The risky lack of variety in agriculture and, hence, our daily diet has triggered food brands and companies to announce a broad reformulation of their total product portfolio for people and the planet (Juniper et al., 2021).

So, sustainability and more regenerative agriculture will be a new key element in the food product design. But what are the objective criteria to determine when a new food product is sustainable? Often, a low environmental, economic and social impact is described as sustainable.

The regular list of sustainable practices ranges from the use of more sustainable farming, better packaging and softer manufacturing processes to logistics and working conditions. What is sometimes forgotten in this context are the economic consequences for the producers. Even so, these are the key drivers for business decisions. The "people, planet, profit" (3P) concept captures this element by reminding us that without profit, companies cannot invest in better people and planet practices.

For many companies, it is a regular practice to include a sustainability assessment in the product design phase. These tools and calculations focus on the environmental impacts, e.g., carbon, water or ecological footprint; the nutritional footprint; or a Life-Cycle Sustainability Assessment (LCSA). They are designed as E2E tools. Multinational food companies have started to share a validated online calculator with their suppliers in the farming sector to optimize the impact of E2E food making. The "Cool Farm Tool" (CFT) developed by Unilever, the University of Aberdeen and the Sustainable Food Lab (Cool Farm Alliance,) is set up as *"a pre-competitive community of collaboration with the mission to accelerate the transition to regenerative and sustainable agriculture by providing globally aligned, credible and science-based metrics, tools and resources that create* (Cool Farm Tool, n.d.).

Such sustainability assessments are often complex and time- and labour-intense. Many different, often volatile factors require evaluation. And all this does not improve speed to market, one of the key success criteria in FMCG markets. And they add costs! For the operating company, it's therefore important to ensure that these investments into the future health of the business are not damaging the actual commercial performance. On one side, more sustainable ingredients and materials can be more expensive, they are more difficult to source, might have a higher risk of crop failure and require more costly distribution. But on the other side, they can also improve operational parameters like water usage, carbon emissions or waste management.

Hence, tools and targets for more sustainable products and operations need to be integrated into the product development process from the beginning. In the early phases of an NFPD project, the impact calculation may not be fully accurate, but it should give an idea about the main levers influencing the score. Companies have, therefore, started to include a scoring scheme for sustainability aspects in project milestones discussions, like the gate decision in stage-gated processes. Through that, sustainability becomes a leading design principle for the team and especially the packaging engineers.

The E2E perspective is important, as it might allow, for example, the acceptance of an increase in GHG in an early step of the process, if later the savings elsewhere outweigh the effect.

The generation of a packaging material may initially create more GHG but then can be reused several times by consumers or trade. Glass is an example here. It is similar for the costs: Food products can use – at least temporarily more expensive ingredients from sustainable sources if a cost saving somewhere on the way from farm to fork can compensate for the oncosts, without the need to increase consumer prices or live on a lower business margin. Often such a change helps the wider transformation of the business. Can oncosts on the agronomical side, raw or packaging material, be compensated for by new investments into supply chain or distribution, so that the materials can be handled more economically? Or is a switch to better recyclable packaging material possible, but only with a shelf-life reduction, that challenges the existing distribution models? But if the manufacturer, distributor and trade can agree on a 3-month shelf-life reduction, will the consumer notice, react or complain? Or will a shorter shelf life underline the foodiness, so the change can happen without massive oncost, margin dilution or price increase?

For the composition of food products, a move to more sustainable practices can mean that the 100,000 tons of corn, wheat, rice, potatoes, chicken, pork and beef, etc., must be successively replaced by much less used material from beans, pulses, cereals, sprouts, grains, tubbers, mushrooms, nuts and seeds. Today, "seventy-five percent of the global food supply comes from just 12 plant and five animals species. And three of them create nearly 60 percent of calories from plants in the entire human diet" (Bracher, 2022). Many, often forgotten, plant variants have been identified in the meantime, allowing more sustainable farming while having a good nutritional profile. Suggested alternatives include cereals, like finger millet, buckwheat and fonio, or nuts and seeds like hemp and flax. For most of those crops, sufficient supply and the necessary large-scale farming practices must still be developed and sufficient volume must be commercially available at a reproducible quality.

For the food product developers, the immense task ahead is then to replace over time all those non-sustainable ingredients in their existing product range with the new, better ones. And the necessary exchange is not a simple ingredient replacement. Most of the ingredients originated from those crops will be different. Taste and texture may need adjustment, basically all physical and chemical properties need to be understood and controlled and shelf life and processing behaviour must be established. And consumer acceptance must be established. Most of those "future 50 foods" have the potential to replace the common overused variants broadly, but for many, the knowledge about the functional properties and applications are scattered.

Companies that have started the exchange program immediately saw that this is not a rejuvenation program for just a couple of years or three. Adding a few moringa leaves or some flour to existing food products can already be a full reformulation program but is in fact only a "baby step" towards the change needed. The required shift might take several years of hard and continuous development effort, but the journey will create valuable learnings from early successes, but also from failure and disappointments. It would be very desirable if students of food science and technology would get early hands-on experience during their academic training on how to use these rather unknown, but truly "future fit" ingredients.

And sometimes it might simply not be possible to replace ingredients in a way that the consumer cannot detect the difference. Then, loyal consumers of the "original" need an explanation of why their favourite food is now different. Whether a pure rational argumentation that the product is now more nutritious and better for the planet is sufficient to keep consumers loyal is a valid question. An interesting thought in this context was mentioned in a 2023 Mintel report, which mentioned "weatherproofed provisions" as a trend in NFPD. Here, the more abstract environmental problem becomes personalized as a help for people to endure tougher climate and weather conditions (Jenny, 2023). Examples are niche ideas like Japan's "after sauna meals", drinks with cooling aftertaste, self-heating hot snacks and meals to save energy, ultraviolet (UV) protectants, products containing marine collagen and other heating or cooling botanicals out of traditional Chinese medicine.

For others, especially the environmentally more sensitized younger generations, the rational arguments may be sufficient. But these have to be honest, quality arguments. Many consumers are fatigued by the massive waves of claims, promises, logos, brand stories, links, etc., that are flooding the markets. Digital tools like apps allow consumers to judge the "eco-friendliness" of food products by scanning the product barcode (Open Food Facts, n.d.). People use these services, but how much this will drive change in consumer behaviour remains to be seen.

GLOBAL VERSUS LOCAL SOURCING

Sustainable ways to grow farm-based ingredients is a key aspect. A second one is how much transport and shipment is acceptable in an environmentally responsible business. This is a question for food companies sourcing many items from many places. The distance a food item travels from production to consumption has a significant environmental impact, especially on the greenhouse gas emissions. So, must the foods business of the future prioritize local farming and production? The intuitive answer is "yes" and only a second view shows a clearer picture.

The GLAMUR research initiative of the European Commission took 3 years to assess the sustainability of different food chains. What they saw was first that "local" and "global" were weak classifications. Both describe the extreme end of a continuum in which all food products and constituents are placed. Rather than measuring "local" and "global" sustainability as alternative choices, the assessment must focus on the specific strengths and weaknesses of a particular supply chain, including complementarities and improvement potential (Brunori et al., 2016).

No doubt, measures of food mileage show that eating locally produced food reduces the carbon footprint of the food system and can mitigate climate change: A study published in 2022 in *Nature Food* calculated that global food miles generate about 3.0 gigatonnes of carbon dioxide equivalent (GtCO2e) per year; that's 19% of total food system emissions. But surely, not everything consumers want can be grown locally. The truth is that fruits and vegetables have the highest food miles emissions, since they often require refrigeration and are often consumed out of season. And even more, locally grown fruits and vegetables need a lot of water and energy that is becoming scarce in some traditional growing areas. And it does not seem likely that consumers will quickly turn to "in season" consumption. A company making products with chocolate, coffee beans, tropical nuts and palm oil, etc., will need a different solution to cut their environmental impact than stopping to source from outside, contrary to global tomato-based product makers or meat sources. Using a local fishery can cut mileage and is beneficial if natural stocks are not depleted. So like always, a simple black-and-white approach is not helpful. Differentiated strategies are needed to book progress. Several companies and services can help today's enterprises to optimize their material logistics without making the business inviable or uncompetitive (https://www.top10erp.org/erp-software-comparison/best-fit/food-and-beverage?).

FINAL WRAP-UP

It's time to put it together and try some educated guesses about the key characteristics of new food product in the years to come.

First, food consists of macronutrients, which were the heroes in the late 19th century and first half of the 20th century. Fat, the best energy provider, was the first to be accused of causing adverse effects when consumed too much, in a moment when sufficient calories were no longer the main concern. More research split the fat class into good and bad guys, unsaturated and saturated. Local cuisines traditionally using food items rich in unsaturated fats, like the Mediterranean diet, were praised for their health contribution. And it took a long time before the pendulum started to swing back here. The old, rock-solid assumption that too much saturated fat causes cardiovascular disease (CVD) became weaker and weaker (Astrup et al., 2020).

From the mid-20th century, the rise of carbohydrates started. They showed no correlation with cardiovascular health, but they were well known, bulky and filling, very versatile in application, everywhere available and not expensive, and contained only half of the calories of fat. The food industry did a great job in making them very palatable and cheap. The price of sugar is now where it was in 1920, and for most of the time in between it was even much lower (https://tradingeconomics.com/commodity/sugar). Replacing high-caloric fat with lower-caloric carbohydrates including sugars became a valid option to manage weight. Then science rediscovered that sugar-triggered insulin is a potent blocker of fat catabolism. These findings suggest that some of the core principles in popular weight management guidelines need to be rewritten, at least for those people having high insulin levels.

The agro-industrial sector had huge success with cost-effective starchy crops, like maize, rice, wheat, potatoes and soybean, feeding the constantly growing population. A couple of years later, the environmental consequences of these crops, such as soil depletion, started to cloud the view for many carbs. And on top of that now, there are new health concerns.

In light of the overweight and obesity endemic, nutritional profiles like a low glycaemic index (GI) and/ or a low insulinic response are becoming important design elements. The focus on physiological responses like the activation of fat store degradation might be much more effective than simple calorie counts.

The enabling "better ingredients" contain more amylose, lipid, protein, dietary fibre and resistant starch (RS). And the food product developers have to find ways to make products with those ingredients that fully match the sensorial expectation, with a minimum of additives. Agronomists and food processors must find ways to bring new, or ancient, varieties into the major food chain using more of the total fruit, grain or kernel. The "*foods of the Anthropocene*" must score well on environmental and nutritional scales (Willett et al., 2019).

While the nutritional view about fats is shifting back, the proteins are the stars on stage. And with plant-based foods, consumers have a wide choice of meat alternatives without mixed feelings about their carbon footprint or the well-being of farm-based animals. Now there are countless food products enriched with protein on the market. The protein flood may have already approached its zenith. There isn't hardly any product category not pumping up their offerings with extra protein. And while the factual nutritional benefits get marginal, the "extra of healthy proteins" has more become an excuse for additional snacking, while experts start to raise their voices about the health consequences of too much protein, like dehydration, kidney disease, digestive problems, misbalanced nutrient uptake, cancer or recently the impact of branched-chain amino acids on insulin levels and diabetes (Bloomgarden, 2018).

Macronutrients are not good or bad for health! The benefits of a balanced diet are undeniable. And there isn't a single food product or category that covers all needs. For new food products a question is how much "the single product" needs to take the responsibility to reflect a healthy balance. Not all drinks need extra protein, milk products do not need extra carbs and not all fat needs to be unsaturated or "extra virgin".

Having said that, there is no doubt that food product developers must keep confirmed "perceived baddies" (fat, sugar, salt, etc.) at bay. New design solutions will have to better disconnect "liking" and "palatability" from the adverse health effects associated with "ultra-processed" or "too energy dense" foods. If that is not happening voluntarily, higher taxes, advertising limitations and market entry barriers will build more pressure.

While we need to push science and technology in foods to provide solutions for these challenges, sometimes the answer may already exist. Endless efforts have been made to find salt (sodium) alternatives. Up to now, the solution of getting to at least 30% reduction has not been found. But the long-scolded monosodium glutamate (MSG) may become an alternative (see Chapter 2 on MSG). It is perhaps more effective than all other substances and very cheap. If that potential becomes confirmed, opinion leaders must help to overcome the negative consumer perception that has been built up based on 50 years of old "bad science". Still, food designers in industry and food service must find ways to minimize the salt in a formulation or dish to what is needed for the "same liking".

Food "goodness" must focus on the total composition. A food item is not just a "carrier" for an extra portion of something "healthy". That's the role of supplements! Food and eating are all about the emotions beyond the mechanical "nutritional intake". Treating food as "nutritional carriers" dismisses the potential. Which extra benefit good food could bring is best decided in conjunction with the active and the food composition and a few key principles.

The food matrix should not contain anything that compromises the delivery of the "beneficial active". Adding an extra portion of essential iron to a very popular tea initially sounds great: On one side, it is a popular "carrier", affordable also in lower-income groups that are more often suffering from a lack of essential iron. But, when the tea catechins are complexing the iron, the promise is not kept. The nutrient is not reaching the body but precipitates in the cup. Addition of vitamin C can change the picture, but the antioxidant itself is vulnerable to heat. Decades of experimental work luckily provide a lot of best practices for the nutritional enrichment of popular foods and beverages (Hurrell RF. Ensuring the Efficacious Iron Fortification of Foods: A Tale of Two Barriers. Nutrients (Hurrell, 2022).

Many more of these synergistic and antagonistic effects between the different food constituents are known and provide great guidance for future product development: Vitamin D and calcium! Iron and vitamin A, but not calcium! Micronutrients and dietary fibre, but not all fibres!

Dietary fibres are another class of food constituents that have not reached their full potential. The evidence about their positive role in overweight and obesity, blood sugar, insulin response and other non-communicable diseases (NCDs) and the whole microbiome is still increasing. Here the task for new food product development is to make them tasty and palatable, without compromising their beneficial effects.

A "future fit" product design approach is the key to commercial success. And the "hottest" battlefield right now are proteins, including the shift away from animal based to new sources. The triumph of plant proteins in many food categories is not only the echo of people being more conscious about the origin of their food products but also about what it is doing to their body and what it means for the environment.

How quickly alternative proteins can displace animal protein will not be determined by nutritional and environmental superiority but the sensorial satisfaction. Biochemically, the challenges here are the different protein structures: The more globular plant-based proteins need to be transformed to mimic the texture of fibrous animal muscle tissue. In dairy products, the casein micelles build structures that give conventional cheese properties that plant-based proteins do not have. And finally, cooking meat creates iconic flavours, while the heating of plant-based protein ingredients often leads to off-flavours (Zhao, 2024).

Lots of progress has been made in the past years, often using specific ingredients and processing functionalities to close the gap. Winning solutions of the future will have to use less "industrial design", "additive functionality" and fewer industry "tricks" to create matching solutions. These improvements are under way and touch the whole vertical value stream from crop choice, harvest location and protein extraction procedures to food mileage and household waste management.

Speaking about food processing, the sometimes-voiced blunt condemnation of food processing, as such, is not sustainable. Restraining the possibilities of science and technology to make positive contributions to future food is counterproductive. The resulting food products and their quality, not the processing, should be judged in the light of the latest knowledge (Knorr & Augustin, 2021). Food processing makes the nutritional value of a food item worse or better or leaves it unchanged. Today, we have so many easy possibilities to judge the nutritional quality of what goes into the stomach that we do not need guesses about the impact of pretreatment.

Front-of-pack label systems like the "Nutri-Score" again make the composition "centre of the plate". Attempts are made to provide a simple, user-friendly guidance of the total product. And they aim to acknowledge the diversity in healthy eating. Taking nutritional responsibility does not mean all food must have the intent to convert to an "average nutrition profile". A particular food format must take responsibility where it makes sense. A good pizza must limit the energy and sodium content and maybe increase the number of vegetables. But it does not need to come with the energy content of spinach or a higher protein content in its base or sauce. And not everything that is loved must become a nutritional hero. Ice cream should continue to reduce sugar and maybe fat content, but if manufacturers are then able to create a creamy texture through the stabilization of air in the matrix, that is great processing and not the destruction of a natural profile.

An impactful nutrient score will have to constantly review the basic algorithms to reflect the latest evidence about whole grain, sugar and salt content, fat and oil types, and additives and sweeteners. The scoring systems, whatever they will be, should not aim to condemn "less nutritious" food products and push them out of the diet but should create a motivation for suppliers to constantly make them better and for the consumer to look for the superior eating experience.

What will change over the next years is the usage of the "additive toolbox" available to make food products. While there is yet no conclusive evidence that "artificial" additives make food generally less healthy, the opinion pendulum swings towards natural ingredients: Colourants, emulsifiers, preservatives, flavours, stabilizers, etc. The natural, non-chemical, synthesized or modified alternatives are "functionally" getting better and better; can carry a clean, natural image; and are often also environmentally friendly. Herbs and spices look like a very promising category here. When smartly added into ethical food items, they often do not even need taste and aroma adaptation.

The focus on key elements in food products will at least stay, but the role of different food constituents will change. A short summary about the future potential of various key ingredients in foods is given in Table 18.1.

TABLE 18.1 Expected change in attractiveness of key food ingredients and technology

ATTRIBUTE	*TREND*	*RATIONALE AND SCOPE*
Salt	(– –)	Increasing salt reduction initiatives globally Persisting health concern Lesser need as taste enhancer and preservative
Simple Sugar	(– –)	NCD concerns about simple sugars, low reputation Increasing focus on role in energy intake, insulinic effects More natural no- or low-caloric sweeteners as available replacers Increasing regulatory and tax pressure
Fibre	(+++)	More alternatives as complex carbs, low blood sugar, low insulin Positive effects on metabolome and immunomodulation Lowering risk of CVD Wide practical history in wide variety of food products
Protein	(++)	Presently a nutrient hero with wide applications Many non-animal protein alternatives entering market Longer-term focus on naturally protein containing food items Foody, clean non-animal origin
Fat	(+)	Revised reputation in human nutrition, lesser CVD concerns Broad role as taste and aroma carrier Positive health image with unsaturated fats Deep functional understanding in food products New "light" solutions appearing, using new technologies
Plant based	(+++)	Strong growth supported by new alternative sources New alternatives with superior functionality and ability to mimic meat/dairy Cleaner technologies for making plant protein More locally available protein alternatives Wider range of key crops, revival of "ancient" variants
Cultured meat	(+)	Growth in high-end gastronomy (niche) Limited application as hybrid meat with plant based Price not competitive compared to alternatives No main market farm-based meat alternative
Insect protein	(+/–)	Niche, no major role as farm-based meat replacement Missing meat image, no leading advantage versus plant based Sensorially fully matching solution still missing Neophobia hurdle
Mushroom protein	(++)	Closer match of taste and texture with animal proteins Dominating patents expired, new innovators entering market Additional potential as functional ingredient and taste modulator Low perception as good protein source
Pre-/pro-/syn-/para	(++)	Still growing evidence list about health benefits A main road to personalized nutrition (my microbiome) Wide application potential and large-scale manufacturing in place
Sweetener	(+/–)	Less artificial sweeteners, non-caloric More natural no- or low-caloric sweeteners
Artificial Additives	(– – –)	Increasing science focus on role in NCDs More natural alternatives with better functionality emerging Public view about health of artificial solutions deteriorating

(Continued)

TABLE 18.1 (Continued) Expected change in attractiveness of key food ingredients and technology

ATTRIBUTE	*TREND*	*RATIONALE AND SCOPE*
Functional food	(+++)	More options with better research benefits Stronger enabling science and technology Increasing consumer demand and awareness Possibility to shift part of health care costs to consumers Regulation in place, becoming more innovation friendly
Algae	(+)	Alternative "natural" food additive Additional functional health perspective Hero ingredient in Asian cuisine Sensorial impact in mass food items still limiting Production costs and remaining safety questions
Fermented food	(++)	Widely accepted as "soft" food process technology Many opportunities for biosciences to optimize product and processes Key ingredient in the "top dishes" of many traditional cuisines New evidence for health benefits Global potential for cultural heroes (kimchi, sauerkraut, tofu)
GM food	(++/–)	Combination with traditional food technologies Powerful biotechnology with more breakthroughs (Crispr/Cas) New varieties with solid environmental and health benefits Consumer and regulatory resistance beginning to fade

The expected relative strength of existing and emerging ingredient- or technology trends for new food products. Classification ranges from (+++) = most positive gradually to (---) most negative.

Farm-based meat will not disappear, but it will become more premium. It will be carefully prepared, consciously eaten with mindful pleasure, but less often, in dishes where its sensorial strength (especially texture and aroma) can shine. Mince-meat meals, top dishes in most kitchens of the world, sausages, burgers, etc., do not need animal protein to match existing meat variants. More people will become flexitarians rather than vegans or vegetarians.

Lab-grown meat has started to enter the first markets. That does not mean it will have easy market entrance everywhere. The technical solutions to make it more price competitive are premature. The intrinsic "foodiness" perception of the tissue is low. More transparent processing distances it further from farm-based food. Hiding these aspects will increase suspicion. Safety remains a watch out point, and the environmental benefits need better proof. In the short term, a niche appearance in high-end restaurants for experimental customers seems likely.

Insect protein may get much cheaper to make, has a similar good nutrition profile to meat and is more environmentally friendly. But it is an unknown safety territory, with various new sources of contamination. Most importantly, strong consumer objections in many countries remains a major obstacle to the adoption. With more plant alternatives getting better and better, the appetite to switch to insect protein might not grow much.

Mushroom-based meat replacements seem to have the best precondition to catch a great share of meat-free product growth. In a sense, they have not been able to "bring their rubber to the road" in the past 30 years. It will be exciting to see how the consumer perception shifts in the coming years.

And mixed solutions may offer an additional opportunity to balance the pros and cons of the different sources and create a convincing alternative, if consumers do not object to this "engineering approach".

Generally, nutrition and environmental credentials will not be a possible "added benefit" for successful new products. They will be a mandatory, essential core element that has to be provided without impacting the great eating experiences. It will be "and -and" not an "and-or". And those intangible "good for people and planet" benefits will need a convincing, touching, and credible story line. Particularly for the big companies and their brands, it will be important to communicate openly the superiority of their products to reduce the mistrust consumer have built up in the past. The shoppers must sense that the promises are open and honest and no longer the outcome of smart legal interpretations, with a tendency to overpromise and greenwash. It might be much more credible to talk about "true intentions", "work in progress" and "hurdles and opportunities" than blunt achievement statements. Future products will not only be "safe by design" but "healthy and sustainable by design".

There shouldn't be any doubt that the future of agronomy, the food industry and all distribution channels will be more sustainable. The very few scientific callers in the desert have been joined by major parts of the consumers, especially in the younger generation. It is good to see that basically all companies have made strong, measurable, meaningful and time-bound commitments to become more sustainable. They will be held accountable for their promises. And only their true impact here will build their licence to operate in the future.

Some of these necessary changes will be possible without structural oncosts for the food makers. New technologies, smarter biological processes, efficiency gains, reduced waste and better use of waste streams should help to manage "the cost of making", as they did in the past. That is where the food sector is traditionally outstanding, keeping prices stable and often low, while maintaining margin and overall product performance.

But that will not work everywhere. Especially not if the cost structure is more inclusive, calculating the full E2E costs ("the real costs"): Some of the very successful industrial processes and technologies will become less wanted, like chemical synthesis or modification. Trying to hide the "non-food" aspect from consumers will not be a viable option. Openly explaining the "making of" can help to build trust in the technology. And great honest storytelling is also needed to convince consumers to accept the little extra charge or a functional compromise.

Taste and aroma are probably not the prime area to test that willingness. But whether the consumer always needs a shelf life of 12, 15, 18, or 24 months at room temperature for their products is a valid question. Particularly, as for the massive amount of plastic packaging globally, no truly game changing technology is visible up to now. If the consumer can accept a compromise, maybe a little shelf-life reduction, cost-competitive materials exist to help to reduce, reuse and recycle plastic packaging.

The previous section talked a lot about what a successful new food product must deliver. This discussion focussed on the content. The other problem in innovation is "how to create that content" and make successful innovations. Sources for inspiration are in a lot of different places. But in most cases, that will not simply give an executable the product brief. They need to be evaluated and translated into a product idea, based on the insights generated. There is no doubt that innovation speed and turnover in foods will further accelerate. More new ideas will come to more markets more quickly. Doing things more quickly means to be aware that this means higher risk for mistakes. New technologies like AI will help to improve the predictability and create a more reliable outlook. But still, the management processes used to create new food products need to be fit for purpose. They need to fit the company strategy, the innovation culture and the project DNA.

An idea that is based on foresight and envisioned to create a new product platform or breakthrough might take longer to finish all steps to launch and can benefit from a well-controlled stage-gated process to manage return on investment. The business case here must take a wider and longer perspective to promote the idea. A line or range extension can benefit a lot from quick, agile decision making with high-level autonomy for the team.

And both, insight and foresight have moments on the way to market, where conscious decision making under uncertainty helps move things forward.

So many things have changed and moved on in the past years in new food product development, and this is beyond trends and technologies, project management and operating models.

The wider view on successful innovation mechanics must speak about the role of R&D in that business and in relation to other functions in this process. Young professionals joining enterprises as food developers, technologists, nutritionists, packaging engineers, etc., should make stronger contributions to the future of the business they join, becoming more strategic partners, not "just" brilliant idea executors.

This means that R&D cannot wait for the other functions like marketing and market research to conclude for them the results of their work. They need to be active partners in the identification and translation of a consumer need into a food product or service. R&D contributions also have to leave the compartmental responsibility and reach out to supply chains to have an understanding of which product design can be made under which circumstances.

There is so much more business responsibility for food developers. But one cannot simply ask for more business responsibility; the other functional teams in the company must welcome such extended contributions and trust the input. That happens best through active, knowledgeable and valuable contributions all along the development process. If a key ingredient looks too expensive to match the cost target given in a business case, it might sometimes be possible to cut the amount needed or find a cheaper alternative. But now and then, it's not possible without damage to the proposition. Expecting the shopper to pay a higher price or asking the business to accept a lower margin normally doesn't work either. A much deeper look into how else the consumer want can be met can be the better way forward. This requires a willingness to engage in the analytics of market research, going from basic appreciation to operational, so that the assumptions made are fully understood and maybe jointly improved. A deeper financial understanding of the underlying business case offers opportunities to look at various more indirect cost items differently to shape a business case successfully. Understanding the product manufacturing or material sourcing and storage can help to identify reserves and optimization. Sometimes a small design change in the product creates a big advantage in production or distribution. Assuming more sales or faster turnover can make a case favourable. But that cannot be just brought into the calculation for the gatekeeping meeting, hoping that these numbers will truly come later. It needs a robust and convincing argumentation from R&D to promote an alternative approach. Here at least a good understanding of the E2E financial calculations helps to see the elements in the mix that can be influenced. If the calculated marketing expenditures are seen as high compared to the material and manufacturing costs, a great product demo idea, which can be activated at point of sales or in the social media, can help to convince people and cut marketing costs.

This book, therefore, spent time to explore those spaces, where R&D can engage more actively for the benefit of the overall business success. With all these new powerful developments in science and technology, R&D is normally the only function in a company able to understand why and how a new opportunity would add value for the business! Taking out salt from all products under one famous brand sounds like a wonderful contribution to public health. But if the product performance is suffering, shoppers may turn to the competitor, the financial situation of the brand or company may deteriorate, and the well-intended contribution to public health will not happen, as no one will be buying the new product. To find the "good" idea and split it from the "bad" and "ugly" one requires the honest wish to understand the capabilities of the business. What is good for one company or organization may be detrimental for another. And the application of new technical and scientifically knowledge is often not a "what" question, but a "how" one. And that "how" normally needs to be answered by R&D.

Finally, successful new food products are not only based on great insights but also on foresight. We can reliably analyse the present and predict the nearer future with some confidence. The key task of people in innovation and product development is to design products or services that will be successful in the future: That can be in 6 months with the next launch date or in 6 years in a different world. Sometimes that's easy ("Which new ice cream flavours will be popular next summer?"); sometimes it's more difficult ("Will double emulsion change all dressings?"). But what innovators do is use the insights from the present (and sometimes the past) to imagine products for the future. These people often work in R&D!

I would like to close with a figure I recently found back in a conversation on "LinkedIn" that Sterling Crew, President of the Institute of Food Science and Technology, shared (see Figure 18.1). It illustrates the discrepancy between the jobs food scientists and technologists are doing and the perception of the public of what they are doing.

As with most appealing statements, the take-home message accentuates the differences a bit too much to make the point clear. This one is an encouragement for everybody working in those professions or planning to do so to actively speak about the job essentials openly and correct the impression that industrially made food is basically the direct upscaling of kitchen food preparation and cooking the layperson has in mind.

FIGURE 18.1 What food scientists and technologists do, and what the public might think they do. (From LinkedIn, n.d.)

REFERENCES

Astrup, A., Magkos, F., Bier, D. M., Brenna, J. T., De Oliveira Otto, M. C., Hill, J. O., King, J. C., Mente, A., Ordovas, J. M., Volek, J. S., Yusuf, S., & Krauss, R. M. (2020). Saturated Fats and Health: A Reassessment and Proposal for Food-Based Recommendations. *Journal of the American College of Cardiology*, *76*(7), 844–857. https://doi.org/10.1016/j.jacc.2020.05.077

BCG. (2023, May 30). *Whetting-consumers-appetite-for-sustainable-foods*. https://www.bcg.com/publications/2023/whetting-consumers-appetite-for-sustainable-foods

Bloomgarden, Z. (2018). Diabetes and Branched-Chain Amino Acids: What Is the Link? *Journal of Diabetes*, *10*(5), 350–352. https://doi.org/10.1111/1753-0407.12645

Brunori, G., Galli, F., Barjolle, D., Van Broekhuizen, R., Colombo, L., Giampietro, M., Kirwan, J., Lang, T., Mathijs, E., Maye, D., De Roest, K., Rougoor, C., Schwarz, J., Schmitt, E., Smith, J., Stojanovic, Z., Tisenkopfs, T., & Touzard, J. (2016). Are Local Food Chains More Sustainable than Global Food Chains? Considerations for Assessment. *Sustainability*, *8*(5), 449. https://doi.org/10.3390/su8050449

Club of Rome. (2024, January 30). *The limits to growth - Club of Rome*. https://www.clubofrome.org/publication/the-limits-to-growth/

Cool Farm Tool. (n.d.). *The Tool – Cool Farm Tool | An online greenhouse gas, water, and biodiversity calculator.* https://coolfarm.org/the-tool/

European Union. (n.d.). *Data.europa.eu.* https://data.europa.eu/data/datasets/s2241_505_eng?locale=en0

Food and Agriculture Organization. (n.d.). *What is agrobiodiversity?* http://www.fao.org/docrep/007/y5609e/y5609e01.htm#bm1

Hurrell, R. F. (2022). Ensuring the Efficacious Iron Fortification of Foods: A Tale of Two Barriers. *Nutrients, 14*(8), 1609. https://doi.org/10.3390/nu14081609

Jenny, Z. (Director). (2023). *Food and drink trends global: What food and drink consumers want and why: Now, next and in the future.* https://insights.mintel.com/rs/193-JGD-439/images/Mintel_2023_Global_Food_and_Drink_Trends.pdf

Jeurissen, R. (2000). Cannibals With Forks: The Triple Bottom Line of 21st Century Business. *Journal of Business Ethics, 23*(2), 229–231. https://doi.org/10.1023/a:1006129603978

Juniper, T., Gregory, P., Drewnowski, A., & Redmond, A. (2021). 50 foods for healthier people and a healthier planet. In *FUTURE 50 FOODS.* https://www.wwf.org.uk/sites/default/files/2019-02/Knorr_Future_50_Report_FINAL_Online.pdf

Knorr, D., & Augustin, M. (2021). Food Processing Needs, Advantages and Misconceptions. *Trends in Food Science & Technology, 108*, 103–110. https://doi.org/10.1016/j.tifs.2020.11.026

LinkedIn. (n.d.). *LinkedIn Login, Sign in | LinkedIn.* https://www.linkedin.com/search/results/content/?heroEntity

Mack, O., Khare, A., Krämer, A., & Burgartz, T. (2016). *Managing in a VUCA World.* Springer.

Meadows, D. H., Randers, J., & Meadows, D. L. (1972). The Limits to Growth (pp. 101–116). In: L. Robin, S. Sörlin, & P. Warde (eds.) *The Future of Nature: Documents of Global Change.* New Haven: Yale University Press. https://doi.org/10.12987/9780300188479-012

Nestlé Global. (n.d.b). *Sustainability at Nestlé.* https://www.nestle.com/sustainability

Open Food Facts. (n.d.). *Eco-Score: The environmental impact of food products.* https://world.openfoodfacts.org/eco-score-the-environmental-impact-of-food-products

Pariona, A. (2019, June 7). *What are the world's most important staple foods?* WorldAtlas. https://www.worldatlas.com/articles/most-important-staple-foods-in-the-world.html

PWC. (2023, February 16). *Consumer-markets/consumer-insights-survey-Feb-2023.* https://www.pwc.com/gx/en/industries/consumer-markets/consumer-insights-survey-feb-2023.html

Ritchie, H., Rosado, P., & Roser, M. (1970, January 1). *Environmental impacts of food production.* Our World in Data. https://ourworldindata.org/environmental-impacts-of-food

Schreefel, L., Schulte, R., De Boer, I., Schrijver, A. P., & Van Zanten, H. (2020). Regenerative Agriculture – The Soil Is the Base. *Global Food Security, 26*, 100404. https://doi.org/10.1016/j.gfs.2020.100404

Sterling, B., & Charlebois, S. (2017, January). *Building social license to operate within the food system: (Learning from the Lessons of Others).* https://www.researchgate.net/publication/312630805_Building_Social_License_to_Operate_within_the_Food_System_Learning_from_the_Lessons_of_Others

Sustainability for All. (n.d.). https://www.activesustainability.com/sustainable-development

The 17 Goals. (n.d.). https://sdgs.un.org/goals

Unilever. (2024). *Our strategy and growth action plan.* https://www.unilever.com/our-company/strategy

Willett, W., Rockström, J., Loken, B., Springmann, M., Lang, T., Vermeulen, S., Garnett, T., Tilman, D., DeClerck, F., Wood, A., Jonell, M., Clark, M., Gordon, L. J., Fanzo, J., Hawkes, C., Zurayk, R., Rivera, J. A., De Vries, W., Sibanda, L. M., & Murray, C. J. L. (2019). Food in the Anthropocene: The EAT–Lancet Commission on Healthy Diets from Sustainable Food Systems. *Lancet, 393*(10170), 447–492. https://doi.org/10.1016/s0140-6736(18)31788-4

WBCSD. (n.d.). *Staple crop diversification.* Retrieved July 13, 2024, from https://www.wbcsd.org/contentwbc/download/12605/187147/1

Zhao, D. (2024). Plant-Based Protein Processing and Texture Properties. In: Du, X., & Yang, J. (eds) *Flavor-Associated Applications in Health and Wellness Food Products.* Springer, Cham. https://doi.org/10.1007/978-3-031-51808-9_12

Index

Note: Page numbers in *italics* and **bold** refer to figures and tables, respectively.

G

J

K

N

Q

V

W

X

Y

Z